OCR

D0208550

# Dreams, Stars, and Electrons

---

# Dreams, Stars, and Electrons

SELECTED WRITINGS OF
LYMAN SPITZER, JR.

*Edited by Lyman Spitzer, Jr., and
Jeremiah P. Ostriker*

PRINCETON UNIVERSITY PRESS
PRINCETON, NEW JERSEY

Copyright © 1997 by Princeton University Press
Published by Princeton University Press, 41 William Street,
Princeton, New Jersey 08540
In the United Kingdom: Princeton University Press, Chichester, West Sussex

Library of Congress Cataloging-in-Publication Data

Spitzer, Lyman, Jr., 1914–
  Dreams, stars, and electrons : selected writings of Lyman Spitzer, Jr. / edited by Lyman
Spitzer, Jr., and Jeremiah P. Ostriker.
      p.      cm.
  Includes bibliographical references.
  ISBN 0-691-03702-7 (CL : alk. paper).—ISBN 0-691-02797-8 (pbk. : alk. paper)
  1. Interstellar matter.   2. Stellar dynamics.   3. Space science.   4. Plasma physics;
Controlled fusion.
  I. Ostriker, J. P.   II. Title.
QB790.S664   1997
523.1'125—dc20                                                                          96-16473
                                                                                        CIP

This book has been composed in Times Roman

Princeton University Press books are printed on acid-free paper
and meet the guidelines for permanence and durability of the
Committee on Production Guidelines for Book Longevity of the
Council on Library Resources

Printed in the United States of America by Princeton University Press
1 3 5 7 9 10 8 6 4 2
1 3 5 7 9 10 8 6 4 2
(Pbk.)

*To my wife—*
*my cherished companion,*
*my constant support*

*—L. S.*

———————————————

# CONTENTS

## PART TWO: *Stellar Dynamics*

## PART THREE: *Space Science*

# PART FOUR:  *Plasma Physics; Controlled Fusion*

# PART FIVE:  *Miscellaneous*

# INTRODUCTORY REMARKS

LYMAN SPITZER'S contributions to the physical sciences include seminal and fundamental advances in four major fields—the interstellar medium, stellar dynamics, plasma physics, and space astronomy. The sheer volume of work is staggering, with four research monographs and more than one hundred articles in refereed scientific journals in over a half century of active research. Spitzer's trademark has been incisive physical insight, coupled with an ability to formulate and accurately solve appropriate model problems. The impact of his work is strengthened by a crisp and lucid style of exposition. He has invariably discovered, at the outset of an investigation, which are the important physical effects to be modeled carefully and which processes can be ignored in the initial assay. This is a skill that cannot easily be taught, but the readers of this book will come away with a vision of how a remarkable scientific mind works.

Spitzer has chosen big problems to work on, and somehow, for him, complexities unravel and the fundamental simplicities become apparent. In the area of plasma physics, Spitzer, following H. Alfvén, helped to establish the physical and mathematical foundations of this important discipline in the 1950s. Spitzer recognized early the importance of determining the thermal, electrical (the "Spitzer conductivity"), and mechanical transport coefficients in a fully ionized gas. His pioneering studies in basic plasma physics culminated in the volume "Physics of Fully Ionized Gases" (1956, second edition in 1962), which became a classic, oft cited text, influential to the education of successive generations of plasma physicists.

His work in initiating the study of controlled thermonuclear fusion for the practical generation of energy is well known. Spitzer proposed to the Atomic Energy Commission in 1951 the building of a machine, a "Stellarator," which would be "designed to obtain power from the thermonuclear reactions between deuterium and either deuterium or tritium." Progress since that early time, in several countries, has been continuous, and the cherished point of energy "break even" seems now in sight.

In theoretical astrophysics, Spitzer essentially founded as a discipline the study of "interstellar matter," and made major contributions to stellar dynamics. For the interstellar medium, he was early to suggest that the brightest stars in spiral galaxies have formed recently from the gas and dust there. He also noted the presence and importance of interstellar magnetic fields, the likelihood of a multiphase medium (with hot, warm, and cool components), and the significance of dust grains. His numerous contributions to our knowledge of interstellar space and his broad understanding of the subject were codified in his 1978 monograph, "Physical Processes in the Interstellar Medium," which is still a standard graduate text in this field. In stellar dynamics, he made major contributions to defining "relaxation" and to showing how this process caused a dense stellar system ultimately but inexorably to approach a singular state, the approach accelerated by a distribution of stellar masses but retarded by the presence of binaries. His many contributions to this field were summarized in the 1987 volume, "Dynamical Evolution of Globular Clusters."

Spitzer's seminal contributions to space astronomy are legendary. In 1946, he proposed in a report under Project Rand, entitled "Astronomical Advantages of an Extra-Terrestrial Observatory," the development of large space telescopes. He outlined the advantages to be gained due to greater angular resolution (overcoming astronomical "seeing" problems), to the increased wavelength coverage available, and to the stability of a low-gravity environment. All of these benefits have been realized to some extent with present satellite experiments, with Spitzer a major contributor to their realization. Under his direction, a group of Princeton scientists developed the extremely successful *Copernicus* (32-inch) ultraviolet satellite. Launched in 1972, it made a number of significant astronomical discoveries, among them an accurate value for the cosmologically important ratio of deuterium-to-hydrogen in interstellar space. The currently very productive Hubble Space Telescope, now returning incomparable pictures of the cosmos, was in a quite literal sense Spitzer's "brainchild." He played major roles in shepherding it through many difficult stages of its existence from the earliest planning to its recent refurbishment. Spitzer continues to sit as an elder statesman on the Space Telescope Institute Council, providing wise guidance for this extraordinarily important scientific venture.

In addition to purely scientific skills, Spitzer's vigorous personality, sound judgment, and basic human decency propelled him to positions of leadership at a variety of levels. At Princeton, where he was Chair of the Department of Astrophysical Sciences and Director of the Observatory for a third of a century (1947–79), he built one of the world's leading institutions for astronomical education and research, with an almost unique atmosphere for research. In collaboration with Martin Schwarzschild, his brilliant colleague, Spitzer established a pleasant and mutually supportive environment in which a generous interest of each scientist in other's research leads to increased productivity and originality, as well as the cross-fertilization exemplified best by Spitzer's own work. Also at Princeton, Spitzer was the originator and early leader of the institution that became the Princeton Plasma Physics Laboratory, a major national facility.

As a national scientific administrator, he served as Director of the wartime Sonar Analysis Group (1944–46), President of the American Astronomical Society (1958–60), and Chair of the Space Telescope Institute Council (1981–90), and held other major national leadership positions on numerous committees, commissions, etc., which have guided the scientific life of the nation. Spitzer's service to his country was recognized by medals for scientific achievement and national service by NASA in 1972 and 1976, and by the U.S. National Medal of Science in 1980.

Among Spitzer's international honors are the Karl Schwarzschild Medal (Germany) in 1975, the Gold Medal of the Royal Astronomical Society (England) in 1978, and the Jules Janssen Medal of the Société Astronomique de France (1980), and culminated with the Crafoord Prize of the Royal Swedish Academy of Sciences in 1985, which is equivalent to the Nobel Prize in those areas of science excluded from the Nobel award.

Spitzer continues his scientific work at Princeton, primarily with senior research staff analyzing recent results of the Hubble Space Telescope. His current research focuses on using this premier instrument to help understand the "diffuse matter in space," the title of one of his most influential books. My first task as an eager, young postdoctoral fellow at Princeton in 1965 was to check and proofread this text. I learned firsthand the characteristics of Spitzer's style—clarity, logic and an unswerving concentration on fundamentals, coupled with an incredible capacity for the hard, detailed work of science. Universally admired as a scientist and cherished as a friend and mentor by many, Spitzer's life and work will serve as a model for generations to come.

Jeremiah P. Ostriker  
Princeton  
August 1996

# PREFACE

LATE in 1993, Jerry Ostriker asked if I might be interested in the publication of a volume presenting a selection of my papers. As a consulting editor for Princeton University Press he had been responsible for publishing, in English, the scientific papers of Ya. B. Zeldovich; each paper was accompanied by a commentary discussing briefly the background and the subsequent development of that research field. Jerry pointed out to me that such a book of my own selected papers would make available to astronomers some articles which had never been published in their entirety or which had appeared in publications not easily available. Also, inclusion of individual commentaries would increase the usefulness of the book, especially for scientists not previously exposed to some of the research fields discussed. In any case, I knew I would enjoy re-educating myself in the fields treated in each paper, to allow me to include in some commentaries useful updates on subsequent research programs. So I offered my enthusiastic support in the preparation of this volume.

Which papers should be included? To keep the sales price down and make the book as widely available as possible, Jerry and I agreed that we should aim at a volume less than five hundred pages in length. This decision did not permit us to include more than a fraction of my published works. The first papers I chose to include were those which seemed most relevant to present research activities. Hence, for example, I omitted the 1939 paper which cast grave doubt on the tidal-encounter theory of planet formation. Also omitted are a number of joint papers for which my own contribution seemed, in retrospect, relatively minor. On the other hand, a few survey papers are included, partly to give an adequate background for specialized research work in the same field, partly to give material of possible value in the history of astronomy.

Jerry had urged me to include a few articles which were not primarily scientific, but which might give some insight concerning Princeton astronomers, myself included. The last five papers in the book represent my response; these contain relatively little discussion of astrophysical theory or of the data which theory attempts to explain.

In some papers of substantial length, one or more sections have been omitted to save space. Each such omission is indicated in a footnote (†) at the beginning of the paper and is usually mentioned in the accompanying commentary. A few typographic errors in the equations have been corrected, indicated in each case by a footnote (†). Except for obvious typographic slips, any errors corrected in the text are also so indicated.

Whenever feasible, the commentaries include brief status reports on those research topics which seem relevant to the papers which they accompany. Of course, space limitations restrict these summaries to simplified statements of what seems to be our present knowledge, with little discussion of conflicting opinions. The references cited usually represent a very small fraction of the important papers available, but might be helpful in starting more thorough studies. There are virtually no references to papers appearing after the end of 1994, when systematic writing of the commentaries was completed.

A list of my published papers is included at the end of this volume, including: (A) books, (B) technical papers, and (C) nontechnical and (somewhat) more popular papers. Brief comments during discussions at scientific meetings have been omitted, as have internal memoranda relating to my various activities; otherwise the list is complete as far as I know.

As with several previous books of mine, I am much indebted to the Institut d'Astrophysique in Paris, and to its Director, Dr. Alain Omont, for providing the ideal environment needed for a concentrated three-month effort on this book. I am grateful to many astronomical friends and colleagues for their helpful suggestions on preliminary drafts of the various commentaries, in particular to J. N. Bahcall, J. Bergeron, D. H. DeVorkin, B. T. Draine, J. L. Greenstein, E. B. Jenkins, D. C. Morton, D. E. Osterbrock, B. D. Savage, M. Schwarzschild, S. R. Shapiro, W. Stodiek, S. D. Tremaine, J. W. Tukey, and D. G. York. Special thanks go to J. P. Ostriker for a careful reading of all this material and for a wide variety of important scientific comments. I am indebted to our daughter, Lydia, for the commentary accompanying the summary of my early personal history in Paper 32. My wife has been particularly helpful in suggesting stylistic changes to give a clearer, less pedagogical prose.

*Princeton University Observatory*
*June 1995*

# PART ONE

# Interstellar Matter

# THE DYNAMICS OF THE INTERSTELLAR MEDIUM. I.
# LOCAL EQUILIBRIUM

(REVISED VERSION: ASTROPH. J. 93, 369, 1941 AND 94, 232, 1942)

## *Commentary*

THIS paper had its genesis in 1938–1939, when I was a postdoctoral fellow at the Harvard College Observatory. I remember well the excitement occasioned by Bethe's pioneering work[1] that year, showing that the energy radiated by the brighter stars was generated by nuclear fusion, via the carbon-nitrogen-oxygen cycle, converting hydrogen to helium. Martin Schwarzschild and I had lively discussions of this result, whose consequence seemed to be that supergiant stars could not have been shining at their present rate since the formation of our Galaxy, which must be considerably older than some $10^9$ years, the age of the Earth. Since we knew of no apparent means for a star to go into a low-luminosity state of suspended animation, we concluded that these extremely luminous stars must have formed relatively recently; the dust clouds in spiral galaxies provided a likely site for the birth of these stars.

As a young theorist I was attracted to sweeping theoretical explanations, an inclination which I suspect was strengthened by my intense respect for Eddington. The popular books which he and Jeans wrote, presenting broad surveys of astronomy, gave me a lifelong fascination with the big problems of the universe—its formation, structure, and evolution. Eddington's classic monograph, Internal Constitution of the Stars, seemed to me a model of physical insight, using intuitively plausible approximations to obtain a general picture of how stars are built. I was disappointed that my personal contacts with him, when he was my nominal mentor at Cambridge University in 1935–1936, were not particularly helpful or inspiring. However, this experience did not in the least diminish my admiration for his creative genius in theoretical astrophysics.

Attracted as I was by broad, inclusive theories, I assembled the arguments given in this paper, and even aspired to develop this point of view as an entirely deductive theory, with the results of Table 1 all following inexorably from the difference of angular momentum between flattened spirals and globular star systems. By the time I moved to Yale the following year, I realized that our basic information on the properties of the interstellar medium was grossly inadequate for any portion of this grandiose theoretical project. So I embarked on a long-range program to analyze various physical processes in interstellar space, a project in which I have been engaged, off and on, for much of my career.

Paper #1 was my first attempt to publish in this field. In September 1940, I sent copies of this draft to two of my senior astronomer friends, whom I greatly respected. After tactful comments on the "new approach to interstellar problems," they each devoted some eight pages to proposed clarifications and other improvements. One of them added that "the metaphysical introduction is not suitable," and proposed that this material be deleted and perhaps published at the end of my proposed series of papers on interstellar matter. Naturally, I was disappointed by the recommendation that the observational summary in Table 1, together with the accompanying text, be deleted. I had already omitted the one item which seemed to me somewhat speculative; this is the suggestion that the continuing formation of young stars from interstellar gas in spiral galaxies provided a plausible explanation for some of the observations. Nevertheless, I followed the advice, with §1 to 3 appearing first,[2] and with an expanded discussion of radiation pressure appearing a few months later.[3] World War II interrupted my series of technical papers on interstellar matter. The concepts put forward in Table 1 were published in 1948—see Paper #3.

I am pleased that this early discussion is being published here. While Table 1 refers to differences between the two extreme types of stellar systems, the text a few paragraphs later states that as regards points 4, 5, and 6 in the Table, E galaxies are generally similar to the globular systems (essentially globular clusters). Similarly Sa and Sb galaxies differ from Sc types more in degree than in kind. Thus, the lower three rows of Table 1 describe differences of physical content between spiral and elliptical galaxies.

These contrasts are now well established. Baade's justly celebrated paper[4] on stellar populations has greatly strengthened the observational evidence on the difference of maximum stellar luminosity between elliptical and spiral galaxies. His observations of neighboring spiral galaxies have supported the association of dust lanes with supergiant stars, an association fully confirmed in many modern studies. Interstellar matter is now known to be present in many elliptical galaxies also, but generally with a density much too low to form stars at a significant rate (see Commentary, Paper #2). As regards the frequency of supernovae (point 6), a recent analysis[5] indicates that the relatively few supernovae observed in E galaxies are all of type Ia, not the Ib and II types which are conspicuous in spiral galaxies and are the only ones attributed to the death of massive stars. While the detailed origin of these differences is obscure (as are the marked differences of rotational velocities recently found between different elliptical galaxies), the sweeping picture presented in Table 1 seems now to be substantiated in broad outline.

The argument that the brightest stars in spiral galaxies must have been formed relatively recently from the interstellar matter present depends on the ages of supergiant O and B stars; I was well aware that our knowledge of stellar structure and evolution was inadequate to convince other astronomers of these low ages. By the 1960s this situation had entirely changed; stellar evolution studies had established[6] ages of 10 to $20 \times 10^9$ years for stars in globular clusters and from $10^6$ to $10^7$ years for the bright O and B supergiants in Galactic clusters and associations. There seemed to be general agreement that these supergiant stars had formed recently from interstellar gas and dust. More recent analyses[7] give about the same ages.

The conclusion that supergiant O and B stars in our own Galaxy have formed recently, and that presumably others are forming currently, gave an enormous boost to theoretical studies of star formation. Observations of physical conditions in star-forming regions can provide the theorist with crucial "observational guidance" (a phrase from S. Chandrasekhar). Such guidance became even more dominant with the development of infrared astronomy many decades later, but was important in the 1940s also. A preliminary approach to a star formation scenario, from that early time, appears in Paper #3 in this volume.

The sections of Paper #1, following the introduction, explore a number of physical aspects of the interstellar medium, especially those relevant to the various dynamical processes involved in star formation. For example, the relaxation and equipartition times found for interstellar particles are sufficiently short that the gas may usually be treated as a fluid, with atomic mean free paths short compared to relevant dimensions. The mutual force between grains of dust, in the presence of an ambient radiation field, can be a significant factor leading to condensations in the interstellar medium. These discussions in 1939 were very much handicapped, of course, by lack of the extensive information now available on this medium. In particular, the large difference in kinetic temperature between H II regions and the denser H I clouds[8] was unknown then. The 100-to-1 ratio of gas to dust (by mass) in interstellar space (Paper #12 in this volume) had not yet been measured, and the presence[9] of solid particles with radii down to some 5 Å was unsuspected.

Of the results given in Paper #1, one which is particularly affected by this new information concerns the electric charge on small solid dust particles, or grains. For radiation shortward of 1500 Å the photoelectric efficiency can[10] exceed 0.1. This large enhancement of photoelectric emission over what I had assumed in 1939 tends to give grains a positive charge, while for the smallest grains the photon absorption is so slight that negative charges can appear, with a single electron adhering to the grain about half the time.[11] These processes have an important effect on the heating of warm H I gas by photoelectrons from the small particles (see Commentary #4).

As evident from gaps in the numbering of the sections and equations, some discussions in the draft paper have been omitted from the version given here. Two items were dropped from the published version[2,3] because they did not seem of sufficient interest. These items, omitted here also, concern: (1) dynamical properties of the interstellar medium if the mass in solid particles were to exceed that in separate atoms—at the end of §2; (2) condensation of gas to form solid particles—in a final §7. Also omitted here are §3 on the viscosity of the interstellar gas and a brief discussion in §4 of motion by a solid particle near a massive, luminous star; these two topics are better treated in the published version.[2,3] Apart from these omissions, the only changes in the original version have been the modifications noted in Table 2 and one change in notation in §4.

### REFERENCES

1. Bethe, H. A.—*Phys. Rev.* **55**, 434 (1939).
2. Spitzer, L.—*Astroph. J.* **93**, 369 (1941).
3. Spitzer, L.—*Astroph. J.* **94**, 232 (1941).
4. Baade, W.—*Astroph. J.* **100**, 137 (1944).
5. Cappellaro, E. et al.—*Astron. Astroph.* **273**, 383 (1993).
6. Sears, R. L., Brownlee, R.—In: Stars and Stellar Systems VIII, eds. L. H. Aller and D. B. McLaughlin, Chicago Univ. Press, 619 and 630 (1965).
7. Schaller, G., Schaerer, D., Meynet, G., and Maeder, A.—*Astron Astoph. Supp.* **96**, 269 (1992); VandenBerg, D. A.—*Astroph. J. Supp.* **51**, 29 (1983).
8. Kulkarni, S. R. and Heiles, C.—In: Interstellar Processes, eds. D. J. Hollenbach and H. A. Thronson, Jr., Dordrecht: Reidel, 87 (1987).
9. Allamandola, L. J., Tielens, A. G. G. M., and Barker, J. R.—*Astroph. J. Supp.* **71**, 733 (1989).
10. Draine, B. T.—*Astroph. J. Supp.* **36**, 595 (1978).
11. Bakes, E. L. O. and Tielens, A. G. G. M.—*Astroph. J.* **427**, 822 (1994).

# *Paper*

ABSTRACT

Observations suggest that many differences between late-type spiral galaxies and globular systems may have some connection, direct or indirect, and that the presence of an interstellar medium may be important in theories of galactic structure. As a preliminary to a theoretical treatment of such possibilities, the known properties of interstellar matter are here discussed. Interstellar lines are shown to originate in regions of hydrogen ionization, provided that there is one hydrogen atom per $cm^3$ in space and that hydrogen is much more abundant than other elements. Non-hydrogenic atoms are known to be highly ionized; electron collisions will charge interstellar dust particles to a potential of $-2.2 \times 10^{-4}T$ volts, where $T$ is the electron temperature. Equipartition of kinetic energy between dust and atoms is established in $6 \times 10^4$ years, provided that the radius of the particles is not greater than $10^{-5}$ cm. The interstellar medium will behave as a perfect gas.

The general radiation field of the Galaxy, homogeneous in galactic longitude, is shown to give rise to an inverse-square force between two metallic dust particles in the galactic plane that is $1.4 \times 10^4/d^2$ times the force of gravity between them, where $d$ is the density of matter in the particle. Similar forces between dust and atoms and between two atoms are shown to be unimportant.

Recent work,[1] though by no means precise, indicates that the average density of interstellar matter is about $10^{-24}$ gm/$cm^3$ in order of magnitude, that a substantial fraction of this is in the form of compound particles with diameters ranging around $10^{-5}$ cm,[2,3] and that the rest includes neutral and ionized atoms[4] and molecules.[5] The distribution of the small "dust" particles is very spotty, such particles apparently occurring in clouds in the galactic plane.

The influence of this medium on galactic and stellar evolution has not been extensively discussed. Hoyle and Lyttleton[6] have advanced a theory of stellar accretion, but their work has been criticized by Atkinson.[7] Gamow[8] has alluded briefly to the formation of stars from clouds of absorbing matter, but this work has not been quantitative, and overlooks some of the important aspects of the problem. The influence of absorbing matter on the general problems of galactic structure, as

TABLE 1

| Sc Spirals | Globular Systems |
|---|---|
| 1. High Angular Momentum | No Angular Momentum |
| 2. Extremely flattened spheroids | Spheres |
| 3. Spiral Arms—Symmetry of rotation through 180° about axis | Symmetry about any axis |
| 4. Interstellar dust and atoms | No traces of any interstellar matter |
| 5. Maximum stellar luminosity between $10^4$ and $10^5 L_\odot$. | Maximum stellar luminosity less than $10^3 L_\odot$ |
| 6. Supernovae frequent | Supernovae rare |

distinct from the obvious effect on the observations relevant to these problems, has apparently not been stressed.

The observational evidence suggests, however, that the presence of interstellar matter is perhaps connected in some way with other properties of stellar systems. Table 1 lists the most important differences between the two extreme types of stellar systems, the late-type Sc spirals, and the globular nebulae (and clusters). The first two entries are of course the simplest and the most obviously related. The presence of spiral arms has been attributed by Lindblad[9] to the type of gravitational potential which exists in a highly flattened disc. The existence of absorbing matter in late-type spirals is evident from the photographs of these objects. Absorbing matter has not yet been detected in globular clusters or in nebulae, or even in elliptical nebulae. Irregular nebulae are observed to have patches of obscuring matter, but these systems are peculiar in many respects. Any absorbing matter which exists in the typical globular or elliptical systems is certainly not concentrated into the dark lanes seen in the late-type spirals.

The maximum luminosity of stars in the Galaxy—excluding novae—corresponds to an absolute magnitude between $-5$ and $-7$. Stars brighter than $-2$ are lacking from globular clusters.[10] The globular and elliptical nebulae are not resolved into stars; if it is assumed that the stars are unobscured, this fact sets a limit of about $-2$ for the absolute magnitude of the brightest stars in the closest elliptical nebulae, M32.[12] Since the surface brightnesses of all elliptical systems are roughly comparable, and since no stars are resolved in any of them, we may take $-2$ as an upper limit of the absolute magnitude for the stars in such systems.

The significance of this upper limit for stellar luminosity arises from its relation to current theories of stellar energy. Bethe[12] has shown that the carbon-nitrogen nuclear transmutation cycle suffices for the energy of the main-sequence stars. The chief difficulty in the problem of stellar energy generation lies in reconciling the continued existence of the brightest stars with the time scale of $2 \times 10^9$ years. The maximum energy available from the transmutation of a gram of hydrogen is $0.008c^2$, or roughly $7 \times 10^{18}$ ergs. Hence any star which has a luminosity much greater than 100 ergs/gm sec cannot have radiated at that rate for $2 \times 10^9$ years without calling on additional sources of energy, which nuclear physicists have not yet found.

Since a star's luminosity increases as its hydrogen is consumed, we may take 300 ergs/gm sec as the present critical rate of energy generation. This corresponds to a luminosity of about $10^3 L_\odot$, or an absolute magnitude of $-2.9$, and a mass of $7M_\odot$. The inference is that in globular and elliptical systems the Bethe theory is wholly adequate, but that we must find some explanation for the brightest stars in the spiral galaxies.

The relative frequency of supernovae given as the last entry in Table 1 refers to the fact that of the 18 supernovae observed before May 1938, only one of these occurred in an elliptical system.[13] If we take Hubble's[14] estimate that the elliptical nebulae constitute 17 percent of the total, we find that the probability of a supernova in any one spiral system is three times as great as the corresponding probability for an elliptical or globular system. With so few observations, however, this conclusion is highly tentative.

The Sc spirals and the E0 nebulae represent the two extremes. The intermediate types show intermediate properties in some respects; all the elliptical systems, however, are similar to the globular ones as regards points 4, 5, and 6 in Table 1.

Any theory of nebular structure must presumably explain the differences between the two types of systems. The observational data are too inadequate for any exhaustive attempt in this direction. But present theoretical knowledge is sufficient for a preliminary evaluation of the role which interstellar matter may play in this respect. We know that interstellar matter exists on a large scale in spiral systems. Can it account for any of the other characteristic differences between spiral and elliptical systems?

A knowledge of the characteristics of the interstellar medium is necessary for a discussion of this question. An attempt has accordingly been made in the present investigation to determine the properties of interstellar matter with respect to ionization, equipartition of kinetic energy, radiation pressure, and the condensation of atoms to form larger particles. Many of the results are not new, but are gathered together, apparently for the first time, in a survey which may be of value independently of the rest of the work. Subsequent papers will deal with the behavior of the medium in the gravitational and radiation fields of a galaxy, and with the frequency, nature, and possible evolution of condensations within the medium.

# I. Ionization

It has been realized for some time that interstellar gases, despite the low energy density of ionizing radiation, are doubtless in a high state of ionization. Also the velocity of the free electrons should be relatively high, corresponding to the color temperature of the ionizing radiation. A precise determination of $n_e$, the number of electrons per cm$^3$, is important partly because it reveals the general level of ionization, and partly because it sets a lower limit to the density of interstellar atoms.

One of the most complete investigations of this problem has been carried through by Dunham.[15] The intensity of interstellar H and K relative to λ4227 gives a ratio of 3500 Ca$^+$ atoms to every Ca atom. This fact, together with the energy density of interstellar radiation computed by Dunham, gives an electron density of about 10 cm$^{-3}$. The ionization formula used neglects the fact that ionization occurs only from the ground state, and yields a value for $n_e$ about ten times too large. A comparable value of 3 hydrogen atoms per cm$^3$ has been found by B. Strömgren[16] from an analysis of Struve and Elvey's[17] data on Hα emission in Cygnus and Cepheus. The discovery[5] of lines from interstellar CH, NaH, and CN is a verification of so high a density of hydrogen; we may therefore assume a density of $1.6 \times 10^{-24}$ gm/cm$^3$ for the interstellar atoms.

As Strömgren has pointed out,[16] the ionization of hydrogen is strongly affected by the heavy absorption of radiation beyond the Lyman limit. The result of Strömgren's analysis is that within a distance of 26 parsecs from a B0 star and 140 parsecs from an O5 star the hydrogen is very largely ionized, provided that one hydrogen atom per cm$^3$ is assumed. Outside this region the hydrogen is largely neutral, and radiation beyond the Lyman limit is almost completely absorbed. The rough agreement of these numerical values with the radii of observed emission regions provides a further check on the assumed density of hydrogen.

These two regions, which according to Strömgren are rather sharply separated, will be referred to here as H II, the region of hydrogen ionization, and H I, the region of neutral hydrogen. If there are a hundred hydrogen atoms to every nonhydrogenic atom the electron pressure will depend almost entirely on the ionization of hydrogen. Hence $n_e$ should be a hundred times as great as H II as in H I, and neutral atoms should show a corresponding concentration in H II relative to H I.

The increase of ionizing radiation near B and O stars present in H II will tend to offset this effect. If we accept Strömgren's estimate, however, that roughly one-tenth of interstellar hydrogen is ionized, then the average intensity of radiation between 1000 Å and 3000 Å cannot be more than 5 to 10 times as great in H II, on the average, as in H I. Since, then, the ratio of neutral to ionized atoms should be more than ten times as great in H II as in H I, it is evident that the observed interstellar lines arise primarily in H II; Dunham's determination of $n_e$ should be taken to refer to this region only.

If this interpretation is correct, the number of Na and Ca atoms per $cm^3$ must be ten times as great as that given by Dunham. This still leaves hydrogen ten to twenty times more abundant by mass than other atoms, excluding helium.

Small particles will also be highly ionized. The influence of cosmic rays discussed by Corlin[18] in the case of meteorites is quite negligible for particles of radius $10^{-5}$ cm. Since the total number of primary cosmic ray particles entering the upper atmosphere of the earth is 0.09 per $cm^2$ per sec,[19] a dust particle of so small a radius will be hit only once every hundred years, and will be little affected by the collision in any case. During the same period more than $3 \times 10^4$ electrons would collide with a neutral particle of this size, even if the electron density were as low as $10^{-3}$ per $cm^3$.

It is rather the simple processes of electron collision and photoelectric ionization that are responsible for the ionization of interstellar particles. The charge $Q_d$ on the dust particle must be such that a steady state exists. The cross-section $\pi \sigma_i^2$, for the capture of an ion of charge $Q_1$ mass $m_i$, and velocity $v_1$ relative to the particle is given by

$$\pi \sigma_i^2 = \pi \sigma^2 \left( 1 - \frac{2 Q_d Q_i}{m_i v_i^2} \right). \tag{1}$$

The total number of captures per second for ions of this type, which we shall denote by $\Gamma_i$, may be found by multiplying (1) by $n_i v_i$ and averaging over a Maxwellian velocity distribution, the average extending only over those values for which $\sigma_i^2$ is positive. The integration is elementary and we find

$$\Gamma_i = 4 \left( \frac{\pi}{3} \right)^{1/2} \frac{n_i \sigma^2 E_i^{1/2}}{m_i^{1/2}} \begin{cases} 1 - \gamma_i, & \gamma_i < 0, \\ e^{-\gamma_i}, & \gamma_i > 0. \end{cases} \tag{2}$$

where

$$\gamma_i = \frac{3Q_iQ_d}{2\sigma E_i} = \frac{3VZ_i}{2\epsilon_i};\tag{3}$$

$E_i$ and $\epsilon_i$ denote the mean kinetic energy of particle $i$ in ergs and electron volts, respectively; $V$ is the electrical potential of the dust particle in volts.

The number of photoelectrons ejected per second will be roughly $\pi\sigma^2 Uc\zeta/h\nu_t$, where $U$ is the energy density of radiation per cm³, $\nu_t$ is the threshold frequency, and $\zeta$ is the fraction of all quanta traversing the sphere which are effective in photoionizing it. The fraction $\zeta$ will be a product of the fraction of the total radiation which has a frequency greater than $\nu_t$ and the efficiency of such radiation in ejecting electrons. Since as we shall see below the particles will be negatively charged, in general, all the photoelectrons will escape.

The assumption of a steady state requires that the total charge gained by the particle per second must vanish. If we assume equipartition of kinetic energy, $E_i$ must be independent of $i$ and may be replaced by $E$. If we assume also that protons are the only positive ions present, we find from (2)

$$1 + \gamma + K = \left(\frac{m_p}{m_e}\right)^{1/2} e^{-\gamma},\tag{4}$$

where subscripts $e$ and $p$ refer to electrons and protons, respectively; $\gamma$ is equal to $\gamma_e$ and $-\gamma_p$, where these quantities are defined in (3); $K$ is defined by

$$K = \frac{(6\pi)^{1/2}Uc\zeta}{4h\nu_t n_e\left(\overline{v_p^2}\right)^{1/2}}.\tag{5}$$

The quantity $K$ is the ratio of the number of electrons ejected per second photoelectrically to the number of protons which would be captured per second by direct collision if the particle were neutral. For most substances the photoelectric threshold lies between 1000 Å and 2000 Å. With Dunham's value[15] of $5.2 \times 10^{13}$ ergs/cm³ for $U$, and a root mean square proton velocity of $1.6 \times 10^6$ cm/sec, corresponding to a temperature of 10,000°, we see that $K$ is less than $10\zeta/n_e$. Since only about $3 \times 10^{-2}$ of the total radiant energy in space is of wave length shorter that 2000 Å, and since the efficiency of radiation in producing photoelectrons is rarely as high as $10^{-3}$, we see that $\zeta$ will be less than $10^{-5}$ in all probability. Hence in H II $K$ will be much less than unity and may be neglected.

If we then set $m_p/m_e$ equal to 1840, (4) has the numerical solution

$$\gamma = 2.51. \tag{6}$$

Even if $K$ is as great as 10, the value of $\gamma$ is one half that given by (6); we shall assume that (6) is generally valid.

If we combine (6) and (3) we find that the change $Q_d$ on the particle is such that the potential energy of an electron on the surface is $2/3 \times 2.51$ times the mean kinetic energy of the electron. If this mean kinetic energy equals $\epsilon$ electron volts, and if $V$ is the potential of the particle, we have simply

$$V = -1.67\epsilon. \tag{7}$$

If the electron velocities corresponds to a temperature $T$, $\epsilon$ is given by

$$\epsilon = 1.29 \times 10^{-4}T. \tag{8}$$

If $T$ is 10,000°, a dust particle will therefore be charged to a potential of $-2.2$ volts. Unless the electron density is much below $10^{-3}$, in which case dust particles will be positively charged, we may conclude that any small compound particle will be charged to a potential of about $-2$ volts, independently of its composition or its radius. If the radius $\sigma$ of the particle is $10^{-5}$ cm, this corresponds to a charge of $140e$.

## II. EQUIPARTITION OF KINETIC ENERGY

The rate at which energy is equalized between particles of different mass has been calculated[20] for the gravitational case. The resultant formulae may be carried over to the electrostatic case provided that the gravitational constant $G$ is replaced by $Q_1Q_2/m_1m_2$, where $Q_1, Q_2, m_1$, and $m_2$ are the charges and masses of the two interacting particles. If the kinetic energies of the two particles are each distributed according to the Maxwellian law, we find for the time rate of change of $E_1$, the mean kinetic energy of particle 1.

$$\frac{dE_1}{dt} = \beta n_2 \frac{E_2 - E_1}{(E_2/m_2 + E_1/m_1)^{3/2}}, \tag{9}$$

where $n_2$ is the number of particles of type 2 per $cm^3$; $\beta$ is defined by

$$\beta = \frac{2(3\pi)^{1/2}Q_1^2Q_2^2}{m_1 m_2} \log\left(1 + \frac{v^4\mu^2 R_e^2}{Q_1^2 Q_2^2}\right). \tag{10}$$

In the argument of the logarithm in (10) $v^2$ may be taken as the larger of the two mean square velocities; $R_e$ is the maximum distance at which an encounter is effective; $\mu$ is the reduced mass. If $Q_1$ is greater than $Q_2$, $R_e^3$ is approximately equal to $Q_1/Q_2 n_2$, or to $1/n_1$, whichever is the smaller.

Let us take $m_i$ equal to $A_1 m_0$, where $m_0$ is the mass of unit atomic weight, and $Q_1$ equal to $Z_1 e$, where $e$ is the electronic charge. The logarithmic term becomes

$$\log\left(\frac{v^4\mu^2 R_e^2}{Q_1^2 Q_2^2}\right) = 32.9\left(1 + \frac{1}{7.1}\log\frac{\epsilon}{Z_1^2 Z_2^2} + \frac{1}{21}\log_{10}\frac{Z_1}{Z_2 n}\right). \tag{11}$$

The correction term in parentheses is small; if we neglect it we have form (9) and (10),

$$\frac{d\epsilon_1}{dt} = 4.15 \times 10^{-6} n_2 \frac{Z_1^2 Z_2^2}{A_1 A_2} \frac{\epsilon_2 - \epsilon_1}{(\epsilon_1/A_1 + \epsilon_2/A_2)^{3/2}}, \tag{12}$$

where $\epsilon_i$ is again the mean kinetic energy of particle $i$ in electron volts, and is numerically equal to $E_1/1.59 \times 10^{-12}$.

Let particle 2 be the particle of greater velocity; i.e., let $\epsilon_2/A_2$ exceed $\epsilon_1/A_1$. If $\epsilon_2$ is relatively constant then equation (12) may be integrated directly. This will be the case if: 1. the energy density $n_2 \epsilon_2$ of particle 2 exceeds $n_1 \epsilon_1$ (a mass abundance $n_2 A_2$ greater than $n_1 A_1$ implies this of course); 2. the energy $\epsilon_2$ is kept constant by other means; or 3. the deviations from equipartition are small. If we neglect $\epsilon_1/A_1$ in the denominator of (12) and assume, then, that $\epsilon_2$ is constant, we integrate to find

$$\epsilon_2 - \epsilon_1 = Ce^{-t/t(1,2)} \tag{13}$$

where $C$ is a constant and $t(1,2)$ is the time of equipartition of particle 1 with a faster particle 2; $t(1,2)$ is given by

$$t(1,2) = 2.4 \times 10^5 \frac{A_1 \epsilon_2^{3/2}}{A_2^{1/2} Z_1^2 Z_2^2 n_2} \text{ sec.} \tag{14}$$

If the faster particle 2 is also the more massive, (13) is valid only until such time as the less massive particle 1 has attained a velocity comparable with that of 2. If the faster particle has a smaller energy density than the slower, $\epsilon_2$ will change relatively much more than $\epsilon_1$, (13) is no longer valid, and the approach to equipartition is not exponential.

Another time of relevance is the relaxation time $\tau$, defined as the time in which deviations from a Maxwellian velocity distribution fall to $1/e$ of their initial value. From the formula in the gravitational case[21] we find, with the same substitutions as before,

$$\tau = 3.6 \times 10^5 \frac{A^{1/2}\epsilon^{3/2}}{Z^4 n} \text{ sec.} \tag{15}$$

This value refers to equipartition between similar ions only. It is approximately equal to the value of $t(1, 1)$ found from (14), a check on the calculation of both quantities.

The velocities of protons are at least equal to those of dust particles. Hence a mass abundance of dust less than that of ionized hydrogen implies that the energy density of dust is less than that of hydrogen. In this case (14) is applicable and the dust will come into equipartition with protons in a time $t(d, p)$, where d refers to dust particles, and p to protons. In H I the hydrogen atoms are not ionized, and (14) applies only if the dust velocities are not greater than a few kilometers a second. Values of $\tau$ and $t(1, 2)$ for typical interstellar particles in both H I and H II are given in Table 2.

The different particles represent typical cases for the population of interstellar space. While the number of calcium atoms per cm³ is not likely to be as high as $10^{-3}$, that of all atoms other than hydrogen and helium is probably of this order of magnitude. The material density of a dust particle is perhaps higher than unity; we shall use the symbol $d$ to denote this density of matter within the particle to avoid confusion with $\rho$, the mass of matter actually present in a cubic centimeter of interstellar space.

The values of $t(1, 2)$ in Table 2 refer to interactions with the next lightest particle present, except for dust particles in H II where the rate of equipartition with protons is much greater than with $Ca^{++}$ atoms. Collisions with neutral H atoms are not important in this connection unless such atoms are $10^4$ times more numerous than the heavier ones.

Since $Z^2/A$ varies as $1/\sigma d$ for a dust particle, it is evident that dust of all sizes will rapidly come to equipartition with protons, and that in fact equipartition of energy will be established between all the interstellar particles within $2 \times 10^3$ years if the hydrogen is ionized, and within

TABLE 2

|        |        | Electrons | Protons[†] | Ca$^{+++}$[†] | Dust |
|--------|--------|-----------|-----------|-----------|------|
| H I | $n$ | $2.0 \times 10^{-3}$ | — | $10^{-3}$ | $10^{-10}$ |
|     | $\rho$ | $1.8 \times 10^{-30}$ | — | $6.6 \times 10^{-26}$ | $4.2 \times 10^{-25}$ |
|     | $\tau$ | 0.19 years | — | 6.7 years | $2.2 \times 10^{4}$ years |
|     | $t(1,2)$ | | — | $2.4 \times 10^{3}$ years | $5.5 \times 10^{4}$ years |
| H II | $n$ | 1.0 | 1.0 | $10^{-3}$ | $10^{-10}$ |
|     | $\rho$ | $9.1 \times 10^{-28}$ | $1.7 \times 10^{-24}$ | $6.6 \times 10^{-26}$ | $4.22 \times 10^{-25}$ |
|     | $\tau$ | 3.4 hours | 150 hours | 6.7 years | $2.2 \times 10^{4}$ years |
|     | $t(1,2)$ | | 0.48 hours | 0.11 years | $1.4 \times 10^{3}$ years |

The quantities $n$ and $\rho$ are the space densities in number of particles per cm$^3$ and gm/cm$^3$, respectively, for each type of particle. The values of $\tau$ for each particle give the time required to establish a Maxwellian distribution of velocities at energies corresponding to 10,000° by interaction with similar particles only. The quantity $t(1,2)$ is the time required for equipartition of energy to be established by interaction with lighter particles. It is assumed that the radius of a dust particle is $10^{-5}$ cm, and that the density of matter within the particle is unity.

[†] Some of the numerical values for protons and Ca$^{++}$ ions have been changed to agree with the more accurate values in the published version.[1]

$6 \times 10^4$ years if it is not. The random velocities of the larger particles will be very slow; if $d$ equals unity and $\sigma$ equals $10^{-5}$, a particle should have a velocity of 32 cm/sec at 10,000°, or less than a mile an hour. The mean free path of such articles will be roughly two astronomical units, about the same as that for electrons in H II.

It is usually assumed that the temperature of the entire assembly will be dictated by the electron temperature; this in turn will be determined by the ionizing radiation. The rate at which the temperature changes as external circumstances change, as stars of different colors and luminosities pass by, is in turn determined by the rate of ionization. The mean life of a free electron—the interval between photoejection and capture —is roughly $2 \times 10^5$ years if there is one proton in every cubic centimeter. Hence in H II, where half the kinetic energy is in the motion of the electrons, the general velocity temperature will follow any changes in the ionizing radiation.

In H I, however, the number of positive ions is much less and the mean life is greater by a factor of $10^2$ or more; since the electronic energy is a small fraction of the total kinetic energy, if neutral hydrogen is present in abundance, the general temperature will change little if at all in $10^8$ or $10^9$ years from this cause. Collisions between atoms and dust will probably lead to a slow decrease of kinetic energy in such a case, provided that the atoms are not actually captured by the particles.

It is probable, however, that the temperature in H I is only slightly less than that in H II, about 20,000–30,000°, the color temperature of the early B and O stars. This result depends, however, on an assumed abundance of 100 hydrogen atoms to each nonhydrogenic one, an assumption which may be seriously in error.

## IV. Radiation Pressure

We discuss here an effect of radiation pressure that has apparently not been considered. Two absorbing particles in a homogeneous radiation field will be forced towards each other by radiation pressure, since the shadow of each on the other will produce an uncompensated force along the line joining the two particles. If the energy density within a frequency interval $d\nu$ is $U_\nu d\nu$, and the absorption cross-section is $\pi\sigma^2 Q_\nu$, we have a resultant radiation-pressure force between the two particles given by

$$f_{r\nu} = \pi\sigma_1^2 Q_{1\nu} \times \pi\sigma_2^2 Q_{2\nu} U_\nu / (4\pi r^2). \tag{24}$$

The subscripts 1 and 2 refer to the two particles. If we integrate (24) over all $\nu$ to find the total force $f_r$, then the ratio of this force to the force of gravity between the two particles, which we denote by $f_g$, becomes, in the case of two similar particles,

$$\frac{f_r}{f_g} = \frac{9UK}{64\pi\sigma^2 d^2 G}; \tag{25}$$

$\sigma$, $d$, and $G$ have the same meanings as before; $U$ is the integrated energy density of radiation and

$$K = \overline{Q_\nu^2} = \frac{1}{U} \int_0^\infty U_\nu Q_\nu^2 d\nu. \tag{26}$$

Metallic particles are primarily responsible for absorption as opposed to scattering. For such particles $Q$ is approximately equal to $2\pi\sigma/\lambda$, provided that $\sigma$ is less than $\lambda/2\pi$. Thus as $\sigma$ decreases, (25) approaches a limiting value which is independent of $\sigma$. With Dunham's value $5.2 \times 10^{-13}$ erg/cm$^3$ for $U$[15], (26) yields

$$\frac{f_r}{f_g} = 1.4 \times 10^{-5} \frac{1}{d^2} \left(\frac{1}{\lambda^2}\right). \tag{27}$$

The average value of $1/\lambda^2$ integrated over Dunham's energy distribution $U_\nu$ is approximately $2.8 \times 10^9$ for no absorption; if the corresponding data for absorption are taken $K$ is only 0.4 as great as in the case of no absorption. To find a lower limit in (27) we replace $1/\lambda^2$ by $10^9$, $d$ by 7.8, and find that $f_r/f_g$ equals 230. If $\sigma$ is greater than $5 \times 10^{-6}$ the approximation for $Q$ is no longer valid and $f_r/f_g$ will be less than this value. If $d$ is unity, however, $f_r/f_g$ is increased to $1.4 \times 10^4$

Nonmetallic particles will give scattering rather than absorption; the ratio of the scattering to the total extinction—scattering plus absorption —is called the albedo of the absorbing medium. Determinations of this quantity are somewhat discordant,[31] but the most probable value at present is about 0.35,[32] giving equal amounts of absorption and scattering. Unless the scattering efficiency per gram of matter is very much less than that of absorption per gram of metallic particles, we see that there exists an attraction between the particles of interstellar space which amounts to at least a hundred times the force of gravity between them, on the average. This force, which for small optical depths behaves mathematically in the same way as gravitational attraction, is of great importance in the formation of condensations within the medium.

The similar force between dust and atoms cannot be of much importance in the distribution of interstellar matter. Let us consider the extreme case in which all the line radiation is absorbed by a dust cloud and as a result atoms have been attracted to and concentrated in the cloud. The excess pressure of atoms within the cloud is given by $(1 + Z)nkT$, where $n$ is the number of atoms per $cm^3$ within the cloud minus the corresponding number outside, $Z$ is the average number of free electrons per atom, and $k$ is the usual Boltzmann constant. If these atoms are held within the cloud by radiation pressure $P_r$ we must have

$$P_r = \tfrac{1}{3}\zeta U = (1 + Z)nkT, \tag{28}$$

where $\zeta$ is the fraction of the total radiation which atoms are capable of absorbing, and $P_r$ is the pressure on the atoms arising from this source. Since $\zeta$ is less than $10^{-3}$, in general, we have

$$n < \frac{1.3}{(1 + Z)T}. \tag{29}$$

Since $T$ is about 10,000° and may be considerably more, the increase in the concentration of singly-ionized atoms from this source amounts to $7 \times 10^{-5}$ atoms/$cm^3$. Inasmuch as the average interstellar abundance of Na and Ca atoms is already of this order, radiation can at most double the concentration of such atoms.

For a dust particle, on the other hand, we may take $Z$ equal to 150, and $\zeta$ equal to unity. At a temperature of $10^4$ the value of $n$ given by (29) is $9 \times 10^{-4}$, corresponding to a maximum density of about $10^{-18}$ gm/cm$^3$.

The general radiation field is sensibly constant with galactic longitude, owing to relatively heavier absorption in the direction of the galactic center, but varies greatly with galactic latitude. Under such circumstances two scattering particles in the galactic plane will scatter radiation out of the plane; this will decrease the intensity of radiation between them and will give rise to a force tending to push the two particles together. This is of no importance for atoms, for the reasons discussed above. For dust particles this will increase the initial mutual attraction between two particles which both scatter and absorb radiation, and will provide also a force between two scattering nonabsorbing particles. This assumes that the line joining the atoms is parallel to the galactic plane; if not, the force may be repulsive. When a condensation has developed appreciable optical depth, however, the effect vanishes, as the radiation will diffuse through the dust and develop an isotropic radiation field.

## REFERENCES

1. Seares, F. H.—*Publ. Astron. Soc. Pacific* **52**, 80 (1940); Ann. d'Astroph. **1**, 21 (1940).
2. Schalen, C.—Uppsala Medd. No. 64 (1935).
3. Greenstein, J. L.—Harvard Circular No. 422 (1937).
4. Merrill, P. W., Sanford, R. F., Wilson, O. C., and Burwell, C. G. *Ap. J.* **86**, 274 (1937).
5. McKellar, A.—*P.A.S.P.* **52**, 187 (1940).
6. Hoyle, F. and Lyttleton, R. A.—*Proc. Cambridge Philos. Soc.* **35**, 592 (1939).
7. d'E. Atkinson, R.—*Month. Not. RAS* **100**, 500 (1940).
8. Gamow, G.—The Birth and Death of the Sun, (New York: Viking Press), p. 197 (1940).
9. Lindblad, B.—*Ap. J.* **92**, 1 (1940).
10. I am indebted to Dr. Shapley for this information.
11. Hubble, E. P.—The Realm of the Nebulae (New Haven: Yale Univ. Press), p. 139 (1936).
12. Bethe, H. A.—*Phys. Rev.* **55**, 434 (1939).
13. Baade, W.—*Ap. J.* **88**, 285, (1938).
14. Hubble, E. P.—ref. 11, p. 55.
15. Dunham, T. H., Jr.—*Proc. Am. Philos. Soc.* **81**, 277 (1939).
16. Strömgren, B.—*Ap. J.* **89**, 526 (1939).
17. Struve, O. and Elvey, C. T.—*Ap. J.*, **88**, 364 (1938).

18. Corlin, A.—*Zs. f. Ap.*—**15**, 259 (1938).
19. Bowen, I. S., Millikan, R. A., and Neher, H. V.—*Phys. Rev.* **53**, 217 (1938).
20. Spitzer, L., Jr.—*Month. Not. RAS* **100**, 396 (1940).
21. Spitzer, L., Jr.—ref. 20, formulae (11), (15), and (17a).
31. Struve, O.—*Ap. J.* **85**, 194 (1937); Henyey, L. G.—ibid. **85**, 255 (1937); Greenstein, J. L. and Henyey, L. G.—ibid. **89**, 647 (1939).
32. I am indebted to Dr. Greenstein for this value.

# THE DYNAMICS OF THE INTERSTELLAR MEDIUM. III.

## GALACTIC DISTRIBUTION

(ASTROPH. J. 95, 329, 1942)

## *Commentary*

THE purpose of this work was to show that the observed absence of gas in globular and elliptical systems, emphasized in Paper #1, was a natural and direct result of the dynamical structure of these systems. The physical arguments presented are straightforward and show that gas can be coextensive with stars in a massive such galaxy only if the gas is very hot. Equation (34) for the maximum mass of cool gas and equation (41) for the density distribution in a plane one-dimensional system were new in 1941, but are valid, of course, only for the idealized models considered. In Paper #17 a variant of the former equation is used to give the maximum mass in relatively heavy stars within the core of a globular cluster.

The principal conclusion of Paper #2, that most of the volume of elliptical galaxies must be relatively free of interstellar matter, is fully consistent with present evidence as far as cool (100 K) or warm ($10^4$ K) gas is concerned. Traces of such gas are certainly present in many ellipticals, and with more sensitive detection methods the fraction of ellipticals found to contain such traces increases more and more. Thus a survey of elliptical galaxies at 21 cm detected emission[1] in only 5 percent of the E systems observed, while CCD measures of selective extinction (with B, V, and I pass-bands) indicated dust lanes and patches[2] in 80 percent[2]. Ionized warm gas was found from H$\alpha$ and [NII] emission in 57 percent of the E galaxies examined[2]. The total mass of these gaseous components in an E system is very low, in the range $10^4$ to $10^6 M_\odot$ for the cold gas presumably associated with the dust[1,2] and $10^3$ to $10^4 M_\odot$ for the warm gas, found mostly in the central regions.[2,3]

These masses are tiny fractions, less than $10^{-5}$, of the total stellar mass in a luminous E galaxy. The apparent dust distribution in these systems often shows very little correspondence with the star light, suggesting that this cold gas was recently acquired, perhaps in a merger with another galaxy or an intergalactic cloud. Such interactions may have important effects on galaxy evolution. The warm ionized gas can probably be attributed in part to material ejected by stars which would be expected to give the greatest effect near the center, where the stellar density is highest. We may conclude that the observed trace gases do not weaken the theoretical inference in Paper #2 that cold (or warm) gas cannot be widely distributed in pressure equilibrium throughout a typical E galaxy.

Another conclusion in Section 1, that the gas in ellipticals cannot be hot, is evidently overstated. The cooling analysis here, based on constant $n_e$, assumes an interstellar gas density of 1 H atom $cm^{-3}$; this is a reasonable value for the disk of our Galaxy, but is less so for ellipticals. In fact, the X-ray data indicate a proton density of about 0.1 $cm^{-3}$ in centers of E systems.[4] At this density the rate of radiation will be decreased below the value found above by a factor $1/100$, and the cooling time, increased by a factor 10, giving acceptable values for these quantities. A more accurate conclusion of the analysis in Paper #2 would have been: (A) a hot gas, with rms random velocities comparable to those of the stars, can exist in hydrostatic equilibrium throughout an elliptical galaxy; (B) to keep such a gas hot would require either an atomic density well below 1 $cm^{-3}$ or injection of heating energy at a rate substantially greater than any then known. The possibility of measurable X-ray fluxes from elliptical galaxies could have been an additional conclusion!

We now know that hot gas is present in most of the brighter ellipticals with a temperature[4] of about $10^7$ K, corresponding to a dispersion of 300 km $s^{-1}$ for proton velocities in the line of sight. The total mass of this gas is in the range $10^9$–$10^{10} M_\odot$. The relevant heating sources are now believed to include supernova explosions, stellar winds, and infalling intergalactic gas. Since the random velocities of stars and atoms in this hot gas are comparable, the gas should be co-extensive with the galaxies, according to the theory in Paper #2, in general agreement with observations.

The discussion of the density distribution in NGC 3115 at the end of Paper #2 is, of course, highly tentative. However, the possible presence of a non-luminous disk in the galaxy's mid-plane seems an interesting suggestion. More recent studies of this galaxy, now classified as S0, provide much more accurate and detailed data than were available in 1941. These data have been used[5] to construct a dynamical model, with separate ratios of mass density to luminosity density assumed for the disk and for the surrounding spheroid. This model does not fit the observations at all well, and more extreme assumptions as to mass-luminosity ratios, such as a dark, thin disk in the equatorial plane, for example, might be required.

## REFERENCES

1. Bregman, J. N., Hogg, D. E., and Roberts, M. S.—*Astroph. J.* **387**, 484 (1992).
2. Goudfrooij, P., Hansen, L., Jorgensen, H. E., and Nørgaard-Nielson, H. U.—*Astron. Astroph. Supp.* **105**, 341 (1994).
3. Burbidge, E. M. and Burbidge, G. R.—*Astroph. J.* **142**, 634 (1965).
4. Fabbiano, G.—*Annu. Rev. Astron. Astroph.* **27**, 87 (1989).
5. Capaccioli, M., Cappellaro, E., Held, E. V., and Vietri, M.—*Astron. Astroph.* **274**, 69 (1993).

# 2

## Paper

### ABSTRACT

Measurements of the rotational velocity in NGC 3115 show that the random stellar velocities in this elliptical galaxy are of the order 200 km/sec; other elliptical systems with masses of $10^{10}$ $M_\odot$ will also contain stars with the same velocity dispersion. The random velocities of interstellar atoms and dust particles in equilibrium, however, are less than 20 km/sec. Analysis of electron captures and free-free transitions, taking into account the variation of the $g$-factors with energy, shows that atoms of appreciable abundance will, in fact, reach such a low-velocity equilibrium within $10^7$–$10^8$ years; collisions between dust particles will similarly reduce dust velocities within $10^5$–$10^6$ years. Unless interstellar particles are subject to unknown forces, the velocities of these particles in the more massive elliptical and spherical galaxies must be much less than those of the stars.

It follows that any interstellar matter in such stellar systems must be highly concentrated to the center or to the equatorial plane. For a rigorously spherical system the total mass of such matter cannot exceed $4 \times 10^{-4}$ times the mass of all the stars. In an elliptical system the mass is unrestricted, but if the mass of interstellar matter exceeds that of the stars, then half of such matter must lie within 3 parsecs from the equatorial plane. In any case the light observed from spherical and elliptical systems of large mass must be direct starlight, not diffuse or scattered light.

The equatorial layer of interstellar matter in an elliptical galaxy may in theory contain as much or more matter than the stars in the system. The determination by Oort of the stellar density distribution in the elliptical nebula NGC 3115, and in particular the high-luminosity gradient which he found close to the equatorial plane, suggest that a large fraction of the mass of this system may be in the form of dark matter in the equatorial plane. Matter in so dense a layer would presumably be quite different in its physical state from interstellar matter in our own Galaxy and would perhaps condense into faint stars of low random velocity. Further observations are necessary to decide whether a rotating massive disk of dark matter may play an important part in the structure of some elliptical systems.

The presence of interstellar atoms and dust particles in spiral galaxies is known from direct observation. The question naturally arises as to whether interstellar particles may be present also in elliptical and globular galaxies. The diffuse, nonstellar appearance of most elliptical systems has frequently led to the suggestion that most of the light from such galaxies might consist of diffuse radiation, scattered by small particles.

Direct observations on this point are difficult and by themselves not very conclusive. Fortunately, the equilibrium of interstellar particles in a galactic system is susceptible to rather simple theoretical analysis. Given the forces that act on the particles, it should be possible to demonstrate which equilibrium configurations are possible. It has been shown in the first two papers of this series[1] that the interstellar medium behaves as a perfect gas and that the direct effects of radiation pressure may be neglected for most purposes. The problem is therefore reduced to the equilibrium of a gas in a gravitational field.

The most important quantity in such a problem is the ratio of the random stellar velocities to the velocities of the interstellar particles. Stellar velocities in elliptical galaxies are unfortunately rather uncertain. For one system, NGC 3115, direct measurements of the rotational velocity by Humason[2] are available; the observed values range from 80 km/sec near the center to 450 km/sec farther out. It is clear that since this nebula has a considerable extension perpendicular to its galactic plane, the random stellar velocities must be an appreciable fraction of the rotational velocities and are probably in the neighborhood of 200 km/sec.

For other systems the only relevant data are the random velocities of nebulae in clusters, which indicate a mass of roughly $10^{10}$ M$_\odot$ or more for an elliptical system.[3] Since the radius[4] containing half the mass of an elliptical system is at most 1000 parsecs, the virial theorem gives a mean square stellar velocity of 200 km/sec. Nebular masses are so uncertain, however, that this value for the random stellar velocities cannot be assumed for all elliptical systems.

Nevertheless, it is of considerable interest to investigate the effects associated with such high galactic masses. The following analysis will therefore be developed for galaxies in which the random stellar velocities are assumed to be of the order of 200 km/sec. The results of the analysis may be applied with some certainty to NGC 3115. It is highly

[1] L. Spitzer, Jr., *Ap. J.*, **93**, 369, 1941; **94**, 232, 1941.
[2] *Report of the Director, Mt. W. Obs. 1936–37*, p. 31.
[3] S. Smith, *Ap. J.*, **83**, 23, 1936; F. Zwicky, *Ap. J.*, **86**, 217, 1937.
[4] E. Hubble, *The Realm of the Nebulae*, p. 178, Yale University Press, 1936.

probable that they are relevant to many other elliptical and globular galaxies as well. Whether or not most of the elliptical and globular galaxies have such high masses and such high random stellar velocities cannot be decided until more observational data are available.

In contrast to the assumed velocities of 200 km/sec for stars, the equilibrium random velocities of interstellar particles are quite low, some 20 km/sec for protons and electrons at 10,000°, less for other atoms, and negligibly small for dust particles. Since equilibrium will be reached in much less than $10^9$ years, we may infer that the root mean square velocity of interstellar particles does not exceed 20 km/sec. It follows that the root mean square stellar velocities in the elliptical galaxies considered here are some ten or more times as great as the velocities of interstellar particles.

A necessary consequence of this large difference in velocities is that any interstellar matter in such a system must be almost entirely concentrated toward the equatorial plane or, in the case of a spherical system, toward the center. A factor of ten in velocities corresponds to a factor of one hundred in kinetic energy per unit mass. Hence a star in the equatorial plane of an elliptical galaxy will have at least one hundred times as much energy per unit mass as any atoms or dust particles in the vicinity and will clearly be able to rise much farther from the equatorial plane before the gravitational attraction of the galaxy pulls it back.

The analysis proceeds in three separate parts in the following sections. First, the rate of dissipation of energy for atoms or dust particles at high velocities must be determined in order to exclude the possibility that the interstellar medium could have remained at a temperature of millions of degrees during $10^9$ years. Second, the equilibrium of a gas within a spherical gravitational field is examined, and a maximum total mass is found for the interstellar matter in a spherical galaxy. In the case of a typical globular cluster the situation is quite different, since the mean square stellar velocities are comparable with the atomic ones. It is doubtful, however, whether some of the smaller clusters have a sufficient gravitational force to hold atoms at all.

In the third and last section the equilibrium of dust and atoms in systems possessing angular momentum is investigated. Here again the relative concentration of the interstellar medium may be evaluated, and as in a spherical system it may be shown that the amount of interstellar matter throughout most of a massive elliptical galaxy must be quite small. The effect which a thin, dense equatorial layer of dark matter can produce on the distribution of stars is also examined. A comparison of the predicted effects with observational data in NGC 3115 suggests that this galaxy may, in fact, possess a large amount of dark matter concentrated to its equatorial plane.

## 1. Dissipation of Energy

In the general case of a gas at high temperature the dissipation of energy will proceed in may ways. To obtain an upper limit for the time required to dissipate large energies, we shall here consider the two types of interstellar media which are the most likely to remain at an initially high temperature—a medium consisting wholly of hydrogen atoms and one composed exclusively of dust particles. In each case the rate of loss of energy will be computed; possible mechanisms for an offsetting increase in energy will be considered later. Since dust and atoms are considered wholly separately, the following results are independent of the equipartition of kinetic energy between dust and atoms which appears when both types of particles are present and which was discussed quantitatively in paper I. It is clear that if both dust and atoms are present together the rate of energy dissipation will be increased from the values found below. This strengthens the conclusion that interstellar particles cannot long remain at high velocities.

A medium composed wholly of hydrogen atoms will be considered first. It the average kinetic energy per atom exceeds the ionization energy $E_0$, the gas will be largely ionized, since the cross-section for ionization by electronic or atomic impact is much greater than that for electron capture. The resultant assembly of protons and free electrons will come to equipartition of kinetic energy fairly rapidly. It was shown in paper I that for protons and electrons with an assumed density of 1 atom per $cm^3$ and a kinetic temperature of 10,000°, deviations from equipartition of energy fell to $1/e$ their initial value in 0.48 years. Since this time varies as the cube of the velocity, it is evident that even for velocities of 600 km/sec, corresponding to a temperature of 14,000,000°, the time of equipartition is only $2.7 \times 10^4$ years. We may therefore use with confidence the appropriate Maxwellian velocity distributions for both protons and electrons.

A gas of protons and electrons may lose energy by radiation in two ways—electron captures and free-free transitions. Probabilities for these processes are known from wave-mechanical theory. It is frequently assumed in astrophysical investigations that the Gaunt "g-factor" is equal to unity. This is a valid approximation for low energies but may be badly off when the energies are very high. Accurate values of $g$ have been exhaustively determined by Menzel and Pekeris.[5] The equations given by Bethe,[6] based on the analyses of Sommerfeld, Stobbe, and others, are less complete but are in a more convenient form for the

[5] *M.N.*, **96**, 77, 1935.
[6] *Handb. d. Phys.*, **24**, Part I, 488, 1933.

present purpose. The rate of loss of energy will be computed from these equations for two processes—recapture of electrons directly in the ground state and radiation of energy in free-free transitions. If other processes are important, they will, of course, increase the rate of energy dissipation.

Let $E$ and $v$ be the energy and velocity, respectively, of an electron, and let $E_0$ be the ionization energy of hydrogen. The energy radiated in a single recombination into the ground state will be a function of the initial relative velocity $V$; if $V$ were the same for all encounters, the number of such recombinations per second would be $\pi\sigma_c^2 n_p n_e V$, where $\pi\sigma_c^2$, a function of $V$, is the cross-section for the capture process, and $n_p$ and $n_e$ are the numbers of protons and electrons per cm³, respectively. Since the protons have a root mean square velocity only 1/43 as great as the electrons, we may without serious error replace the relative velocity $V$ by the electron velocity $v$; with this approximation the energy radiated in a single encounter becomes $E + E_0$. To find the rate of change of $\overline{E}$, the mean energy of protons and electrons, the energy loss per unit time must be averaged over all values of $v$ and divided by $n_p + n_e$, or $2n_e$. This yields the equation

$$\frac{d\overline{E}}{dt} = -\frac{3}{2}\left(\frac{2}{\pi}\right)^{1/2} \frac{n_p}{m^{1/2}\overline{E}^{3/2}} \int_0^\infty E(E + E_0)e^{-3E/2\overline{E}}\pi\sigma_c^2 \, dE, \quad (1)$$

where $m$ is the mass of the electron.

The capture cross-section $\pi\sigma_c^2$ for the lowest quantum state may be written as

$$\pi\sigma_c^2 = A \frac{E_0^{5/2}}{(E + E_0)E^{3/2}} \, \vartheta_1\left(\frac{E_0}{E}\right), \quad (2)$$

where

$$A = \frac{2^6 e^2 h}{3m^2 c^3} = 1.44 \times 10^{-21} \text{ cm}^2. \quad (3)$$

If we let $k^2$ equal $E_0/E$, then $\vartheta_i(E_0/E)$ in equation (2) may be written in the form

$$\vartheta_1(k^2) = \frac{2\pi k}{1 + k^2} \frac{\exp(-4k \cot^{-1}k)}{1 - \exp(-2\pi k)}. \quad (4)$$

When $k$ is very large or very small we have the expansions

$$\vartheta_1(k^2) = \frac{2\pi e^{-4}}{k}\left(1 + \frac{1}{3k^2}, \cdots\right), \tag{5}$$

$$\vartheta_1(k^2) = 1 - \pi k \cdots . \tag{6}$$

Since the Gaunt $g$-factor is given by

$$g = 4(3)^{1/2}\left(\frac{E_0}{E}\right)^{1/2} \vartheta_1\left(\frac{E_0}{E}\right), \tag{7}$$

it is evident that when $E_0/E$ is large, $g$ is equal to its familiar value, $8\pi 3^{1/2}e^{-4}$, or 0.797. When $E_0/E$ is small and $\vartheta_1$ is unity, the value of $g$ is very much less than unity. Values of $\vartheta_1(E_0/E)$ are given in Table 1.

Equations (2) and (4) may be substituted into equation (1) to give an integral formula for $d\overline{E}/dt$. Since this integral can apparently not be evaluated in a simple analytical form, we may approximate by taking $\vartheta_1(E_0/E)$ outside the integral sign with its value for some appropriate value of $E$. This obviously gives asymptotically correct results as $\overline{E}$ approaches infinity and $\vartheta_1(k^2)$ becomes equal to unity. If this procedure is also to give correct results when $E_0/\overline{E}$ is large, and $\vartheta_2(k^2)$ is approximately given by equation (5), then it is readily shown that the appropriate value of $E$ is $2\overline{E}/3\pi$. This value is relatively small, since, when $E_0/\overline{E}$ is not large, most of the energy losses come from the recapture of low-velocity electrons. With this approximation, then $d\overline{E}/dt$ becomes

$$\frac{d\overline{E}}{dt} = -\frac{3An_p E_0^{5/2}}{2^{1/2}m^{1/2}\overline{E}} \vartheta_1\left(\frac{3\pi E_0}{2\overline{E}}\right). \tag{8}$$

This method of procedure, although correct when $\overline{E}$ is very large or very small, is not very accurate when $E_0/\overline{E}$ is near unity; the resultant error in equation (8) should be less than some 20 per cent at the most, however, which is as accurate as we shall need here. Since $E_0$ varies at the square of the ionic charge, it is evident that a few highly ionized elements will very greatly increase the rate of radiation.

### TABLE 1
#### Correction Factor $\vartheta_1$

| $E/E_0$ | $\vartheta_1$ | $E/E_0$ | $\vartheta_1$ | $E/E_0$ | $\vartheta_1$ |
|---|---|---|---|---|---|
| 0.01 | 0.012 | 1.0 | 0.14 | 100 | 0.74 |
| .02 | .016 | 2.0 | .20 | 200 | .81 |
| .04 | .023 | 4.0 | .29 | 400 | .86 |
| .06 | .029 | 6.0 | .34 | 600 | .88 |
| .08 | .034 | 8.0 | .39 | 800 | .89 |
| .10 | .038 | 10.0 | .42 | 1,000 | .90 |
| .20 | .054 | 20.0 | .52 | 4,000 | .95 |
| .40 | .080 | 40.0 | .63 | 10,000 | .97 |
| .60 | .102 | 60.0 | .68 | 40,000 | .98 |
| 0.80 | 0.120 | 80.0 | 0.72 | 100,000 | 0.99 |

To visualize conveniently the order of magnitude of the effects involved, one may define a time of dissipation $T_d$ by the relationship

$$T_d = \frac{\overline{E}}{\left|\dfrac{d\overline{E}}{dt}\right|}.\tag{9}$$

The quantity $T_d$ so defined is the length of time required for the complete dissipation of energy by radiation if the rate of dissipation remained constant as $\overline{E}$ decreased. Since actually $|d\overline{E}/dt|$ increases as $\overline{E}$ decreases, the actual time required for the cooling of a gas at high temperature is somewhat less than $T_d$. If we insert numerical values into equation (8) and substitute into equation (9), we have for the time of dissipation arising from recombinations in the ground state

$$T_d(\text{rec}) = \frac{6.6 \times 10^4}{n_p \vartheta_1 (4.7 E_0/\overline{E})} \left(\frac{\overline{E}}{E_0}\right)^2 \text{ years.}\tag{10}$$

Values of $T_d(\text{rec})$ for $n_p$ equal to unity and for various root mean square proton velocities are given in Table 2.

The free-free transitions may be similarly treated, except that here we must integrate over all possible transitions as well as over all velocities.

TABLE 2
Values of the Time of Dissipation

| $v$ Km/Sec | 100 | 200 | 400 | 600 | 1000 |
|---|---|---|---|---|---|
| $\bar{E}/E_0$ | 3.82 | 15.3 | 61 | 137 | 382 |
| $T_d(\text{rec})$ | $8.0 \times 10^6$ | $6.0 \times 10^7$ | $5.3 \times 10^8$ | $2.6 \times 10^9$ | $1.4 \times 10^{10}$ |
| $T_d(f-f)$ | $5.3 \times 10^6$ | $1.1 \times 10^7$ | $2.1 \times 10^7$ | $3.2 \times 10^7$ | $5.3 \times 10^7$ |
| $T_d(\text{dust})$ | $1.1 \times 10^6$ | $5.5 \times 10^5$ | $2.8 \times 10^5$ | $1.8 \times 10^5$ | $1.1 \times 10^5$ |

*Note:* Values of $T_d$ give the time in years required to dissipate the entire kinetic energy of the assembly if the rate of dissipation remained constant and if no other process were acting.

The exact equations become very complicated, and it will suffice to take the asymptotic form for small $E_0/\bar{E}$, which becomes

$$\frac{d\bar{E}}{dt} = \frac{3}{8\pi} \left( \frac{3}{\pi} \right)^{1/2} \frac{An_p E_0}{m^{1/2}\bar{E}^{3/2}} \int_0^\infty E e^{-3E/2\bar{E}} \vartheta_2 \left( \frac{E_0}{E} \right) dE, \quad (11)$$

where the various symbols have the same meanings as before. If we again let $k^2$ equal $E_0/E$, we have

$$\vartheta_2(k^2) = 2\pi k \int_1^\infty \frac{1}{1 - \exp(-2\pi uk)} \log \frac{u+1}{u-1} \frac{du}{u^2}. \quad (12)$$

These results may be derived either from the formula given by Bethe, which is valid only for large electron energies, or from the asymptotic result given by Menzel and Pekeris,[7]

$$g = \frac{2 \cdot 3^{1/2} l}{1 - \exp(-2\pi l)} \log \frac{l+k}{l-k}, \quad (13)$$

where $k^2$ equals $E_0/E$ is usual and $l^2$ equals $E_0/E'$; $E$ and $E'$ are the initial and final energies of the electron, respectively. The quantity $u$ in equation (12) is the ratio of $l$ to $k$.

When $E$ is large, $2\pi(E_0/E)^{1/2}$ is small, and for most of the relevant range of integration in equation (12) the denominator may be replaced by $2\pi uk$. The resultant integral may then be evaluated exactly, and we

[7] Ref. 5, eqs. (1.47) and (1.49).

find that $\vartheta_2$ is unity. Equation (11) can then be integrated, and we have

$$\frac{d\overline{E}}{dt} = -\frac{An_p E_0 \overline{E}^{1/2}}{2\pi(3\pi)^{1/2} m^{1/2}}. \tag{14}$$

This is equivalent to a formula given by Menzel,[8] provided that $\bar{g}$ is set equal to $2 \cdot 3^{1/2}/\pi$, or 1.10. Thus we see that the average value of $g$ is very nearly unity, despite the fact that $g$ varies from a small quantity to infinity over the range of integration. When $\overline{E}$ is small, $g$ may be set equal to unity, as may be seen from the series expansions given by Menzel and Pekeris.[5] We may therefore assume that equation (14) is valid for intermediate velocities as well as for high velocities.

If we define $T_d$ as in equation (9), the time of dissipation arising from free-free transitions becomes

$$T_d(f-f) = 2.7 \times 10^6 \, \frac{1}{n_p} \left(\frac{\overline{E}}{E_0}\right)^{1/2} \quad \text{years.} \tag{15}$$

Values of $T_d(f-f)$ are shown in Table 2 for various values of the root mean square proton velocity; the number of protons per $cm^3$ is again set equal to unity. It is evident from this table that, for most velocities, recaptures in the ground state are of negligible importance in the dissipation of energy. This is presumably true of recaptures in other lower states as well. The condition for continuity at the series limit indicates that recaptures in the higher quantum states will behave in the same way as the free-free transitions. Such recaptures will, of course, decrease the time of dissipation even further from the values given in Table 2.

Lastly, the dissipation of energy by direct encounters between dust particles must be computed. Let us consider an encounter between two particles of equal mass, $m_d$, such that the center of gravity of the two particles is at rest and their original relative velocity is $V_1$. Let the perpendicular distance between the original paths be equal to $p_1$. Then the total initial energy and angular momentum of the two particles is $\frac{1}{4}m_d V_1^2$ and $\frac{1}{2}p_1 m_d V_1$. If $V_2$ is the relative velocity of separation and $p_2$ the value of $p$ after collision, the energy loss per particle becomes

$$\Delta E = \tfrac{1}{8}m_d(V_1^2 - V_2^2) \tag{16}$$

[8] *Ap. J.*, **85**, 330, 1937, eq. (30); a factor $K/h$ has been omitted from this equation.

while the conservation of angular momentum yields

$$p_1 V_1 = p_2 V_2. \tag{17}$$

The elimination of $V_2$ from equations (16) and (17) gives

$$\Delta E = \tfrac{1}{8} m_d V_1^2 \left( 1 - \frac{p_1^2}{p_2^2} \right). \tag{18}$$

It is obvious that the total loss of energy must be the same if the center of gravity is not at rest, and we may therefore use equation (18) for the energy loss in every collision.

The maximum loss of energy occurs when $p_2$ is a maximum. Since $p_2$ can scarcely exceed $2\sigma$, the geometrical diameter of each particle, we may replace $p_2$ by $2\sigma$ in equation (18) and multiply the resultant expression by $\zeta$, the ratio of the actual energy loss to the maximum loss possible. To find the rate of decrease of the mean kinetic energy $\bar{E}$, we must multiply $\Delta E$ by $2\pi p_1 dp_1 V_1 n_d$, where $n_d$ is the number of dust particles per cm³, and then integrate over $p_1$ from zero to $2\sigma$. This yields

$$\frac{d\bar{E}}{dt} = -\frac{\pi}{4} \bar{\zeta} \sigma^2 m_d V_1^3 n_d, \tag{19}$$

where a mean value of $\zeta$ has been taken.

Finally, we must determine the mean value of the factor $V_1^3$ in equation (19). If the distribution of dust-particle velocities is assumed to have spherical symmetry, then

$$\overline{V_1^2} = 2\overline{v_d^2}, \tag{20}$$

where $\overline{v_d^2}$ is the mean square velocity of the individual particles. If the distribution of dust velocities were Maxwellian, we should have

$$\overline{V_1^3} = \frac{2^{7/2}}{3^{3/2}\pi^{1/2}} \left( \overline{V^2} \right)^{3/2}. \tag{21}$$

Although there is apparently no reason why the velocities of very rapidly moving dust particles should obey the Maxwellian distribution, equation (21) should provide an adequate approximation to the true state of affairs, particularly since the numerical factor in equation (21) should not differ much from unity in any case. If we let $m_d$ equal

$4\pi\sigma^3 d/3$, where $d$ is the density of matter within the particle, and let $n_d$ equal $\rho_d/m_d$, where $\rho_d$ is the density of dust in grams/cm$^3$, we may combine equations (19), (20), and (21) to find

$$\frac{1}{\bar{E}}\frac{d\bar{E}}{dt} = -\frac{4}{(3\pi)^{1/2}}\frac{\rho_d\bar{\zeta}}{\sigma d}(\overline{v_d^2})^{1/2}, \tag{22}$$

where now $\bar{\zeta}$ denotes the average over $V_1$ as well as over $p_1$. With the same definition of $T_d$ as before, we find that if $\sigma$ equals $10^{-5}$ cm, $d$ equals 7.8, and $\rho_d$ is $1.7 \times 10^{-24}$, equation (22) gives

$$T_d(\text{dust}) = \frac{1.1 \times 10^{12}}{\bar{\zeta}(\overline{v_d^2})^{1/2}} \text{ years.} \tag{23}$$

The quantity $\bar{\zeta}$ is not likely to be very small when $\overline{v_d^2}$ is large. For the rapid collisions we are concerned with here, the energy of impact is very much greater than the binding energy of the particles, which is normally the source of the elastic forces. Hence such rapid collisions will probably be completely inelastic, and two such particles will in all likelihood vaporize when their velocity of impact is several hundred kilometers per second. To find a lower limit on $T_d$, however, $\bar{\zeta}$ has been set equal to 0.1 in the computation of the values in Table 2.

It is evident from Table 2 that the interstellar matter of appreciable density will have lost its high energy in $10^7$ years or less. There is no evidence for any mechanism which can avert this loss of kinetic energy. Ionization by radiation provides at most an energy comparable to $E_0$ for each such recombination; ionization by electron impact is likely to be more frequent, and this involves no net change in $\bar{E}$. Interaction with stars will be small, particularly at such high velocities. For a particle to be deflected 90° by a star of solar mass, moving with a relative velocity of 200 km/sec, the undisturbed path of the particle must come within 4 solar radii of the sun's surface. Even though this distance may be considerably increased by radiation pressure from the brighter stars, such close encounters are rare and may be neglected.

The role of cosmic rays is somewhat more obscure, since little is known about the numbers of low-energy particles, which would be of importance in this connection. Particles of the energy and abundance observed terrestrially may be shown to be unimportant. The number of particles entering the earth's atmosphere per cm$^2$ per sec is[9] 0.09. Since

[9] I. S. Bowen, R. A. Millikan, and H. V. Neher, *Phys. Rev.*, **53**, 217, 1938.

the cross-section of a light nucleus of cosmic-ray impact is roughly $10^{-25}$ cm$^2$, the mean life between encounters is $10^{18}$ years. Since the mean energy acquired per encounter is at most $10^{10}$ e.v., this corresponds to a gain of 1 e.v. per particle in $10^8$ years, which is insufficient to offset a loss of several hundred e.v. per particle in $10^7$ years.

If the mass of interstellar matter in a galaxy is comparable with that of the stars, it is difficult to find any mechanism which will maintain the particles of the medium in a state of high kinetic energy. An energy loss of 400 e.v. in $10^7$ years, corresponding to an initial proton velocity of roughly 300 km/sec, is a loss of 1 erg per gram per second; and to maintain this in a medium of appreciable mass requires an energy source comparable to that of all the stars. There is no evidence whatever for such an additional source of energy. We may conclude that, on the basis of present theory, the mean square velocity of interstellar particles cannot possibly be as high as 100 km/sec if the density of such matter is appreciable; in all probability interstellar matter is in kinetic equilibrium with a velocity temperature between 5,000° and 20,000°, and with velocities of not more than 20 km/sec.

This analysis refers, of course, to purely random velocities. These results, therefore, do not exclude the possibility that the interstellar medium could be supported by large-scale currents with velocities much greater than the microscopic random velocities of the individual particles. If such currents are assumed to be irregular, as in the case of turbulent stellar atmospheres, the energy of macroscopic motion would soon be dissipated in all probability. When two such currents, with velocities far exceeding the velocity of pressure waves in the medium, encountered each other, the directed macroscopic velocities would tend to be converted into random microscopic motions. The only current system which could apparently be stable in $10^9$ years or more would be a simple rotation of the medium about the center of the system—a case which is, of course, relevant in an elliptical galaxy. With this exception one may assume that large-scale currents play no direct part in the equilibrium of the interstellar medium.

## 2. Spherical Systems

The equilibrium of interstellar matter in a spherical galaxy is simplified when the mean square velocity of the interstellar particles is much less than that of the stars. As we shall see below, the total mass of the interstellar medium in such a case must be much less than that of the stars and will therefore not affect appreciably the structure of the stellar system. The problem therefore reduces to the equilibrium of a gas in a potential field which arises partly from a known distribution of stars, partly from the gas itself.

The specific analysis is complicated by two factors. In the first place, the structure of the spherical galaxies is not well known. Second, the mass of the medium complicates the equations. General limits will therefore be discussed for the most part. Where effects produced by the precise structure of a galaxy are of interest, the Emden polytrope for $n = 5$ will be used. This is known to give a good approximation of globular clusters, and even for galaxies it should give some indication of the general effects to be expected when the mass of the system is highly concentrated toward the center.

First, let use derive the form of the virial theorem which will be useful. The equation of motion in the $x$-direction for a particle of mass $m_p$ may be written

$$m_p \frac{d^2 x}{dt^2} = F_x. \tag{24}$$

A subscript $p$ will be used throughout to denote interstellar particles, either dust or atoms. If we multiply equation (24) by $x$, we find

$$\tfrac{1}{2} m_p \frac{d^2 x^2}{dt^2} - m_p \left( \frac{dx}{dt} \right)^2 = x F_x. \tag{25}$$

If it is assumed that the system comprises a large number of identical particles in a steady state, we may average over all the particles, in which case the second derivative in equation (25) vanishes. The contribution of the atomic interaction forces to $x F_x$ may be neglected, on the average, compared to the external and gravitational forces. If we assume also that both the distribution of velocities and potential field possess spherical symmetry, then equation (25) becomes

$$\overline{v_p^2} = -\overline{rF_r}/m_p, \tag{26}$$

where $F_r$ is the radial force, and $\overline{v_p^2}$ is the mean square velocity of the particles in question.

For a gas of small mass in a cluster of uniform stellar density $\rho_{s0}$ the radial force is given by

$$\frac{F_r}{m_p} = -\frac{4\pi G \rho_{s0}}{3} r. \tag{27}$$

The assumption that $\rho_{s0}$ is constant is legitimate when the medium is concentrated toward the core of the system. From equation (26) we have in this case

$$\overline{v_p^2} = \frac{4\pi G\rho_{s0}}{3}\overline{r_p^2},\tag{28}$$

where a subscript $p$ is used to indicate that $r^2$ as well as $v^2$ is averaged over all the particles (atoms or dust) in the medium.

For the stars the force per unit mass is less than that given by equation (27), since the density decreases outward. From equation (28) and the corresponding inequality for $\overline{v_s^2}$, we have, therefore, the inequality

$$\frac{\overline{r_p^2}}{\overline{r_s^2}} < \frac{\overline{v_p^2}}{\overline{v_s^2}}.\tag{29}$$

When we come to extend this result to a specific stellar structure, such as the Emden polytrope $n = 5$, we find that $\overline{r_s^2}$ is infinite. The analysis is therefore best given in terms of $r_{s(1/2)}$ and $r_{p(1/2)}$, the radii containing half the stellar mass and half the mass of the medium, respectively. The analysis involves a determination of the mean potential energy of the polytrope.[10] The final result is

$$\frac{r_{p(1/2)}}{r_{s(1/2)}} = 0.414\left(\frac{\overline{v_p^2}}{\overline{v_s^2}}\right)^{1/2}.\tag{30}$$

We may infer that the ratio of the size of the interstellar system to that of the stellar one is in general somewhat less than the ratio of the root mean square velocities.

The neglect of the total mass $M_p$ of the interstellar particles is valid, provided that the density of the medium is everywhere less than that of the stars. When this is not the case, when $\rho_p$ is greater than $\rho_{s0}$, it is evident that the gravitational attraction of the medium on itself will be greater than that of the stars on the medium. We may replace equation

[10] It is sometimes stated that the potential energy of the polytrope $n = 5$ is infinite, since a factor $n - 5$ appears in the usual formula for the total potential energy. But the radius $R$ of the system is also infinite in this case; the energy is, in fact, finite, and may be determined by a direct integration.

(28) by the more general formula

$$\overline{v_p^2} = \frac{4\pi G\rho_{s0}}{3}\,\overline{r_p^2} + \beta\,\frac{GM_p}{\left(\overline{r_p^2}\right)^{1/2}}, \tag{31}$$

where $\beta$ is a constant of unit order of magnitude. The second term in equation (31) is the one which usually appears in applications of the virial theorem and is equal but opposite in sign to the potential energy per unit mass of a sphere of gas in its own gravitational field.

It is evident from equation (31) that if $M_p$ is fixed there is a value of $\overline{r_p^2}$ such that $\overline{v_p^2}$ is a minimum. Similarly, if $\overline{v_p^2}$ is fixed, there is a value of $\overline{r_p^2}$ at which $M_p$ is a maximum. This maximum value of $M_p$ is given by

$$\{M_p(\text{max})\}^2 = \frac{1}{9\pi\beta^2\rho_{s0}}\left(\frac{\overline{v_p^2}}{G}\right)^3, \tag{32}$$

which may be written as

$$\left(\frac{M_p(\text{max})}{M_s}\right)^2 = \frac{4}{27\beta^2}\left(\frac{\overline{v_p^2}\,r_{s(1/2)}}{GM_s}\right)\frac{3M_s}{4\pi r_{s(1/2)}^3\,\rho_{s0}}. \tag{33}$$

It is clear that $3M_s/4\pi r_{1/2}^3$ is twice the mean stellar density interior to $r_{s(1/2)}$; this will be less, as a rule, then the central density $\rho_{s0}$. Since $GM_s/r_{s(1/2)}$ is roughly equal to $\overline{v_s^2}$ and $\beta$ is approximately unity, we have

$$\frac{M_p(\text{max})}{M_s} \leq 0.38\left(\frac{\overline{v_p^2}}{\overline{v_s^2}}\right)^{3/2}. \tag{34}$$

If we turn again to the polytrope $n = 5$, we find that in this special case $3M_s/4\pi r_{s(1/2)}^3$ is equal to $0.447\rho_{s0}$, while $GM_s/r_{s(1/2)}$ is equal to $2.60v_s^2$. These two factors reduce the constant in equation (34) from 0.38 to 0.060. This indicates that equation (34) gives a generous upper limit on $M_p$. If, in accordance with the results of section 1, we assume that the ratio of root mean square velocities for stars and particles is roughly equal to ten, it follows from equation (34) that the maximum possible mass of interstellar matter in a spherical galaxy is at most $4 \times 10^{-4}$ of the stellar mass and that any such matter must be highly concentrated toward the center of the galaxy.

The existence of a maximum mass for the medium may be explained simply. Let us consider an assemblage of interstellar particles whose mean square velocity is fixed and whose distribution in the cluster satisfies the equations of equilibrium; let us examine the sequence of configurations which is obtained as the total mass $M_p$ is gradually increased. When $M_p$ is zero the value of $\overline{r_p^2}$ will satisfy equation (28). As the mass of the medium is gradually increased, the self-attraction of the medium begins to become important. At first this may be offset by a contraction in the medium, reducing the attraction of the stars on the interstellar matter. Eventually, however, a point is reached at which a further contraction increases the self-attraction exactly as much as it decreases the attraction of the stars. At this point no increase in $M_p$ is possible without also increasing $\overline{v_p^2}$—this is the point of maximum mass for a fixed mean square velocity. When $M_p$ has a value less than this critical one, equation (31) will have two relevant roots for $\overline{r_p^2}$, one corresponding to a diffuse configuration in which the gravitational attraction of the stars is predominant, the other, a dense system subject primarily to its own self-attraction.

The relevance of this maximum mass is restricted by the fact that the analysis holds only for systems possessing no angular momentum whatever. Most spherical galaxies probably have some angular momentum, and in such a case interstellar matter could contract to the equatorial plane, forming a flattened, rapidly rotating disk of dark matter of the type discussed in the next section. The total mass of such matter could not be very great, however, since the gravitational attraction of a very massive layer of dark matter would impart some ellipticity to the observable stellar distribution. In any case we may conclude that interstellar dust and atoms play a wholly subordinate role in the constitution of spherical stellar systems of large mass and large random stellar velocity.

In systems of small mass, such as globular clusters, the random stellar velocities are of the same order as the velocities of the particles. In such systems the total mass of dust particles present could in theory be very large, provided that not more than a small number of atoms was present. An appreciable number of atoms would slow down the dust particles as is shown in paper I and lead eventually to a collapse of the medium, unless the mass of atoms present exceeded sufficiently the mass of dust to support the dust particles.

For a primarily gaseous medium, on the other hand, a new restriction appears. If the mean square atomic velocities were slightly greater than those of the stars and $M_s$ were assumed to be much greater than $M_p$, the atoms would apparently leave the system entirely. According to equation (26), the average value of $rF_r$ must vary directly with the mean square velocity. While it is easy to find a distribution of matter for which this average will be very much decreased, given the gravitational field of a particular stellar system, there is a limit to the amount this quantity can be increased. Since most of the stars are near the region in which $rF_r$ is a maximum, the average value of this quantity for any distribution of matter will not much exceed its average value for the stars. For the polytrope $n = 5$, for instance, the maximum value of $rF_r$ is only 1.31 times its average value for the entire polytrope. It follows that interstellar particles with a mean square velocity some 20 or 30 percent greater than the velocity of the stars and a total mass considerably less than that of the stars cannot possibly be in equilibrium within a cluster.

A disequilibrium of this sort would not lead to an immediate escape of the medium, provided that the mean square particle velocities were less than the mean square velocity of escape. In fact, the medium would expand adiabatically to reach a new equilibrium with lower mean square velocities of the particles. But in the case of atoms the kinetic temperature is presumably determined essentially by the color temperature of the ionizing radiation and other factors; if these various factors tended always to give the atoms a mean square velocity appreciably greater than that of the stars and if the gravitational attraction of the medium upon itself were unimportant, such an atomic medium would gradually expand and leave the cluster. If the mass of the interstellar atoms exceeded that of the stars, this difficulty would not arise. In such a case the medium would hold itself together under its own attraction, and the observed stellar system would be imbedded in a globular atomic cluster of much greater mass.

## 3. Elliptical Systems

The analysis of the preceding section must be modified for systems which possess angular momentum and axial symmetry. In such a case the interstellar medium will be confined to the equatorial plane, forming a highly flattened disk imbedded in the more extended elliptical aggregation of stars. If the initial distribution of the medium is assumed to be the same as that of the stars, then, as the random velocities of the interstellar particles decreased, there would be some contraction toward the galactic center as well as toward the equatorial plane. The constancy of angular momentum during such a contraction would lead to an increase of the centrifugal force, which would remain proportional to the inverse cube of the distance from the axis of rotation. A new equilibrium would therefore be reached in which the increased centrifugal force just balanced the attraction toward the galactic center; the medium would, of course, be revolving about the galactic center more rapidly at any point than the stars in the immediate neighborhood.

The relative contraction necessary to attain such an equilibrium may be computed on various assumptions; the ratio of the equilibrium and initial distances from the axis of rotation for a given element of matter is roughly equal to the ellipticity of the nebula or to the fourth root of the ellipticity, depending on whether the potential is that of a point mass or of a homogeneous spheroid. This subject is intimately connected with the uncertain problem of the origin of the galaxies. We shall therefore not consider the problem of the density distribution of interstellar matter with increasing distance along the equatorial plane but rather the distribution in directions perpendicular to this plane. Such an analysis rests on firm ground, since it depends only on the conditions of equilibrium and not on any assumed initial properties.

In the following analysis, distance from the equatorial plane will be denoted by $z$; the distance of a point from the axis of rotation will be denoted by $R$. As before, subscripts $p$ and $s$ will be used to distinguish quantities pertaining to the interstellar particles—dust or atoms—from those pertaining to the stars. Regions near the center of elliptical systems will not be discussed here, since these will be closely analogous to spherical galaxies; we shall consider, rather, the structure of the galaxy when $R$ is relatively large.

The analysis depends on whether $F_z$, the gravitational force in the $z$-direction arises primarily from the central mass of the system or from the neighboring matter in the equatorial plane. If the former is the case, then both particles and stars will move in the same external potential field, which for small values of $z/R$ varies as $z^2$. It is clear that in such a case

$$\frac{\overline{z_p^2}}{\overline{z_s^2}} = -\frac{\overline{v_p^2}}{\overline{v_s^2}}. \tag{35}$$

The total mass of the interstellar medium per unit area of the equatorial plane, as well as the total stellar mass per unit area, is quite unrestricted in such a case, provided only that the total density is always less than the central mass of the galaxy, $M_c$, divided by $4\pi R^3$; this is the usual condition that the attraction of the central mass be greater than the attraction of the local matter.

When the total density of stars and medium is greater than $M_c/4\pi R^3$, however, we may neglect the rest of the galaxy and as a first approximation consider only the one-dimensional problem, in which the gravitational potential $\Phi$ is determined by the equation

$$\frac{d^2\Phi}{dz^2} = 4\pi G(\rho_s + \rho_p), \tag{36}$$

while the condition of equilibrium gives

$$\frac{d}{dz}\left(\tfrac{1}{3}\rho_s \overline{v_s^2}\right) = -\frac{d\Phi}{dz}\rho_s, \tag{37}$$

together with an equation for the interstellar particles, identical with equation (37) except that a subscript $p$ replaces the subscript $s$. One may assume that $\overline{v_s^2}$ and $\overline{v_p^2}$ are not functions of $z$. This neglects primarily the dispersion in stellar masses, since the chief effect of such a dispersion is to increase $\overline{v_s^2}$ with $z$.

The solution of these equations depends on $\rho_{p0}$ and $\rho_{s0}$, the densities of particles and stars when $z = 0$. Let us assume first that $\rho_{p0}$ is much less than $\rho_{s0}$. Then we may neglect $\rho_p$ in equation (36); if we divide equation (37) by $\rho_s$, then differentiate with respect to $z$, and use equation (36) to eliminate $d^2\Phi/dz^2$, we have

$$\frac{d}{dz}\left(\frac{1}{\rho_s}\frac{d\rho_s}{dz}\right) = -\frac{12\pi G}{v_s^2}\rho_s. \tag{38}$$

If we make the substitutions

$$\rho_s = \rho_{s0}\Lambda(\xi), \tag{39}$$

$$z = \xi\left(\frac{\overline{v_s^2}}{12\pi G\rho_{s0}}\right)^{1/2}, \tag{40}$$

the equation assumes a dimensionless form. With the boundary conditions that $\Lambda(0) = 1$ and that $\Lambda'(0) = 0$, the solution of equation (38) becomes

$$\Lambda(\xi) = \operatorname{sech}^2(2^{-1/2}\xi). \tag{41}$$

The mass $M_s(\xi)$ per unit area between $-\xi$ and $+\xi$ is

$$M_s(\xi) = \left(\frac{2\overline{v_s^2}\rho_{s0}}{3\pi G}\right)^{1/2}\tanh(2^{-1/2}\xi). \tag{42}$$

From equations (40) and (42) it follows that $z_{s(1/2)}$, the value of $z$ such that half the stellar mass per unit area lies between $z_{s(1/2)}$ and $-z_{s(1/2)}$, is given by

$$z_{s(1/2)} = 0.274 \frac{M_s}{\rho_{s0}} \tag{43}$$

where $M_s$ equals $M_s(\infty)$ and is simply the total stellar mass above and below a unit area of the equatorial plane. Since $\tanh(\infty) = 1$, equation (42) gives $M_s$ in terms of $\overline{v_s^2}$ and $\rho_{s0}$.

For the interstellar particles, on the other hand, we may neglect the change in $\rho_s$ and set $d\Phi/dz$ equals to $4\pi G\rho_{s0}z$. The equation of hydrostatic equilibrium for the particles then yields

$$\log \frac{\rho_p}{\rho_{p0}} = -\frac{6\pi G\rho_{s0}}{\overline{v_p^2}} z^2. \tag{44}$$

The mass per unit area leads to the usual error function; if we determine the value of $z_{p(1/2)}$ by integrating equation (44) and if we take $z_{s(1/2)}$ from equation (43), substituting from equation (42) for $M_s$, we have

$$\frac{z_{p(1/2)}}{z_{s(1/2)}} = 0.87\left(\frac{\overline{v_p^2}}{\overline{v_s^2}}\right)^{1/2}. \tag{45}$$

Here, as in equation (35) and in the globular galaxies, the ratio of the dimensions of the two systems is roughly equal to the ratio of the root mean square velocities.

The chief difference between the elliptical and the globular systems is the absence of any maximum mass for the interstellar medium. No matter how great the mass of interstellar matter, it is always possible to make $z_{p(1/2)}$ so small that the mean square velocity is arbitrarily small. If $\rho_{p0}$ should exceed $\rho_{s0}$, however, equation (44) is not longer applicable. To find the distribution of the medium in this case, a new solution of the equations of equilibrium must be found. If $\rho_{p0}$ is considerably greater than $\rho_{s0}$, then $\rho_p$ will exceed $\rho_s$ in most of the region in which the density of interstellar matter is appreciable. One may in such a case assume that the distribution of the medium is independent of the presence of stars and set $\rho_s$ equal to zero in equation (36) for the gravitational potential $\Phi$. The resultant equations for $\rho_s$ are identical with equations (38)–(43)—derived for $\rho_s$ on the assumption that $\rho_p$

could be neglected—provided that a subscript $p$ is substituted for the subscript $s$ throughout.

When the equilibrium of the stars is considered in this case, however, the neglect of $\rho_s$ in the determination of $\Phi$ will not yield an adequate approximation, if $M_s$ exceeds $M_p$. In the general case a new solution of equation (38) must be found for $\rho_s$ such that $\Lambda'(0)$ equals a finite value, determined by the gravitational attraction of the thin layer of interstellar dust and atoms. In the limiting case in which $M_p$ is greater than $M_s$, we may, in fact, neglect $\rho_s$ in the determination of $\Phi$. For all values of $z$ this gives $d\Phi/dz = 2\pi GM_p$. In such a case we have for the stars the usual isothermal distribution in a constant-force field, which gives

$$\log \frac{\rho_s}{\rho_{s0}} = -\frac{6\pi GM_p}{\overline{v_s^2}} z. \tag{46}$$

The value of $z_{s(1/2)}$ in this case may be found by a simple integration of equation (46).

To compare the results for interstellar matter and for stars in the case when $M_p$ exceeds $M_s$, we take the value of $z_{p(1/2)}$ found when equations (42) and (43) are applied to dust and atoms rather than to stars and divide this by the value of $z_{s(1/2)}$ found from equation (46) above. The resultant ratio is

$$\frac{z_{p(1/2)}}{z_{s(1/2)}} = 1.57 \frac{\overline{v_p^2}}{\overline{v_s^2}}. \tag{47}$$

We see that when the mass of interstellar matter exceeds that of the stars, the relative concentration varies as the energy rather than as the velocity. It is obvious that for intermediate cases, in which $\rho_{p0}$ exceeds $\rho_{s0}$ but $M_p$ is less than $M_s$, the ratio of $z_{p(1/2)}$ to $z_{s(1/2)}$ will follow a relationship intermediate between equations (45) and (47).

Numerical values for $z_{p(1/2)}$ may be computed from these results. If the ratio of root mean square velocities is set equal to ten as before and if $z_{s(1/2)}$ equals 200 parsecs, the value of $z_{p(1/2)}$ equals 18 parsecs when $\rho_{p0}$ is less than $p_{s0}$, and 3.1 parsecs when the total mass of interstellar matter is greater then that of the stars. If the mass of interstellar matter is an appreciable fraction of the mass of the system, $z_{p(1/2)}$ will clearly not exceed a few parsecs.

The density of the medium in this latter case would be at least $10^{-20}$ gm/cm$^3$, and possibly very much greater. With so great a density it is doubtful whether the electron temperature could remain as high as

10,000°, and it is likely that the root mean square velocity would fall, increasing the density even further. Such a medium would probably be unstable and would perhaps condense into stars, meteorites, or large dark bodies. But these processes of condensation would not increase $\overline{v_p^2}$, and the thin massive layer of matter, surrounded by a vastly more extended stellar envelope, would still remain. The probable state of matter in the equatorial plane will not be discussed in detail here.

It should be pointed out that such a dark layer is equivalent in its effect to a distribution of very faint dwarf stars with small peculiar motions. In fact, the hypothetical equatorial layer of interstellar matter discussed here might quite possibly condense into stars of absolute magnitude $+15$ or fainter. The random velocities of stars formed in this way would obviously be small. From an observational standpoint it would be very difficult to distinguish a layer of such stars from a layer containing bodies roughly the size of meteorites; the phrase "dark matter" will be used to denote all such possibilities.

If a massive equatorial layer of this type should exist within an elliptical galaxy, it would lead to directly observable results. The mass deduced from the rotational velocities of the stars would be greater than the mass derived from the luminosity of the system. More important, the distribution of stars on each side of the equatorial plane would follow equation (46) rather than the quite different density function in equation (41). The former distribution is striking in its finite derivative at the equatorial plane. Unfortunately, this effect would be observable only if the plane of the galaxy were parallel to the line of sight from earth.

These characteristics are remarkably close to the description of NGC 3115, analyzed in detail by Oort.[11] The luminosity distribution in this elliptical system of class E7 is completely in disaccord with the mass distribution as deduced from the observed constant angular velocity.[2] Furthermore, the logarithmic gradients of luminosity density in the $z$-direction have a nearly constant slope very near to the equatorial plane. The values of $\log \rho$ as a function of $z$ for various values of $R$ are shown in Figure 1. These were taken from Figure 7 in Oort's paper; $R$ and $z$ are expressed in seconds of arc. The curved dashed line $A$ shows the distribution to be expected from equation (41), when the stars themselves are primarily responsible for the gravitational attraction. The straight dashed line $B$ represents the distribution given by equation (46), when the equatorial layer of dark matter is responsible for most of the gravitational force on the stars.

[11] *Ap. J.*, **91**, 273, 1940.

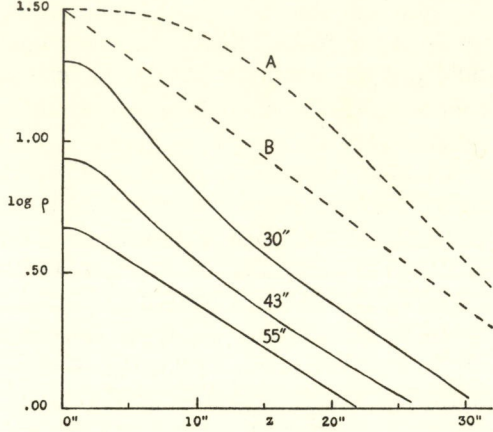

FIG. 1. Stellar density distribution in NGC 3115. The logarithm of the observed luminosity density, $\rho$, is plotted against $z$, the perpendicular distance from the equatorial plane in seconds of arc. The two dashed curves are the theoretical results if $\rho$ is a function of $z$ only, for two cases: A. A uniform isothermal gas under its own gravitational attraction. B. The same gas under the attraction of a massive equatorial layer of dark matter.

It will be noted that the observed curves in Figure 1 correspond to case $B$ more closely than to case $A$. The observed logarithmic gradients depart from linearity for very small values of $z$, but a slight inclination of the equatorial plane to the line of sight would account for this effect. The curvature evident for larger values of $z$ could be explained partly by a dispersion in stellar velocities and partly by the deviation of the potential field from the one-dimensional case discussed here. It seems difficult to account for the observational values of log $\rho$ when $z$ is moderately small, if only luminous stars are assumed to be present. The evidence for an equatorial layer of dark matter in NGC 3115 is suggestive but not conclusive. Further examination of elliptical galaxies, and particularly of the density gradients near the equatorial plane, would be necessary to decide the issue.

The central conclusion of the present paper, however, does not concern the possible existence of such a dark layer. The chief result is that, in NGC 3115 and probably in most of the elliptical and globular galaxies as well, the amount of interstellar matter throughout most of the volume of these systems—in the regions surrounding most of the stars—must be negligible. Unless there are unsuspected forces acting on interstellar particles, the light which comes from elliptical and globular systems of large mass must be almost entirely direct starlight, not diffuse or scattered light.

This view is supported by the failure to find space reddening or interstellar absorption lines in these objects. The lack of interstellar emission lines in elliptical objects[12] is also consistent with this view, although the absence of early-type stars provides an alternative explanation for these observations. The chief support for these conclusions, however, is the essentially theoretical argument presented here. Better determinations of nebular masses would decide to what extent these results may be applied to all elliptical and globular galaxies.

[12] N. U. Mayall, *Pub. A.S.P.*, **51**, 282, 1939.

# THE FORMATION OF STARS

(PHYSICS TODAY 1, 6, 1948)

## *Commentary*

THIS paper, designed for the general scientific reader, is focused not so much on the detailed processes of star formation as on the evidence regarding when and where stars have formed. Thus the chief emphasis is on the point of view put forward in Table 1 of Paper #1, including: the nuclear evidence that supergiant O and B stars must have formed late in the life of our galaxy; observational evidence that these stars are born only in regions characterized by gas and dust; and certain relevant properties of the interstellar medium. The long delay between Papers #1 and 3 resulted, of course, from my absorption in sonar research during World War II. During this interval, Baade's important 1944 paper was published, showing conclusively that the maximum stellar luminosity in elliptical systems was about $10^3$ $L_\odot$, just as in globular clusters; this result extended firmly to elliptical galaxies the picture proposed in Paper #1. Paper #3 was published at about the same time as a paper[1] I had presented at a Harvard symposium in 1946. The account included here gives a fuller description of my general picture.

The star formation process itself is also discussed in Paper #3. A principal suggestion is that the radiative force which pushes grains together (as discussed in Paper #1) accelerates the initial stages of interstellar condensation. Since this process would concentrate solid grains with respect to gas, it would produce stars with a relatively high abundance of elements heavier than hydrogen and helium; these heavier elements tend to accumulate preferentially in grains. A few years later, when it became clear that the relative abundance of the heavier elements was much the same in young stars as in the interstellar medium as a whole (gas plus grains), this process was dropped from scenarios for the birth of stars.

In the decades since this paper appeared, tremendous advances in both observation and theory have made possible the creation of a broad new field of research on how stars form. New detectors of microwave radiation have permitted extensive studies of molecules deep in molecular clouds, and subsequently infrared measures have given data on the thermal emission from dust in these clouds. Thus extensive data have accumulated on the physical properties—density, temperature, and motions—within star-forming regions,[2, 3] whose dusty mantles hid them from sight in visible and ultraviolet light. Measures of interstellar magnetic fields have been made,[4] chiefly via Zeeman splitting or Faraday rotation. Theoretical computations[3] of various phases in star formation have elaborated earlier work on thermal stability[5] magneto-hydrodynamics,[6] ambipolar diffusion,[7] accretion discs,[8] and other topics. A number of different scenarios for star formation have been proposed and analyzed.[3]

The simplified models developed[3] in this burgeoning field have fitted many features of the observations, but are necessarily approximate. In consequence, the distribution of initial stellar masses, for example, remains unexplained. As the power of available computers continues to increase, it may become feasible some day to simulate star formation numerically with sufficient realism to give a conclusive picture of just what happens as a star is born.

## REFERENCES

1. Spitzer, L.—In Centennial Symposia, Harvard Obs. Monograph No. 7, 87 (1948).
2. Myers, P. C.—In: Interstellar Processes, eds. D. J. Hollenbach and H. A. Thronson, Jr. (Dordrecht: Reidel) 71 (1987).
3. The Physics of Star Formation and Early Stellar Evolution, eds. C. J. Lada and N. D. Kylafis (Dordrecht: Kluwer) (1991).
4. Heiles, C.—In: Interstellar Processes, eds. D. J. Hollenbach and H. A. Thronson, Jr. (Dordrecht: Reidel) 171 (1987).
5. Field, G. B.—*Astroph. J.* **142**, 531 (1965).
6. Alfvén, H.—Cosmical Electrodynamics (Oxford: Clarendon Press) (1950).
7. Mestel, L. and Spitzer, L.—Monthly Notices, Roy. Astr. Soc. **116**, 503 (1956).
8. Lüst, R.—*Zs. Naturforsch.* **7a**, 87 (1952).

## *Paper*

If research in astronomy had stopped in 1913, our knowledge of stellar evolution today would be in a satisfactory state. At that time astronomers had a plausible theory of a star's life cycle. Einstein's theory of relativity, advanced only a few years before, showed that mass and energy were interchangeable. It was therefore natural for astronomers to assume that stars were formed as large massive bodies which through successive century after century continued to radiate away matter. Ultimately most of the matter in a star, according to this picture, would be radiated away as light and heat. In this way all the stars, despite their large differences in mass, formed part of the same evolutionary sequence.

Unfortunately, this simple, sweeping, and satisfying picture became discredited by additional information, both astronomical and physical. On the astronomical side, evidence began to accumulate that the universe has not lasted long enough for most stars to radiate away must of their matter. The expansion of the universe, the presence of uranium on the earth, the existence of certain relatively transitory clusters of stars, all indicate that something happened about three billion years ago. If the universe was not created then, it was certainly very extensively reorganized; some sort of cosmic explosion apparently took place at that time. Since the sun, a fairly typical star, would require many hundreds of billions of years to radiate an appreciable fraction of its mass, its total mass has obviously not changed appreciably within this last few billion years.

On the physical side, nuclear physicists have learned a great deal about the specific processes by which matter can be converted into energy. The only known process of importance which can liberate energy inside a star is the combination of four hydrogen atoms to form a helium atom. Calculations carried out by the nuclear physicist, Professor Hans Bethe, show that in the stars this process occurs through the catalytic action of carbon and nitrogen nuclei. Since four hydrogen atoms weigh 0.7 percent more than one helium atoms, the additional mass is released as energy and can be radiated by the star. Even if a star is originally all hydrogen, the total mass radiated can evidently not exceed a very small fraction of the mass of the star.

As a result of these findings we now know that the universe has apparently not lasted long enough in its present form for stars to radiate much of their mass, and in any case there seems to be no physical process by which a star could radiate away most of its matter even if there were time enough. We are forced to conclude that the present variety of stars in the sky is the result of the original method of star formation rather than of any evolutionary process. And the formation of stars in general is still a closed book, since the explosion of the universe a few billion years ago has so far defied any attempts at detailed analysis. It is even possible that the basic laws of nature may have been quite different at that time. Thus our research in the direction of general stellar evolution reminds one of Browning's philosopher, who had "...written three books on the soul, Proving absurd all written hitherto And putting us to ignorance again."

## SUPERGIANT STARS

While the original of the universe is still beyond our understanding, some progress has been made in explaining the origin of a certain class of stars, which may have been created relatively recently. A supergiant star is one which radiates light and heat some ten thousand times as strongly as our own sun. There are not many of these stars, but in a galaxy of many billions of lesser stars they stand out in the same way that a searchlight stands out from a swarm of fireflies. These stars are burning their candle at both ends and they cannot last very long, astronomically speaking. Within a mere hundred million years, such a star must burn all its hydrogen into helium. There is no known way in which a star can remain dark for a long period of time and then suddenly start shining. We conclude that these supergiant stars have formed within the last hundred million years—less than a tenth of the age of the universe.

Of course, it is possible that nuclear physicists have overlooked some important process by which a star can radiate a much larger fraction of its mass then the hydrogen-into-helium process liberates. This does not seem very likely, since the energies with which the atoms hit each other inside a star average only a few thousand electron volts—a small fraction of the energies developed in such atom-busting devices as the cyclotron and synchrotron—and since the nuclear reactions produced at low energies have been fairly well explored in the laboratory.

If it is assumed that these stars have in fact been formed within the last hundred million years, the mechanism for this formation is a problem which astronomers may hope to investigate with some hope of success. Within this interval, conditions in the universe have apparently not changed very much and an examination of the universe about us may actually indicate how supergiant stars have formed in the past, and may even be forming at the present time.

## CLOUDS—THE CLUE

The clouds of matter which float about between the stars are an obvious source of material for star formation. Recent investigations show that these clouds are in fact so closely associated with supergiant stars that a physical connection between them seems very likely.

In brief, the observations indicate that supergiant stars are found only in those aggregations of stars where interstellar clouds of matter are also present. More specifically, observations of stellar galaxies, each one a million or so light years away and each, like our own galaxy, containing many billions of stars, show that supergiant stars are found only in spiral galaxies. These spiral systems, like the huge galaxy in which our sun is located, are flattened, disk-shaped systems some hundred thousand light years in diameter, each one rotating about an axis of perpendicular to the plane of its disk. A typical spiral galaxy is shown in Fig. 1. The characteristic feature of these systems, after which they are named, is the presence of a pair of arms which apparently come out of the central nucleus and wind around the system.

In the elliptical galaxies—which are not rotating so rapidly, are not so flattened, and show no spiral structure—no supergiant stars are found. In fact, long-exposure plates at the Mt. Wilson Observatory have shown that the stars in these systems have a sharp upper limit on their brightness; no star greater than the critical brightness can be found, while below this critical brightness myriads of stars appear on the photographic plate. This result is in marked contrast to the observed brightness of the stars in spiral galaxies, where there are always one or two brightest supergiant stars, a number of less bright supergiants, and a gradually increasing number of fainter and fainter stars. This sharp upper limit on the brightness of stars in elliptical galaxies is just what one would expect if no new stars had been formed since the beginning of the universe, and if the brightest ones had burned up all their fuel and gone out.

Detailed examination of galaxies also indicates that clouds of matter between the stars are found only in spiral systems. In elliptical galaxies the vast stretches between the stars are very nearly empty, but in flattened spiral galaxies like our own there is about as much matter between the stars as there is inside the stars. This association between obscuring clouds and supergiant stars is strengthened by the fact that in the closest galaxy, the great nebula in Andromeda, supergiant stars are observed to occur in exactly those regions where the obscuring clouds are most prominent. Thus the observational evidence indicating a physical connection between clouds and supergiant stars is very strong.

FIG. 1. A typical spiral galaxy. Both supergiant stars and interstellar clouds of obscuring matter are found only in these huge, flattened rotating systems. (Yerkes Observatory Photo)

Before we can accept the hypothesis that supergiant stars have in fact formed from these clouds we must investigate whether or not there is some process which could cause interstellar matter to condense into stars. In this way we are led to consider the physical nature of the stuff between the stars, and the forces which operate on it. Thirty-five years ago the very existence of interstellar matter was not fully realized but recently extensive information on this topic has been obtained.

### Atoms in Space

The dominant constituents of interstellar matter are believed to be individual atoms. These atoms absorb or emit light of particular wavelengths, which can be measured accurately by use of the spectroscope. In some regions, where the gas is at a high temperature, bright emission lines of hydrogen, oxygen, and nitrogen are observed. Measurements of the intensities of these lines show that the density of the interstellar gas is about one hydrogen atom in each cubic centimeter, with other elements present as slight impurities. The interstellar medium is a much better vacuum than is ever obtained in a terrestrial laboratory. If a fly were to breathe a single breath into a vacuum chamber as big as the Empire State Building, the resulting density of the air would still be much greater than the density of the interstellar gas.

In other regions of space the interstellar gas is cool, and no emission lines are produced. Instead, the atoms absorb the light from distant stars, producing absorption lines at particular wavelengths. The absorption lines of the abundant gases, hydrogen, helium, nitrogen, oxygen, etc., when these are cool, lie far out in the ultraviolet where they cannot be detected. Interstellar absorption lines of sodium, calcium, titanium, and iron lie within the observable spectrum and have been observed in the spectra of bright stars a few thousand light years away. These lines are very sharp, and can usually be distinguished from the lines produced by the atoms in a stellar atmosphere, where the high temperature and pressure give wide lines.

Recent work has been concerned with the detailed distribution of interstellar gas. Measurement of the strongest absorption lines, with the most powerful spectrographs available at the 100-inch telescope of the Mt. Wilson Observatory, shows that a single line is frequently made up of several components. Each separate component is produced by absorption in a single cloud of gas, the different components being separated in wavelength by the difference in Doppler effect produced by the different cloud velocities. These clouds, each one about twenty light years across, are moving through space at speeds of some ten to twenty miles a second. A more detailed understanding of the nature of these clouds is desirable before one can discuss in detail how interstellar matter can form new stars. Further work along these lines is now in progress.

### Solid Particles

In addition to the separate atoms drifting about in space, small solid particles, or grains, are also present. Each grain is about one hundred-thousandth of an inch in diameter; ten thousand placed end to end would make a line about as long as a period on this page. Since the size of these grains is just about equal to the wavelength of visible light, these particles are of the size which is most effective in absorbing and scattering light waves. These particles are responsible for the general obscuration produced by the clouds shown in Fig. 1. Particles of smaller size are presumably also present, but these do not produce such a noticeable effect, and can therefore not be detected.

The properties of these particles have been determined from accurate measurements of the obscuration which they produce in light of different wavelengths. This obscuration is greater for blue light than for red light, which proves that the particles cannot be much *larger* in size than the wavelength of light. On the other hand, the obscuration varies inversely only as the first power of the wavelength, instead of as the fourth power which is observed for scattering by the molecules of the atmosphere. From this one can conclude that the grains are not very much *smaller* in size than the wavelength of light. In this way a particle size of about the wavelength of light has been determined. From the fact that the grains seem to scatter more than they absorb it seems likely that they are dielectric rather than metallic in composition. If, as seems likely, these grains were produced by the sticking together of individual atoms, the enormous abundance of hydrogen relative to other elements would be expected to produce solid hydrogen compounds, in particular, ordinary ice. However, impurities of all other elements would also be present.

Studies of the distribution of these grains have indicated that the clouds in which these grains are concentrated are apparently identical with the gaseous clouds already described. Thus whatever pushes atoms into clouds also pushes the grains together.

## Forces in Interstellar Space

To discuss in detail how stuff in space can condense to form new stars we must determine the physical conditions of matter in space. In particular, we must combine the observational evidence described above with our knowledge of basic physical principles to investigate the different forces that are at work on the different particles. Only in this way can we predict how the interstellar medium will behave under various widely different conditions.

In the immense vacuum between the stars, an interstellar particle spends most of its time moving in a straight line without interruption. Occasionally, one of two things may happen to it: an encounter with another interstellar particle, or an encounter with a light wave, or photon. The information which physicists have obtained on such processes is not so complete as astronomers would like, but is sufficient for an approximate evaluation of the effects which these various collisions will produce.

The collisions of the interstellar atoms and grains with each other help determine the temperature of matter in space. In most cases, the collisions are elastic and the kinetic energy of the different particles is exchanged back and forth; as a result, the distribution of velocities corresponds to that in thermal equilibrium at some particular temperature. Photoemission of energetic electrons from hydrogen atoms and grains, on absorption of photons, tends to keep the temperature high, but inelastic collisions between atoms and grains tend to give a low temperature. Near a very hot and very bright star the gas will be heated up to about 10,000°K, but in other regions a temperature of about 100°K seems likely. This difference of temperature between different regions is believed to produce cosmic currents, or winds, in the same way as the winds on earth are produced.

In some cases the interstellar particles stick to each other on collision. Thus atoms stick together to form molecules, molecules stick together to form larger molecules, and grains grow by slow accretion. This process was analyzed during the war by a number of Dutch astronomers, who were able to show that the interstellar grains have probably been formed by this evolutionary process within the last few billion years. More accurate physical information on collisions between particles at low energies is required to make this theory more quantitative.

Collisions between grains and photons are important in star building. It is well known that light exerts pressure. Since starlight in a galaxy comes from all directions in the galactic plane, a single grain will be knocked this way of that by photon collisions, without any net motion resulting. However, when several grains are present, the shadow of each one on the other unbalances the radiative force, and photons striking from the opposite sides push the grains toward each other. As a result, there is an effective force of attraction between grains which is several thousand times as great as the gravitational force between them.

## STAR FORMATION

The further we go away from observational data the more uncertain our theories become. The mechanism of star formation, which is the ultimate objective of much of the work described above, is still in a rather speculative state. However, putting all the above information together does provide a reasonable preliminary picture for the process by which stars can be formed from interstellar matter.

The process may be assumed to start with an interstellar gas, formed at the same time as the rest of the universe. The first step in the process is then the slow condensation of interstellar particles from the gas. After these particles have reached a certain size, the radiative attraction between them forces them together and they drift toward each other, forming an obscuring cloud in a time of about ten million years. In a cloud, where the density of grains is high, the temperature tends to be low. In the surrounding region the high temperature produces high pressure, and the low-temperature, low-pressure cloud therefore becomes compressed. In this way the density of gas within a cloud will be increased, corresponding to the observed result that a cloud of grains is also a cloud of gas.

Currents produced by differences of temperature and also by the general rotation of the galaxy will tend to tear some of these clouds to pieces. On the other hand, the forces of condensation will pull them together, and some clouds may be expected to go on contracting. The radiative force becomes ineffective when the clouds become so opaque that light does not penetrate into them very far. At this point gravitation takes over and tends to produce a further contraction. In this stage a cloud has a diameter of a light year or less. Small opaque clouds of this type, called globules, have been known for some time, and are shown in Fig. 2.

One of the chief problems concerns the angular momentum of this prestellar globule, or protostar. According to Newton's laws of motion, the angular momentum, which is proportional to the product of the radius and the rotational velocity, remains constant; as the radius decreases the rotational speed increases. Since the radius of a typical cloud is some ten million times the radius of a supergiant star, this increase in rotational speed can be quite impressive. Unless some way can be found to dispose of the angular momentum, a protostar would hurl itself to pieces by centrifugal force. The possibility that turbulent motions in the gas may carry the angular momentum away has been explored by several German astronomers. However, the turbulent velocities involved would exceed the velocity of sound in the interstellar gas, and physical information about this type of turbulence is virtually nonexistent. In this country the possibility has been advanced that a galactic magnetic field might produce electrical eddy currents in a rotating protostar, which would then damp out the angular momentum.

An interesting variant of this star-building picture has been proposed by Dr. Fred Whipple, one of the astronomers who has contributed most to this theory of star building. He suggests that a condensing cloud may

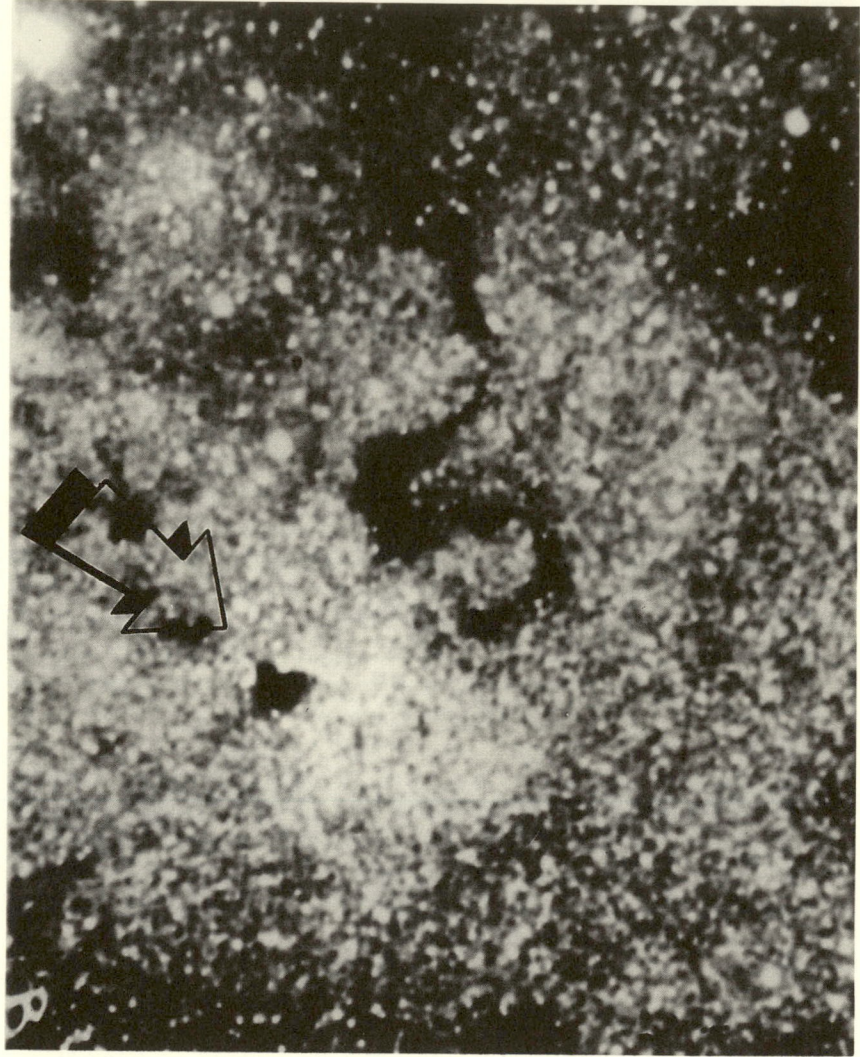

Fig. 2. The compact opaque globule may be a protostar, contracting to form a new star.

have produced our solar system. In view of the widespread general interest in the formation of the solar system, such a bold extrapolation of these theoretical concepts back to conditions several billion years ago is naturally of much significance.

It is evident that the picture of star formation which has been described here is still in a formative stage. The work in progress is being carried out cooperatively by a number of astronomers all over the world. Perhaps when the 200-inch telescope probes further into the secrets of space, and when further progress in experimental and theoretical physics increases our understanding of the processes at work between the stars, we may then outline with more assurance the detailed steps by which supergiant stars may be forming almost before our very eyes.

# THE TEMPERATURE OF INTERSTELLAR MATTER. I

(Astroph. J. 107, 6, 1948)

## *Commentary*

INFORMATION on the kinetic temperature is a vital input to all discussions of how the interstellar gas moves and evolves. Hence in 1946 I began a systematic analysis of the various physical processes which combine to determine this temperature in equilibrium. This program led to three publications. Paper #4 was the first of these and treated a variety of processes which increase the energy of random motions; that is, which heat the gas. The second[1] considered cooling processes (conversion of kinetic energy into radiation), while the third[2] computed approximately the equilibrium value of the temperature $T$ at which the heating and cooling are equal. A particularly noteworthy result of these three papers was about a hundredfold difference of $T$ between H II regions and H I clouds; the former are characterized by nearly complete ionization of H, while in the latter, the H atoms are mostly neutral. In H II gas, surrounding luminous early-type stars, $T$ is typically about 10,000 K, while in the separate, usually dusty, H I clouds a value of at most 100 K for $T$ is typical.

This temperature difference, which strongly affects many of the various interactions occurring in the interstellar medium, results from simple physical principles. In H II gas, heating results whenever a proton captures an electron, which is subsequently ejected on absorption of an ultraviolet stellar photon and carries off much more kinetic energy that it had before capture. Cooling results whenever an electron loses energy by collisional excitation of a heavy ion, which then radiates the excitation energy. In thermal equilibrium these two processes balance at $T \approx 10,000$ K, the exact value depending[3] on the abundances of the heavy ions relative to H and, of course, on the collisional cross-sections.

In H I gas the radiation which can ionize H atoms is heavily absorbed, and other heating processes are less by orders of magnitude. Hence the temperature falls far enough so that most electrons have kinetic energies too low to excite most of the heavy ions present. The 10,000 K kinetic temperature of an ionized interstellar gas was pointed out by

Eddington, but the much lower temperature of H I clouds has not been generally realized. Observational confirmation[4] of this low H I temperature was obtained in the 1970s, first by comparison of emission and absorption in the 21 cm line of neutral H, and next by the relative population of $H_2$ molecules in the $J = 1$ and $J = 0$ levels, with no electronic or vibrational excitation.

In the fifty years since the appearance of Paper #4 and its two accompanying articles, the principal features of thermal equilibrium theory have changed relatively little. Detailed computations, on the other hand, have greatly altered, as accurate excitation cross-sections have been obtained (replacing 1947 values whose uncertainties ranged up to an order of magnitude). In addition, high speed computers have become available to calculate equilibrium temperatures in increasingly complex situations, including heating by X rays and cooling by molecules.

One aspect in which qualitatively new elements have been added to the theory is the proposal of additional heating processes in H I gas. Two developments have pushed in this direction. First, as calculations became more precise and realistic, it was increasingly evident that the processes being considered led to equilibrium temperatures of some 40 K in dusty H I clouds; these values, about half those observed, are low enough to reduce markedly the excitation of the $C^+$ fine-structure level, a principal coolant for $T$ less than 100 K. Second, increasingly refined observations of the low-density H I gas between the clouds indicated a kinetic temperature of a few thousand degrees K, comparable to that in H II regions.

An early attempt to account for warm H I gas, with these higher temperatures at relatively lower densities, was based on the assumption that low-energy cosmic rays, with energies between 10 and 100 MeV and high particle fluxes, permeated the Galaxy (see the final paragraph below). The cosmic rays actually reaching the Earth have energies exceeding 1000 MeV, but the flux of these particles is so low that their heating effect is negligible. The theory of interstellar heating by low-energy cosmic-ray protons had several attractive features,[5] but various effects which it predicted for interstellar molecules disagreed[6] with observations.

A later, more successful, attempt is based on the relatively high interstellar abundance of very small particles, with radii less than 10 Å and containing less than 1000 atoms (assumed to be predominantly carbon). Such tiny grains (or large molecules) can be heated to a temperature exceeding 100 K when they absorb a single photon, an effect which is important in explaining the observed infrared emission from interstellar clouds.[7] These very small particles are believed to show much stronger photoelectric emission than do the larger grains considered earlier. A preliminary analysis[8] of the heating produced by the photoelectrons from these small particles suggest that this effect may well provide the additional heating required for warm H I regions, through other possible mechanisms could also contribute.

The hot phase of the interstellar gas, with $T \approx 10^6$ K is not considered here (see Paper #7). Theory indicates that these high temperatures do not require a thermal equilibrium, but result from dynamical transients, such as supernova explosions and high-velocity stellar winds.

As with Paper #1, some of the text has been omitted from Paper #4. At the end of the sixth section, on interstellar grains, the original paper contains a detailed discussion of the photoelectric properties of different bulk materials for radiation at visible and near-ultraviolet wavelengths. Since this discussion has been superseded by a systematic, more up-to-date survey,[9] all this text has been deleted. Also omitted is a final section showing that the heating rate by cosmic rays reaching the Earth is negligible, but pointing out that particles of lower energy might be present and could make a dominant contribution to interstellar heating.

## REFERENCES

1. Spitzer, L.—*Astroph. J.* **109**, 337 (1949).
2. Spitzer, L. and Savedoff, M. P.—*Astroph. J.* **111**, 593 (1950).
3. Menzel, D. H. and Aller, L. H.—*Astroph. J.* **94**, 30 (1941).
4. Spitzer, L.—In: Physical Processes in the Interstellar Medium, New York: J. Wiley, Sect. 3.4a, b (1978).
5. Field, G. B., Goldsmith, D. W., and Habing, H. J.—*Astroph. J. Lett.* **155**, L149 (1969).
6. Spitzer, L. and Jenkins, E. B.—*Annu. Rev. Astron. and Astroph.* **13**, 133 (1975).
7. Puget, J. L. and Léger, A.—*Annu. Rev. Astron. and Astroph.* **27**, 161 (1989).
8. Bakes, E.L.O. and Teilens, A.G.G.M.—*Astroph. J.* **427**, 822 (1994).
9. Draine, B. T.—*Astroph. J. Supp.* **36**, 595 (1978).

# 4

## *Paper*[†]

### ABSTRACT

The processes which tend to increase the equilibrium kinetic temperature of interstellar matter are considered in detail. The equations for the energy gained by photoelectric ionization of $H$ and other atoms, by the photodetachment of electrons from $H^-$, and by photoelectric emission from solid grains are put into a form suitable for numerical computation. Since the electrical charge on the grains enters into the problem, the various physical processes, such as the sticking probability of electrons, which determine the magnitude of the charge on a grain, are also discussed, and the quantitative physical information available on these processes is summarized. The kinetic energy transferred from cosmic rays to the much-less-energetic atoms and grains is treated briefly. The processes which tend to decrease the equilibrium kinetic temperature and the final equilibrium temperatures to be expected will be discussed in a subsequent paper.

Since thermodynamic equilibrium is not even remotely attained in interstellar space, there is no uniquely defined temperature which characterizes interstellar matter. It was pointed out by Eddington,[1] however, that the frequent elastic encounters between the interstellar particles would establish a Maxwellian velocity distribution, with equipartition of kinetic energy among the various interstellar atoms. Interactions between the atoms and the small solid grains[2] responsible for the extinction of starlight will bring[3] the kinetic energy of the grains also into equipartition with the energy of the interstellar particles. The specific deviations from the Maxwellian distribution which may be expected have recently been investigated quantitatively.[4] A numerical application of these general results shows that in interstellar space, as well as in the somewhat denser nebulae, such deviations are wholly

---

[†] Discussions of photoelectric properties and cosmic-ray heating are omitted here.

[1] *Proc. Roy. Soc. A*, **111**, 424, 1926 (Bakerian Lecture).

[2] In accordance with discussions among several workers in the field, the solid interstellar particles are referred to as "grains"; the terms "dust" and "smoke," which imply a certain type of origin for the interstellar grains, are not used here, but are reserved for discussions of the evolution of solid particles.

[3] L. Spitzer, Jr., *Ap. J.*, **93**, 369, 1941.

[4] D. Bohm and L. H. Aller, *Ap. J.*, **105**, 131, 1947.

negligible. Thus the distribution of particle velocities corresponds very closely to the distribution in thermodynamic equilibrium at some particular temperature, which may be called the "interstellar kinetic temperature."

The classic discussion of interstellar temperature is due to Eddington.[1,5] He pointed out that electrons, when ejected from an atom by photoionization, have an average kinetic energy corresponding to the color temperature of the ionizing radiation—about 10,000°–20,000° for the ultraviolet radiation capable of ionizing the atoms present. Since free-free transitions of electrons in the fields of the positive ions reduce the electron energy, a value of 10,000° was suggested by Eddington as the kinetic temperature of interstellar matter.

A somewhat different result obtained by Rosseland,[6] who found that the kinetic temperature of an interstellar medium composed entirely of hydrogen was a smaller fraction—about one-sixth—of the color temperature of the exciting radiation. This result depends in part on the assumption that the radiation emitted by the interstellar gas at each frequency may be found from Kirkhoff's law and is thus equal to the absorption coefficient at that frequency multiplied by the Planck function for black-body radiation at the kinetic temperature. While this assumption is justified in thermodynamic equilibrium, its validity in interstellar space is open to serious question. In particular, as compared with thermodynamic equilibrium, the ground state of an atom in interstellar space is enormously overpopulated relative to the ionized state. The absorption coefficient at frequencies beyond the ionization limit is proportional to the number of these neutral atoms, while the emission is proportional to the number of the ionized atoms and electrons. It follows that the intensity of radiation emitted by captures in the ground state is enormously less than that found from Kirkhoff's law. One may conclude that Rosseland's analysis is not directly relevant to the exact determination of interstellar kinetic temperature.

In accordance with Eddington's results, a kinetic temperature of about 10,000° has usually been assumed in most of the subsequent theoretical work on interstellar matter. The importance of this temperature is evidenced by the variety of papers in which the results depend critically on this assumption. Among the subjects discussed in these investigations are: the condensation of interstellar atoms to form grains;[7] the ionization equilibrium of interstellar metallic atoms;[8,9] the ioniza-

[5] *Internal Constitution of the Stars* (Cambridge University Press, 1926), p. 371.

[6] *Theoretical Astrophysics* (New York: Clarendon Press, 1936), p. 321.

[7] B. Lindblad, *M.N.*, **95**, 20, 1934; *Nature*, **135**, 133, 1935; D. ter Haar, *B.A.N.*, No. 361, 1943 (*Ap. J.*, **100**, 288, 1944).

[8] O. Struve, *Proc. Nat. Acad. Sci.*, **25**, 36, 1939.

[9] T. H. Dunham, Jr., *Proc. Amer. Phil. Soc.*, **81**, 277, 1938.

tion equilibrium of interstellar hydrogen;[10] the dissociation equilibrium of interstellar molecules;[11, 12, 13] the origin of short-wave radiation observed from the galactic center;[14, 15] the charge on solid grains,[3, 16] and the formation of concentrations of grains, which subsequently grow into stars.[17] In some of these references no explicit discussion of the kinetic temperature is given, but it is implicitly assumed that this temperature is equal to the color temperature of the dilute ionizing radiation, which is taken equal to 10,000° or 15,000°.

The possibility that collisions between electrons and molecules might seriously reduce the kinetic temperature in regions where the interstellar medium is relatively dense has been suggested[18] in connection with a theory of accretion of interstellar matter by stars. The quantitative analysis of inelastic collisions between electrons and molecules will be discussed in the second paper of the present series. It has also been pointed out[3] that collisions between interstellar atoms and grains tend to reduce the kinetic temperature. A detailed discussion of this mechanism has recently been given in an important paper by F. Hoyle.[19] However, Hoyle's treatment is concerned primarily with the properties of a hypothetical interstellar medium whose density exceeds $5 \times 10^{-22}$ gm/cm$^3$; the analysis is presented in connection with the general structure of spiral galaxies. The present discussion concerns primarily the properties of interstellar matter in the neighborhood of the sun, to which Hoyle's analysis is not directly relevant.

In the steady state the kinetic temperature of matter is determined by the condition that the kinetic energy gained per cubic centimeter per second must just equal the corresponding energy lost. Thus an investigation of the kinetic temperature must deal with two types of processes: those which convert other types of energy, chiefly radiant energy from the stars, into kinetic energy of random motion, and those which dissipate kinetic energy by converting it into other forms, chiefly the far-infrared radiation emitted by the grains and the visible radiation emitted as forbidden emission lines by the atoms. The present paper, the first of a series on interstellar kinetic temperatures, investigates the

[10] B. Strömgren, *Ap. J.*, **89**, 526, 1939.
[11] A. McKellar, *Pub. A.S.P.*, **52**, 187, 1940.
[12] P. Swings, *Ap. J.*, **95**, 270, 1942.
[13] H. A. Kramers and D. ter Haar, *B.A.N.*, No. 371, 1944.
[14] L. G. Henyey and P. C. Keenan, *Ap. J.*, **91**, 625, 1940.
[15] G. Reber and J. L. Greenstein, *Observatory*, **67**, 15, 1947.
[16] B. Jung, *A.N.*, **263**, 425, 1937.
[17] F. L. Whipple, *Ap. J.*, **104**, 1, 1946.
[18] F. Hoyle and R. A. Lyttleton, *Proc. Cambridge Phil. Soc.*, **35**, 405, 1939; **36**, 424, 1940.
[19] *M.N.*, **105**, 287, 1945.

processes which contribute to the kinetic energy in the steady state. Collisions between interstellar clouds and similar time-dependent processes may also contribute to the kinetic energy; the discussion of such nonequilibrium effects is postponed to a subsequent paper.

The first section of the present paper treats in a general way the various processes which can convert radiant energy from the stars into kinetic energy of the interstellar particles. In all processes a compound particle is disrupted, the absorbed energy being divided among the component particles. Since in a steady state as many compound particles are formed by encounters as are disrupted by photon absorption, it is convenient to regard these processes as superelastic collisions, i.e., two components particles collide, emit a quantum, and adhere together for a short time but subsequently dissociate with the absorption of a quantum, the final kinetic energy exceeding the initial energy. The following sections apply these general equations to a number of specific types of superelastic collisions—i.e., electron capture by protons, metallic ions, neutral hydrogen atoms, and solid grains and also capture of atoms by one another. Some of the processes which have been investigated are relatively unimportant in most of interstellar space but might become dominant under some conditions. In the section dealing with grains the factors affecting the electrical charge on the grain are also treated, since this charge affects the number of electrons captured by the grain. A final section analyzes the transfer of energy from cosmic rays to the interstellar medium. A second paper, now in preparation, discusses the various inelastic encounters which tend to dissipate the kinetic energy of the interstellar particles. In a subsequent discussion these various results will be used to determine the kinetic temperature of interstellar matter.

It should be emphasized that many of these results are quite preliminary. Adequate theoretical or experimental research is lacking on many of the important physical phenomena. For example, if the radiation from the brighter stars contains a very great excess of ultraviolet radiation, the kinetic temperature of interstellar space will obviously be affected. In addition, the unexplained presence of cosmic rays suggests that unsuspected processes may play a significant role in interstellar problems. It may be of interest, however, to follow out as far as possible the consequences of known physical processes and to see whether these consequences are in agreement with the known observational facts.

## I. General Formulae for Superelastic Collisions

The photoelectric effect is the primary source of kinetic energy in interstellar space. By this process light from the stars is converted into electron kinetic energy, and this energy is rapidly divided among the other particles in the medium. When an electron is captured and subsequently re-emitted on capture of a photon, the kinetic energy is usually increased. This difference between the kinetic energy after photoemission and the kinetic energy before capture may be called the "net energy gain." This quantity, multiplied by the probability that a free electron will be captured per second and averaged over all initial electron velocities and over all frequencies of the absorbed photons, is the average net energy gain per second per free electron and will be denoted by the symbol $G_e$. The following additional subscripts will be used to denote the type of particle with which the electron is colliding: $p$ for protons; $H$ for neutral hydrogen atoms; $i$ for ions of atoms other than hydrogen and helium, especially calcium, sodium, etc.; $m$ for molecules; and $g$ for grains. Thus $G_{ep}$ will denote the average net energy gain per free electron per second in electron-proton encounters resulting in electron capture and subsequent photoemission. The total energy gain per second per free electron, resulting from all these processes, will then be the sum of all the $G_e$'s.

The number of each type of particle per cubic centimeter will be denoted throughout by $n$, with the subscripts $e$, $p$, $H$, $i$, $m$, and $g$ referring, as before, to electrons, protons, neutral hydrogen atoms, ions of atoms other than hydrogen and helium, molecules, and grains, respectively. The velocity of an electron will be denoted by $v$, its kinetic energy before capture by $E_1$, and after capture and re-emission by $E_2$; $E_{2p}$, for example, will be used to denote the kinetic energy of a photoelectron ejected after capture by a proton, while $\overline{E_{2p}}$ will denote the average value of this quantity. The quantity $\overline{E_1}$, the average kinetic energy of an interstellar particle, is assumed to be the same for all types of particles, in accordance with the results reached in a previous paper.[3] It will be assumed throughout that the velocity of an electron relative to an atom, molecule, dust grain, etc., may be set equal to the electron velocity $v$ and that all the kinetic energy of photoelectric emission is carried away by the electron. In view of the very small relative mass of the electron, this assumption is sufficiently accurate. The capture cross-section of each type of particle for electrons will be denoted by $\sigma$, with the appropriate subscripts; this choice of notation represents a change from previous work,[3] where $\sigma$ was taken as the radius of a grain.

First, we develop a general formula for $G_e$, which may be applied to all types of photoelectric emission. If the relative velocity of the two types of particles were the same in all encounters and if the kinetic energy before capture and after photoejection were also constant from one encounter to the next, $G_e$ would be given by the simple formula

$$G_e = n\sigma v(E_2 - E_1),\qquad(1)$$

where $n\sigma v$ is the number of electron captures per second per free electron by the type of particle in question, with a particle density $n$ per cubic centimeter. For each specific process, all quantities in equation (1), except $v$, will carry a subscript indicating the type of particle which captures the electron. Since, in fact, $v$ has a Maxwellian distribution and since $\sigma$ and $E_1$ may both vary with $v$, we must average $G_e$ over all velocities; in addition, $E_2$ must be averaged over all frequencies in the rotation field. If we denote averages by horizontal bars, we have

$$G_e = n\left(\overline{\sigma v} \times \overline{E_2} - \overline{\sigma v E_1}\right).\qquad(2)$$

More specifically, the fraction of electrons whose velocities lie between $v$ and $v + dv$ is given by $P(v)\,dv$, where

$$P(v) = \frac{4}{\pi^{1/2}} L^{3/2} v^2 e^{-Lv^2}.\qquad(3)$$

The quantity $L$ is related to the kinetic temperature, $T$, by the relationship

$$L = \frac{m_e}{2kT},\qquad(4)$$

where $m_e$ is the mass of the electron. With this explicit function for $P(v)$, we may write down the two averages

$$\overline{\sigma v} = \frac{4L^{3/2}}{\pi^{1/2}} \int_0^\infty \sigma(v) v^3 e^{-Lv^2}\,dv,\qquad(5)$$

and

$$\overline{\sigma v E_1} = \frac{2m_e L^{3/2}}{\pi^{1/2}} \int_0^\infty \sigma(v) v^5 e^{-Lv^2}\,dv,\qquad(6)$$

where $\frac{1}{2}m_e v^2$ has been substituted for $E_1$.

To find $\overline{E_2}$ we must know the absorption coefficient $\kappa(\nu)$ for photoelectric emission and $U(\nu)$, the density of radiant energy per cubic centimeter per frequency interval. Since $\kappa(\nu)$ is the ratio of the energy absorbed per particle per second to the energy flux per square centimeter per second, the total energy absorbed in a frequency interval $d\nu$ per second per particle will be $\kappa(\nu)cU(\nu)\,d\nu$. The corresponding number of photons absorbed is obtained on dividing this absorbed energy by the energy $h\nu$ of each individual photon. It should be noted that $\kappa(\nu)$ represents only that part of the absorption which leads to the emission of one photoelectron per absorbed quantum. In the general case a fraction $q(\nu)$ of this absorbed energy will appear as kinetic energy of the photoelectron. Some of the absorbed energy is required to increase the potential energy of the electrons—i.e., to eject them; in the case of solid grains some of the absorbed energy also goes into the thermal energy of the grain. The average energy of a photoelectron is obtained by dividing this total kinetic energy given to electrons by the number of photoelectrons emitted per second. Thus we have

$$\overline{E_2} = \frac{\displaystyle\int_{\nu_0}^{\infty} q(\nu)\kappa(\nu)U(\nu)\,d\nu}{\displaystyle\int_{\nu_0}^{\infty} \frac{\kappa(\nu)U(\nu)\,d\nu}{h\nu}};\tag{7}$$

the integration extends from the photoelectric threshold frequency, $\nu_0$, up to infinity. When the ionizing radiation is dilute black-body radiation, with a color temperature $T_c$, then $\overline{E_2}$ is usually about equal to $kT_c$. Thus for each process it will be convenient to introduce an effective color temperature, defined by the relationship

$$T_{cj} = \frac{\overline{E_{2j}}}{k};\tag{8}$$

for each specific process, $j$ is replaced by the appropriate subscript. Thus for the ionization of $H$ atoms, the effective color temperature is denoted by $T_{cp}$. In the following sections the above equations will be used to compute $G_e$ for a number of processes.

## II. Protons

First, we consider encounters between electrons and protons and compute $G_{ep}$. This process has been considered in some detail by Menzel and others in a series of papers.[20, 21] dealing primarily with the electron temperatures in planetary nebulae. Electron captures have also been discussed by Cillié[22] in connection with the Balmer decrement in

[20] D. H. Menzel, *Ap. J.*, **85**, 330, 1937.
[21] J. G. Baker, D. H. Menzel, and L. H. Aller, *Ap. J.*, **88**, 422, 1938.
[22] G. Cillié, *M.N.*, **92**, 820, 1932.

gaseous nebulae. Since Cillié's work is not concerned with kinetic temperatures and since the viewpoint in the present paper is quite different from that of Menzel and his co-workers and may give added insight into the nature of the problem, this subject will be treated again here.

The cross-section for capture of an electron, whose velocity is $v$, in the $n$th quantum state of hydrogen may be written as $\sigma_{np}(v)$. This cross-section may be found from equation (23) of Menzel's basic paper.[20] The total number of emissions per second per frequency interval, given by Menzel's $E_{\kappa h}/h\nu$, is equal to the total number of electron captures per corresponding velocity interval; hence we have

$$\frac{E_{\kappa n}}{h\nu}\, d\nu = \sigma_{np} n_p n_e v P(v)\, dv. \tag{9}$$

The quantity $v$ may be eliminated by the equation

$$h\nu = \tfrac{1}{2}mv^2 + \frac{h\nu_0}{n^2}, \tag{10}$$

where $\nu_0$ is the frequency of the Lyman series limit. Combining equations (3), (4), (9), and (10) with Menzel's equation (23) and substituting for Menzel's $K$ from his equation (6), we have, finally,

$$\sigma_{np} = A\frac{\nu_0}{\nu}\,\frac{h\nu_0}{\tfrac{1}{2}m_e v^2}\,\frac{g}{n^3}, \tag{11}$$

where $\nu$ is given again by equation (10) and where

$$A = \frac{2^4}{3^{3/2}}\,\frac{he^2}{m_e^2 c^3} = 2.11 \times 10^{-22}\ \text{cm}^2. \tag{12}$$

Throughout the following we shall set the Gaunt factor, $g$, equal to unity. Menzel and his coworkers have investigated the error introduced by this assumption and find that it is small in general—usually less than about 20 percent. In any hydrogenic atom, such as $He^+$, or $Li^{++}$, where the nuclear charge is $+Ze$, equations (11) and (12) are still valid, provided that one takes into account the direct proportionality between $\nu_0$ and $Z^2$.

If now we introduce equation (11) into equations (5) and (6) and make the substitution

$$u = Lv^2 = \frac{m_e v^2}{2kT}, \tag{13}$$

we find for the partial cross-section the following average:

$$\overline{\sigma_{np}v} = \frac{2A}{(\pi L)^{1/2}} \frac{1}{n^3} \frac{hv_0}{kT} \int_0^\infty \frac{e^{-u}\, du}{\dfrac{1}{n^2} + \dfrac{kT}{hv_0}u}. \tag{14}$$

If we introduce the quantities

$$\beta = \frac{hv_0}{kT} = \frac{158{,}000°}{T}, \tag{15}$$

$$Ei(x) = \int_x^\infty e^{-w} \frac{dw}{w}, \tag{16}$$

and let

$$w = u + \frac{\beta}{n^2}, \tag{17}$$

we find, on summing over all quantum levels $n$,

$$\overline{\sigma_p v} = \frac{2A}{(\pi L)^{1/2}} \beta \phi(\beta) \tag{18}$$

where

$$\phi(\beta) = \sum_{n=1}^\infty \frac{\beta}{n^3} e^{\beta/n^2} Ei\left(\frac{\beta}{n^2}\right). \tag{19}$$

Similarly, we find from equation (6),

$$\overline{\sigma_{np}vE_1} = \frac{m_e A}{\pi^{1/2}L^{3/2}} \frac{\beta^2}{n^3} \int_0^\infty \frac{ue^{-u}}{u + \dfrac{\beta}{n^2}}. \tag{20}$$

If we make the substitution (17) as before and sum over all $n$, equation (20) yields

$$\overline{\sigma_p v E_1} = \frac{m_e A}{\pi^{1/2} L^{3/2}} \beta \chi(\beta),$$ (21)

where

$$\chi(\beta) = \sum_{n=1}^{\infty} \frac{\beta}{n^3} \left\{ 1 - \frac{\beta}{n^2} e^{\beta/n^2} Ei\left(\frac{\beta}{n^2}\right) \right\}.$$ (22)

Values of the functions $\phi(\beta)$ and $\chi(\beta)$ are given in Table 1. These were found by direct summation up to $n = 10$, with the remaining terms approximated by an integral. These functions change so slowly with $T$ that their values at any intermediate temperature may be found by simple interpolation. It may be noted that the function $\phi$ is equal to $\beta$ times the function $G_{T_e}$ used by Baker, Menzel, and Aller.[21]

The functions $\phi(\beta)$ and $\chi(\beta)$ have a simple physical significance when $\beta$ is large. As $\beta$ increases, $\phi(\beta)$ approaches the ratio between the number of electron captures in all states and the corresponding number of captures in the ground state. When $\beta$ exceeds 20, for example, $\phi(\beta)$ is within 5 percent of this ratio; but with decreasing $\beta$ the agreement is less close. When $\beta$ lies between 0.5 and 1.0, with $\phi(\beta)$ between 0.70 and 0.96, the total number of captures in all levels is between 1.5 and 1.6 times the number in the ground level. Similarly, for large $\beta$, $\chi(\beta)$ approaches the ratio of the kinetic energy radiated per second by electron captures in all levels to the corresponding energy radiated by captures in the ground level. When $\beta$ lies between 0.5 and 2.0, $\chi(\beta)$ is less than unity, and this ratio lies between 1.3 and 1.5. Evidently the relative importance of captures in excited levels changes extremely slowly with changing temperature.

The average kinetic energy of a captured electron is obtained on dividing equation (21) by equation (18), which yields $kT\chi(\beta)/\phi(\beta)$. Thus the ratio $\chi(\beta)/\phi(\beta)$ gives the average energy of the captured electrons, in units of $kT$. Since the mean kinetic energy of free electrons is $3kT/2$, the ratio of the electron energy lost per capture to the mean kinetic energy is $2\chi(\beta)/3\phi(\beta)$, and this quantity is listed in the last column of Table 1. The relatively low energy of the captured electrons results from the increase of capture cross-section with decreasing velocity. If no dissipation of energy occurred except through electron captures by protons, the kinetic temperature would adjust itself so that the average energy of a captured electron would equal $kT_{cp}$, the average energy of an electron emitted photoelectrically (see eq. [8]);

TABLE 1
Values of $\phi(\beta)$ and $\chi(\beta)$*

| $\beta$ | $T$ | $\phi(\beta)$ | $\chi(\beta)$ | $2\chi(\beta)/3\phi(\beta)$ |
|---|---|---|---|---|
| 0.5 | 316,000° | 0.70 | 0.35 | 0.33 |
| 1.0 | 158,000 | 0.96 | 0.55 | .38 |
| 2.0 | 79,000 | 1.26 | 0.82 | .43 |
| 5.0 | 31,600 | 1.69 | 1.24 | .49 |
| 10.0 | 15,800 | 2.02 | 1.56 | .52 |
| 20.0 | 7,900 | 2.36 | 1.89 | .54 |
| 50.0 | 3,160 | 2.82 | 2.34 | .55 |
| 100.0 | 1,580 | 3.16 | 2.67 | .56 |
| 200.0 | 790 | 3.51 | 3.01 | .57 |
| 500.0 | 316 | 3.98 | 3.47 | .58 |
| 1,000.0 | 158 | 4.32 | 3.81 | .59 |
| 2,000.0 | 79 | 4.67 | 4.16 | .59 |
| 5,000.0 | 31.6 | 5.13 | 4.63 | .60 |
| 10,000.0 | 15.8 | 5.47 | 4.97 | .61 |
| 20,000.0 | 7.9 | 5.82 | 5.32 | 0.61 |

* As $\beta$ increases, $\phi(\beta)$ approaches the ratio:

$$\frac{\text{Number of electrons captured in all levels of H}}{\text{Number of electrons captured in the ground level}},$$

while $\chi(\beta)$ approaches the ratio:

$$\frac{\text{Kinetic energy radiated by captures in all levels of H}}{\text{Kinetic energy radiated by captures in the ground level}}.$$

For all $\beta$, $(2\chi(\beta))/(3\phi(\beta))$ equals the ratio:

$$\frac{\text{Average kinetic energy of the electrons captured}}{\text{Average kinetic energy of the free electrons}}.$$

hence in such a case the kinetic temperature would exceed the effective color temperature $T_{cp}$ by a factor between 1.1 and 2.0. Actually, free-free transitions, which will be discussed subsequently, reduce the kinetic temperature more nearly to $T_{cp}$.

To evaluate numerically equation (2) for $G_{ep}$, it is necessary to determine $\overline{E_{2p}}$, the average kinetic energy of the ejected photoelectron. When an electron is ejected from an H atom in the ground state, the absorbed energy is $h\nu$, while the kinetic energy of the electron is $h(\nu - \nu_0)$, where $\nu_0$ is again the frequency of the Lyman limit. Thus we may write for this case

$$q(\nu) = \frac{h(\nu - \nu_0)}{h\nu} = 1 - \frac{\nu_0}{\nu}. \tag{23}$$

The atomic absorption coefficient $\kappa(\nu)$ for an H atom in the ground state may, to a first approximation, be written as

$$\kappa(\nu) \propto \frac{1}{\nu^3}. \tag{24}$$

Examination of the exact formula[23] shows that equation (24) is a close approximation for electron energies of practical interest.

The energy density $U(\nu)$, which also appears in equation (7) for $\overline{E_2}$, will depend on position in the Galaxy. In interstellar space as a whole the average values given by Dunham[9] are probably the best available. The integral in equation (7) has been evaluated numerically for Dunham's values computed on the assumption of selective absorption. The effective color temperature $T_{cp}$, defined in equation (8), is then found directly from $E_{2p}$ and is equal to 27,000°. The corresponding value found when no absorption is assumed is 32,000°, but this value is presumably less realistic than the previous one.

Near a single bright B star, this computed average value is no longer relevant. Close to a star the radiation will be approximately dilute black-body radiation at the color temperature $T_c$, where $T_c$ is essentially the surface temperature of the star. In this case an explicit formula may be obtained for $T_{cp}$. If equations (23) and (24) are substituted in equation (7), and $U(\nu)$ is replaced by a constant times $B_\nu(T_c)$, the Planck function for the radiant intensity, we find

$$\overline{E_{2p}} = \frac{\displaystyle\int_{\nu_0}^{\infty} \left(1 - \frac{\nu_0}{\nu}\right) \frac{d\nu}{e^{h\nu/\kappa T_c} - 1}}{\displaystyle\int_{\nu_0}^{\infty} \frac{1}{h\nu} \frac{d\nu}{e^{h\nu/kT_c} - 1}}. \tag{25}$$

If we introduce the quantity $\beta_c$, defined as

$$\beta_c = \frac{h\nu_0}{kT_c} = \frac{158,000°}{T_c}, \tag{26}$$

[23] H. Bethe, *Handb. d. Phys.* (J. Springer, 1933), **24**, Part I, 477.

equation (25) may be written in the form

$$\overline{E_{2p}} = kT_c \psi(\beta_c) = kT_{cp} \tag{27}$$

where use has been made of equation (8) and where

$$\psi(\beta) = \frac{\int_\beta^\infty \dfrac{dy}{e^y - 1}}{\int_\beta^\infty \dfrac{dy}{y(e^y - 1)}} - \beta. \tag{28}$$

Evidently $\psi(\beta_c)$ is the ratio of the effective color temperature, $T_{cp}$, to the actual color temperature, $T_c$. If the quantity $(e^y - 1)^{-1}$ in both integrands in equation (28) is expanded and the integrals are evaluated term by term, one finds

$$\psi(\beta) = \frac{\sum\limits_{n=1}^\infty \dfrac{1}{n} e^{-n\beta}}{\sum\limits_{n=1}^\infty Ei(n\beta)} - \beta. \tag{29}$$

Values of $\psi(\beta)$, computed from equation (29), are given in Table 2, together with the resultant values of the effective color temperature. It will be noted that $\psi(\beta_c)$ exceeds 0.80 as long as $\beta_c$ exceeds 3; thus the effective color temperature is within 20 percent of the actual color temperature of the ionizing radiation if this latter temperature is less than about 50,000°.

On the basis of a theory developed by B. Strömgren,[10] the values given in Table 2 are always to be used whenever $G_{ep}$ is appreciable. According to this picture, H atoms are ionized only in spherical regions surrounding each star, outside of which the hydrogen is almost completely neutral. The regions of ionized hydrogen are called H II regions, while the regions of neutral hydrogen are called H I regions. If the density of hydrogen in the galactic plane is about 1 atom per cubic centimeter, the H II regions are primarily confined to the immediate vicinity of the early-type stars. If this picture is correct, the only relevant values of $T_{cp}$ are those near O and B stars and may be found from the top few lines of Table 2. In most other regions the hydrogen is neutral, and photoelectric ionization of hydrogen will be negligible. However, there is some reason to suspect that interstellar atoms are predomi-

TABLE 2
Values of $\psi(\beta_c)$ and $T_{cp}$

| $\beta_c$ | $T_c*$ | $\psi(\beta_c)^\dagger$ | $T_{cp}^\ddagger$ | $\beta_c$ | $T_c*$ | $\psi(\beta_c)^\dagger$ | $T_{cp}^\ddagger$ |
|------|----------|------|----------|-------|---------|-------|---------|
| 0.5  | 316,000° | 0.449 | 142,000° | 5.0   | 31,600° | 0.868 | 27,400° |
| 0.75 | 211,000  | .537 | 113,000  | 10.0  | 15,800  | .922  | 14,600  |
| 1.0  | 158,000  | .599 | 94,600   | 20.0  | 7,900   | .957  | 7,560   |
| 1.5  | 105,000  | .686 | 72,000   | 50.0  | 3,160   | .981  | 3,100   |
| 2.0  | 79,000   | .739 | 58,400   | 100.0 | 1,580   | .990  | 1,560   |
| 3.0  | 52,700   | .808 | 42,600   | 200.0 | 790     | 0.995 | 786     |
| 4.0  | 39,500   | 0.844 | 33,300  |       |         |       |         |

* $T_c$ is the color temperature of radiation from a star.
† $\psi$ is the ratio of $T_{cp}$ to $T_c$.
‡ $T_{cp}$ is the effective color temperature for photoionization of H.

nantly concentrated in dust clouds,[24,25] between the clouds the density of hydrogen may be less than 1 atom per cubic centimeter and may be so small that most of the hydrogen outside the clouds is ionized,[26] i.e., that the entire interstellar gas, outside of dust clouds, is an H II region. In such a case $T_{cp}$ may be set equal to 27,000°, except near O and B stars, where the values in Table 2 are applicable.

Now, if equations (18), (21), and (27) are substituted in equation (2), as applied to electron–proton encounters, and equations (4), (12), and (15) are used to eliminate $L$, $A$, and $\beta$, respectively, we have

$$G_{ep} = \frac{2.85 \times 10^{-27}}{T^{1/2}} n_p \{ T_{cp} \phi(\beta) - T\chi(\beta) \} \, \text{erg/sec.} \quad (30)$$

Equation (30) gives the average net gain of kinetic energy per free electron per second resulting from electron-proton encounters.

## III. ATOMS OTHER THAN HYDROGEN

Next, the gain in kinetic energy resulting from electron captures by ions other than protons may be considered. In regions where the hydrogen is ionized (H II regions), the low abundance of most other atoms makes

[24] L. Spitzer, Jr., *Pub. A.A.S.*, **10**, 235, 1941.
[25] J. H. Oort and H. C. van de Hulst, *B.A.N.*, Vol. **19**, No. 376, 1946.
[26] The importance of this possibility has been emphasized by B. Strömgren in correspondence.

their contribution relatively negligible. Thus the value of $G_e$ corresponding to electron capture by O III and other highly ionized atoms need not be considered. The only other type of atom which may add an important contribution to $G_e$ in H II regions is helium. The density of helium relative to hydrogen is difficult to estimate; the available data for planetary nebulae, the sun, and $\tau$ Scorpii suggest[27] about 1 or 2 He atoms for each 10 atoms of hydrogen. In H I regions He atoms will not contribute an appreciable $G_e$, but in H II regions the photoelectric ionization of these atoms may increase the kinetic temperature somewhat. If this contribution from He atoms is comparable to that from hydrogen, the He atoms will be ionized only within a certain radius from the hot star, and this radius may be substantially less than the radius of the region of ionized hydrogen. The neglect of He atoms will give a lower limit for the temperature of H II regions. While no detailed analysis of the contribution from He atoms will be made here, the approximate equations derived in this section may be applied to He atoms as well as to others.

In H I regions, where hydrogen is neutral, the ionization of elements other than hydrogen and helium must be considered. The most important atoms are those which are cosmically abundant and which have a low ionization potential, such as Na and Ca atoms. As a first approximation, any atom may be treated by the equations developed above for hydrogen. The applicability of the hydrogenic formulae for other atoms has been investigated by Bates, Buckingham, Massey, and Unwin.[28] These authors point out that captures in highly excited states are predominantly captures in states of high angular momentum, which become hydrogen-like most rapidly with increasing $n$. Especially at low temperatures, at which captures in the highly excited states become increasingly important relative to captures in the ground state, the hydrogenic formulae should become valid.

This general expectation is supported by recent computations[29] of transition probabilities for the Na atom. From the oscillator strengths given for the transitions from the various states of the Na atom up to the continuum, the probabilities of electron capture in these various states may be determined from Milne's formula,[28,30] based on the principle of detailed balancing in thermodynamic equilibrium. For captures in the ground 3s state and in the excited 4s state, the cross-sections so determined are less than 0.1 times the hydrogenic values.

[27] L. H. Aller and D. H. Menzel, *Ap. J.*, **102**, 263, 1945.

[28] *Proc. Roy. Soc. London*, *A*, **170**, 322, 1939.

[29] M. Rudkjöbing, *Pub. Copenhagen Obs.*, No. 124, 1940.

[30] E. A. Milne, *Phil. Mag.*, **47**, 209, 1924.

For capture of slow electrons in the excited 3p and 4p levels the cross-sections are equal to 1.17 and 1.06, respectively, times the hydrogenic values. Although similar computations of $f$-values for the $Ca^{++}$ ion[31] do not show such rapid convergence toward the hydrogenic values, one may tentatively adopt the hydrogenic-capture cross-sections as the best available general estimate, except for states of zero angular momentum. Since captures in such s states are relatively unimportant, we may safely use the hydrogenic formulae for all excited states. Certainly, for temperatures as low as 100°, which may be expected in H I regions and for which electron captures in very highly excited states become dominant, use of the hydrogenic formulae should give accurate results for the rate of electron capture.

In summing the capture cross-section over all excited states, a problem arises concerning the ground state. If this ground state is an $ns$ (or an $ns^2$) level, the captures to the excited $np$ (or $nsnp$) levels may be important. Since the hydrogenic formulae derived in the previous section do not distinguish between states with different angular momentum but with the same total quantum number, it is necessary to include the captures in all the states with a particular value of $n$ in order to take into account the captures in the $p$, $d$, or $f$ levels. For atoms such as Na and Ca, the error thus introduced will not be serious, owing to the relative unimportance of the $s$ levels. As a rough working rule, if half or less of the states with total quantum number $n_1$ are filled when the atom is ionized but unexcited, the capture cross-section may be determined by adding the values of $\sigma_{np}$ for all $n$ equal to or greater than $n_1$; but if more than half these states are filled, the sum begins with the capture cross-section for $n$ equal to $n_1 + 1$. The symbol $k$ will be used to denote the total quantum number $n$ for which the hydrogenic cross-section should be used. On this basis, superelastic collisions between electrons and $He^+$ atoms are described by exactly the same formulae as are the corresponding collisions between electrons and protons; thus for $He^+$ as for $H^+$, $k$ equals 1. For capture of electrons by singly ionized atoms from lithium through nitrogen, however, $k$ equals 2, while for atoms from oxygen through titanium, $k$ equals 3, and for atoms from vanadium through silver, $k$ equals 4. A more accurate method of interpolation between the values for adjacent $k$'s could, of course, be used, but the relatively crude scheme described here should provide an adequate first approximation.

---

[31] L. Green, informal communication.

On the basis of this approximation, we have, from equations (18) and (19),

$$\overline{\sigma_i v} = \frac{2A}{(\pi L)^{1/2}} \beta \phi_k(\beta) \tag{31}$$

where

$$\phi_k(\beta) = \sum_{n=k}^{\infty} \frac{\beta}{n^3} e^{\beta/n^2} Ei\left(\frac{\beta}{n^2}\right). \tag{32}$$

Similarly,

$$\overline{\sigma_i v E_1} = \frac{m_e A}{\pi^{1/2} L^{3/2}} \beta \chi_k(\beta), \tag{33}$$

where

$$\chi_k(\beta) = \sum_{n=k}^{\infty} \frac{\beta}{n^3} \left\{ 1 - \frac{\beta}{n^2} e^{\beta/n^2} Ei\left(\frac{\beta}{n^2}\right) \right\}. \tag{34}$$

Values of the functions $\phi_k(\beta)$ and $\chi_k(\beta)$ are given in Table 3.

For capture of an electron by a singly ionized atom, $\beta$ is given by equation (15) in the preceding section. For capture by an atom which has lost $Z$ electrons, one must take into account the fact that $v_0$ varies as $Z^2$. Equations (31)–(34) are still valid in this case, provided that we use the following definition for $\beta$:

$$\beta = \frac{158{,}000 Z^2}{T}. \tag{35}$$

It may be noted that, while these equations may be used to determine the rate of electron capture, given the kinetic temperature and the number of electrons and ions per cubic centimeter, they cannot be used to determine the rate of photoionization and the ionization equilibrium resulting. A captured electron will rapidly cascade to the ground state, from which the probability of photon capture may be very different from its hydrogenic value.

To evaluate $G_{ei}$ it is necessary to determine $\overline{E_{2i}}$, defined in equation (7), or the effective color temperature $T_{ci}$, defined by equation (8). The detailed value of these quantities depends on the absorption coefficient $\kappa(\nu)$ for the transition in question. Recent studies[32,33] have cast some

[32] S. Chandrasekhar, *Ap. J.*, **102**, 23, 1945.
[33] Ibid., p. 395.

TABLE 3
Values of $\phi_k(\beta)$ and $\chi_k(\beta)$

| $\beta$ | $T$ | $\phi_2$ | $\phi_3$ | $\phi_4$ | $\phi_5$ | $\phi_6$ | $\chi_2$ | $\chi_3$ | $\chi_4$ | $\chi_5$ | $\chi_6$ |
|---|---|---|---|---|---|---|---|---|---|---|---|
| 0.5 | 316,000° | 0.24 | 0.12 | 0.075 | 0.052 | 0.038 | 0.082 | 0.034 | 0.018 | 0.011 | 0.008 |
| 1.0 | 158,000 | 0.36 | 0.20 | 0.125 | 0.087 | 0.065 | 0.15 | 0.063 | 0.034 | 0.021 | 0.014 |
| 2 | 79,000 | 0.54 | 0.31 | 0.20 | 0.14 | 0.11 | 0.27 | 0.13 | 0.082 | 0.058 | 0.044 |
| 5 | 31,600 | 0.84 | 0.52 | 0.36 | 0.25 | 0.21 | 0.49 | 0.27 | 0.17 | 0.12 | 0.09 |
| 10 | 15,800 | 1.11 | 0.73 | 0.52 | 0.39 | 0.31 | 0.71 | 0.41 | 0.27 | 0.19 | 0.14 |
| 20 | 7,900 | 1.41 | 0.98 | 0.74 | 0.58 | 0.47 | 0.97 | 0.61 | 0.41 | 0.29 | 0.21 |
| 50 | 3,160 | 1.84 | 1.37 | 1.09 | 0.89 | 0.75 | 1.37 | 0.94 | 0.68 | 0.51 | 0.38 |
| 100 | 1,580 | 2.17 | 1.69 | 1.38 | 1.16 | 1.00 | 1.69 | 1.23 | 0.96 | 0.75 | 0.61 |
| 200 | 790 | 2.51 | 2.02 | 1.70 | 1.47 | 1.28 | 2.02 | 1.54 | 1.25 | 1.02 | 0.86 |
| 500 | 316 | 2.98 | 2.49 | 2.16 | 1.92 | 1.72 | 2.44 | 1.98 | 1.65 | 1.42 | 1.22 |
| 1,000 | 158 | 3.32 | 2.82 | 2.50 | 2.25 | 2.05 | 2.81 | 2.32 | 1.99 | 1.75 | 1.55 |
| 2,000 | 79 | 3.67 | 3.17 | 2.84 | 2.59 | 2.39 | 3.16 | 2.66 | 2.33 | 2.09 | 1.89 |
| 5,000 | 31.6 | 4.13 | 3.63 | 3.30 | 3.05 | 2.85 | 3.63 | 3.13 | 2.80 | 2.55 | 2.35 |
| 10,000 | 15.8 | 4.47 | 3.97 | 3.64 | 3.39 | 3.19 | 3.97 | 3.47 | 3.14 | 2.89 | 2.69 |
| 20,000 | 7.9 | 4.82 | 4.32 | 3.98 | 3.73 | 3.53 | 4.32 | 3.82 | 3.48 | 3.23 | 3.03 |

doubt on previous computations of this function, and an elaborate calculation of $T_{ci}$ for different atoms is therefore not appropriate at the present time.

For an approximate evaluation of $T_{ci}$, we use the relationship that $T_{ci}$ tends to be approximately equal to the color temperature $T_c$, provided that $h\nu_0/kT_c$ is large compared to unity. This relation, which was proved by Eddington,[5] is based on the assumption that the absorption coefficient at the series limit is finite—a legitimate assumption for the photoelectric ionization of atoms and molecules. The color temperature of interstellar radiation may be found from the values of the energy density given by Dunham.[9] The ratio of the energy densities at 1000 and 2000 A, corresponding to photon energies between 6 and 12 electron-volts, corresponds to black-body radiation at 17,000°, if Dunham's values, corrected for the effect of selective absorption, are used. This result may be compared with the effective color temperatures of 13,000° and 24,000°, found for the photoionization by integrated starlight of H-like atoms whose ionization potentials are 6 and 12 volts, respectively; in such atoms the absorption coefficient beyond the series limit varies as the inverse cube of the frequency, and the effective color temperature is found from equations (7) and (8) by a direct graphical integration based on Dunham's values. As a conservative estimate, designed to give an upper limit on the kinetic temperature in H I regions, we shall finally set $T_{ci}$ equal to 20,000°. In H II regions near O and B stars, where a greater effective color temperature of the ionizing radiation might be expected, superelastic collisions between electrons and ionized atoms other than hydrogen or helium may be neglected.

Finally, then, we find for $G_{ei}$ the following equation, applicable to any ions,

$$G_{ei} = \frac{2.85 \times 10^{-27} n_i}{T^{1/2}} \{T_{ci}\phi_k(\beta) - T\chi_k(\beta)\} \text{ ergs/sec.} \quad (36)$$

The quantity $k$ is the total quantum number $n_1$ for a captured electron in the ground state, provided that the number of electrons with this total quantum number in the ground state of the ion is no greater than $n_1^2$; otherwise $k$ equals $n_1 + 1$.

## IV. NEUTRAL HYDROGEN ATOMS

Next, the contribution of neutral hydrogen must be considered. There is an appreciable probability that an H atom will capture an electron, forming an $H^-$ ion. It may be readily shown that such an ion will be neutralized rapidly by photon absorption, with the result that the relative number of $H^-$ ions is relatively small. However, the number of electrons captured per second by H atoms is sufficiently great so that this process may actually be more important as a source of kinetic energy in H I regions than is the process of electron capture by atomic ions.

The capture cross-section $\sigma_H$ may be computed from the atomic absorption coefficient for the $H^-$ ion. From the condition that in thermodynamic equilibrium the number of electrons captured with a velocity $v$ must equal the number of electrons ejected with the same velocity, one finds the equation

$$\frac{\sigma_H}{\kappa(\nu)} = \frac{h^2\nu^2}{2m^2v^2c^2}, \tag{37}$$

where $\kappa(\nu)$ is the atomic absorption coefficient for an $H^-$ ion. Equation (37) is a special case of Milne's formula,[30] referred to above. From the values of $\kappa(\nu)$ (denoted by $\kappa_\lambda$) given by Chandrasekhar,[33] values of $\sigma_H$ have been computed and are given in Table 4 for different values of $v$. For values of $T$ less than 1000°, corresponding to wavelengths between 14,000 A and 16,550 A, an approximate formula,[34] based on a series expansion of the free-electron wave function, was used to compute $\kappa_\lambda$; these approximate values may be in error by as much as 10 percent. The table in the second column also lists values of the temperature at which the root-mean-square velocity equals the values in the first column. For comparison with electron captures by protons, the table also gives values of $\sigma_{1p}$, computed from equation (11) for $n$, $g$, and $Z$ all equal to 1. The number of captures in the fourth-quantum level of hydrogen, corresponding to the captures in the lowest unoccupied levels of the $Ca^+$ ion would be less than a fourth of the values of $\sigma_{1p}$ listed in the table. Since the number of H atoms in H I regions exceeds the number of metal ions by at least $10^3$, it is evident that electron captures of H atoms will be much more numerous than those by other ions, provided that the kinetic temperature is not less than 1000°. Hence, at temperatures greater than 1000°, $G_{ei}$ may be neglected as compared to $G_{eH}$, but at lower temperatures $G_{ei}$ may become more important. Electron captures by other neutral atoms, such as oxygen, are presumably negligible, owing to the relatively low abundance of such atoms compared with hydrogen.

[34] S. Chandrasekhar, informal communication.

TABLE 4

Cross-Sections for Electron Capture

| $v$ | Velocity Temperature $T$ | Protons $\sigma_{1p}$ | H atoms $\sigma_H$ |
|---|---|---|---|
| $0.126 \times 10^7$ cm/sec | 3.48° | $63,700 \times 10^{-22}$ cm² | $0.0231 \times 10^{-22}$ cm² |
| 0.178 | 6.95 | 31,800 | .0326 |
| 0.252 | 13.9 | 15,900 | .0461 |
| 0.308 | 20.9 | 10,600 | .0566 |
| 0.359 | 28.3 | 7,800 | .0659 |
| 0.401 | 35.3 | 6,260 | .0733 |
| 0.490 | 52.7 | 4,200 | .0897 |
| 0.698 | 107 | 2,060 | .124 |
| 0.952 | 199 | 1,110 | .167 |
| 1.34 | 389 | 566 | .230 |
| 1.65 | 599 | 368 | .292 |
| 1.93 | 819 | 268 | .347 |
| 2.17 | 1,040 | 212 | .421 |
| 3.18 | 2,220 | 98.0 | .554 |
| 4.10 | 3,700 | 58.0 | .636 |
| 4.64 | 4,730 | 44.7 | .657 |
| 5.01 | 5,520 | 38.3 | .661 |
| 5.13 | 5,790 | 36.2 | .663 |
| 5.80 | 7,390 | 27.8 | .656 |
| 6.49 | 9,260 | 22.1 | .632 |
| 6.92 | 10,500 | 19.4 | .619 |
| 7.74 | 13,200 | 15.3 | .585 |
| 8.47 | 15,800 | 12.2 | .554 |
| 9.15 | 18,400 | 10.5 | .527 |
| 9.67 | 20,600 | 9.00 | .502 |
| 10.9 | 26,100 | 6.73 | .456 |
| 12.9 | 36,600 | 4.45 | .384 |
| 15.5 | 52,800 | 2.81 | .303 |
| 19.6 | 84,500 | 1.46 | .203 |
| 28.9 | 184,000 | 0.437 | 0.0814 |

Since the cross-section $\sigma_H$ varies only slightly with the electron velocity $v$ over the region of interest, a computation of the detailed averages given in equations (5) and (6) would scarcely be worthwhile. Instead, an approximate method will be used to determine $G_{eH}$. We may write

$$\overline{\sigma_H v} \approx \sigma_H(\bar{v}) \times \bar{v} \tag{38}$$

and

$$\overline{\sigma_H v E_1} \approx \sigma_H(\bar{v}) \times \overline{v E_1} \approx \sigma_H(\bar{v}) \times \tfrac{1}{2} m_e \bar{v}^3, \tag{39}$$

where $\bar{v}$ denotes the root-mean-square electron velocity.

The quantity $\overline{E_{2H}}$, as defined in equation (7), must also be known before $G_{eH}$ can be computed. Since all the excess energy of the photon, above the photoelectric threshold energy, $h\nu_0$, appears as kinetic energy, $q(\nu)$ is given again by equation (23), where $\nu_0$ is now the threshold for negative hydrogen absorption, corresponding to a threshold wavelength of about 16,500 A. The absorption coefficient $\kappa(\nu)$ may be taken from Chandrasekhar.[33] The energy density $U(\nu)$ may be taken from Dunham's values[9] computed without regard to absorption. If, as before, we equate $\overline{E_{2H}}$ to $kT_{cH}$, the value of $T_{cH}$ computed on these assumptions is about 11,000°, corresponding to a mean excess energy of about 1 electron-volt for the absorbed photon. This is the value that would result if all the radiation in space had a wavelength of about 7000 A.

This value of $T_{cH}$ is almost certainly too high; the presence of selective absorption will increase $U(\nu)$ in the infrared relative to the visible and thus reduce the average energy of an absorbed photon. Observations of interstellar absorption do not extend sufficiently far in the infrared to allow an evaluation of this effect; for example, Dunham's computed values of $U(\nu)$ extend only to 10,000 A. For an accurate determination of $T_{cH}$, it would be necessary to recompute $U(\nu)$ on different assumptions concerning the selective absorption in space. Since $G_{eH}$ does not usually play a significant role in interstellar space, an upper limit on $T_{cH}$ will suffice for most purposes, and we shall accordingly set $T_{cH}$ equal to 11,000°. With this assumption, equation (2) for $G_{eH}$ becomes

$$G_{eH} = 1.40 \times 10^{-10}\sigma_H(\bar{v})n_H T^{1/2}(7,300° - T). \qquad (40)$$

Values of $\sigma_H(\bar{v})$ may be read from Table 4.

## V. MOLECULES

Photoionization and photodissociation of molecules must next be considered. Apparently little is known about the capture cross-section of ionized molecules for electrons. On the other hand, the abundance of the observed molecules is believed[35] to be low, with values of about $10^{-6}$ CH and CN molecules per cubic centimeter. This is considerably greater than the observed densities of Na and $Ca^+$ atoms. However, Na and Ca atoms are believed to be largely ionized, while the relative numbers of $CH^+$ and CH atoms, for example, seem to be about the same, since their absorption lines are apparently of comparable intensities. Thus the number of hydride molecules per cubic centimeter is

---

[35] T. H. Dunham, Jr., *Pub. A.A.S.*, **10**, 123, 1941.

apparently only a very small fraction of the corresponding number of interstellar atoms. As pointed out by P. W. Merrill,[36] it is even possible that the observed molecular lines are circumstellar and that the actual abundance of molecules between the stars is even lower than $10^{-6}$ per cubic centimeter. In any case, superelastic encounters between electrons and such molecules and between the component atoms making up a molecule may evidently be neglected.

Molecular hydrogen may possibly be much more abundant in H I regions of interstellar space that are the observed molecules CH and CN. However, the ionization energy of $H_2$ is 15.3 volts, 1.8 volts more than that of atomic hydrogen. Since ultraviolet radiation beyond the Lyman limit will be almost completely absorbed in H I regions, any $H_2$ molecules present will remain neutral and will therefore not contribute to the kinetic energy of the assembly. Photodissociation of $H_2$ molecules in H I regions is equally unlikely, since the energy required for this process is 14.5 volts when the molecule is in its ground state.

## VI. SOLID GRAINS

Another source of kinetic energy in interstellar space is the photoelectric ionization of small solid particles, or grains. This effect may again be treated by the general equations developed at the beginning of this paper. Of all the electrons which strike a grain and stick to it, a fraction $1 - \alpha$ will be ejected by the photoelectric effect. Since the number of electrons captured by a grain must be very nearly equal to the number lost or neutralized, the remaining fraction $\alpha$ may be assumed to attach themselves to positive ions which strike the grain; electrons leaving in this way will soon be ejected when the atom absorbs a photon. Thus, as before, each encounter in which the electron sticks to the grain may be regarded as a superelastic collision, which may increase the kinetic energy of the assembly.

This point of view neglects the positive ions which stick to the solid grains, neutralizing electrons captured by the grain. However, this virtual disappearance of ions and electrons from the interstellar gas will have a relatively minor effect on the kinetic temperature of the remaining particles, an effect which will be appreciable only during a time so great that at least half the atoms originally present become incorporated into the growing dust grains. Especially since the predominant H atoms adhere only rarely to the grains, this small effect is generally negligible and will not be considered here.

[36] *Pub. A.S.P.*, **58**, 354, 1946.

On this basis, the computation of $G_{eg}$ for grains is a straightforward process but is somewhat involved, since the properties of photoelectric emission are not simple. One source of complexity is that the value of $G_{eg}$ depends markedly on the charge of the grain, which, in turn, depends on the kinetic temperature. Thus it will be necessary to compute the potential of the grains for each assumed temperature and then, with the resultant value of $G_{eg}$ and with the $G_e$'s found for other processes also, to solve for the kinetic temperature. This process must be repeated until the temperature found equals the assumed temperature. In this section, the formulae for the three average quantities occurring in equation (2) will be given first. Next, general formulae for the total number of photoelectrons emitted per second and for the resultant potential of the grain will be derived. Finally, the values of the various physical quantities occurring in these equations will be discussed.

The average product $\overline{\sigma_g v}$ which occurs in equation (2) is the number of electrons captured by a grain per second per unit electron density. A similar quantity, the total number of electrons captured by a grain per second, denoted by $\Gamma$, has been obtained in a previous paper.[3] This formula for $\Gamma$ must be multiplied by the sticking probability, $\xi_e$ for an electron, to take into account the fact that not all electrons may adhere to the grain; the appropriate values of $\xi_e$ for metallic and nonmetallic grains are discussed in subsection $a$ below. If also the previously derived value of $\Gamma$ is divided by $n_e$ to yield $\overline{\sigma_g v}$, and if $3kT/2$ is substituted for the kinetic energy $E$, we have

$$\overline{\sigma_g v} = 2\xi_e \left( \frac{2\pi k}{m_e} \right)^{1/2} a^2 T^{1/2} \times \begin{cases} 1 - \gamma, & \gamma < 0, \\ e^{-\gamma}, & \gamma > 0, \end{cases} \quad (41)$$

where $a$ is the radius of the solid grain (the symbol $\sigma$, previously used[3] for this quantity is here used for the cross-section) and where

$$\gamma = -\frac{eV}{kT}; \quad (42)$$

the quantity $V$ is the electrostatic potential of the grain, in e.s.u, while $e$ is the electronic charge in the same units. Evidently $\gamma$ is a measure of the potential energy of an electron at the surface of the grain; thus, if the grain is negatively charged, $\gamma$ is positive, while for a positively charged grain $\gamma$ is negative.

The total kinetic energy lost per second by collisions of electrons with a grain, per unit electron density, is readily found by methods similar to those formerly used[3] and is given by the expression

$$\overline{\sigma_g v E_1} = 2\left(\frac{2\pi}{m_e}\right)^{1/2} \xi_e k^{3/2} T^{3/2} a^2 (2 + |\gamma|) \times \left\{\begin{matrix} 1, & \gamma < 0, \\ e^{-\gamma}, & \gamma > 0. \end{matrix}\right\} \quad (43)$$

Equation (43) also has been multiplied by $\xi_e$ to allow for the fact that not all the electrons will stick to the grain when they strike it.

The average value of $E_{2g}$ for this process must also be evaluated. When the electrons leave the grain by attaching themselves to a positive ion, the average kinetic energy with which they are subsequently ejected from the atom will be the same as that already computed for superelastic encounters between electrons and atoms. Thus for a fraction $\alpha$ of the electrons captured by a grain, $\overline{E_{2g}}$ will be $kT_{ci}$, where $T_{ci}$ is about 20,000° in H I regions; in H II regions $T_{cp}$, lying between 20,000° and 60,000°, replaces $T_{ci}$. It is fortunate that this process usually has a minor effect on the computed temperature, since, as shown below, its probability is most uncertain.

It is also conceivable that electrons might leave a grain by attaching themselves to a neutral H atom which strikes the grain. However, an electron attached to an otherwise neutral H atom has a total energy of only $-0.75$ electron-volts, as compared to the considerably lower energy of electrons in most solids. Hence this process seems improbable and will not be considered further.

It remains to compute $\overline{E_{2g}}$ for the fraction $1 - \alpha$ of the captured electrons which are ejected from the grain by photoelectric absorption. Results will first be obtained for a grain with no electrical charge. Equation (7) is applicable in the present case, provided that we define $\kappa(\nu)$ as the energy absorbed photoelectrically by a grain per unit intensity of incident radiation. When radiation of energy density $U(\nu)$ per frequency interval is present in space, an amount $\pi a^2 c U(\nu)$ passes through the geometrical cross-section $\pi a^2$ of a grain per second. A fraction $Q(\nu)$ of this radiation is absorbed, but only a fraction $\chi(\nu)$ of the absorbed photons leads to the emission of a photoelectron. This quantity $\chi(\nu)$ is called the "photoelectric efficiency." The photoelectric absorption coefficient $\kappa(\nu)$ therefore becomes

$$\kappa(\nu) = \pi a^2 Q(\nu) \chi(\nu). \quad (44)$$

Equation (7) depends also on the quantity $q(\nu)$, the fraction of the photon energy which appears as kinetic energy of the emitted electron. Since an electron must be given a potential energy $h\nu_0$ to escape from a solid, where $\nu_0$ is the threshold energy, the "available" energy for a free electron is, as before, $h(\nu - \nu_0)$. However, some of this available energy is lost by collisions as the electron leaves the solid, and additional energy is also required to eject electrons whose kinetic energy in the solid is less than the maximum zero-point energy. As a result of these two factors, the kinetic energy of the photoelectrons produced by photons of frequency $\nu$ will, on the average, be a fraction $q_1$ of the available energy. We may therefore write

$$q(\nu) = q_1\left(1 - \frac{\nu_0}{\nu}\right). \tag{45}$$

The subsequent discussion of the factors determining $q_1$ indicates that this quantity does not change much with frequency; hence $q_1$ will be assumed constant.

If equation (44) and (45) are substituted in equation (7), we have, for an uncharged grain,

$$\overline{E_{2g}} = q_1 \frac{\int_{\nu_0}^{\infty}\left(1 - \frac{\nu_0}{\nu}\right)Q(\nu)\chi(\nu)U(\nu)\,d\nu}{\int_{\nu_0}^{\infty} Q(\nu)\chi(\nu)U(\nu)\frac{d\nu}{h\nu}}. \tag{46}$$

In accordance with equation (8), a color temperature $T_{cg}$ may be defined so that $kT_{cg}$ equals $\overline{E_{2g}}$. If we define $I_1$ and $I_2$ as follows:

$$I_1(\nu_0) = \int_{\nu_0}^{\infty} Q(\nu)\chi(\nu)U(\nu)\frac{d\nu}{h\nu}, \tag{47}$$

$$I_2(\nu_0) = \int_{\nu_0}^{\infty} Q(\nu)\chi(\nu)U(\nu)\frac{d\nu}{h\nu_0}, \tag{48}$$

then $\overline{E_{2g}}$ and $T_{cg}$ may conveniently be expressed in the form

$$\overline{E_{2g}} = kT_{cg} = q_1 h\nu_0\left(\frac{I_2 - I_1}{I_1}\right). \tag{49}$$

The physical significance of the integral $I_1$ will become apparent below in the discussion following equation (54).

If the grain is charged, these results must be modified. A positive charge will attract some electrons back to the grain and prevent them from escaping. Thus the surface acts as though the effective photoelectric threshold were increased by an amount $eV/h$. The effective value of $\chi(\nu)$ will also be modified. Since $\chi(\nu)$ for an uncharged grain is not known very accurately, it will here be assumed that the effective value of $\chi(\nu)$ is equal to the appropriate value for an uncharged grain with a threshold equal to $\nu_0 + eV/h$. Thus equation (49) may be used in this case, provided that this effective photoelectric threshold is used.

If the charge on the grain is negative, two effects appear. In the first place, the kinetic energy of the photoelectron is increased by the electrostatic repulsion, the increase in $\overline{E_{2g}}$ amounting to $|eV|$. In the second place, the threshold frequency will be diminished by the electrical field. For plane metallic surfaces of pure metal this reduction of threshold has been thoroughly studied. On the assumption that the threshold energy results from the presence of an image force, the reduction of the photoelectric threshold may be shown to be proportional to the square root of the accelerating electrical field; these results are in moderately good agreement with observation.[37] This theory, if applied to small grains, indicates that the photoelectric threshold energy is reduced by $|eV|/Z_g^{1/2}$, where $Z_g$ is the number of electrons on the grain. For a potential of 1 volt, $Z_g$ is in the neighborhood of 100, and the photoelectric threshold is reduced by only a tenth of a volt, which is quite negligible.[38] For surfaces contaminated with gas, the simple theory is no longer applicable; but, as long as the potential of the grain does not exceed 1 or 2 volts, yielding electrical fields of less than 200,000 volts per centimeter, the reduction of the photoelectric threshold produced by the electrical field is not likely to be large, and we shall therefore neglect it here.

Finally, then, if we take into account both photoemission and neutralization of positive ions, we have for a charged grain the following equation:

$$\overline{E_{2g}} = akT_{ci} + (1 - \alpha)kT_{cg}, \tag{50}$$

where

$$kT_{cg} = q_1 h \nu_{0e} \left[ \frac{I_2(\nu_{0e}) - I_1(\nu_{0e})}{I_1(\nu_{0e})} \right] + \begin{cases} 0, & \gamma < 0 \\ \gamma kT, & \gamma > 0 \end{cases}, \tag{51}$$

---

[37] A. L. Hughes and L. A. DuBridge, *Photoelectric Phenomena* (New York: McGraw-Hill Book Co., 1932), sec. 6–15.

[38] A more exact analysis by H. C. van de Hulst (*Recherches astr. de l'Obs. d'Utrecht*, in preparation), which takes into account the curvature of the grain surface, gives essentially similar results.

$\gamma$ is given by equation (42), and $\nu_{0e}$ is related to $\nu_0$, the photoelectric threshold of an uncharged grain, by the equation

$$\nu_{0e} = \nu_0 + \begin{cases} \dfrac{-\gamma kT}{h}, & \gamma < 0 \\ 0, & \gamma > 0 \end{cases}. \tag{52}$$

Next the total number of electrons emitted will be considered, together with the other factors determining the electrical charge on the grain. The denominator of equation (7), when applied to the present case, gives the total number of photoelectrons emitted by the grain per second. As before,[3] we may define $K$ as the ratio of the photoelectrons emitted per second to the number of positive ions striking an uncharged grain per second; this latter quantity is obtained from equation (41), with $\gamma$ and $\xi_e$ set equal to 0 and 1, respectively, with $m_e$ replaced by the ionic mass $m_i$, and with the ion density $n_i$ inserted as a factor. Evidently, $K$ will depend on the charge of the grain. The value for an uncharged grain will be derived first. We have

$$K = \frac{(6\pi)^{1/2} c}{4 n_i \left(\overline{v_i^2}\right)^{1/2}} \int_{\nu_0}^{\infty} \frac{Q(\nu)\chi(\nu)U(\nu)\,d\nu}{h\nu}, \tag{53}$$

where the mean square ionic velocity $\overline{v_i^2}$ has been substituted for $3kT/m_i$. Equation (53) is a more complete version of equation (7) in a previous paper.[3]

A positive charge on the grain will attract some electrons back to the grain and prevent them from escaping. As before, this effect may be taken into account by using a greater effective threshold $\nu_{0e}$, given again by equation (52). The decrease in threshold produced by a negative charge on the grain will again be neglected. In terms of the integral $I_1$, defined in equation (47), we have, finally, for a charged grain,

$$K = 2.06 \times 10^6 \frac{A_i^{1/2}}{n_i T^{1/2}} I_1(\nu_{0e}), \tag{54}$$

where $A_i$ is the atomic weight of the positive ion. The quantity $I_1$ appearing in equations (49) and (54) may be regarded as the effective number of photons per cubic centimeter of interstellar space; i.e., the number of photoelectrons emitted from a grain per second equals $I_1$

multiplied by the velocity of light and by the geometrical cross-section of the grain.

From the value of $K$ given in equation (54) it is possible to determine the charge on the solid grain. By an obvious generalization of equation (6) in a previous paper,[3] we find

$$\xi_i(1 + \gamma) + K = A_i^{1/2}\xi_e\left(\frac{m_0}{m_e}\right)^{1/2}e^{-\gamma}, \quad \gamma > 0;$$

$$\xi_i e^{\gamma} + K = A_i^{1/2}\xi_e\left(\frac{m_0}{m_e}\right)^{1/2}(1 - \gamma), \quad \gamma < 0;$$

(55)

where the quantity $\xi_e$ is again the sticking probability for the electron, while $\xi_i$ is the probability that an ion will leave with an electron captured from the grain (or stick to the grain); $A_i$ is the atomic weight of the predominant positive ion present, while $m_0$ is the mass of unit atomic weight.

From the terms in equation (55) it is a simple matter to find $\alpha$, the ratio of the number of positive ions neutralized per second to the number of electrons captured per second. We find

$$\alpha = \frac{\xi_i}{\xi_e A_i^{1/2}\left(\dfrac{m_0}{m_e}\right)^{1/2}} \times \begin{cases} \dfrac{(1 + \gamma)}{e^{-\gamma}}, & \gamma > 0 \\[2ex] \dfrac{e^{\gamma}}{(1 - \gamma)}, & \gamma < 0 \end{cases} = \begin{cases} \dfrac{\xi_i(1 + \gamma)}{\xi_i(1 + \gamma) + K}. \\[2ex] \dfrac{\xi_i e^{\gamma}}{\xi_i e^{\gamma} + K}. \end{cases}$$

(56)

This completes the mathematical formalism required to determine $G_{eg}$. Equation (55), together with equations (54) and (52), determines $\gamma$, the ratio of the potential energy of an electron on a grain to $kT$, which is two-thirds of the average kinetic energy of a free particle; the quantity $\gamma$ affects all the other processes. Equation (56) determines $\alpha$, which must be substituted in equation (50) for $\overline{E_{2g}}$. Of the two temperatures in equation (50), $T_{eg}$ is found from equation (51), while $T_{ci}$ equals 20,000° in H I regions; in H II regions $T_{cp}$, found from Section II above, replaces $T_{ci}$. Finally, equation (41) and (43) determine $\overline{\sigma_g v}$ and $\overline{\sigma_g v E_1}$, which, together with $\overline{E_{2g}}$, must be substituted in equation (2) for $G_{eg}$. If we substitute for the numerical values, this final equation becomes

$$G_{eg} = 2.70 \times 10^{-10} \xi_e n_g a^2 T^{1/2}$$

$$\times \begin{cases} (1 - \gamma)\{\alpha T_{ci} + (1 - \alpha)T_{cg}\} - (2 - \gamma)T, & \gamma < 0 \\ e^{-\gamma}\{aT_{ci} + (1 - \alpha)T_{cg} - (2 + \gamma)T\}, & \gamma > 0 \end{cases},$$

$$(57)$$

where $T_{eg}$, $\gamma$, and $\alpha$ are determined by equations (51), (55), and (56), respectively.[†]

To apply these results it is necessary to specify the constants $\xi_e$, $\xi_i$, $q_1$, and $\nu_0$, together with the functions $\chi(\nu)$ and $Q(\nu)$. All these quantities depend on the composition of the grains, concerning which nothing very definite is known. If the grains form in interstellar space from initially small nuclei, different atoms sticking to the grain on impact, a rather conglomerate grain is to be expected, composed largely of the hydrides of carbon, nitrogen, and oxygen. A mixture of metallic atoms, such as sodium, magnesium, and some iron, should be present, however. Whether such impurities could impart some metallic properties to the grains is a matter of conjecture. In any case a surface layer of adsorbed H and He atoms should be present, so that, even if the grains were made of pure metal, their properties would be those of a contaminated surface rather than those of a pure metallic surface. In view of these many uncertainties, it is not possible to specify exact values for any of the above quantities.

---

[†] In equation (57) a typographic error has been corrected, with $10^{-11}$ replaced by $10^{-10}$.

# CONTINUOUS EMISSION FROM PLANETARY NEBULAE

(WITH J. L. GREENSTEIN)

(ASTROPH. J. 114, 407, 1951)

## *Commentary*

DURING my biennial visits to Pasadena and the Mt. Wilson Observatory (see Commentary, Paper #6), I profited frequently from luncheon discussions with Jesse Greenstein, who was adept at finding observational results that seemed to call for theoretical interpretation. One day in Pasadena, Jesse mentioned to me an apparent excess emission at short wavelengths in the continuous emission spectra of some gaseous nebulae. Our subsequent discussions led to Paper #5, analyzing the possibility that two-photon emission by H atoms in the metastable 2s state might be a source of such emission. The primary results in this paper were obtained from calculations based on established quantum-mechanical analyses. These results are: the spontaneous two-photon emission probability from the 2s to the 1s level of neutral H; also, the frequency distribution of the emitted photons. As far as I know, ours were the first definite numerical values published for these quantities.

A number of similar but more precise calculations appeared in ensuing years. According to a relatively recent work[1] the correct value of $A_{2s, 1s}$ is 8.2249 s$^{-1}$, less by 0.026% than the value in Paper #5. For the frequency distribution function, the corrections found in the relative values are of order 0.5%.

A modification now required in Paper #5 is the inclusion of proton collisions in the discussion of 2s-2p transitions. In Paper #5 we assumed these were negligible, but later work showed[2] that protons are about an order of magnitude more effective than electrons in changing angular momentum by transitions between hydrogen levels of the same total quantum number, $n$. Thus in equation (23), the constant $8.2 \times 10^{-6}$ should be replaced[3] by $6.5 \times 10^{-5}$ for $T = 10^4$ K, $n_e = n_p$. This change reduces, of course, the fraction of radiative recombinations which result in two-photon emission.

The third section below is devoted to an evaluation of the mean free path of L$\alpha$ photons. This discussion was included to show when H excitation by L$\alpha$ could increase $X_C$, the fraction of electron-proton recombinations leading to two-photon emission. The result was negative for photons remaining within the H II zone. According to present concepts, if a L$\alpha$ photon produced in an H II region travels through much H gas, it is unlikely to be converted into two-photon emission, but instead will tend to be absorbed by dust grains, either directly within the surrounding gas or after scattering from the adjacent H I region back into the H II zone. When deexcitation by protons is also taken into account, the overall effective value of $X$, the ratio of two-photon jumps to radiative recombinations, is not likely to be increased above its value ($\approx 1/3$) for direct radiative capture.

The treatment of L$\alpha$ scattering in Section 3 above is apparently the first analysis of this problem to include not only frequency redistribution (noncoherent scattering) in the Doppler core of the line profile but also the nearly coherent scattering in the resonance wings. The calculations here ignored the slight Doppler shifts of frequency in the line wings. This approximation introduced appreciable errors for large optical thickness and in a subsequent analysis[4] was replaced by a more realistic treatment. Extensive theoretical work in later years has led to inclusion of the exact frequency distribution and of the important absorption by dust. The detailed numerical results obtained[5] have been used in interpreting L$\alpha$ observations of galaxies.[6]

Recent analyses indicate that two-photon emission makes a small contribution to the diffuse interstellar light,[7] an appreciable one to the emission from planetary nebulae,[8] and a major one to the blue continuum of Herbig-Haro objects.[9] In this last, the gas immediately behind the driving shock is hot, but H atoms are still predominantly neutral; in this region the two-photon emission from collisionally excited H atoms much exceeds the recombination emission.

Evidently two-photon emission is now a standard ingredient for model calculations of emission from a rarefied gas. For recombination radiation, such computations generally rely on a later tabulation[10] of emitted intensities; the accuracy of these and other later values is limited by an uncertainty of some 20 percent in the 2s-2p collision rate coefficients used. Section IV of Paper #5, which is omitted here, gives a similar tabulation, ignoring the proton collisions discussed above. In the limit of low $n_e$ the intensities tabulated in this Section IV, for $X = 0.32$, are essentially identical with more recent work and were the first to portray the effect of two-photon emission on the spectra of emission nebulae, reducing the Balmer discontinuity and flattening the ultraviolet spectrum at $\lambda < 3000$ Å.

## REFERENCES

1. Nussbaumer, H. and Schmutz, W.—*Astron. Astroph.* **138**, 495 (1984).
2. Pengelly, R. M. and Seaton, M. J.—Monthly Notices, *Roy. Astr. Soc.* **127**, 165 (1964).
3. Osterbrock, D. E.—Astrophysics of Gaseous Nebulae and Active Galactic Nuclei; Mill Valley, CA: Univ. Sci. Books, Sect. 4.3 (1989).
4. Osterbrock, D. E.—*Astroph. J.* **135**, 195 (1962).
5. Hummer, D. G. and Kunasz, P. B.—*Astroph. J.* **236**, 609 (1980).
6. Chen, W. L. and Neufeld, D. A.—*Astroph. J.* **432**, 567 (1994).
7. Hurwitz, M., Bowyer, S., and Martin, C.—*Astroph. J.* **372**, 167 (1991).
8. Miller, J. S. and Mathews, W. G.—*Astroph. J.* **172**, 593 (1972).
9. Dopita, M. A., Binette, L., and Schwartz, R. D.—*Astroph. J.* **261**, 183 (1982).
10. Brown, R. L. and Mathews, W. G.—*Astroph. J.* **160**, 939 (1970).

# Paper[†]

## Abstract

Simultaneous emission of two photons by an H atom in the metastable 2s level is considered as a source of a continuum in planetary nebulae. Detailed calculations show that the probability of two-photon emission is 8.23 $\sec^{-1}$ and that the intensity of radiation by this process increases markedly with increasing frequency. About 32 percent of electron captures lead directly to the 2s level and then to two-photon emission, since collisional de-excitation proves unimportant. Transitions from the 2p to the 2s level, induced by collisions with free electrons, convert a part of the remaining L$\alpha$ quanta into this continuous radiation, conversion occurring after a quantum of L$\alpha$ radiation has been scattered about $10^{10}$ times, on the average, in a region of ionized H. A neutral hydrogen region may surround the H II region. In such an H I envelope, about $10^{13}$ scatterings are required for conversion, but the density of neutral H is so much greater that most conversion by collision probably takes place there. The fraction of L$\alpha$ so converted may vary over a wide range, depending on the physical conditions.

The theory is used to predict the total emission from an ionized hydrogen gas. The Balmer jump is reduced, and the decrement of the continua shortward of the Paschen and Balmer limits is also reduced. A bluish continuum is to be expected in the region from $\lambda$ 6000 to $\lambda$ 3646. Our analysis applied to the ultraviolet observations now available results in a reduction of electron temperatures by about 25 percent. The available wide-slit observations indicate the possibility that a bluish "visual continuum" may exist.

The continuous spectrum of planetary nebulae in the visual region was first measured by Page,[1] and has been analyzed more recently by Aller and Minkowski[2] and by Page.[3] These observations show that longward of the Balmer continuum there exists continuous radiation, with appreciable strength and with a nearly constant intensity per unit wavelength interval between $\lambda$ 3000 and $\lambda$ 4800.

Attempts to explain this continuous spectrum have been, so far, uniformly unsuccessful. Previous suggestions have been summarized by

[†] Section IV, Relative Intensity of Two-Photon Emission, has been omitted here.
[1] *M.N.*, **96**, 604, 1936.
[2] Unpublished.
[3] *Ap. J.*, **96**, 78, 1942.

Greenstein and Page,[4] who show that emission of radiation in the formation of $H^-$ ions cannot explain the data. The present paper investigates a different source of continuous radiation, based on the simultaneous emission of two quanta from a hydrogen atom in the metastable 2s level. An appreciable fraction of electrons captured by protons will reach the 2s level on their way down to the ground state. In addition, L$\alpha$ radiation may be converted into two photons of this visual continuum. The absorption of L$\alpha$ radiation will excite an H atom to the 2p level; and, during the brief interval before the L$\alpha$ quantum is re-emitted, a collision with a free electron may induce a transition from the 2p to the 2s level. All one-photon transitions from the 2s down to the 1s level are forbidden, and, unless a collision with another free electron induces a transition back to the 2p level, the excited electron will jump down to the 1s level, emitting two photons. Evidently, the sum of the energies of these two photons equals the energy of the L$\alpha$ photon.

While the probability that an L$\alpha$ photon will be converted into two photons is relatively small, a single L$\alpha$ photon is scattered an enormous number of times before it can escape from the planetary nebula. Moreover, the Zanstra process converts most of the stellar energy beyond the Lyman limit into L$\alpha$ photons. Thus it seems possible that some of the energy emitted by the central star in a planetary nebula can be converted into continuous radiation by two-photon emission.

When this work was nearly complete, it was found that this process had already been considered by Minkowski and Aller,[5] who rejected it because the predicted color distribution was too blue, and, more recently, by A. Y. Kipper;[6] no details of Kipper's work are apparently available in this country.

## I. Probability of Two-Photon Emission

The general theory of two-photon processes has been given by M. Goppert Meyer,[7] and a detailed application of the theory in the case of the 2s-1s transition in H has been given by Breit and Teller.[8] Since Breit and Teller's numerical computations were approximate and did not consider at all the change of intensity with frequency, more detailed computations are required. Let the frequencies of the two photons

[4] *Ap. J.*, **114**, 106, 1951.
[5] Informal communication.
[6] *A.J.U.S.S.R.*, **27**, 321, 1950.
[7] *Ann. d. Phys.*, **9**, 273, 1931.
[8] *Ap. J.*, **91**, 215, 1940.

emitted by $y\nu_{12}$ and $(1 - y)\nu_{12}$, where $\nu_{12}$ is the frequency of an L$\alpha$ photon; evidently

$$\nu_{12} = \tfrac{3}{4}cR, \tag{1}$$

where $c$ is the velocity of light and $R$ is the Rydberg constant for H. Let $A(y)\,dy$ be the probability that a photon is emitted with a frequency in the range $\nu_{12}\,dy$. From equation (6.2) in the paper by Breit and Teller, we have

$$A(y) = \frac{9a^6cR}{2^{10}}\,\psi(y), \tag{2}$$

where $a$ is the fine-structure constant $2\pi e^2/hc$, and

$$\psi(y) = y^3(1 - y)^3 \left| \sum_{m=2}^{\infty} R_{mp}^{1s} R_{mp}^{2s} \left( \frac{3}{1 + 3y - 4/m^2} + \frac{3}{4 - 3y - 4/m^2} \right) \right.$$

$$\left. + \int_0^{\infty} C_{1s} C_{2s}\,dx \left( \frac{3}{1 + 3y + 4x^2} + \frac{3}{4 - 3y + 4x^2} \right) \right|^2. \tag{3}$$

The quantities $R_{mp}^{ns}$ and $C_{ns}$ are radial quantum integrals defined by Breit and Teller.

Values of $\psi(y)$ have been computed in detail, with $R_{mp}^{ns}$ taken from the tabulation by H. Bethe,[9] and $C_{ns}$ from the paper by M. Stobbe;[10] since Stobbe's tables were not sufficiently complete, new values of these functions were computed from his formulae. The resultant values of $\psi(y)$ are given in Table 1. Since $\psi(1 - y)$ equals $\psi(y)$, no values are given for $y$ greater than 0.5. The emissivity $j_\nu$ per unit frequency interval is proportional to $h\nu A(y)$, and therefore varies as $y\psi(y)$; values of the relative emissivities are given in the last column.

The familiar Einstein coefficient $A_{2s,1s}$ for the two-photon transition is given by

$$A_{2s,1s} = \frac{1}{2} \int_0^1 A(y)\,dy = \frac{9a^6cR}{2^{11}} \int_0^1 \psi(y)\,dy. \tag{4}$$

[9] *Handb. d. Phys.* (Berlin: J. Springer, 1933), **24-1**, 442.
[10] *Ann. d. Phys.*, **7**, 661, 1930.

TABLE 1

Relative Probabilities and Intensities of Two-Photon Emission

| $y$ | $\lambda(A)$ | Probability $\psi(y)$ | Emissivity $y\psi(y)$ | $y$ | $\lambda(A)$ | Probability $\psi(y)$ | Emissivity $y\psi(y)$ |
|------|--------|-------|--------|------|------|-------|-------|
| 0.00 |        | 0     | 0      | 0.30 | 4052 | 4.546 | 1.363 |
| 0.05 | 24,313 | 1.725 | 0.0863 | 0.35 | 3473 | 4.711 | 1.649 |
| 0.10 | 12,157 | 2.783 | 0.2783 | 0.40 | 3039 | 4.824 | 1.929 |
| 0.15 | 8105   | 3.481 | 0.5222 | 0.45 | 2702 | 4.889 | 2.200 |
| 0.20 | 6078   | 3.961 | 0.7922 | 0.50 | 2431 | 4.907 | 2.454 |
| 0.25 | 4862   | 4.306 | 1.077  |      |      |       |       |

The factor of $\frac{1}{2}$ is required, since there are two photons, and each pair is counted twice. Numerical integration yields the result

$$\int_0^1 \psi(y)\, dy = 3.770. \tag{5}$$

Inserting numerical values into equation (4), we find

$$A_{2s,1s} = 8.227 \text{ sec}^{-1}, \tag{6}$$

a value close to the upper limit found by Breit and Teller; an increase in $a$ above the value used by Breit and Teller is partly responsible for this relatively high value of $A_{2s,1s}$.

## II. EXCITATION OF 2s LEVEL

We consider, now, the processes by which an electron can reach the 2s state under conditions prevailing in planetary nebulae. The first and simplest process is that in which an electron reaches the 2s state by electron capture, either by direct capture in this state or by capture in a higher state, with subsequent cascading downward to the 2s state.

We first compute the probability $X_{r,n}$ that an electron, on recombination with a proton in the level of total quantum number $n$, passes through the 2s state on its way down. let $\Gamma_{n'l',nl}$ be the number of electrons jumping from the level $n'l'$ to the level $nl$ per second per cubic centimeter; jumps down from and up to the free state will be represented by $\Gamma_{f,nl}$ and $\Gamma_{n'l',f}$ respectively. Let $\Gamma^*_{n'l',nl}$ represent the corresponding quantity in thermodynamic equilibrium with the same density of protons and electrons and at a temperature corresponding to the mean kinetic energy of protons and electrons. Evidently, from the principle of detailed balancing,

$$\Gamma^*_{n'l',nl} = \Gamma^*_{nl,n'l'}.\tag{7}$$

Also, since the electron velocity distribution is Maxwellian, we have

$$\Gamma^*_{f,nl} = \Gamma_{f,nl}.\tag{8}$$

We wish to compute the fraction of electrons, captured in the level of total quantum number $n$, which are captured in the level of angular momentum $l$. From equations (7) and (8) we see that this fraction, $y_{nl}$, is given by

$$v_{nl} = \frac{\Gamma_{f,nl}}{\Sigma_l\,\Gamma_{f,nl}} = \frac{\Gamma^*_{nl,f}}{\Sigma_l\,\Gamma^*_{nl,f}}.\tag{9}$$

To a first approximation the ratio on the right-hand side of equation (9) is given by the ratio of $gf$ values. The use of an integrated $f$ value for transitions to the continuum neglects the shape of the radiation spectrum and the detailed form of $df/dv$ for continuous absorption, but should give an adequate first approximation. Equation (9) then becomes

$$y_{nl} = \frac{g_{nl}f_{nl,f}}{\Sigma_l\,g_{nl}f_{nl,f}}.\tag{10}$$

Evidently in the case $n = 2$ we have

$$X_{r,2} = y_{20}.\tag{11}$$

For captures of electrons in levels of higher $n$, we must consider subsequent transitions. We may write, in general,

$$X_{r,n} = \frac{\Sigma_l\,z_{nl}y_{nl}}{\Sigma_l\,y_{nl}},\tag{12}$$

where $z_{nl}$ is the fraction of electrons captured in the level $n,l$ which cascade down to the level $2,0$. For the level $n = 3$ we have

$$z_{3l} = \begin{cases} 0 & \text{for} \quad l = 0 \\ 1 & \text{for} \quad l = 1 \\ 0 & \text{for} \quad l = 2. \end{cases}\tag{13}$$

Transitions down to the ground level ($n = 1$) are neglected, since any photons emitted in this way will be immediately reabsorbed. For higher levels, more complicated results are obtained. The relative probabilities of two competing downward transitions may be determined directly from the transition probabilities given by Bethe.[9] For $n = 4$, for example, we have

$$z_{4l} = \begin{cases} 0.42 & \text{for} \quad l = 0 \\ 0.74 & \text{for} \quad l = 1 \\ 0.26 & \text{for} \quad l = 2 \\ 0.00 & \text{for} \quad l = 3. \end{cases} \tag{14}$$

If we now substitute these results for $z_{nl}$ in equation (12) and use in equation (10) the value of $f_{nl, f}$ given by Bethe, we find

$$X_{r, n} = \begin{cases} 0.38 & \text{for} \quad n = 2 \\ 0.45 & \text{for} \quad n = 3 \\ 0.38 & \text{for} \quad n = 4. \end{cases} \tag{15}$$

For greater values of $n$, the value of $X_{r, n}$ decreases gradually. However, the number of electron captures on the $n$th level varies as $1/n^3$ for binding energies less than the kinetic energy of the free electron. In the planetary nebulae, where the electron temperature corresponds to a mean kinetic energy of about 1 volt, captures on levels about $n = 5$ may therefore be neglected. The fraction of electrons captured which reach the 2s state should be somewhere between 0.30 and 0.35; we shall assume a mean value of 0.32. Collisional de-excitation may be taken into account in the manner discussed below, with the result that, if $T$ is 10,000°, $X_r$, the weighted mean of $X_{r, n}$, becomes

$$X_r = \frac{0.32}{1 + 8.2 \times 10^{-6} n_e}. \tag{16}$$

A second mechanism for reaching the 2s state is radiative excitation of the 2p state, followed by a collision with a free electron, inducing a transition to the 2s state. The low probability for this collisional process is offset by the very high number of times that a quantum of L$\alpha$ radiation will be absorbed and re-emitted before it leaves the nebula. On the assumption that all excitation is by radiative absorption of L$\alpha$ quanta, we now wish to compute the ratio between two-quantum jumps from 2s to 1s and one-quantum jumps from 2p to 1s. This ratio, which we denote by $\zeta$, gives the probability that a quantum of L$\alpha$ radiation will be converted into two photons when it is absorbed by an H atom.

This ratio clearly depends on the probability of collisionally induced transitions, which has been considered in detail by Breit and Teller.[8] Their computations may readily be modified to include the Lamb shift[11] of the 2s level relative to the $2p_{1/2}$ level; this shift has only a small effect, since the energy of the transition is in any case very small compared to the energy of the incident electron.[12]

With obvious modifications, the equations $S'$ and $S''$ by Breit and Teller may be combined to yield for the transition probability $C_{2s,2p}$

$$C_{2s,2p} = \frac{6n_e h^2}{\pi v m^2} \left\{ \ln \frac{mv^2}{|E_{2s} - E_{2p_{1/2}}|} + 2\ln \frac{mv^2}{|E_{2a} - E_{2p_{3/2}}|} \right\}, \quad (17)$$

where $n_e$ is the electron density per cubic centimeter; $m$ is the electron mass; and $v$ is the velocity of the free electron in centimeters per second. In wave numbers,

$$|E_{2s} - E_{2p_{1/2}}| = 0.035 \text{ cm}^{-1}, \quad (18)$$

$$|E_{2s} - E_{2p_{1/2}}| = 0.365 \text{ cm}^{-1}. \quad (19)$$

Substituting numerical values and replacing $1/v$ by the harmonic mean at temperature $T$, we have

$$C_{2s,2p} = 6.21 \times 10^{-4} \frac{n_e}{T^{1/2}} \ln(5.7T)\left[1 + \frac{0.78}{\ln(5.7T)}\right]\text{sec}^{-1}. \quad (20)$$

The second term in brackets on the right-hand side of this equation is small compared to unity and will be neglected here. Collisional transitions induced by collisions with neutral H atoms may be neglected if $n_H$ does not exceed $10^3 n_e$; the cross-section for such encounters will be several orders of magnitude less than the $10^{-12}$ cm$^2$ predicted for electron collisions.

The relative populations of the 2s and 2p levels are readily computed from the condition that statistical equilibrium exists, i.e., that the number of electrons jumping out of each level equals the number jumping in. We neglect simulated two-quantum emissions. The quantity

[11] *Phys. Rev.*, **72**, 241, 1947.
[12] This situation is in marked contrast to that prevailing in a constant electrical field, where the Lamb shift decreases the radiative transition probability from 2s to 1s by a factor of about 1000, according to G. Luders, *Zs. f. Naturforsch.*, **5a**, 608, 1950.

$\zeta$ equals the ratio of populations in the 2s and 2p levels, multiplied by the ratio $A_{2s,1s}/A_{2p,1s}$. After some analysis we obtain

$$\zeta = \frac{g_{2s}C_{2s,2p}}{g_{2p}A_{2p,1s}} \frac{1}{1 + C_{2s,2p}/A_{2s,1s}}. \tag{21}$$

The ratio $g_{2s}/g_{2p}$ is $\frac{1}{3}$, and, if we insert numerical values from equation (20), taking the value of $A_{2s,1s}$ from equation (6), we have

$$\zeta = \frac{3.31 \times 10^{-13} n_e \ln 5.7T}{T^{1/2} + 7.5 \times 10^{-5} n_e \ln 5.7T}. \tag{22}$$

When $T$ is 10,000° K, a standard value for most planetary nebulae, equation (22) yields

$$\zeta = \frac{3.62 \times 10^{-14} n_e}{1 + 8.2 \times 10^{-6} n_e}. \tag{23}$$

For values of $n_e$ between $10^3$ and $10^4$ per cubic centimeter—typical values for most planetaries—$\zeta$ is about $10^{-10}$. Of about $10^{10}$ L$\alpha$ quanta absorbed by an H atom, one will give rise to a two-quantum jump. If a region of neutral H surrounds the planetary, $n_e$ in such a region will be less by a factor of $10^3$, and $\zeta$ will equal about $10^{-13}$.

### III. Mean Free Path of L$\alpha$ Quantum

We have seen that some 30–35 percent of the electrons captured by a proton produce two-quantum emission. The others each produce, among other things, a quantum of L$\alpha$ radiation; and, if the nebula is sufficiently thick optically, these quanta will also be converted into the visual two-photon continuum. Here we consider the mean free path of an L$\alpha$ quantum, on the average, before it is converted into two quanta by this process.

The situation is idealized by the assumption that at some distance from the central star, the density of neutral H is constant in space. With this assumption, the necessity for solving the diffusion equation is eliminated; and the simplified analysis of Brownian motion may be applied. Let $s_\nu$ be the mean free path of a photon before absorption and let the absorption coefficient of neutral H for this photon be $a_\nu$. If

$n_H$ is the density of neutral H, then we have

$$s_\nu = \frac{1}{n_H a_\nu}.$$ (24)

If the probability of conversion into two-quantum radiation is $\zeta$ per absorption, then the photon will travel $1/\zeta$ mean free paths, on the average, before it is so converted. The directions of successive paths will be uncorrelated, and the mean square distance $l_\nu^2$ will increase proportionally to the number of paths traveled; hence $l_\nu^2$ will equal $s_\nu^2$ multiplied by $1/\zeta$.

To obtain a realistic picture, we must take into account the Doppler change of $\nu$ in successive paths, depending on the thermal motion of the absorbing atom and on the angles between the absorbed and emitted photon. This type of noncoherent scattering has been considered by Henyey[13] and applied to planetary nebulae by Zanstra.[14] The detailed correlation of frequencies between the absorbed and subsequently re-emitted photon will be ignored in this first approximation, and only the statistical distribution of frequencies will be considered; this distribution may be assumed to follow the Maxwellian distributions of H-atom velocities. The mean square distance traveled must be averaged over this distribution of frequencies, and we have the basic equation

$$l^2 = \frac{1}{\zeta n_H^2} \int_0^\infty \frac{\phi(\nu)\, d\nu}{a_\nu^2},$$ (25)

where $\phi(\nu)\, d\nu$ is the fraction of atoms emitting quanta in the frequency range $d\nu$. From the usual Doppler formula and the Maxwellian velocity distribution we have

$$\phi(\nu) = \frac{1}{\pi^{1/2} b} e^{-(\Delta\nu/b)^2},$$ (26)

where $\Delta\nu$ is the difference in frequency from the undisplaced frequency and

$$b^2 = \frac{2kT\nu^2}{m_H c^2}.$$ (27)

[13] *Proc. Nat. Acad. Sci.*, **26**, 50, 1940.
[14] *B.A.N.*, Vol. 11, No. 401, 1949.

If a Doppler profile of the line-absorption coefficient were assumed and $a_\nu$ therefore varied as $\phi(\nu)$ for all $\Delta\nu$, the integral in equation (25) would diverge; physically the $L\alpha$ radiation would leak out of the nebula in the far wings of the profile. Actually, only the center of the line profile is given by the usual Doppler formula, and for large $\Delta\nu$ the resonance wings dominate. To an adequate approximation we may write

$$a_\nu = \frac{\pi e^2 f}{mc} \times \begin{cases} \phi(\nu) & (\Delta\nu \leq b) & (28a) \\[2ex] \phi(\nu) + \dfrac{\gamma}{\pi(\Delta\nu)^2} & (\Delta\nu > b), & (28b) \end{cases}$$

where $e$ and $m$ are the electronic charge and mass and $\gamma$ is the damping constant, numerically equal to $A_{2\mathrm{p},1\mathrm{s}}/4\pi$.

If equation (28) is substituted in equation (25), we find that the integral comes mostly from values of the integrand for which the Doppler wings and the resonance wings are about equal. If we let $w$ be the value of $(\Delta\nu/b)^2$ at which these two contributions to $a_\nu$ are equal, then from equations (26) and (28) we see that $w$ satisfies the equation

$$we^{-w} = \frac{\gamma}{\pi^{1/2}b}. \tag{29}$$

In the present instance $\gamma$ is several orders of magnitude less than $b$, and $w$ is moderately large. If we define a new variable, $u$, by the relation

$$u = \left(\frac{\Delta\nu}{b}\right)^2 - w, \tag{30}$$

then the integral in equation (25) becomes

$$\int_0^\infty \frac{\phi(\nu)\,d\nu}{a_\nu^2} = \frac{m^2 c^2 b^2 e^w}{\pi^{3/2} e^4 f^2 w^{1/2}} \int_{-w}^\infty \frac{e^u(1 + u/w)^{3/2}\,du}{(1 + e^u + u/w)^2}. \tag{31}$$

In deriving equation (31) we have used equation (28b) for all values of $\Delta\nu$; this diminishes somewhat the value of the integral for small $\Delta\nu$, but this region of $\Delta\nu$ contributes a negligible amount to the integrand in any case. When $w$ is infinitely great, the integral in equation (31) equals unity. For $w = 10$, the integral differs from unity by about 10 percent, a difference which we shall neglect. We have, combining equations (24) and (31),

$$l^2 = \frac{m^2 c^2 b^2 e^w}{\pi^{3/2} e^4 f^2 n_H^2 w^{1/2} \zeta}.$$  (32)

For L$\alpha$ radiation, $f$ is 0.416, and $\gamma$ is $4.97 \times 10^7$ sec$^{-1}$. If we set $T$ equal to 10,000° K, $b$ is $1.06 \times 10^{11}$. The corresponding value of $w$ found from equation (29) is 10.60. Substituting numerical values in equation (32) and making use of equation (23) for $\zeta$, we obtain

$$l = 2.4 \times 10^3 \frac{(1 + 8.2 \times 10^{-6} n_e)^{1/2}}{n_e^{1/2} n_H} \text{ parsecs.}$$  (33)

If $l$ found from equation (33) is small compared to the radius $R$ of the planetary nebula, then all the L$\alpha$ radiation will be converted into two-photon emission. If, on the other hand, $l$ much exceeds $R$, the previous analysis is not strictly applicable. In this case the quanta will escape from the nebula after they have traveled a distance $R$, on the average, from the point of origin. Since the number of scatterings required to travel a distance $R$ varies as $R^2$, the number of scatterings experienced by a photon before escape will be less than the number $1/\zeta$ required for conversion into two-photon radiation by the fraction $(R/l)^2$, where $l$ is the mean free path computed from equation (33) on the assumption that $R$ is effectively infinite. If we denote by $X_c$ the fraction of L$\alpha$ radiation so converted by collisions, we have, approximately,

$$X_c = \left(\frac{R}{l}\right)^2.$$  (34)

It should be noted that, if the H atoms are distributed in a filamentary system, with regions of high density embedded in regions of lower density, the mean value of $1/n_H^2$ will be increased, and $l$ will exceed the value given in equation (33).

The fraction $X_c$ of L$\alpha$ radiation converted into two-photon emission by collisions can be evaluated from a comparison between the size of the planetary and the free path, $l$, given by equation (33). Observations give only the average $n_e$ in the main body of the nebula; $n_H$ is variable, amounting to about $10^{-3} n_e$ in the main body and increasing at the outer boundary as $n_e$ decreases. The estimate of the mean value of $n_e^{1/2} n_H$ must be based on a definite model for the nebula. Page and

Greenstein[15] identify the visible portion of the nebula with the ionized hydrogen region (Strömgren sphere) surrounding the central star. They find that the observed radii $R$ of planetary nebulae agree with those predicted on the basis of Strömgren's theory,[16] using the observed $n_e, T_e$, and the $T_s, R_s$ of the exciting star. We adopt the model used by Strömgren, that of a homogeneous sphere of pure hydrogen of density $n$, surrounding a star which radiates as a black body.

First, we consider the fraction of $L\alpha$ radiation converted within the H II region; we denote this fraction by $X_{cII}$. Equation (33) is valid only if $n_e$ and $n_H$ are constant. However, for approximate results we may use this equation for actual nebulae, introducing the following mean value of $n_e^{1/2} n_H$:

$$\overline{n_e^{1/2} n_H} = \frac{n^{3/2}}{R} \int_0^R x^{1/2} (1 - x)\, dr, \tag{35}$$

where $x$, the fraction of H ionized, is given by the Strömgren theory.[16] We will assume that $R$ equals $s_0$; since $n_e$ is less than $10^5$, we will drop the correction term in the numerator of equation (33). Since $R$ is much smaller than $l$, we have, from equation (34),

$$X_{cII}^{1/2} = \frac{R}{l} = \frac{n^{3/2}}{2.4 \times 10^3} \int_0^R x^{1/2} (1 - x)\, dr. \tag{36}$$

The integrand is known as a function of $r$ from Strömgren's differential equation (12) for $1 - x$ less than 1 and from his approximation formula (17) for the region $r$ about equal to $s_0$. A scale parameter, $a$, exists which measures both the thickness of the layer in which hydrogen becomes neutral and the fraction of hydrogen neutral near the star. The combination $as_0$ is independent of the properties of the star:

$$as_0 = aR = \frac{1}{6.3n}. \tag{37}$$

The unit of length is 1 parsec for $R$; $n$ is the total number of hydrogen atoms and ions per cubic centimeter. The fraction of neutral hydrogen

[15] Ap. J., **114**, 98, 1951.
[16] Ap. J., **89**, 526, 1939.

in the inner part of the sphere is given by

$$1 - x = \frac{a}{1 - (r/R)^3}\left(\frac{r}{R}\right)^2, \qquad (1 - x) < 1. \qquad (38)$$

Thus $n_H$ depends mainly on $a$. The mean $(1 - x)x^{1/2}$ has been obtained by numerical integration; it is $2.7a$ for $a = 10^{-2}$ and $3.8$ for $a = 10^{-3}$. At $r/R = 0.7$ the value of $(1 - x)x^{1/2}$ from equation (38) is $0.75a$. The mean ionization determined by integration is seriously influenced by the large value of $1 - x$ in the rim of the ionized region but is not changed in order of magnitude. The maximum value of $(1 - x)x^{1/2}$ is 0.38, but this value is encountered in a shell only $0.007R$ thick. The order of magnitude of the mean free path should be correct, in spite of the large variation of $1 - x$.

Combining equation (36) and (37) and the results of the integrations, we obtain

$$X_{cII} = 3.2 \times 10^{-8}n, \quad \text{for} \quad a = 10^{-2}, \qquad (39a)$$

$$X_{cII} = 6.4 \times 10^{-8}n, \quad \text{for} \quad a = 10^{-3}. \qquad (39b)$$

The value of $a$ can be computed from the observational data. For a low-temperature and density ($n_e = 750$) nebula, NGC 40, the compilation by Page and Greenstein results in $a = 1.5 \times 10^{-3}$; for NGC 7009, $n_e = 6800$ and $a = 0.7 \times 10^{-3}$. Thus expression (39b) is sufficiently good; $X_{cII}$ is $4.1 \times 10^{-5}$ for NGC 40 and $5.0 \times 10^{-4}$ for NGC 7009. Certain nebulae with apparently large $n_e$, such as NGC 6790 and IC 4997, would have $X_{cII}$ equal to $1.5 \times 10^{-3}$ and $4.6 \times 10^{-3}$, respectively. Thus the known planetaries convert an insignificant fraction of L$\alpha$ radiation into two-photon emission by means of collisions within the H II region.

Page and Greenstein[15] pointed out that, if a planetary nebula has been expanding for a sufficiently long period, the visible disk or shell will be only the ionized core of the gas. The apparent boundary of the H emission is then the edge of the Strömgren sphere, and the neutral gas may extend far beyond. Without necessarily adopting the hypothesis of continuous ejection of matter from the star, we may still assume that the gas is expanding radially outward to large distances from the visible boundary at a uniform velocity. Then the density $n(r)$ varies as $1/r^2$, and the total number of atoms per square centimeter outside the visible boundary is $Rn(R)$. Thus in approximate results we can replace this H I region by a uniform layer of thickness $R$. In such a region the free electrons come primarily from $C$ and $Mg$, which will be largely ionized.

The electron density $n_e$ may be set equal to $n_H/2000$, and we have

$$\overline{n_e^{1/2}n_H} = 2.2 \times 10^{-2}n^{3.2}. \tag{40}$$

We let $X_{cI}$ be the fraction of $L\alpha$ radiation converted into two-photon emission in the $HI$ shell outside the planetary. Then, by use of equations (33) and (34), we have

$$X_{cI}^{1/2} = \frac{R}{l} = 0.9 \times 10^{-5}n^{3/2}R. \tag{41}$$

The density of a typical planetary may be taken as $n = 5000$ and $R = 0.1$ parsecs; then $X_{cI} = 0.10$, i.e., about one-tenth of $L\alpha$ is converted into two-photon emission. This is much larger than the $X_{cII}$, which would be near 0.001, for these standard conditions. We can expect $X_{cI}$ to vary from zero to about $\frac{1}{4}$, depending on physical conditions in this outer envelope. If we take into account the probable fall of temperature in $HI$ regions and the consequent increase of $\zeta$ and decrease of $b$, the value of $X_{cI}$ is somewhat increased. If we let $X/(1 - X)$ be the ratio of the energy of two-photon radiation to $L\alpha$ radiation, evidently $X$ may vary from about 0.32 to nearly 1 in planetaries of differing density. A more refined theory of radiative transfer, taking into account the large spatial variations of $n_H$ through the nebula, both in $HI$ and $HII$ regions, would be required for a more accurate and detailed calculation of $X_c$.

We are grateful to Drs. O. C. Wilson and R. Minkowski for a number of illuminating discussions on this problem.

# A COMPARISON OF THE COMPONENTS IN INTERSTELLAR SODIUM AND CALCIUM*

### (WITH P. McR. ROUTLY)

### (ASTROPH. J. 115, 227, 1952)

## *Commentary*

WHEN Martin Schwarzschild and I came to Princeton, in 1947, we both felt that we should undertake some observational research, partly to supplement the theoretical studies we were pursuing and partly to provide a proper balance in the graduate program we were planning to develop. We persuaded the University that this objective could best be achieved by sending each of us in alternate years to Pasadena for a term of observing (see Paper #31). The alternative, providing a large telescope in the lowlands of New Jersey, was not appealing either scientifically or financially. So we began the custom of observing at the Mt. Wilson Observatory, where fortunately we were usually alloted time on the 100-inch telescope. Sometimes a graduate student would come with one of us, and help obtain observations for a Ph.D. thesis. For me, this arrangement continued more or less regularly until 1962, after which my involvement with space astronomy prevented further such visits.

Paper #6 above was one of the chief results of this observational program. The scientific background summarized in the Introduction of this paper has given me a strong interest in comparing the individual Na I and Ca II components of interstellar absorption features. This program, started in earlier visits, became the principal objective of my Pasadena stay in 1950, when I was accompanied by Paul Routly, one of our graduate students. He assisted me in obtaining spectrograms and began the analysis which led to his Ph.D. thesis some six months later.

The primary result of this investigation appears in figure 1, indicating that the column-density ratio between Na I and Ca II atoms in an individual absorption component depends strongly on $|v|$, the absolute value of the component's radial velocity (in the Local Standard of Rest). As $|v|$ increases from $< 10$ km s$^{-1}$ to $> 20$ km s$^{-1}$, this ratio decreases

* This article is based on observations made by the senior author as guest investigator at the Mount Wilson and Palomar Observatories. The work was supported in part by a contract with the Office of Naval Research.

by about one and a half orders of magnitude on the average, with a very substantial scatter. Later surveys, with larger data bases, have obtained very similar plots.[1,2]

A theoretical discussion of these results in a final section of Paper #6 mentions the possibility that exchange of atoms between gas and dust grains might be responsible for the effect shown in figure 1. However, we rejected this explanation because we saw no plausible physical reason why Na and Ca atoms should differ in this respect. Instead we invoked a possible difference of ionization between clouds of low and high $|v|$. Such a difference might be produced by an appreciable difference of $T$, with a resultant difference in the rate of ionization by electron collisions. While this process could be important in some situations, it appears to be a somewhat minor effect overall. Hence, Section VI, which discusses the theory of this effect, has been omitted here, though the resulting final conclusion in section VII to which it led has been retained.

Some two years later the now generally accepted interpretation of this effect was summarized in my Russell Lecture.[3] Strömgren's detailed analysis[4] had confirmed earlier that in normal dense interstellar clouds Ca was under-abundant with respect to Na, in comparison with the Sun, for example. The same theory indicated to us that in the lower-density, high-velocity clouds the Na/Ca ratio is more nearly normal. Also, in a contemporary discussion,[5] the acceleration of clouds to high velocity was attributed to "explosions" of interstellar gas, caused by the intense burst of ionizing and heating ultraviolet radiation associated with the birth of a new, massive star. Thanks to these two developments, it seemed theoretically appealing to assume a general picture of Ca atoms selectively locked up in grains (i.e., "depleted" from the gas) in normal, dense clouds, but evaporating to some extent in the disruptive environment of cloud acceleration.

In more recent years, ultraviolet spectroscopic data obtained from space vehicles has shown that this same picture applies also to a wide variety of other elements in the interstellar gas. In particular, the *Copernicus* satellite (see Papers #9 and 11) had sufficient spectroscopic resolution to measure separately both the high-velocity and the low-velocity components produced by various elements along several lines of sight. With these data the low-$|v|$ components showed relative chemical abundances which were quite different from the usual "cosmic" (i.e., Solar and meteoritic) values, with different depletions for different elements.[6] (Since most of the observed ultraviolet features were produced by atoms in their predominant interstellar ionization stage, no corrections for other stages of ionization were needed.) However, the components at high $|v|$ yielded more nearly cosmic abundance ratios of elements,[6] with little or no depletion for $|v| \approx 100$ km/s$^{-1}$. Early observations with the more powerful Hubble Space Telescope have confirmed these results.[7]

While the various processes involved in grain growth and disruption have been extensively studied,[8] many uncertainties remain. No one synthesis of dust grain evolution appears generally consistent with the extensive evidence.[9]

### REFERENCES

1. Siluk, R. S. and Silk, J.—*Astroph. J.* **192**, 51 (1974).
2. Vallerga, J. V., Vedder, P. W., Craig, N., and Welsh, B. Y.—*Astroph. J.* **411**, 729 (1993).
3. Spitzer, L.—*Astroph. J.* **120**, 1 (1954).
4. Strömgren, B.—*Astroph. J.* **89**, 526 (1939).
5. Oort, J. H. and Spitzer, L.—*Astroph. J.* **121**, 6 (1955).
6. Jenkins, E. B.—In: Interstellar Processes, eds. D. J. Hollenbach and H. A. Thronson, Jr., Dordrecht: Reidel 533 (1987).
7. Spitzer, L. and Fitzpatrick, E. L.—*Astroph. J.* **409**, 299 (1993).
8. Seab, C. G.—In: Interstellar Processes, eds. D. J. Hollenbach and H. A. Thronson, Jr., Dordrecht: Reidel 491 (1987); Jones, A. P., Tielens, A.G.G.M., Hollenbach, D. J., and McKee, C. F.—*Astroph. J.* **433**, 797 (1994).
9. Draine, B. T.—In: Evolution of the Interstellar Medium, ed. L. Blitz, San Francisco: Astr. Soc. Pacific 193 (1990).

# 6

## Paper[†]

### ABSTRACT

Radial velocities and equivalent widths of the interstellar D lines, including a number of faint components, were obtained for some twenty stars from high-dispersion spectrograms taken at the Mount Wilson Observatory. Corresponding measures in Ca II were also obtained for four stars.

This material, together with additional measures obtained previously by other investigators, was employed in a comparison between Na and Ca. For each component $N$(Na II) and $N$(Ca II), the numbers of absorbing Na I atoms and Ca II ions in the line of sight, were found whenever possible, as were also the values of the Doppler widths $b$(Na I) and $b$(Ca II). For strong components, the data indicate that $b$(Ca II) in kilometers per second is about 1.5 times $b$(Na I), in qualitative agreement with Wilson's analysis; but for weaker lines these Doppler widths are more nearly equal. On the average, a marked decrease occurs in $N$(Na I)$/N$(Ca II) as the residual radial velocity of the component increases in absolute value; in low-velocity clouds Na I tends to be more abundant than Ca II, while in high-velocity clouds the reverse is true. This result apparently explains why the strong saturated lines of Na I and Ca II do not lie on the same curve of growth [$b$(Ca II) $> b$(Na I)].

Consideration is given to the possibility that differences of ionization between high-velocity and low-velocity clouds may explain this variation of abundance ratio with cloud velocity. In a rapidly moving cloud the kinetic temperature should be increased, since the intercloud gas in front of the cloud will be compressed and heated. At a temperature somewhere between $5000°$ and $10,000°$, collisional ionization of Na I by electrons would reduce the ratio $N$(Na I)$/N$(Ca II) by the observed factor. However, it is uncertain whether the temperature can be so high, and other mechanisms may be important.

## I. INTRODUCTION

Indication of the existence of interstellar calcium clouds was first obtained in 1936 by C. S. Beals,[1] who found evidence of complex structure in the interstellar K lines of $\epsilon$ and $\zeta$ Orionis and $\rho$ Leonis. During the next decade, W. S. Adams much extended the high-

---

[†] Section VI, Theoretical Discussion, has been omitted here.
[1] *M.N.*, **96**, 661, 1936.

dispersion observations of interstellar calcium and showed that, on the average, one star in every two possessed multiple components. In 1943 Adams published a provisional list on the structure of the H and K lines in fifty stars of early type[2] and in 1949 extended the list to three hundred stars.[3] His work includes visual estimates of intensity for the various K-line components, together with accurate determinations of their radial velocities. The following year quantitative measures of equivalent widths of the components of K and H were obtained from Adams' plates by Spitzer, Epstein, and Li Hen.[4]

Owing to greater observational difficulties, the high-dispersion data on Na I have not kept pace with those on Ca II. Sanford, Merrill, and Wilson[5] showed that the interstellar D lines in $\rho$ Leo and in several stars in Orion were clearly double and that the radial velocities of the components agreed well with the corresponding values obtained from H and K. The existence of interstellar sodium clouds, directly confirmed by this work, had been deduced in 1937 by Merrill and Wilson[6] in an analysis of D-line measurements by Merrill, Sanford, Wilson, and Burwell.[7] Although the spectrograms were not of sufficiently high dispersion to show faint components, Merrill and Wilson, in order to explain the data, were forced to the view that the interstellar sodium gas was not distributed homogeneously but was concentrated in the form of clouds. More recently, Merrill and Wilson[8] published preliminary equivalent-width measures of strong and faint D-line components in $\rho$ Leo and $\chi^2$ Ori, along with similar data for Ca II; in five other stars the weaker components of the D lines could not be seen on the spectrograms.

A number of puzzling differences result from a comparison between the interstellar lines of Na I and Ca II. Wilson[9] demonstrated one such difference in 1939 by combining the previous analyses of Wilson and Merrill[6] for Na I and of Sanford and Wilson[10] for Ca II. Wilson used so-called "ratio-curves" of D2/D1 versus D1 and K/H versus H. For the weakest lines the doublet ratio equals 2 but decreases with growing line strength and becomes nearly constant for line strengths exceeding 0.3 A. The surprising feature of these curves, verifying earlier data of

[2] *Ap. J.*, **97**, 105, 1943.
[3] Ibid., **109**, 354, 1949.
[4] *Ann. d'Ap.*, **13**, 147, 1950.
[5] *Pub. A.S.P.*, **50**, 58, 1938.
[6] *Ap. J.*, **86**, 44, 1937.
[7] *Ap. J.*, **86**, 274, 1937.
[8] *Pub. A.S.P.*, **59**, 132, 1947.
[9] *Ap. J.*, **90**, 244, 1939.
[10] *Ap. J.*, **90**, 235, 1939.

Beals, was that the Ca II ratio leveled off at a value of approximately 1.58, whereas the Na I ratio lay much lower, at 1.20. On fitting these observed ratio-curves by means of theoretical curves of growth, each characterized by a mean velocity parameter or Doppler constant $b$, Wilson show that $b(\text{Na I}) = 0.12$ A and $b(\text{Ca II}) = 0.24$ A, corresponding to mean velocities of 7.5 km/sec for Na I and 22 km/sec for Ca II. Wilson proposed two alternatives: either a physical mechanism operates which permits the small-scale Na I and Ca II velocities to differ considerably in the same region of space, with the heavier atoms possessing the higher velocity, or the application of the curve of growth is invalid. Jentzsch and Unsöld[11] have considered the theoretical aspects of this problem further and have shown clearly that the anomalous behavior of the ratio-curves cannot be explained by a uniform difference between the abundance of Na I and Ca II, by a superposition of velocity distributions common to Na I and Ca II, or by any combination of both.

Other interesting differences appear between interstellar Na I and Ca II when faint components are intercompared. Merrill and Wilson[8] found that, while the most intense component of Na I is considerably stronger than the corresponding component in Ca II, the weaker Na I components tend to appear weaker than the corresponding Ca II components, if, indeed, they are visible at all. Jentzsch and Unsöld[11] have pointed out that exactly the opposite effect would occur if the ratio of Na I to Ca II atoms were the same in all interstellar clouds.

The present investigation was undertaken to extend the high-dispersion interstellar observations on Na I and to carry out a more thorough comparison than has hitherto been possible with the interstellar Ca II data.[3, 4] The six sections which follow are concerned with the observational material, measurement of radial velocities, measurement of equivalent widths, interpretation of the observational data, theoretical discussion, and, finally, a summary of the results obtained.

## II. Observational Data

Observations on interstellar lines in twenty-three stars are presented in this paper. The stars observed are primarily those which showed[3] well-resolved components of H and K. Most of the plates were obtained at the coudé spectrograph of the 100-inch telescope of the Mount Wilson Observatory, chiefly during the summer and fall of 1950, with a few plates obtained in 1947 and 1948. In addition, several plates were kindly taken for this program by Drs. Ira Bowen and O. C. Wilson at the coudé focus of the Hale 200-inch telescope on Mount Palomar. Also

[11] *Zs. f. Phys.*, **125**, 370, 1948.

TABLE 1
Stars Observed for Interstellar Lines

| Star HD | Name | R.A. (1990) | Decl. (1990) | l | b | Sp. | m |
|---|---|---|---|---|---|---|---|
| 21278 | ... | $3^h 20.9^m$ | $+48° 43'$ | 115° | −05° | B5 | 4.9 |
| 24912 | ξ Per | 3 52.5 | +35 30 | 128 | −12 | O7 | 4.0 |
| 36822 | $φ^1$ Ori | 5 29.3 | +09 25 | 163 | −11 | B0 | 4.5 |
| 36861 | $λ^1$ Ori | 5 29.6 | +09 52 | 163 | −11 | O8 | 3.7 |
| 37043 | ι Ori | 5 30.5 | −05 59 | 177 | −18 | O8 | 2.9 |
| 37128 | ε Ori | 5 31.1 | −01 16 | 173 | −16 | cB0 | 1.8 |
| 37468 | σ Ori | 5 33.7 | −02 39 | 174 | − 16 | B0 | 3.8 |
| 37742 | ζ Ori | 5 35.7 | −02 00 | 174 | −15 | B0 | 2.0 |
| 38771 | κ Ori | 5 43.0 | −09 42 | 182 | −17 | cB0 | 2.2 |
| 41335 | ... | 5 59.4 | −06 42 | 181 | −12 | B2e | 5.1 |
| 42087 | 3 Gem | 6 03.7 | +23 08 | 155 | 03 | B2 | 5.8 |
| 91316 | ρ Leo | 10 27.5 | +09 49 | 204 | 54 | cB1 | 3.8 |
| 166937 | μ Sgr | 18 07.8 | −21 05 | 338 | −03 | cB8e | 4.0 |
| 167264 | 15 Sgr | 18 09.3 | −20 46 | 338 | −03 | B0 | 5.4 |
| 169454 | ... | 18 19.6 | −14 02 | 345 | −02 | B0e | 6.8 |
| 175754 | ... | 18 51.7 | −19 17 | 344 | −11 | B0p | 7.0 |
| 184915 | κ Aql | 19 31.5 | −07 15 | 0 | −15 | B0 | 5.0 |
| 190429 | ... | 19 59.8 | +35 45 | 40 | 02 | O | 7.8 |
| 193322 | ... | 20 14.6 | +40 25 | 46 | 02 | O8 | 5.8 |
| 199478 | ... | 20 52.4 | +47 02 | 55 | 01 | cB8e | 5.8 |
| 212978 | ... | 22 23.1 | +39 19 | 63 | −15 | B3 | 6.1 |
| 214680 | 10 Lac | 22 34.8 | +38 32 | 65 | −17 | O9 | 4.9 |
| 214993 | 12 Lac | 22 37.0 | +39 43 | 66 | −17 | B1 | 5.2 |

included are observations by Merrill and Wilson, who had previously published[8] only the sum of the equivalent widths of D2 and D1 (or of K and H) for each component, but who generously made available the detailed measures from their high-dispersion plates. For six stars, sodium observations were available, but the calcium determinations were of low accuracy or were lacking altogether. In such cases, additional Ca II plates were taken by the present authors. Table 1 gives the HD number for all stars observed, together with stellar name, right ascension, and declination for the epoch 1900, galactic longitude and latitude, spectral type, and apparent magnitude. Most of the information was taken from the *Henry Draper Catalogue*.

The three cameras available at the 100-inch coudé have focal lengths of 114, 73, and 32 inches. In 1947 to 1949 the grating available gave a dispersion for these three cameras of 2.84, 4.49, and 10.3 A/mm, respectively, in the second order, used for the observations of K and H. Observations of the D lines were made in the first order, with correspondingly less dispersion. In 1950 a new grating was available; for K and H, observed in the third order, the dispersion was the same as with the old grating in the second order, but for the D lines, observed in the second order, the dispersion was 4.26, 6.73, and 15.3 A/mm, respectively. For the relatively few bright stars in Table 1, the 114-inch camera was used for both Na I and Ca II, while for the fainter stars the 73-inch camera was employed almost exclusively. The 32-inch camera was used only for those few faint stars in which the separation of the components was at least 45 km/sec. The four Palomar plates yielding equivalent-width measures were all taken with the 144-inch camera, with a dispersion in the yellow of 3.41 A/mm.

The 103$a$-D Eastman plate was used for Na I, and the Eastman II$a$-O, usually baked for several days, for Ca II. The exposure times for the Na I observations were long, averaging from 3 to 5 hours; those of the Ca II observations were considerably shorter, resulting from increased emulsion sensitivity and greater stellar flux in the blue. The long exposure times and the relative coarseness of the plate grain in yellow-sensitive emulsions make it more difficult to observe faint components in Na I than in Ca II.

### III. Measurement of Radial Velocities

Measurements of the radial velocities were carried out in the standard way. The reliability of each measure was estimated on an arbitrary scale, ranging in decreasing factors of 2 from 2 to $\frac{1}{8}$, and was based upon a general over-all impression of each plate. These numerical weights were used in combining different measures of velocity for a given component. Probable errors corresponding to these weights were calculated from the differences between values obtained from different plates of the same star and are presented in Table 2 for the Na measures. Because of the variety of cameras used and the lack of sufficient data, it was found necessary to group some of the weights in pairs. The only probable error that could be derived for calcium was 1.3 km/sec, referring to a measure of weight 2 taken with the 32-inch camera. The observational material is rather meager to admit probable error determinations of good quality in all cases. It is felt, however, that the probable errors for the sodium observations of weights 2 and 1, taken with the 114- and 73-inch cameras, are reliable. The approximate correctness of the weighting system is indicated by the fact that the

PAPER 6

TABLE 2
Probable Errors of Sodium Radial
Velocities in Km/Sec

| Weight | Spectrographic Camera | | |
| --- | --- | --- | --- |
| | 114-Inch | 73-Inch | 32-Inch |
| $\left.\begin{array}{c} 2 \\ 1 \end{array}\right\}\cdots\cdots$ | 0.8 | 0.8<br>1.0 | |
| $\left.\begin{array}{c} \frac{1}{2} \\ \frac{1}{4} \end{array}\right\}\cdots\cdots$ | 1.4 | ... | 3.9 |
| $\frac{1}{8}\cdots\cdots\cdots$ | 2.5 | ... | ... |

probable errors in Table 2 are about proportional to the inverse square root of the weights.

The agreement between the radial velocity of a component as measured on the Na I plates and the corresponding component as measured by Adams[3] in Ca II is good. The root-mean-square deviation between the two sets of radial velocities is 2.5 km/sec. In this comparison, Adams' radial velocities have been increased by the constant amount of 1.5 km/sec to account for the fact that Adams used solar wave lengths for H and K, whereas the laboratory values have been adopted in this paper.

Table 3 gives the results of the radial-velocity and equivalent-width measures. Unless otherwise indicated, all the values appearing in the table have been obtained by the present authors. Data on Ca II have been included only in those cases where new measures were carried out. Columns 1 and 2 indicate the HD number of the star and the element under consideration. Column 3 denotes the number of plates taken and the cameras used; the notation "3—114" means that 3 plates were taken with the 114-inch camera. Column 4 gives the order in which the components of the interstellar lines are observed to occur in Ca II. The number 1 is assigned to the most shortward component; the remaining components are then numbered in orderly sequence toward longer wavelengths. In columns 5 and 6 the radial velocities of the D lines and their components (or of the Ca II lines) are listed, together with the weight of each determination. In those cases where the radial velocity of blends was measured, the numbers of the two blended components are given in column 4. In the notes at the bottom of Table 3, for stars denoted by an asterisk, acknowledgement is made to the work of others, suggestions offered as to the possible existence of components in Na I

TABLE 3
Measures of Radial Velocities and Equivalent Widths[A]

| HD No. (1) | El. (2) | Plates (3) | Comp. (4) | R.V. (Km / Sec) (5) | Wt. (6) | D2 or K[†] (A) (7) | Wt. (8) | D1 or H (A) (9) | Wt. (10) |
|---|---|---|---|---|---|---|---|---|---|
| 21278*............. | Na I | 2—73″ | 1 | +1.6 | 1 | 0.197 | 1 | 0.179 | 1 |
|  |  |  | 2 | ... | ... | < 0.014 | ... | ... | ... |
|  | Ca II | 1—73″ | 1 | +1.6 | $\frac{1}{2}$ | 0.061 | $\frac{1}{4}$ | 0.040 | $\frac{1}{2}$ |
| 24912, ξ Per...... | Na I | 2—114″ | 1 | +11.8 | 1 | 0.252 | $\frac{3}{2}$ | 0.198 | $\frac{3}{2}$ |
|  |  |  | 2 | ... | ... | < 0.010 | ... | ... | ... |
| 36822*, φ′ Ori.... | Na I | 2—73″ | 1 | ... | ... | < 0.014 | ... | ... | ... |
|  |  |  | 3 | +24.2 | 1 | 0.29 | 1 | 0.20 | 1 |
| 36861*, λ′ Ori .... | Na I | 2—114″ | 1 | +1.9 | $\frac{1}{8}$ | 0.022 | $\frac{1}{2}$ |  |  |
|  |  |  | 2 | +12.1 | $\frac{1}{8}$ | 0.066 | E | 0.204 | 1 |
|  |  |  | 3 | +25.0 | 4 | 0.207 | 1 |  |  |
|  | Ca II | ... | 1 | ... | ... | 0.073 | 2 | 0.039 | 2 |
|  |  |  | 2, 3 | ... | ... | 0.094 | 2 | 0.053 | 2 |
| 37043*, ι Ori ..... | Na I | 2—114″ | 1 | +4.4 | $\frac{1}{2}$ | 0.072 | 2 | 0.057 | 2 |
|  |  |  | 2 | +25.4 | $\frac{1}{4}$ | 0.055 | 2 |  |  |
| 37128, ε Ori...... | Na I | 1—114″ | 1, 2 | ... | ... | 0.091 | E | 0.047 | E |
|  |  |  | 3 | ... | ... | 0.124 | 1 | 0.074 | 1 |
| 37468, σ Ori ..... | Na I | 1—73″ | 1 | −2.6 | 1 | 0.064 | E | 0.025 | E |
|  |  | 1—114″ | 2 | +22.1 | 2 | 0.183 | 1 | 0.145 | 1 |

TABLE 3—*Continued*

| HD No. (1) | El. (2) | Plates (3) | Comp. (4) | R.V. (Km / Sec) (5) | Wt. (6) | D2 or K[†] (A) (7) | Wt. (8) | D1 or H (A) (9) | Wt. (10) |
|---|---|---|---|---|---|---|---|---|---|
| 37742, ζ Ori...... | Na I | 5—114″ | 1 | −4.0 | 6 | 0.060 | 1 | 0.025 | 2 |
| | | | 2 | +11.9 | 1 | 0.023 | E | 0.018 | E |
| | | | 3 | +22.8 | 7 | 0.145 | 4 | 0.090 | 3 |
| | Ca II | 1—114″ | 1 | ... | ... | 0.017 | 1 | 0.011 | 1 |
| | | | 3 | ... | ... | 0.035 | 1 | 0.022 | 1 |
| 38771*, κ Ori[A] | Na I | ... | 1 | −0.4 | ... | 0.054 | E | 0.025 | E |
| | | | 2, 3 | +22.4 | ... | 0.141 | 4 | 0.094 | 4 |
| | Ca II | ... | 1 | ... | ... | 0.052 | $\frac{1}{4}$ | 0.022 | $\frac{1}{4}$ |
| | | | 2, 3 | ... | ... | 0.043 | $\frac{1}{4}$ | 0.016 | $\frac{1}{4}$ |
| 41335.............. | Na I | 1—73″ | 1 | ... | ... | < 0.006 | ... | ... | ... |
| | | | 2 | +19.9 | 1 | 0.197 | 1 | 0.102 | 1 |
| 42087, 3 Gem..... | Na I | 2—73″ | 1 | −24.0 | $\frac{1}{4}$ | 0.027 | $\frac{1}{2}$ | 0.010 | $\frac{1}{4}$ |
| | | 1—144″ | 2, 3 | +14.1 | 3 | 0.542 | 1 | 0.486 | 1 |
| 91316*, ρ Leo .... | Na I | ... | 1, 2 | −5.8 | ... | 0.124 | 2 | 0.058 | 2 |
| | | | 3 | +23.1 | ... | 0.043 | 1 | 0.012 | $\frac{1}{2}$ |
| | Ca II | ... | 1 | ... | ... | 0.075 | 2 | 0.037 | 2 |
| | | | 2, 3 | ... | ... | 0.050 | 1 | 0.017 | 1 |
| 166937, μ Sgr..... | Na I | 2—73″ | 2 | −6.7 | $\frac{3}{2}$ | 0.202 | $\frac{1}{4}$ | 0.164 | $\frac{1}{4}$ |
| | | | 4 | ... | ... | < 0.014 | ... | ... | ... |

TABLE 3—Continued

| HD No. (1) | El. (2) | Plates (3) | Comp. (4) | R.V. (Km/Sec) (5) | Wt. (6) | D2 or K† (A) (7) | Wt. (8) | D1 or H (A) (9) | Wt. (10) |
|---|---|---|---|---|---|---|---|---|---|
| 167264*, 15 Sgr.... | Na I | 3—73″ | 1 | −9.4 | 3 | 0.281 | 1 | 0.215 | 1 |
| | | | 2 | +5.7 | 2 | 0.032 | $E$ | 0.035 | $E$ |
| | | | 3 | +28.6 | 2 | 0.041 | 1 | 0.032 | 1 |
| 169454............ | Na I | 2—32″ | 1 | −7.9 | 1 | 0.680 | $\frac{3}{2}$ | 0.680 | $\frac{3}{2}$ |
| | | | 2 | +81.7 | 1 | 0.152 | $\frac{3}{2}$ | 0.112 | $\frac{3}{2}$ |
| | Ca II | 1—32″ | 1 | −5.8 | $\frac{1}{2}$ | 0.522 | 1 | 0.382 | $\frac{1}{2}$ |
| | | | 2 | +83.5 | $\frac{1}{4}$ | 0.080 | 1 | 0.111 | $E$ |
| 175754*............ | Na I | 1—32″ | 1 | ... | ... | < 0.020 | ... | ... | ... |
| | | | 2 | −8.3 | $\frac{1}{2}$ | 0.454 | 2 | 0.305 | 2 |
| 184915*, κ Aql.... | Na I | 1—73″ | 1 | −10.8 | $\frac{1}{2}$ | 0.29 | 1 | 0.22 | 1 |
| | | | 2 | ... | ... | < 0.020 | ... | ... | ... |
| 190429............ | Na I | 3—32″ | 1 | ... | ... | < 0.008 | ... | ... | ... |
| | | 1—144″ | 2 | −33.7 | $\frac{3}{2}$ | 0.090 | $\frac{1}{2}$ | 0.091 | $\frac{1}{2}$ |
| | | | 3 | −6.3 | $\frac{3}{2}$ | 0.886 | 1 | 0.783 | 1 |
| | Ca II | 1—32″ | 1 | −77.7 | $\frac{1}{2}$ | 0.125 | 1 | 0.108 | $\frac{1}{4}$ |
| | | | 3 | −10.1 | $\frac{1}{2}$ | 0.406 | 1 | 0.286 | $\frac{1}{4}$ |
| 193322............ | Na I | 2—73″ | 1 | ... | ... | < 0.022 | ... | ... | ... |
| | | | 3 | −11.7 | 1 | 0.475 | $\frac{1}{2}$ | 0.425 | $\frac{1}{2}$ |
| 199478*............ | Na I | 2—73″ | 1, 2 | −12.6 | 3 | 0.535 | 2 | 0.483 | 2 |
| | | 1—144″ | 3 | +22.0 | $\frac{1}{2}$ | 0.022 | $\frac{1}{2}$ | 0.021 | $\frac{1}{2}$ |
| | | | 4 | +43.5 | 2 | 0.081 | 1 | 0.061 | 1 |

TABLE 3—*Continued*

| HD No. (1) | El. (2) | Plates (3) | Comp. (4) | R.V. (Km / Sec) (5) | Wt. (6) | D2 or K† (A) (7) | Wt. (8) | D1 or H (A) (9) | Wt. (10) |
|---|---|---|---|---|---|---|---|---|---|
| 212978 ............ | Na I | 1—144″ | 1 | ... | ... | < 0.008 | ... | ... | ... |
| | | | 2 | − 12.2 | 2 | 0.313 | 1 | 0.213 | 1 |
| | Ca II | 1—73″ | 2 | − 10.9 | $\frac{3}{2}$ | 0.088 | 2 | 0.064 | 2 |
| 214680*, 10 Lac ... | Na I | ... | 1, 2 | − 31.5 | ... | 0.103 | $\frac{1}{4}$ | ... | ... |
| | | | 3 | − 8.5 | ... | 0.237 | 1 | 0.192 | 1 |
| | Ca II | ... | 1, 2 | ... | ... | 0.055 | 2 | 0.021 | 2 |
| | | | 3 | ... | ... | 0.076 | 2 | 0.042 | 2 |
| 214993, 12 Lac .... | Na I | 1—73″ | 1 | ... | ... | < 0.016 | ... | ... | ... |
| | | | 2 | − 10.6 | 1 | 0.240 | $\frac{1}{2}$ | 0.195 | $\frac{1}{2}$ |

*Notes:*

1. Merrill, Sanford, Wilson, and Burwell, *Ap. J.*, **86**, 274, 1937.
2. Merrill and Wilson, elaboration of observations contained in *Pub. A.S.P.*, **59**, 132, 1947.
3. Spitzer, *Ap. J.*, **108**, 274, 1948.

HD 21278   Possible component in Na I at 19 km/sec.
HD 36822   Equivalent widths of component 3 in Na I by 1.
HD 36861   Equivalent widths of Ca II by 2.
HD 37043   Remarkable reversal observed in this star; stronger component in Na I actually the weaker in Ca II.
HD 38771   Radial velocities of Na I by 2; equivalent widths of Na I partly by 2; equivalent widths of Ca II by 2.
HD 91316   Radial velocities of Na I by 2; equivalent widths of Na I and Ca II by 2.
HD 167264 Possible component in Na I at 56 km/sec.
HD 175754 Possible component in Na I at 31 km/sec.
HD 184915 Equivalent widths of component 1 in Na I by 3.
HD 199478 Possible component in Na I at − 47 km/sec.
HD 214680 Radial velocities in Na I by 2; equivalent widths of Na I and Ca II by 2.
† Equivalent widths preceded by a < sign refer to upper limits described in the next section.
[A] In the original printed version, the data for the two Na I components in HD 38771 were interchanged. In this reprinted version the error has been corrected, both here and in Table 5.

not seen in Ca II, etc. The remaining columns in Table 3 will be described in the next section.

## IV. Measurement of Equivalent Widths

The general technique of measuring equivalent widths has been described many times.[12] Two series of calibration spectra were exposed on either side of the stellar spectrum in the usual way, and calibration-curves of the photographic plates were derived from them. Particular care was taken to assure uniform illumination over the slits producing the calibration spectra; and the stellar and calibration spectra exposure times were made as nearly equal as possible, so as to avoid any complications due to reciprocity failure. Special difficulties were encountered in drawing the profiles of the sodium interstellar components on the tracings because of the coarse plate grain. In handling blends between a weak and a strong component, the procedure outlined by Spitzer, Epstein, and Li Hen was followed.[4]

Each individual measure of equivalent width was assigned a numerical weight of 1, $\frac{1}{2}$, or $\frac{1}{4}$; the letter $E$ was reserved to indicate those components which were incompletely resolved from another component. In combining several measures of the same component, the numerical weights were added, but the weight $E$ was not; the result of two $E$ measures was still taken as $E$. Percentage probable errors, corresponding to these weights, were calculated and are given in Table 4. Because of insufficient data, the probable errors were not broken down for the separate cameras; Spitzer, Epstein, and Li Hen found in their work on Ca II that, on the average, the probable errors were independent of the camera focal lengths, since the higher dispersion was used only on the closer stars with weaker interstellar lines. The internal consistency of the weighting system may be checked from the requirement that observations of weight $\frac{1}{4}$ to $\frac{1}{2}$ should have about 1.6 times the probable error of observations of weight 1. From Table 4 it is evident that the probable errors for weights $\frac{1}{2}$ and $\frac{1}{4}$ are relatively too high. It was not thought worthwhile to carry out a revision of the weighting system. A comparison between the values of $\frac{1}{2}(D2 + D1)$, as obtained in this present work, and the few corresponding published values available from Merrill and Wilson and from Spitzer gives a root-mean-square percentage deviation of 16 percent.

In several stars components were visible in Ca II but not in Na I. In each case an upper limit on the equivalent width of the corresponding

---

[12] See, e.g., Williams, *Ap. J.*, **79**, 280, 1934.

TABLE 4
Percent Probable Errors of Equivalent-Width Measures

| Na I | | Ca II | |
|---|---|---|---|
| *Weights* | *Percentage* | *Weights* | *Percentage* |
| 1 | 8.1 | 1 | 5.9 |
| $\frac{1}{2}$ and $\frac{1}{4}$ | 19 | $\frac{1}{2}$ and $\frac{1}{4}$ | 24 |

component of D2 was determined in the following way. On the tracing of the spectrogram, the position at which this faint component of D2 would appear, if it existed, was located. A hypothetical absorption line was then drawn in with the same half-width as the instrumental profile and with the greatest depth which could escape certain detection of the component. The equivalent width of this feature, which depends on the graininess of the plate, the width of the spectrum, etc., was then taken as the upper limit for D2. The half-width of the instrumental contour was determined for each camera from the narrow lines in the iron arc. The values found were 7.6 and 12 km/sec for the 114- and 73-inch cameras, respectively, at the 100-inch coudé, and 8.4 km/sec for the 144-inch camera at the 200-inch. For the 32-inch camera a value of 27 km/sec was assumed, on the assumption that for a particular grating the half-width is inversely proportional to the focal length. Justification of this use of the instrumental contour in determining upper limits is indicated by the fact that, on the average, the widths of the faintest components actually observed are of the same order as the widths of the instrumental profiles.

Intensities of the D lines and their components (also a few new measures for K and H) are given in columns 7 and 9 of Table 3, together with their weights. Values in these columns preceded by a < sign designate the upper limits discussed above.

## V. Interpretation of Observational Data

The method developed by Strömgren[13] has been used to obtain the number of absorbing atoms per square centimeter in each interstellar cloud, together with the dispersion in atomic velocities. Table 2 of Strömgren's paper gives the doublet ratios (D2/D1 or K/H) as functions of $\log w^{(\lambda)}/b^{(\lambda)}$ and $\log N/w^{(\lambda)}$, where $w^{(\lambda)}$ represents the equivalent width of D1 or H, $N$ the number of absorbing Na I atoms or Ca I ions per square centimeter in the line of sight; $b^{(\lambda)}$ is $2^{1/2}$ times the

[13] *Ap. J.*, **108**, 242, 1948.

dispersion in atomic radial velocities and is expressed by Strömgren in angstroms. For purely thermal motions, we have

$$b^{(\lambda)} = \frac{\lambda_0}{c} \left( \frac{2kT}{Am_0} \right)^{1/2}, \qquad (1)$$

where $\lambda_0$ is the wavelength of either D1 or H, $m_0$ is the mass of unit atomic weight, and $A$ is the atomic weight of the atom under consideration. More generally, $b$ would contain a contribution resulting from internal turbulent motions within the cloud.

The values in Strömgren's Table 2 are based on a theoretical curve of growth obtained on the following assumptions: (1) that a single cloud is responsible for the interstellar absorption and (2) that the atoms within the cloud possess a Maxwellian velocity distribution, characterized by the effective velocity parameter $b$. In an extension of his analysis, Strömgren showed that the ratio $N/w^{(\lambda)}$ was not appreciably affected by the number of clouds which entered into the formation of $w^{(\lambda)}$, although the value of $b$ was strongly affected. If the clouds themselves have a Maxwellian distribution of peculiar velocities, characterized by the Doppler constant $b'$, then, for the case in which the clouds produced partially overlapping absorption lines, the effective $b$ should lie somewhere between $b$ and $(b^2 + b'^2)^{1/2}$. In the extreme case, where $N$ clouds give rise to $N$ nonoverlapping components, the effective $b$ would equal $N$ times the value of $b$ for one cloud.

The values of $N$ and $b$ obtained from the present data by Strömgren's method are given in columns 4–7 of Table 5. The Na I doublet ratios, necessary for this method, were derived from Table 3, while the Ca II doublet ratios were calculated primarily from the equivalent-width measures of Spitzer, Epstein, and Li Hen.[4] In those cases where new Ca II measures were determined (Table 3), averages were taken with the values of Spitzer, Epstein, and Li Hen when such were available. The values of $b$ have been converted from angstrom units to kilometers per second.

Table 5 includes only those lines and components for which good determinations of the $N$'s or $b$'s in both Na I and Ca II were possible. A few of the observations were omitted immediately on the grounds of low weight; a weight $\frac{1}{2}$ was considered as the lowest accuracy tolerable for any individual equivalent width. Further restrictions were imposed upon the values of the doublet ratios themselves for the calculation of the $N$'s and the $b$'s. No abundances were determined for components having a doublet ratio less than 1.10, while no $b$'s were computed from doublet ratios greater than 1.80.

TABLE 5

Abundances and Values of $b$ for Na I and Ca II

| HD No. (1) | Comp. (2) | R.R.V. (Km / Sec) (3) | N(Na I) $\times 10^{-11}$ (4) | N(Ca II) $\times 10^{-11}$ (5) | b(Na I) (Km / Sec) (6) | b(Ca II) (Km / Sec) (7) |
|---|---|---|---|---|---|---|
| 21278 | 1 | −2.0 | 192 | 10.6 | 2.34 | 1.74 |
| | 2 | +46.4 | < 0.62 | 1.30[†] | ⋯ | ⋯ |
| 24912 | 1 | +4.3 | 40.4 | 16.7 | 4.43 | 3.63 |
| | 2 | +19.8 | < 0.45 | 1.86[†] | ⋯ | ⋯ |
| 36822 | 1 | −8.9 | < 0.62 | 11.3 | | |
| | 2 | +1.4 | 28.2 | 15.9 | 6.62 | 6.35 |
| | 3 | +10.7 | ⋯ | ⋯ | | |
| 36861 | 1 | −8.4 | 0.98[†] | 8.14 | ⋯ | ⋯ |
| | 2 | + 1.4 } 10.6 | 27.9[†] | 11.9 | ⋯ | ⋯ |
| | 3 | +11.9 | | | | |
| 370043 | 1 | −10.0 | 3.12[†] | 2.04[†] | ⋯ | ⋯ |
| | 2 | +12.9 | 2.32[†] | 10.2 | ⋯ | ⋯ |
| 37468 | 1 | −14.3 | 2.67[†] | 1.67[†] | ⋯ | ⋯ |
| | 2 | +9.1 | 30.2 | 3.72[†] | ⋯ | ⋯ |
| 37742 | 1 | −17.4 | 2.49[†] | 2.96 | ⋯ | ⋯ |
| | 3 | +8.9 | 10.8 | 5.65 | 4.38 | 1.51 |
| 38771 | 1 | −14.9 | 2.31[†] | 3.53 | ⋯ | ⋯ |
| | 2 | + 5.4 } 8.7 | | | | |
| | 3 | +18.2 | 12.7 | 5.77 | ⋯ | ⋯ |
| 41335 | 1 | −16.9 | < 0.27 | 2.04[†] | ⋯ | ⋯ |
| | 2 | +4.6 | 9.5 | 8.27 | ⋯ | ⋯ |
| 42087 | 1 | −35.7 | 1.07[†] | 5.30 | ⋯ | ⋯ |
| | 2 | − 3.6 } 3.0 | | | | |
| | 3 | +11.4 | 351 | 49.0 | 6.62 | 12.9 |
| 166937 | 4 | +39.1 | < 0.62 | 4.65[†] | ⋯ | ⋯ |
| 167264 | 1 | + 2.5 } 3.8 | 54.7 | 68.4 | 5.09 | 5.52 |
| | 2 | +18.9 | | | | |
| | 3 | +40.6 | 5.96 | 11.2 | 0.71 | 4.46 |
| 169454 | 2 | +94.5 | 18.1 | 23.6 | 3.06 | 2.26 |
| 175754 | 1 | −74.6 | < 0.89 | 8.36 | ⋯ | |
| | 2 | + 3.3 } 7.3 | | | | |
| | 3 | +28.0 | 41.1 | 56.5 | 11.2 | 14.4 |
| 184915 | 1 | +0.9 | 39.1 | 16.5 | 5.60 | 4.23 |
| | 2 | +21.3 | < 0.89 | 10.5 | ⋯ | ⋯ |

TABLE 5—*Continued*

| HD No. (1) | Comp. (2) | R.R.V. (Km / Sec) (3) | N(Na I) $\times 10^{-11}$ (4) | N(Ca II) $\times 10^{-11}$ (5) | b(Na I) (Km / Sec) (6) | b(Ca II) (Km / Sec) (7) |
|---|---|---|---|---|---|---|
| 190429 | 1 | −65.2 | < 0.36 | 31.6 | ··· | ··· |
| | 3 | +4.0 | 392 | 100 | 12.2 | 11.3 |
| 193322 | 1 | −26.6 | < 0.98 | 5.39 | ··· | ··· |
| | 2 | −10.1 | | | | |
| | 3 | +0.4 } 2.2 | 256 | 69.1 | 6.11 | 9.07 |
| | 4 | +10.8 | | | | |
| 199478 | 1 | −7.8 } −2.6 | 349 | 48.2 | 6.62 | 17.8 |
| | 2 | +3.0 | | | | |
| | 4 | +54.6 | 10.6 | 9.30 | ··· | ··· |
| 212978 | 1 | −77.9 | < 0.36 | 2.42[†] | ··· | ··· |
| | 2 | −4.7 | 29.4 | 18.4 | 7.13 | 2.94 |
| 214680 | 1 | −28.5 } −24.7 | 4.63[†] | 4.65[†] | ··· | ··· |
| | 2 | −19.5 | | | | |
| | 3 | −3.4 | 45.0 | 9.18 | ··· | ··· |
| 214993 | 1 | −25.6 | < 0.71 | 6.70[†] | ··· | ··· |
| | 2 | −4.0 | 45.6 | 18.7 | 3.97 | 7.56 |

For an appreciable fraction of the observations used in Table 5, the doublet ratios initially exceeded 2. The lines so involved are all relatively weak. Some measures, showing anomalous values of D2/D1, were corrected by assuming the lines to be completely unsaturated and, therefore, to have doublet ratios equal to 2 exactly. The corrected value D′1 was then obtained by equating 3D′1 to the sum of the measured values of D1 and D2. The resulting changes in the original equivalent widths of D1 and D2 were found to be small in all cases. Ca II measures showing anomalous doublet ratios were corrected in the same way. Those abundances in Table 5 derived from equivalent widths altered in the manner described above are followed by dagger signs (†). The assumption of complete unsaturation was also used in one case (HD 36861) to split up a blend and in two other instances to provide values for the equivalent widths of D1 or H where measures of D2 or K only were available. Again the dagger sign was employed to draw attention to the corresponding abundances. Abundances preceded by a < sign are obtained from the upper limits listed in Table 3.

Column 3 of Table 5 gives the residual radial velocities of the components listed in column 2. These values were obtained by taking straight arithmetical averages of the Na I and corresponding Ca II radial velocities[3] and then subtracting the solar motion obtained by Vyssotsky and Janssen.[14] $V = 15$ km/sec, $A = 265°$, and $D = 20°.7$. For interstellar work, the solar motion should be referred to a system of axes rotating around the galactic center with a velocity equal to the mean circular velocity of the solar neighborhood. Since the new solar motion seems to satisfy this requirement better than previous values, it was used in the present paper. No weights were assigned to the residual radial velocities, because Adams did not indicate the accuracy of his individual Ca II measures. In some cases the velocities in column 3 refer to blends of two components. Some of these mean values were derived from direct measurement; in others, estimates were obtained from the individual velocities of the two components, averaged with Adams' visual intensities as weights.

Figures 1 and 2 show the results of Table 5 in graphic form. In figure 1, the values of $N(\text{Na I})/N(\text{Ca II})$ are plotted against the residual radial velocities of the clouds producing the components. It is difficult to state a probable error for the values of $N(\text{Na I})/N(\text{Ca II})$, since the error varies from point to point. By altering D2/D1 in one direction by its probable error, K/H in the opposite direction, recomputing $N(\text{Na I})$ and $N(\text{Ca II})$ through Strömgren's Table 2, and finally taking a grand average, we get the mean probable error as approximately a factor of 2 in either direction. Denoting the residual radial velocity by $V$, we see that figure 1 exhibits two distinct features: first, a sharp peak about $V = 0$, with considerable scatter in the vertical co-ordinate, and, second, a pronounced falling-off for increasing values of $|V|$. While the maximum of the peak at $V = 0$ reaches a value of the order of 18, there exist no points having $|V| > 20$ km/sec for which $N(\text{Na I})/N(\text{Ca II})$ is definitely greater than unity. If we neglect momentarily those points in the vicinity of $V = 0$ for which $N(\text{Na I})/N(\text{Ca II}) < 1$ and which arise in the main from weak lines, a strong correlation is evident between $N(\text{Na I})/N(\text{Ca II})$ and the peculiar velocities of the interstellar clouds. For low-velocity clouds Na I is more abundant than Ca II, while for high-velocity clouds ($> 20$ km/sec) the reverse is true. That these conclusions result from a real velocity effect and are not due to cloud size is clearly shown by plotting $N(\text{Na I})/N(\text{Ca II})$ against $N(\text{Na I})$ or $N(\text{Ca II})$; neither plot indicates any detectable correlation whatever.

Those points lying in the vicinity of $V = 0$ and having $N(\text{Na I})/N(\text{Ca II}) < 1$ tend to weaken the correlation in figure 1. The scatter at $V = 0$ is probably even greater than shown, because a considerable number of the low points involved correspond to upper

[14] *A.J.*, **58**, 56, 1951.

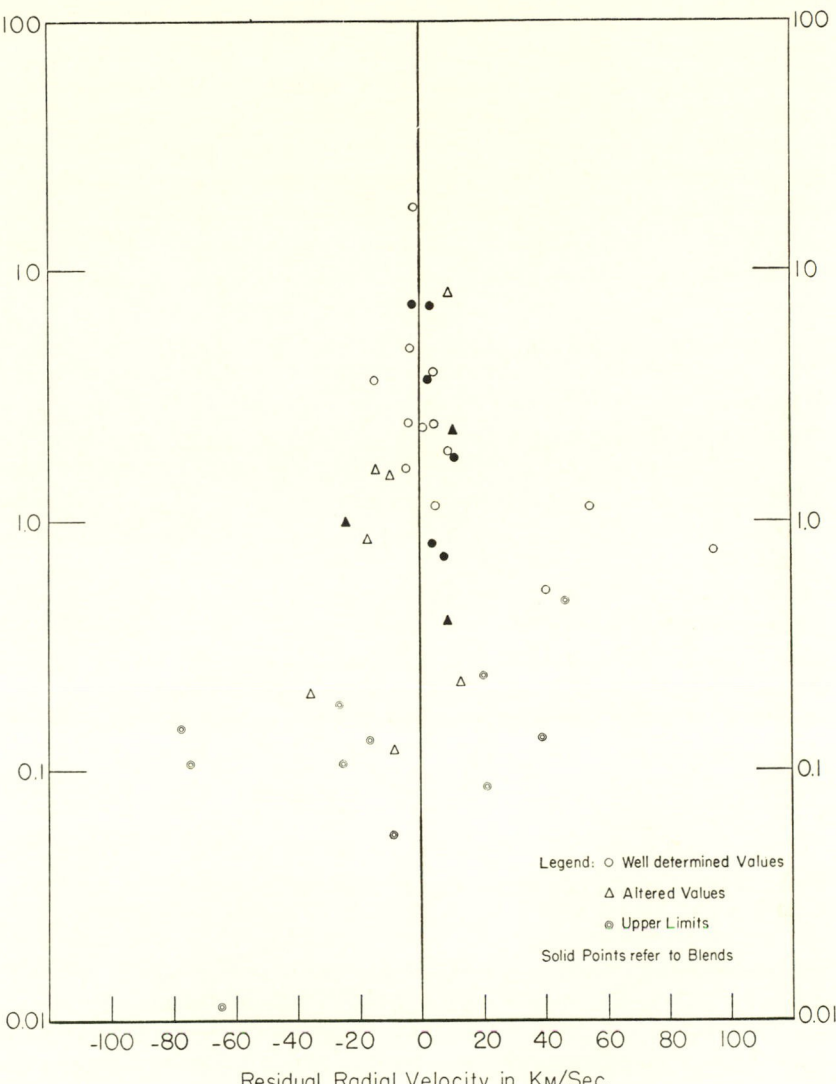

FIG. 1. Plot of log $N(\text{Na I})/N(\text{Ca II})$ versus residual radial velocity.

limits on the Na I abundance. These low-lying points may have a possible origin in clouds which possess high space velocities but have small radial components. The data were considered too scanty to warrant a statistical analysis.

Figure 2 shows the relationship between $b(\text{Ca II})$ and $b(\text{Na I})$. Points

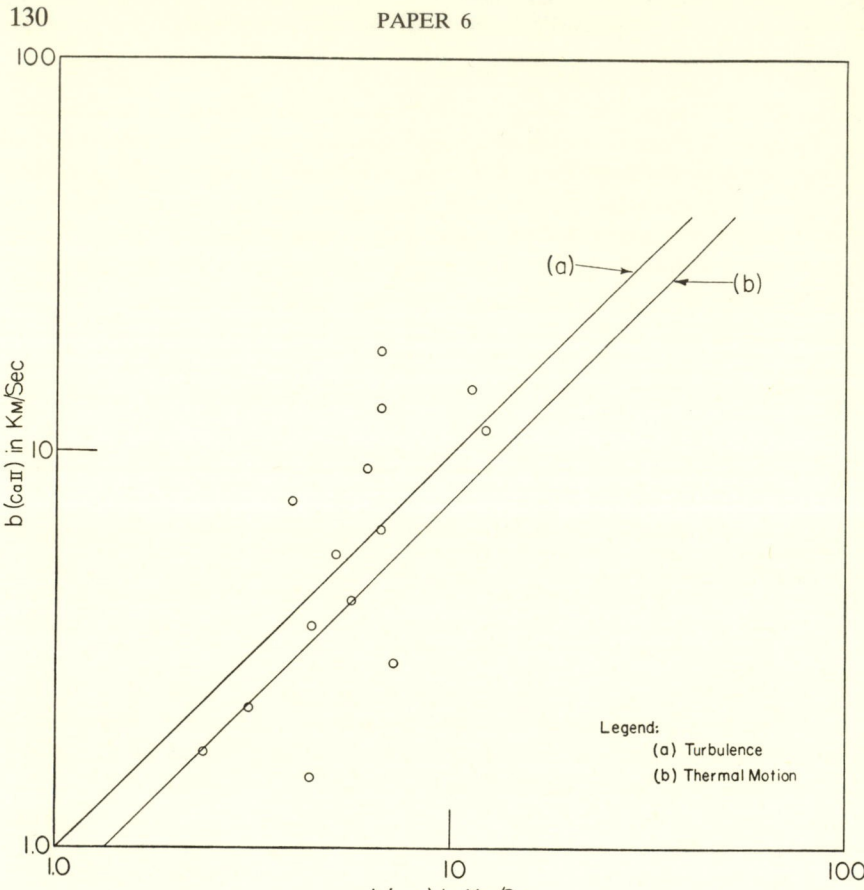

FIG. 2. Plot of log $b$(Ca II) versus log $b$(Na I).

corresponding to the weaker components considered lie in the lower part of the diagram and are relatively uncertain. One point [$b$(Ca II) = 4.5 km/sec, $b$(Na I) = 0.71 km/sec] lies so far from the straight lines shown that it could not even be plotted in figure 2. However, except for this one point, all the weaker lines show $b$'s which are somewhat greater for Na I than for Ca II. Because of the large scatter, it is impossible to decide whether thermal motions [line ($a$) in fig. 2] or turbulent motions [line ($b$)] prevail inside the interstellar clouds producing these components. One may conclude only that the $b$'s for these weaker lines are about equal.

The stronger lines behave quite differently. These lines, corresponding to points in the upper part of the diagram, are more highly saturated, and the $b$'s are more accurately determined. In this region of figure 2, $b$(Ca II) is about 1.5 times as great as $b$(Na I), on the average. Evidently the saturated lines of Na I and Ca II do not lie on the same curve of growth. This result is in qualitative agreement with Wilson's conclusions, deduced from the Na I and Ca II "ratio-curves." However, Wilson's quantitative result, that $b$(Ca II) is 22 km/sec and $b$(Na I) is 7.5 km/sec, exaggerates the difference between the $b$'s. These earlier values were found on the assumption that $b$ is the same for lines of all strengths—an assumption which we now know to be invalid.

A plausible explanation of this difference between $b$(Na I) and $b$(Ca II) may be given in terms of the correlation between $N$(Na I)/$N$(Ca II) and $V$ shown in figure 1. Since the $b$'s represent a mean velocity of the absorbing atoms, they govern the width of the line-absorption coefficient. Let $L_\nu$ denote the total line-absorption coefficient, including the effects of all clouds. Evidently, $L_\nu$ has a highly complex structure. The central part of $L_\nu$ is due to low-velocity clouds having more Na I than Ca II (fig. 1); consequently, the central height of $L_\nu$ for Na I is greater than for Ca II. On the other hand, the wings of $L_\nu$ arise from high-velocity clouds, where Ca II is more abundant than Na I; thus the wings of $L_\nu$ for Ca II are wider than the wings of $L_\nu$ for Na I. The final result is a sharply peaked line-absorption coefficient for Na I and a lower but broader absorption coefficient for Ca II. The effect is illustrated schematically in figure 3, which shows how $b$(Ca II) for highly saturated lines may exceed $b$(Na I).

It might appear possible to verify the result $b$(Ca II) > $b$(Na I) by measuring the half-widths of a number of saturated lines directly from

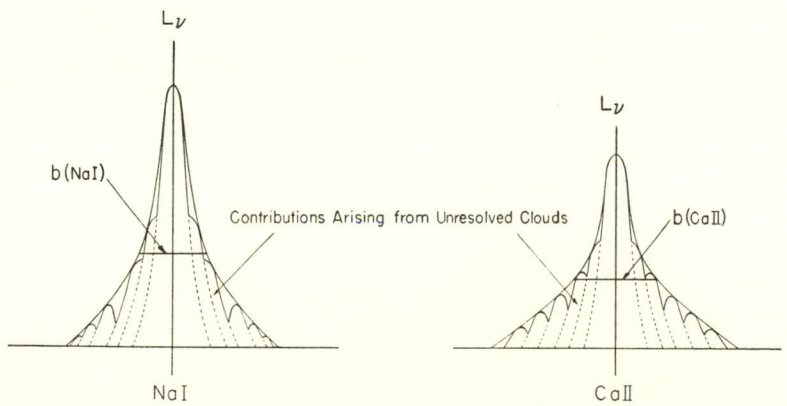

FIG. 3. Diagram of $L_\nu$ for Na I and Ca II showing how $b$(Ca II) > $b$(Na I) for saturated lines.

the tracings. There are two principal reasons why such an approach is not very feasible. First, the lines involved are all highly saturated, and consequently their half-widths bear only an indirect relationship to the half-widths of their corresponding line-absorption coefficients. Second, most of the Ca II and Na I plates were taken with different cameras (114- and 73-inch) which possess different instrumental profiles, and corrections for finite resolving power would be required.

## VII. CONCLUSIONS

The results of the present work may be briefly summarized as follows:

1. In many individual interstellar clouds both sodium and calcium are present.

2. The radial velocities of the Na I components agree with the velocities of the corresponding Ca II features, in accord with conclusion 1.

3. A velocity effect exists, such that Na I tends to be more abundant than Ca II in low-velocity clouds, while in high-velocity clouds the reverse is true.

4. For saturated lines, H and K do not lie on the same curve of growth as the D lines, thus verifying the earlier conclusion of Wilson's that $b(\text{Ca II}) > b(\text{Na I})$; a plausible explanation of this result can be given if use is made of the observed correlation between cloud velocity and the abundance ratio of Na I to Ca II.

5. If the temperature of the high-velocity clouds is higher than some critical value, which lies somewhere between 5000° and 10,000°, collisional ionization of Na I by electrons can explain the relatively low abundance of Na I in these clouds. However, it is uncertain whether the temperature in such a cloud is actually raised so high by contact with the intercloud medium, and other mechanisms may be involved.

# ON A POSSIBLE INTERSTELLAR GALACTIC CORONA*

(ASTROPH. J., 124, 20, 1956)

## *Commentary*

As with Paper #5, this paper had its genesis in a fruitful luncheon discussion with Jesse Greenstein during one of my periodic stays in Pasadena. Jesse told me about the unexpected results obtained by Guido Münch, who observed apparently normal diffuse interstellar clouds at about a kiloparsec or so from the galactic plane. A few years earlier I had emphasized[1] the importance of pressure equilibrium for explaining certain aspects of the interstellar gas. This same point of view applied to the high-$z$ clouds naturally suggested that these clouds might be confined in pressure equilibrium with some extended medium, for which a hot gas seemed the most logical possibility.

Paper #7 explores this possibility, analyzing very briefly certain physical properties of a hot "coronal" gas, with a temperature in the neighborhood of $10^6$ K. Section III includes a calculation of the cooling time of such a gas, which turns out to be probably less than $5 \times 10^9$ years, indicating the likely necessity for a present heat source. This detailed Section III is omitted here, but the more exploratory discussions in the following Section, including a listing of possible heat sources, are retained.

This somewhat speculative theoretical paper, connected only indirectly with solid astronomical facts, had little influence on most other astronomers for some time. However, the prospect of observing the ultraviolet features absorbed by $O^{5+}$ and other highly ionized atoms in a hot coronal gas played an important part in early planning for a satellite spectrometer.[2,3] This prospect actually materialized years later, when the Princeton telescope-spectrometer on the *Copernicus* satellite observed the strong $O^{5+}$ doublet at 1032 and 1038 Å in many stars,[4,5,6] mostly relatively close ones in the galactic disk. Confirmation that hot gas extended to kiloparsec distances from the galactic plane, as sug-

---

* The essential conclusions of the paper were presented to a meeting of Commission 34 (Interstellar Matter) of the International Astronomical Union at Dublin, Ireland, on August 31, 1955.

gested in Paper #7, was obtained subsequently[7] by $N^{4+}$ and $C^{3+}$ observations with the IUE satellite. These absorption-line data indicate that the distribution of highly ionized atoms in the halo is very patchy, a result suggested also for the hot gas by a detailed study[8] of $1/4$ keV X rays.

At present, one of the central research problems[9] in this field is to understand the origin and evolution of the hot gas. Transient heating or cooling through a range of temperatures rather than an isothermal equilibrium seems required to explain why the observed ions exhibit such a wide variety of ionization potentials, ranging from 35 V to produce $S^{3+}$ up to 114 V for $O^{5+}$. A galactic fountain[10] or a supernova produced by an exploding runaway star[11] have been proposed as sources of cooling and recombining hot gas at high $z$. Alternatively, conductive heating and gradual collisional ionization of gas next to a much hotter region (a supernova remnant or a shocked stellar wind) have also been suggested. Active interstellar research programs now underway with the Hubble Space Telescope, together with possible future observations at wavelengths extending much closer to the Lyman limit at 912 Å, should indicate which processes are actually dominant in the hot interstellar gas.

## REFERENCES

1. Spitzer, L.—In: Problems of Cosmical Aerodynamics, IAU-IUTAM Symp., Dayton, Ohio: Central Air Documents Office, p. 31 (1951).
2. Spitzer, L.—In: Earth Satellites as Research Vehicles, Philadelphia: Franklin Inst. (Monograph No. 2) 69 (1956).
3. Spitzer, L. and Zabriskie, F. R.—*Publ. Astr. Soc. Pacific*, **71**, 412 (1959).
4. York, D. G.—*Astroph. J. Lett.*, **193**, L127 (1974).
5. Jenkins, E. B. and Meloy, D. A.—*Astroph. J. Lett.*, **193**, L121 (1974).
6. Jenkins, E. B.—*Astroph. J.*, **219**, 845 (1978).
7. Savage, B. D. and Massa, D.—*Astroph. J.*, **314**, 380 (1987).
8. Snowden, S. L. et al.—*Astroph. J.*, **430**, 601 (1994).
9. Spitzer, L.—*Annu. Rev. Astron. Astroph.*, **28**, 71 (1990).
10. Shapiro, P. R. and Field, G. B.—*Astroph. J.*, **205**, 762 (1976).
11. McKee, C. F. and Ostriker, J. P.—*Astroph. J.*, **218**, 148 (1977).

# 7

## *Paper*[†]

ABSTRACT

The physical conditions in a possible interstellar galactic corona are analyzed. Pressure equilibrium between such a rarefied, high-temperature gas and normal interstellar clouds would account for the existence of such clouds far from the galactic plane and would facilitate the equilibrium of spiral arms in the presence of strong magnetic fields. Observations of radio noise also suggest such a corona.

At a temperature of $10^6$ degrees K, the electron density in the corona would be $5 \times 10^{-4}/cm^3$; the extension perpendicular to the galactic plane, 8000 pc; the total number of electrons in a column perpendicular to the galactic plane, about $2 \times 10^{19}/cm^2$; the total mass, about $10^8 \, M_\odot$. The mean free path would be 4 pc, but the radius of gyration even in a field of $10^{-15}$ gauss would be a small fraction of this. Such a corona is apparently not observable optically except by absorption measures shortward of 2000 A.

Radiative cooling at $10^6$ degrees would dissipate the assumed thermal energy in about $10^9$ years. Cooling by conduction can apparently be ignored, especially since a chaotic magnetic field of only $10^{-15}$ gauss will sharply reduce the thermal conductivity. At $3 \times 10^6$ degrees, near the maximum value consistent with confinement by the Galaxy's gravitational field, radiative cooling is unimportant, and a corona at this temperature might be primeval. The energy source needed at the lower temperatures may be provided by material ejected at high speed from stars or possibly by compressional waves produced by the observed moving clouds. Condensation of cool matter from the corona may perhaps account for the formation of new spiral arms as the old ones dissipate.

Within recent years it has become well established that interstellar gas clouds are found within the "spiral" arms of our Galaxy. Interstellar absorption lines, H II regions around early-type stars, and the 21-cm line of neutral hydrogen all give concordant evidence. The mean density of hydrogen atoms within a spiral arm appears to be about $1/cm^3$. However, the density distribution is nonuniform. According to the present picture—summarized by Spitzer (1954)—within a typical cloud

[†] Section III, Rate of Cooling, has been omitted here.

of neutral hydrogen, where the temperature is about 100° K, the density is about 10 atoms/cm³, as shown by Strömgren (1948). Between the clouds the density is less by a factor of perhaps 100, and the temperature is presumably greater by the same factor. The interstellar gas may be assumed to be at a nearly uniform pressure within the galactic plane, this pressure amounting to $10^{-13}$ dynes/cm².

The possible presence of gas elsewhere in the Galaxy—between the spiral arms and at a great distance from the galactic plane—has not been discussed in detail, largely because of a virtually complete lack of observational information. However, there are a number of observations which suggest that gas may actually be present far from the galactic plane. The present paper provides a brief theoretical discussion of such an "interstellar galactic corona," the reasons for suspecting its presence, its density and kinetic temperature, and its possible importance in the origin of spiral arms and in other contexts as well. At the present time, only a preliminary and tentative analysis is appropriate in so new a field. Hence no great precision is needed, and no firm conclusions can be drawn.

## I. Observational Evidence

The evidence indicating the possible presence of an interstellar galactic corona is of two types. First, the analysis of normal interstellar clouds indicates that regions outside the arms may be permeated by a medium at a pressure about the same as that in the clouds. Second, observations of radio noise suggest that some of the noise originates from a corona surrounding our Galaxy.

### a) Clouds at High Latitude

In some ways the simplest indication of an interstellar galactic corona is obtained from an examination of normal clouds at great distances from the galactic plane. Münch (1956) has found a number of such clouds in a systematic investigation of all accessible B and O stars brighter than 8.0 mag. with a galactic latitude of 30° or more. For example, HD 215733, some 1500 pc from the galactic plane, shows four K components, with a velocity spread of 46 km/sec. In thirteen stars at distances of 500 pc or more from the galactic plane, thirty-five components were observed. While the distances of these clouds cannot be determined directly, the number of components appears to be greater that would be anticipated if the clouds were all within about 200 pc of the galactic plane. Moreover, stars at about 500 pc from the plane tend to show less complex lines than do the more distant stars. Hence it seems probable that an appreciable fraction of these clouds is more than 500 pc from the plane. If the root-mean-square velocity of the interstellar atoms, in the direction perpendicular to the galactic plane, is assumed to equal 10 km/sec, corresponding to a kinetic temperature of about 10000° K for H atoms, the density of the gas at 500 pc should be only 5 percent of the density in the galactic plane.

If a normal galactic cloud, with a density of 10 H atoms/cm$^3$, were placed in such a medium, in which the pressure was only 5 percent of the internal pressure within the cloud, the cloud would expand at the speed of sound, about 1 km/sec for a temperature of 100° K. Hence a cloud 5 pc in radius would double its radius in about $5 \times 10^6$ years, and its pressure would drop to that in the surrounding medium in about $10^7$ years. For comparison, the half-period of oscillation back and forth across the galactic plane (i.e., the time between one passage across the galactic plane and the subsequent passage, in the opposite direction) is about $4 \times 10^7$ years, for amplitudes of 200 pc or less; for clouds rising higher above the galactic plane the corresponding half-period is greater. Thus any cloud at large $z$ is very likely to be in pressure equilibrium with the medium surrounding it.

However, if the density at 500 pc above the plane is only 5 percent of its value in the plane, any cloud will be so rarefied that its absorbing power should be slight. In addition to the reduced number of calcium atoms in the line of sight, with reduced cloud density the electron density will also be reduced, and the fraction of calcium atoms that are singly, instead of doubly, ionized will be diminished accordingly. It is doubtful whether any perceptible absorption line would be produced by a cloud in which the pressure had fallen by a factor of 20 or more. Thus we may conclude that the measures by Münch suggest strongly that the clouds at large $z$ are moving through a medium in which the gas pressure is comparable with that in the galactic plane. A tenuous gas at a high kinetic temperature, in the range of $10^5$ to $3 \times 10^6$ degrees, provides a simple explanation for the continued existence of absorbing clouds far from the galactic plane.

### b) Equilibrium of Spiral Arms

The presence of high-temperature gas between the spiral arms is suggested by the equilibrium of the gas in a spiral arm. This subject has been considered by Chandrasekhar and Fermi (1985), who determined the magnetic field strength from the condition that the gravitational pressure equals the sum of the material and magnetic pressures. However, the root-mean-square velocity of 5 km/sec adopted in their work seems low. The exponential distribution of radial velocities, proportional to $\exp(-v/5)$, found by Blaauw (1952), yields a root-mean-square velocity of $5(2)^{1/2}$ km/sec in one dimension only. If an isotropic velocity distribution is assumed, the three-dimensional root-mean-square velocity, which occurs in the equations of Chandrasekhar and Fermi (1953), becomes $5(6)^{1/2}$, or about 12 km/sec.

With this increased velocity, the kinetic pressure is increased by a factor of 6 and becomes nearly equal to the gravitational pressure. If the difference is attributed to the magnetic pressure, the magnetic field becomes about $3 \times 10^{-6}$ gauss. This determination is quite uncertain, since it depends on the difference between two uncertain and nearly equal quantities. However, if a substantially greater magnetic field is assumed, it is difficult to account for equilibrium. For example, if the magnetic field, $B$, is found from the amplitude of oscillation of the lines of force, following the analysis given in the earlier part of the Chandrasekhar-Fermi paper but utilizing the more realistic value of $v$ found above, $B$ is found to be $3 \times 10^{-5}$ gauss, yielding a magnetic pressure an order of magnitude greater than the gravitational. In this computation $a$, the root-mean-square angular deviation of the lines of force from the direction of the spiral arm, was taken to be 0.13 radian, in accordance with the analysis by Stranahan (1954).

Evidently, the gravitational pressure cannot by itself counterbalance large magnetic pressures. If a magnetic field as great as $10^{-5}$ gauss is assumed, either $B$ must have about the same value both inside and outside the spiral arm, or the gas in the spiral arm must be held in by the pressure of an external gas.

Clearly, the situation will be greatly eased if a spiral arm is assumed to be imbedded in a rarefied gas with a pressure equal to the gas pressure in spiral arms. In such a case there is no difficulty in confining the gas in a spiral arm, especially in view of the possible existence of magnetic fields in the rarefied gas surrounding the spiral arm. Since the magnetic field may actually be less than $10^{-6}$ gauss (see Spitzer 1954), this evidence is certainly not strong, but it is at least suggestive.

### c) Radio Noise

It was suggested by Shklovsky (1952) that much of the cosmic radio noise received on earth might originate in an extended spherical corona surrounding the Galaxy. Support for this suggestion has been obtained by Baldwin (1954) from measures in M31 at a wave length of 3.7 meters. The angular size of the emitting region accompanying this galaxy appears to indicate a spherical corona some 10 kpc in radius, responsible for about two-thirds the total energy radiated by M31 at these wave lengths. Confirming evidence in the case of our own Galaxy has also been obtained by Baldwin (1956). The background of 3.7-meter radiation received at high galactic latitudes has been explained on the assumption of a spherical galactic corona some 10–20 kpc in radius. At most longitudes the observations agree rather well with a theoretical curve computed on this basis. However, the very large deviations observed at about 0° galactic longitude raise some question as to whether lesser irregularities at other longitudes may affect the observed noise levels. Further measurement and analysis are required for conclusive results, but an extended rarefied corona is at least suggested by these data.

Since the radio emission is presumably nonthermal in origin, the radio observations cannot in any case yield values of the density. The interpretation of these radio waves in terms of relativistic electrons moving in a coronal magnetic field, as suggested by Shklovsky (1952), has been discussed by Burbidge (1956). However, the extension of the corona perpendicular to the galactic plane is directly related to the kinetic temperature, and, as we shall see later, a radius between 10 and 20 kpc corresponds to a kinetic temperature of about $10^6$ degrees for ionized H atoms.

### d) Related Suggestion by Pickelner

The possibility of an extended corona of gas has also been discussed by Pickelner (1953, 1955). However, he considers a homogeneous substratum of relatively cool gas, with a density of 0.1 H atoms/cm$^3$ and with turbulent motions of some 70 km/sec, an essentially different picture from the high-temperature corona analyzed here. The chief observational evidence cited by Pickelner is the presence of wide H and K absorption lines, which Spitzer, Epstein, and Li Hen (1950) observed in supergiant B stars and which they attributed to stellar absorption. A re-examination of these data confirms that these wide, shallow absorption features are unquestionably stellar in at least some cases. Of the twelve stars listed by Spitzer, Epstein, and Li Hen as showing this "stellar" absorption, three are known spectroscopic binaries. For two of these, $\chi$ Aur and $\mu$ Sgr, comparison of two plates taken on different dates shows a shift of the sharp interstellar component relative to the wide absorption feature. On the third, $\iota$ Ori, only one plate was available. Furthermore, this wide absorption shows a definite correlation with spectral type, with the strongest absorption in $\chi$ Aur (cB3), 67 Oph (cB7), $\mu$ Sgr (cB8), $\alpha$ Cyg (cA2), and HD 199478 (cB8), and relatively weak absorption for the other eight stars, of types ranging from cB2 to O8. While some of this wide absorption might perhaps be interstellar in origin, rather than truly stellar or possibly circumstellar, the available evidence does not seem to support this possibility.

From a theoretical standpoint, the Pickelner corona seems difficult to maintain. The dissipation of turbulent energy at such a high Mach number would be very rapid, requiring a large source of energy to maintain the motion. While this proposal cannot be excluded, it will not be discussed further in the present paper.

## II. Equilibrium Density

### a) Pressure Equilibrium

In the interstellar medium generally, approximate pressure equilibrium is believed to obtain. The same situation presumably applies to a very hot interstellar corona. Any deviations from hydrostatic equilibrium will produce pressure disturbances traveling at the local sound velocity. At a temperature of $10^6$ degrees, a sound wave moves at 170 km/sec and travels 10000 pc in about $6 \times 10^7$ years. Unless unsuspected additional influences are operative, hydrostatic equilibrium within a corona should be a realistic assumption. We shall assume that pressure equilibrium also exists between such a corona and the normal interstellar medium.

On this assumption the density in the corona is a known function of the temperature. In the normal obscuring H I cloud, the density and temperature are generally assumed to be 10 H atoms/cm$^3$ and 100° K, respectively. Hence, from the familiar gas law and equality of pressure, we obtain for $n_e$ and $n_p$, the density of electrons and protons in the corona,

$$n_e = n_p = \frac{5 \times 10^2}{T}. \tag{1}$$

At the high temperature envisaged, hydrogen will be nearly completely ionized. At a coronal temperature of $10^6$ degrees, $n_e$ is $5 \times 10^{-4}$ atoms/cm$^3$.

### b) Extension Perpendicular to Galactic Plane

The extension of the corona perpendicular to the galactic plane may also be determined from the temperature, if isothermal conditions are assumed. We compute the "coronal height," $z_c$, defined as the value of $z$ at which the density is $1/e$ times its value in the galactic plane, at $z$ equal to zero. In hydrostatic equilibrium we have

$$\frac{d \log n(z)}{dz} = -\frac{mK(z)}{kT}, \tag{2}$$

where $m$ is the mean mass per particle, here equal to half the mass of an H atom, and $K(z)$ is the gravitational acceleration in the $z$ direction, perpendicular to the galactic plane, as a function of $z$, the distance from the galactic plane. We use c.g.s. units in these equations. Equation (2) may be integrated to yield

$$\log \frac{n(z)}{n(0)} = - \frac{\psi(z)}{T},$$  (3)

where

$$\psi(z) \equiv \frac{m}{k} \int_0^z K(z)\, dz.$$  (4)

Evidently $z_c$ is the solution of the transcendental equation

$$\psi(z_c) = T.$$  (5)

Values of $K(z)$ for $z$ up to 500 pc have been determined by Oort (1932). For values of $z$ greater than 4000 pc, Oort computes $K(z)$ from the attraction of a central galactic nucleus containing $1.7 \times 10^{11}$ solar masses and 10000 pc away. With Oort's values of $K(z)$, the values of $\psi(z)$ in Table 1 have been obtained; although the unit of distance in equation 4 is taken to be the centimeter, for convenience in the following tables $z$ is given in parsecs. The values of $z_c$ obtained from equation (5) are given in Table 2, together with the values of the product $n_e z_c$, where the value of $n_e$ is taken at $z$ equal to zero. For $T$ less than $10^5$ degrees K, the corona is restricted to $z$ values of less than 1700 pc. In this range, $K(z)$ is produced by the local material in the galactic plane and is nearly constant for $z$ between 400 and 2000 pc. As

TABLE 1
Values of $\psi(z)$

| $z$(pc) | 100 | 150 | 200 | 400 | 600 |
|---|---|---|---|---|---|
| $\psi(z)$ | $2.18 \times 10^3$ | $4.70 \times 10^3$ | $7.84 \times 10^3$ | $2.18 \times 10^4$ | $3.66 \times 10^4$ |
| $z$(pc) | 1000 | 1500 | 2000 | 4000 | 6000 |
| $\psi(z)$ | $6.85 \times 10^4$ | $1.13 \times 10^5$ | $1.62 \times 10^5$ | $4.06 \times 10^5$ | $7.22 \times 10^5$ |
| $z$(pc) | 10000 | 15000 | 20000 | 40000 | 60000 |
| $\psi(z)$ | $1.40 \times 10^6$ | $2.08 \times 10^6$ | $2.56 \times 10^6$ | $3.48 \times 10^6$ | $3.83 \times 10^6$ |
| $z$(pc) | 100000 | 200000 | 400000 | 1000000 | $\infty$ |
| $\psi(z)$ | $4.12 \times 10^6$ | $4.30 \times 10^6$ | $4.46 \times 10^6$ | $4.53 \times 10^6$ | $4.58 \times 10^6$ |

TABLE 2

Extension of Corona Perpendicular to Galactic Plane

| $T(°K)$ | $3 \times 10^4$ | $10^5$ | $3 \times 10^5$ | $10^6$ | $3 \times 10^6$ | $4.6 \times 10^6$ |
|---|---|---|---|---|---|---|
| $z_c(pc)$ | 520 | 1400 | 3200 | 7500 | 27000 | $\infty$ |
| $n_e z_c (cm^{-2})$ | $2.7 \times 10^{19}$ | $2.2 \times 10^{19}$ | $1.6 \times 10^{19}$ | $1.2 \times 10^{19}$ | $1.4 \times 10^{19}$ | $\infty$ |

a result, $z_c$ varies almost linearly with $T$, and $n_e z_c$ is nearly constant. For greater $z$, the attraction of the central bulge becomes important. For $z$ greater than 10000 pc, $K(z)$ falls off about as $1/z^2$, and $\psi(z)$ approaches a finite value as $z$ approaches infinity. As a result, $z_c$ becomes infinite at a temperature of $4.6 \times 10^6$ degrees K.

For $T$ much greater than $10^6$ degrees, the corona cannot, in fact, be held in by the gravitational attraction of the Galaxy. Analysis of isothermal conditions is somewhat idealized, unless we make the rather drastic assumption that this hypothetical high-temperature gas pervades all intergalactic space. While an intergalactic medium can certainly not be excluded at the present time, there seems little to be gained by considering such a possibility in the complete lack of any observational evidence bearing on the subject. We shall assume, then, that the density of the galactic corona approaches zero for sufficiently large $z$. In the equilibrium configuration, evaporation of coronal protons and electrons into intergalactic space occurs from some limiting surface, and this evaporation keeps the outer surface cool. The gravitational attraction of the Galaxy on this cool layer then provides the pressure needed to confine the inner isothermal regions. If the inner isothermal region of the corona were to be at a temperature of $4 \times 10^6$ degrees, the mass of the cool outer region would have to be comparable with that of the inner region; in fact, it may be shown that the mean temperature of all the coronal atoms cannot much exceed $10^6$ degrees. We shall take $3 \times 10^6$ degrees as a generous upper limit on the coronal temperature.

The total mass of such a corona can be determined from the data in Table 2. For $T$ no greater than $3 \times 10^6$ degrees, the values of $n_e z_c$ do not change rapidly with $T$, and average about $2 \times 10^{19}/cm^2$, corresponding to a mass of about $3 \times 10^{-5}$ gm/cm$^2$, or $6 \times 10^{-5}$ gm in a column 1 cm square in cross-section extending all the way through the Galaxy. For comparison, the total gravitating mass in a column through the galactic plane, as determined by Oort, is about $10^{-2}$ gm/cm$^2$, roughly half of which may be attributed to the interstellar material near the galactic plane. When account is taken of the concentration of this cool gas in spiral arms, we find that a few percent of the total mass of the interstellar material in the Galaxy may be in the corona. If the possibility of much higher coronal densities near the galactic nucleus is ignored, the total mass of the corona is about $10^8 \, M_\odot$.

### c) Mean Free Path and Radius of Gyration

Another quantity which is a simple function of temperature is the mean free path, $l$. This quantity, about the same for electrons as for protons, may be set equal to the root-mean-square velocity, multiplied by the self-collision time, $t_c$, discussed by Spitzer (1956), following the analysis by Chandrasekhar (1942). For the densities and temperatures of interest, the logarithmic term occurring in these expressions is about 30, and we have, approximately,

$$l = \frac{2.0 \times 10^{-15} T^2}{n_e} \text{ pc.} \tag{6}$$

Combining equations (1) and (6), we obtain

$$l = 4.0 \times 10^{-18} T^3 \text{ pc.} \tag{7}$$

We see that the mean free path rises very rapidly with $T$. Even at the maximum temperature of $3 \times 10^6$ degrees K, $l$ is only 108 pc, a small fraction of the coronal extension.

The radius of gyration, $a$, is likely to be much less than the mean free path. For protons at $10^6$ degrees, moving in a magnetic field of $10^{-5}$ gauss, the root-mean-square radius of gyration, $a$, is about $1.3 \times 10^8$ cm, or about $4 \times 10^{-11}$ pc. Even with a vastly smaller magnetic field, the particles would gyrate about the magnetic lines of force many times between collisions. At $10^{-15}$ gauss, for example, $a$ is 0.4 pc for protons and $10^{-2}$ pc for electrons, both at $10^6$ degrees.

### d) Optical Observability

Finally, we discuss how material at this density might be detected by direct optical means. W have already seen that $N$, the number of coronal electrons, or protons, in the line of sight per square centimeter, is about $2 \times 10^{19}$, a rough average of the values of $n_e z_c$ given in Table 2. This value of $N$ is about the number of atoms in the line of sight through 1 cm of air or through the inner corona of the sun, viewed tangentially to the solar limb. The total optical thickness for electron scattering is only $1.3 \times 10^{-5}$, probably much too small to be observed.

While an interstellar corona should emit the same type of radiation as the solar corona, the rate of emission per atom is reduced in about the same ratio as the density, or by a factor in the general neighborhood of $10^{-11}$. It is evident that the intensity of the coronal emission, when reduced by so large a factor, is entirely unobservable.

It may be noted that the Balmer lines would be unobservable even if the recombination coefficient were independent of the temperature. The emission measure of the interstellar corona, defined as $n_e^2 z_c$, where $n_e$ is in cm$^{-3}$ and $z_c$ is in parsecs, is at most 0.1. For H II regions in the Galaxy an emission measure of about 100 is the least that can be detected. Similarly, the thermal radio emission, generated by free-free transitions, is entirely unobservable.

While the interstellar corona can apparently not be observed by its emission spectrum, the number of atoms is sufficient to produce a measurable absorption line. A column of interstellar gas containing $10^{12}$ sodium atoms/cm$^2$ produces measurable D lines. The difficulty with this technique in the case of the solar corona is that at a high kinetic temperature the atoms will mostly be highly ionized, and the ultimate lines will all be in the far ultraviolet. If the state of ionization is like that in the solar corona, the ultimate resonance doublet of Mg x, at 609 and 625 A, should be extremely strong, since the number of magnesium ions in the line of sight in all stages of ionization would be about $10^{16}$/cm$^2$; if 1 percent of these atoms were in the ionized state Mg x, this doublet would be heavily absorbed. However, it is uncertain whether much radiation shortward of the Lyman limit can reach the earth, in view of probable heavy absorption by neighboring H I clouds. The ion O VI, which is isoelectronic to Mg x, has a corresponding doublet at 1038 and 1032 A and might be sufficiently abundant to produce measurable absorption, especially since the ionization potential of O VII has the relatively high value of 739 volts. Similarly, the ultimate lines of N v and C IV, at about 1240 and 1550 A, respectively, might be observable. It would appear that, in principle, an interstellar corona could be detected and analyzed by means of spectroscopic measures from a satellite.

## IV. Problems of Dynamics and Cosmology

The possible existence of an interstellar corona in the Galaxy poses a number of additional problems, besides those discussed in the preceding sections. The more important of these are the following: the possible origin of such a corona, the possible methods of heating, the types of motion possible in such a medium, and the role of such a medium in the formation of spiral arms. These problems cannot be explored with the same definiteness as has been possible for the density distribution and rate of cooling. A brief preliminary exploration of those topics is presented here.

### a) Heating

The problem of origin of such a corona reduces primarily to the problem of heating. Considerable amounts of gas would presumably have been left originally at great distances from the galactic plane when the stars were formed. Also, evaporation of gas from the relatively cool clouds near the plane would provide a steady source of material for such a corona. Hence the central problem in explaining the existence of such a corona is to account for its high temperature.

As we have already seen, the simplest hypothesis is that the corona is itself entirely primeval and was formed at a high temperature some $5 \times 10^9$ years ago. This explanation is possible only if the coronal temperature is substantially in excess of $10^6$ degrees, and perhaps too great to be consistent with confinement of the corona by the gravitational field of the Galaxy. Since the values in Table 2 are somewhat uncertain, however, the hypothesis of a primeval hot corona which has persisted to the present time cannot be excluded.

If the corona is not assumed to be primeval and its temperature is taken to be not appreciably greater than $10^6$ degrees, then the coronal gases must be thermal equilibrium, with the energy gains balancing the losses. We may readily compute $L_{cor}$, the total luminosity of the corona, which must then be balanced by some energy source. If the coronal temperature is taken to be $10^6$ degrees, the cooling time, according to Table 3, is $1.3 \times 10^6$ years. We neglect cooling by conduction, which would reduce slightly the cooling time at this temperature if a magnetic field were entirely absent. Since we have already seen that the mass of the corona may be set equal to $10^8 \, M_\odot$, we obtain

$$L_{cor} = 1.2 \times 10^{39} \text{ erg/sec} = 3.2 \times 10^5 \, L_\odot. \tag{24}$$

This value is a small fraction of the total luminosity of the Galaxy.

We compare $L_{cor}$ with the energy radiated by the cooler interstellar matter in the galactic plane. The H II regions radiate much of the energy emanating from O stars beyond the Lyman limit, which, for the Galaxy as a whole, probably amounts to about 1 percent of the total galactic luminosity, or about $10^8 \, L_\odot$ in order of magnitude. As a check, we compute the energy radiated if 1 percent of the volume of the Galaxy, with an assumed thickness of 200 pc and a radius of 10000 pc, is assumed to contain ionized hydrogen at a temperature of 10000° K and a density of 1 proton/cm$^3$. On capture by a proton, an electron radiates about 15 ev, or $2.4 \times 10^{-11}$ erg. The number of such captures per second at $10000°$ K is $4.3 \times 10^{-13} n_e^2$ (Spitzer 1956). The total energy radiated is $2 \times 10^{41}$ ergs/cm$^3$, or $5 \times 10^7$ $L_\odot$. Evidently, the energy radiated by an interstellar corona at $10^6$ degrees is some two orders of magnitude less than that radiated by the familiar H II regions of the Galaxy.

On the other hand, the energy dissipated by the motions of the interstellar gas is much less. According to Kahn (1955) and Seaton (1955), this energy goes into the H I regions and is mostly responsible for the energy radiated from these regions. The rate of energy dissipation in this process is obtained from the density of kinetic energy and the dissipation time, which in this case is the time between collisions. In the normal interstellar gas, moving with a root-mean-square velocity of 12 km/sec (see Sec. I) and with a mean density of 1 H atom/cm$^3$, averaged over the clouds and the intercloud regions, the density of kinetic energy is $3.6 \times 10^{43}$ ergs/pc$^3$. If collisions between clouds are assumed to dissipate this energy in $10^7$ years and the interstellar material is assumed to fill one-third of a volume 200 pc thick and $10^4$ pc in radius, the rate of dissipation of kinetic energy is $2.4 \times 10^{39}$ ergs/sec, or $6 \times 10^5$ $L_\odot$. Evidently, the radiation of energy from the corona appears to be of the same order of magnitude as the dissipation of kinetic energy by the low-temperature clouds.

Several mechanisms might provide this required energy for heating a galactic corona. One possibility is that compressional waves, produced by moving clouds in the galactic plane, become converted to shock waves in the rarefied gas above and convert their energy into heat. This mechanism has been suggested by Biermann (1946) and Schwarzschild (1948) as a possible means for heating the solar corona. In the interstellar case the ratio of cloud velocity to sound velocity is much greater than in the solar atmosphere, and a greater efficiency of transfer from directed kinetic energy to heat energy may be expected. More information is needed to indicate whether or not this mechanism can actually provide the necessary energy.

Direct heating by stellar radiation seems improbable. Visible light cannot directly produce high temperatures. More energetic photons, with several hundred electron volts of energy, could readily heat a gas to $10^6$ degrees. However, the total flux of radiant energy required would be rather great, since at these frequencies the optical thickness of the corona would not exceed $10^{-3}$ and the required galactic luminosity at wavelengths below 100 A would exceed $10^8$ $L_\odot$. There is no known indication of so great an X-ray flux from the Galaxy.

A mechanism that seems considerably more promising is the high-speed ejection of material from stars. Replenishment of the interstellar medium by stellar ejection has been discussed by Thackeray (1948) and Bierman (1955). According to the authors, the bulk of the emission comes from the early-type stars, which presumably lose matter both by outward motion of the atmosphere at velocities of about 10–100 km/sec and by corpuscular emission at about 1000 km/sec. Much of the emitted gas will presumably strike normal cool clouds in the galactic plane and lose its kinetic energy by increasing slightly the temperature of these clouds. However, early-type stars near the edge of a spiral arm or at considerable distances above the galactic plane may be expected to emit some material which will leave the galactic plane without first striking any substantial amounts of cool gas. This material, moving through the galactic corona, will transfer its kinetic energy to the random thermal energy of the corona.

In addition to the O and B stars, which are primarily confined to the spiral arms, some emission occurs from type II stars, which may already be at considerable distances from the galactic plane. Supernovae, which account for perhaps 1 percent of the total matter ejected from stars, are believed to occur among stars of both populations types.

The energy flow of $3 \times 10^5 \ L_\odot$ required by equation (24) would be produced by a mass of 0.01 $M_\odot$ per year ejected into the corona at a velocity of 600 km/sec. This rate of mass ejection is about that attributed to supernovae and, according to Biermann (1955), is two orders of magnitude less than the probable rate of corpuscular emission from early-type stars. While again no definite conclusions can be drawn at present, this mechanism appears to be a very promising means of keeping a corona hot.

A related source of heating is by means of cosmic-ray particles, with energies of several billion volts per nucleon. However, the results obtained by Spitzer (1948) on heating by cosmic rays may be used to show that this mechanism cannot lead to high temperatures if the flux and spectrum of the energetic particles are the same as observed on earth. We infer that cosmic rays are probably not responsible for heating the corona.

### b) Motions

While the motions of a possible corona are even more hypothetical than the existence of such a system, certain properties of these motions can be explored theoretically. In the unlikely case that the magnetic field is everywhere much less than $10^{-18}$ gauss, electrons and protons move in nearly straight lines over distances comparable with a parsec (see eq. [7]). Any turbulence in this case must have a scale much greater than 1 pc. The motion of a cool cloud through such a medium is difficult to analyze, since the cloud radius may be comparable with the mean free path. The one simplification is that the motions are probably subsonic, since the observed cool clouds are mostly moving less rapidly than the probable sound velocity, which in ionized hydrogen at $10^6$ degrees amounts to 170 km/sec. Heat conduction from the corona to the cloud will form a transition layer surrounding the cloud, and the motions of this layer will complicate the dynamics.

It seems likely that the magnetic field will exceed $10^{-15}$ gauss, in which case radii of gyration of both electrons and protons will be less than 1 pc. In this situation the effective mean free path transverse to the magnetic field is now the radius of gyration, and the gas behaves like a conventional fluid, except that inhomogeneities along the line of force tend to be eliminated rapidly. In this case the gas will flow smoothly around a cool cloud moving through it, with possible generation of turbulent motion in its wake. Conduction of heat to the cool cloud will presumably be negligible. A detailed solution of this hydromagnetic problem will depend, of course, on the geometry of the lines of force, especially if the magnetic field, $B$, is assumed to be so large that the magnetic energy density, $B^2/8\pi$, becomes comparable with the kinetic energy density, $\frac{1}{2}\rho v^2$.

## c) Condensation of Spiral Arms

One interesting feature of a possible galactic corona is that it may provide a mechanism for the formation of new spiral arms. Such a mechanism may be needed to explain why the observed clouds of gas and dust have a spiral structure rather than a circular one. For example, a cloud of gas in the solar neighborhood becomes so distorted by the nonuniformity of galactic rotation that in $5 \times 10^9$ years it becomes a lane with an inclination of only 0.3° to a circular path; i.e., its distance from the galactic center increases by only 5 pc in 1000 pc of distance along the lane. While the 21-cm data obtained by van de Hulst, Oort, and Muller (1954) do, in fact, indicate a nearly circular lane of neutral hydrogen stretching almost a quarter of the way around the Galaxy, the measures suggest an inclination of at least 1°. The outer arms in M31 are also nearly circular, but the data are not sufficiently precise to reveal an inclination of 1° or less.

While the evidence is still insufficient for definite conclusions, the present configuration of spiral arms in our Galaxy and in M31 appears consistent with the hypothesis that from time to time enormous complexes of relatively cool gas are formed in the galactic plane and become sheared by differential galactic rotation into the familiar spiral arms. The gas then condenses both into grains and into new stars, many of which are of early type. In time, as some of the gas condenses into stars, the remainder is swept away in some way, and the arm disappears, its place being taken by a newly formed arm.

The corona may perhaps be tentatively identified as the source of material from which the new arms condense and into which the remnants of old arms evaporate. How can the corona be expected to condense into a cool mass of gas? We have already seen that under some conditions the rate of radiation might be expected to rise with decreasing temperature, with resulting thermal instability. Under such conditions some of the corona would presumably cool and contract, with the remaining coronal gases expanding to take the place of the gas that had contracted. A more detailed study of the ions and their radiating energy levels would be required to indicate whether or not such a mechanism could occur and whether most of the corona would be thermally stable most of the time.

One difficulty with this hypothesis is that the total mass of the corona is at least an order of magnitude less than the total mass of cooler gas. Thus condensation of the entire corona would scarcely provide the material for one spiral arm. However, the condensation will presumably be a gradual process, and, if new material streams into the corona as the condensation proceeds, a spiral arm may be formed with much more material than is in the corona at any one time. Another difficulty is that, if much material is to return to the corona from old spiral arms, the rate of heating must probably be increased by at least an order of magnitude over the estimates given previously. In view of these difficulties and the many uncertainties in this field, the role of an interstellar corona in the evolution of a galaxy is somewhat hypothetical but seems of potentially great importance.

## REFERENCES

Baldwin, J. E., 1954, *Nature*, **174**, 320.
——, 1956, *M.N.*, in press.
Biermann, L., 1946, *Naturwiss.*, **33**, 118.
——, 1955, *Gas Dynamics of Cosmic Clouds: Proceedings of the Cambridge Symposium in July, 1953*, chap. 39, p. 212.
Blaauw, A., 1952, *B.A.N.*, **11**, p. 459 (No. 436).
Burbidge, G. R., 1956, *Ap. J.*, **123**, 178.
Chandrasekhar, S., 1942, *Principles of Stellar Dynamics* (Chicago: University of Chicago Press), chap. ii.
Chandrasekhar, S. and Fermi, E., 1953, *Ap. J.*, **118**, 113.
Cillié, G., 1932, *M.N.*, **92**, 820.
Hulst, H. C. van de, Muller, C. A., and Oort, J. H., 1954, *B.A.N.*, **12**, 177 (No. 452).

Kahn, F. D., 1955, *Gas Dynamics of Cosmic Clouds: Proceedings of the Cambridge Symposium in July, 1953*, chap. 12, p. 60.

Münch, G., 1956, in preparation.[†]

Oort, J., 1932, *B.A.N.*, **6**, 249 (No. 238).

Pickelner, S. B., 1953, *C.R. Acad. Sci. URSS*, **88**, 229; *Pub. Crimean Ap. Obs.*, **10**, 74.

——, 1955, *Mém. Soc. R. Sci. Liége*, **4-15**, 595.

Schwarzschild, M., 1948, *Ap. J.*, **107**, 1.

Seaton, M. J., 1955, *Ann. d'ap.*, **18**, 206.

Shklovsky, I. S., 1952, *Asir. J., U.S.S.R.*, **29**, 418.

Spitzer, L., 1942, *Ap. J.*, **95**, 329.

——, 1948, ibid., **107**, 6.

——, 1954, ibid., **120**. 1.

——, 1956, *Physics of Fully Ionized Gases* (New York: Interscience Publishers).

Stranahan, G., 1954, *Ap. J.*, **119**, 465.

Thackeray, A. D., 1948, *Observatory*, **68**, 22.

Woolley, R. v. d. R. and Allen, C. W., 1948, *M.N.*, **108**, 292.

[†] Published with H. Zirin, *Ap. J.* **133**, 11, 1961.

# ABSORPTION LINES PRODUCED BY GALACTIC HALOS

(WITH J. N. BAHCALL)

(ASTROPH. J. LETT. 156, L63, 1969)

## *Commentary*

IN a scientific discussion during 1968, John Bahcall pointed out to me that if the maximum radii of galaxies were taken to be some 10 to 20 kpc, as suggested by radio and visual observations, the number of galaxies intersected by a line of sight to a distant quasar (at $z \approx 2$, for example) would be less by at least an order of magnitude than the number of separate absorption-line systems seen in the spectrum of a typical quasar. I was familiar with at least the concept of extended gaseous halos around galaxies (see Paper #7), and our conversation led to the suggestion that the extent of such halos might be great enough to account for the observed narrow lines in quasar spectra. Paper #8 appeared shortly afterwards.

With *Copernicus* observations of highly ionized atoms several years in the future, this suggestion was evidently very speculative. The large galactic radii required were not suggested by any direct observational evidence, but seemed physically possible.

These narrow absorption features observed in quasar spectra are generally found in "systems," each a group of lines all with the same redshift, $z$, which is significantly less than $z$ for the quasar emission lines. Paper #8 referred primarily to "metal-line systems," characterized not only by L$\alpha$ of neutral H but also by lines of the more abundant of the heavier elements, such as N, C, Mg, Si, Al, and Fe. In these systems a wide range of ionization states are usually present, including Si II and C IV, for example. Since 1969 extensive information has accumulated on absorption by such atoms in the halo of our Galaxy, and on identification of the specific galaxies responsible for absorption in metal-line systems.

As regards the halo of our own Galaxy, observations with the International Ultraviolet Explorer satellite have shown[1] that highly ionized atoms such as $C^{3+}$ are present at kiloparsec distances from the midplane. Observations of a star in the Large Magellanic Cloud[2] have taken advantage of differential galactic rotation to separate halo absorption at a vertical distance of more than 500 pc from absorption by neighboring disk gas. This absorption by the halo, without the disk, shows[2] marked similarity with absorption in typical metal-line systems. While the correspondence is more qualitative than quantitative, it substantially strengthens the picture proposed in Paper #8.

Detailed identification of distinct galaxies responsible for specific metal-line absorption systems in quasars has been carried out systematically for the pair of Mg II lines at 2800 Å.[3] These features can be observed from the ground if $z > 0.12$. In a sample of 17 Mg II systems with $0.16 < z < 0.90$, 13 of these each had an adjacent galaxy with the same redshift as for the Mg II system, and each was, in all cases, a reasonably bright object. The projected distance of the quasar line of sight from the galactic center varied from $8/h$ to $41/h$ kpc, where $h$ is the Hubble constant in units of 100 km s$^{-1}$ Mpc$^{-1}$. Spherical halos, all with radii of about $40/h$ kpc, provide a good fit with the distribution of distances. The resultant cross-section, multiplied by the number density of galaxies, was found to be consistent with the observed $dN/dz$, the number of such systems per unit $z$. As a result of this and similar studies, the working hypothesis of absorption by extended halos is now widely accepted for formation of metal-line systems.

While the picture advanced in Paper #8 is supported by this recent evidence, some uncertainties remain. Observations toward quasars which show metal-line systems also show[4] in some cases irregular [O II] emission extending from $60/h$ to $270/h$ from the quasar line of sight. The emitting gas, possibly distributed in dwarf irregular galaxies where active star formation is under way, or possibly extending throughout a cluster of galaxies, might contribute enough atoms along the line of sight to produce some of the observed metal-line absorption of quasar light.

There are uncertainties also for these metal-line systems as regards other quantitative details, including[5] the dependence of halo absorption cross-section on atomic species and on red shift (i.e., on age). Observations with the Hubble Space Telescope (HST) can measure most of the relevant atomic species, including Si IV and C IV, at low $z$, where it is easier to identify stellar systems in which the absorbing gases are located. Such data should help[6] to resolve some of these uncertainties and to gain important information, including the distribution and physical properties of gas within galaxies and the intensity of the metagalactic radiation. Larger telescopes on the ground, now obtaining high spectoscopic resolution on very distant quasars, offer the exciting prospect of extending these detailed studies of gaseous galactic halos to much earlier epochs of the Universe.

The origin of L$\alpha$ systems, which make up the "L$\alpha$ forest" observed at $z > 2$, is as yet quite uncertain, despite extensive research.[7] There is evidence that some of these systems, which have a relatively low H column density, are produced in even more extended galactic halos, with effective radii up to $160/h$ kpc.[8] On the other hand, the clustering tendency of the L$\alpha$ clouds at large $z$ is weaker than that of galaxies,[9] suggesting that the L$\alpha$ absorption is produced in separate clouds, not directly associated with any galaxy; such clouds might be confined in pressure equilibrium with an intergalactic medium or might be sub-galactic structures in various stages of gravitational collapse.[10] Again, further HST observations should help to decide between these and other possibilities.

## REFERENCES

1. Savage, B. M.—In: Interstellar Processes, eds. D. J. Hollenbach and H. A. Thronson, Jr., Dordrecht: Reidel 123 (1959).
2. Savage, B. D. and Jeske, N. A.—*Astroph. J.* **244**, 768 (1981).
3. Bergeron, J. and Boissé, P.—*Astron. Astroph.* **243**, 344 (1991).
4. Yanny, B. and York, D. H.—*Astroph. J.* **391**, 569 (1992).
5. Lanzetta, K. M.—In: The Environment and Evolution of Galaxies, eds. J. M. Shull and H. A. Thronsen, Jr., Dordrecht: Kluwer 237 (1993).
6. Bahcall, J. N.—In: Scientific Research with the Space Telescope, IAU Colloquim #54, eds. M. S. Longair and J. W. Warner, NASA CP-2111, p. 215 (1979).
7. QSO Absorption Lines, Probing the Universe, eds. J. C. Blades, D. A. Turnshek, and C. A. Norman, Cambridge: University Press (1988).
8. Lanzetta, K. M., Bowen, D. V., Tytler, D., and Webb, J. K.—*Astroph. J.*, **442**, 538 (1995).
9. Sargent, W.L.W., Young, P. J., Boksenberg, A., and Tytler, D.—*Astroph. J. Supp.* **42**, 41 (1980).
10. Cen, R., Miralda-Escudé, J., Ostriker, J. P., and Rauch, M.—*Astroph. J. Lett.* **437**, L9 (1994).

## Paper

ABSTRACT

We propose that most of the absorption lines observed in quasi-stellar sources with multiple absorption redshifts are caused by gas in extended halos of normal galaxies.

Recent work has established that some quasi-stellar sources have multiple redshift systems in absorption (Bahcall 1968; Bahcall, Greenstein, and Sargent 1968; Burbidge, Lynds, and Stockton 1968; Burbidge 1969; Bahcall, Osmer, and Schmidt 1969). A number of possible explanations have been suggested for this phenomenon (Bahcall et al. 1968; Burbidge et al. 1968; Peebles 1968), but none of the suggestions seem especially plausible when considered in the light of the observed features of the absorption systems. We propose that most of the absorption lines are caused by tenuous gas in extended halos of normal galaxies (see Spitzer 1956 for a review of some earlier work on galactic halos and for a preliminary discussion of the possibility of observing ultraviolet absorption lines formed in such halos).

Our proposal for explaining the observed absorption lines differs from previous suggestions involving galaxies (Wagoner 1967; Shklovsky 1967; Peebles 1968) in that we assume a much larger cross-sectional area for individual galaxies than is indicated by their optical or radio appearance and present a more detailed interpretation of the principal observational features. The probability of intercepting a galaxy along the line of sight is sufficiently large, if galaxies have extended halos, to explain the observations with normal galaxies in standard ($\Lambda = 0$) cosmologies. We first review the observed features of the absorption systems and then show how these features could be caused by galactic halos. We then indicate a possible test of our ideas that could be carried out by means of a satellite-based telescope.

The principal characteristics of the observed absorption systems in PKS 0237-23 and Ton 1530 (see Bahcall 1968; Bahcall et al. 1969) are: (1) for absorption redshifts in the detectable range $z_{abs} \sim 3$ to $z_{abs} \sim 1$ about five absorption systems are found; (2) $z_{em} > z_{abs}$; (3) the special-relativistic velocities $v_{rel}$ between emitter and absorbers range from 0.01 to 0.3; (4) the dispersion in velocities for an individual absorption line is $v_{disp} \sim 10^{-4 \pm 0.5}c$; (5) a wide variety of ionization states of the most abundant elements occurs (they include H I, C II, C IV, Si II, Si III, Si IV, N II, N V, Al II, Al III, and Fe II), but the resonance lines of O I and N I, characteristic of H I regions, are not found; and (6) no lines from metastable levels (e.g., He II $\lambda1640.4$ or C III $\lambda1175.5$) or from excited fine-structure states are definitely identified. In obtaining characteristic 4, we have made use of the fact that even completely dark lines less than about 0.3 Å in width could not have been detected.

The observed large values for the dimensionless ratio $v_{\text{rel}}/v_{\text{disp}}$, which sometimes are greater than $10^3$, lead us to believe that most of the narrow absorption lines are produced in material not associated with the quasi-stellar sources in whose spectra they are detected. A few lines, which correspond to $v_{\text{rel}}/c \lesssim 10^{-2}$, may plausibly be attributed to material associated with the emitting object.

The number $P$ of galaxies intercepted between $z_{\text{em}}$ and $z_c$ (see Bahcall and Peebles 1969; we assume for simplicity $q_0 = \frac{1}{2}$ and $\Lambda = 0$) is

$$P = 2\left[\frac{R_0}{100 \text{ kpc}}\right]^2\left[\frac{N_0}{0.03 \text{ galaxy Mpc}^{-3}}\right]\left[(1 + z_{\text{em}})^{3/2} - (1 + z_c)^{3/2}\right].$$

$$(1)$$

Here $R_0$ and $N_0$ are the local radius and number density of the galaxies, and the Hubble constant, $H_0$, has been set equal to 100 km sec$^{-1}$ Mpc$^{-1}$. We assume that the galactic number density at $z$ satisfies $N(z) = (1 + z)^3 N_0$ but that the average radius is independent of $z$ (at least in the range $z = 1-2$). The value of 0.03 galaxy Mpc$^{-3}$ used as a standard of reference in equation (1) is somewhat arbitrary and is three times smaller than the density of all galaxies (cf., e.g., van den Bergh 1961) if the mean mass per galaxy is taken to be $10^{11}$ $M_\odot$. The reference radius, $R_0$, of 100 kpc is an order of magnitude greater than the galactic radii normally quoted and even exceeds the radius of 50 kpc suggested for the somewhat controversial halo around M31, observed at 158 and 750 MHz, respectively, by Brown and Hazard (1959) and by de Jong (1965). The maximum radii found for M31 by photoelectric photometry and by 21-cm-emission measurements are even smaller; both methods give about 25 kpc according to de Vaucouleurs (1958) and Roberts (1968). However, all these measures represent not a boundary but rather the radius at which the signal produced by the galaxy becomes indistinguishable from the background. The gas density required to produce absorption lines is so low (see below) that it would not be surprising if the maximum radius for absorption of a measurable line were appreciably greater than the maximum radius detected in other ways.

Applying equation (1), we assume $z_{em} = 2.2$ and a reasonable cutoff in detectable redshifts $z_c \sim 1.2$, causing the relative paucity of long-wavelength resonance lines (see Table 1 of Bahcall 1968). We then find $P \sim 5$, in agreement with characteristic 1, if $R_0^2 N_0 \approx (100 \text{ kpc})^2$ (0.03 galaxy $\text{Mpc}^{-3}$). Characteristics 2 and 3 are automatically satisfied by any model in which the absorption is caused by objects between us and the emitter but at cosmological distances from the emitter. Assuming that the internal energies and gravitational potential are comparable, we find $v_{disp} \approx (GM/R)^{1/2}c$, or $v_{disp} \sim 10^{-4}c$ for $R = 10^2$ kpc and $M = 10^{11} M_\odot$. This result is in agreement with characteristic 4. The wide variety of ionization stages seen suggests that either the halo is not isothermal (and that temperatures in the range $2 \times 10^{4°}$ to $2 \times 10^{5°}$ K occur) or else ionization by cosmic rays produces a variety of stages of ionization at any given place. The highly ionized atoms that might be present at temperatures significantly higher than $2 \times 10^{5°}$ K (e.g., O VI $\lambda\lambda 1032.0$, 1037.6) could not have been observed because their resonance wavelengths are detectable only for $z > z_{em}$. Any density less than $10^3$ particles per cubic centimeter is sufficiently small that characteristic 6 is satisfied (see Bahcall 1967).

For the sake of definiteness we assume that the halo has a radius of 100 kpc at a density of $3 \times 10^{-5}$ particle $\text{cm}^{-3}$. A typical column of material is then, in agreement with observation, sufficient to produce dark absorption lines for the allowed transitions of the more abundant ions but not for rarer ions such as Sc III, Ti III, Cu II, Mn II, Ge III, and As III. Numerically, for a line of wavelength about 1500 Å,

$$\tau_{abs} \sim 10^{+6} f_{abs} \left[ \frac{N_{ion}}{N_{total}} \right] \left[ \frac{R_0}{10^2 \text{ kpc}} \right] \left[ \frac{N_{total}}{3 \times 10^{-5} \text{ cm}^{-3}} \right]. \qquad (2)$$

The total mass contained in the halo is $\sim 3 \times 10^6 M_\odot$. Evidently all these numerical estimates are tentative.

Finally, we remark that the existence of a halo of the kind we have postulated around our own and nearby galaxies could be detected with a satellite telescope, which could measure interstellar absorption lines in the region from 1000 to 3000 Å. The OAO C telescope (Rogerson 1963) should be able to make such measures on a few "runaway B stars," some distance from the galactic plane. To reach the early-type stars in the Magellanic Clouds would require a more powerful instrument, which could obtain spectra with a resolution of 0.3 Å on objects as faint as magnitudes 12–14. Later telescopes in the OAO series (apertures of about 40 inches) could have this capability. The potential usefulness of a 120-inch space telescope for research on galactic halos has been discussed elsewhere (Spitzer 1968). Information obtained from such programs could be of central importance for galactic studies, quite apart from its relevance to quasi-stellar sources.

## REFERENCES

Bahcall, J. N., 1967, *Ap. J. (Letters)*, **149**, L47.

——, 1968, *Ap. J.*, **153**, 679.

Bahcall, J. N., Greenstein, J. L., and Sargent, W.L.W., 1968, *Ap. J.*, **153**, 689.

Bahcall, J. N., Osmer, P., and Schmidt, M., 1969, *Ap. J. (Letters)*, **156**, L1.

Bahcall, J. N. and Peebles, P.J.E., 1969, *Ap. J. (Letters)*, **156**, L7.

Bergh, S. van den, 1961, *Zs. f. Ap.*, **53**, 219.

Brown, H. and Hazard, C., 1959, *M.N.R.A.S.*, **119**, 297.

Burbidge, E. M., 1969, *Ap. J. (Letters)*, **155**, L43.

Burbidge, E. M., Lynds, C. R., and Stockton, A. N., 1968, *Ap. J.*, **152**, 1077.

Jong, M. L. de., 1965, *Ap. J.*, **142**, 1333.

Peebles, P.J.E., 1968, *Ap. J. (Letters)*, L121.

Roberts, M., 1968, in *Interstellar Ionized Hydrogen*, ed. Y. Terzian (New York: W. A. Benjamin), p. 617.

Rogerson, J. B., 1963, *Space Sci. Revs.*, **2**, 621.

Shklovsky, I. S., 1967, *Ap. J. (Letters)*, **150**, L1.

Spitzer, L., 1956, *Ap. J.*, **124**, 20.

——, 1968, *Science*, **161**, 225.

Vaucouleurs, G. de., 1958, *Ap. J.*, **128**, 465.

Wagoner, R. V., 1967, *Ap. J.*, **149**, 465.

# SPECTROPHOTOMETRIC RESULTS FROM THE *COPERNICUS* SATELLITE. I. INSTRUMENTATION AND PERFORMANCE

(J. B. ROGERSON et al.)

(ASTROPH. J. LETT. 181, L97, 1973)

## *Commentary*

THE years of effort leading to this paper were motivated in part by the fascinating goal of a large, general-purpose, long-lived telescope in orbit (see Papers #21, 31). In 1955 the Princeton University Observatory began a program in this field, starting first with Martin Schwarzschild's relatively modest and highly successful balloon telescope. Both Martin and I hoped that a succession of larger instruments would lead in time to an observatory in space, orbiting around the Earth.

Like so many U.S. space activities, this one received "full steam ahead" only with the launch of the USSR Sputnik on Oct. 4, 1957. While NASA was being organized, I was asked by the Air Force Cambridge Research Laboratories (AFCRL) if Princeton might be interested in a study contract for an astronomical satellite. My reply was an enthusiastic "Yes." Soon we were using AFCRL funds to support a design study for a spectroscopic telescope, suitable for a satellite of intermediate size. Instrumentation was planned for observations of ultraviolet interstellar lines absorbed by many different species of atoms and molecules, most of which cannot be observed from the ground. My interstellar research had indicated the tremendous increase in our knowledge of interstellar gas that could be obtained from such ultraviolet observations, especially if, as we planned, the spectral resolution were high enough to resolve interstellar from stellar lines. Our other projects supported by AFCRL included: surveys of the most suitable absorption features to observe, fabrication and tests of fused silica egg-crate mirrors, design of high reliability electronics, and tests of gratings and photomultipliers in the vacuum ultraviolet.

In 1960 NASA took over the support of this work as part of the Orbiting Astronomical Observatory (OAO) series of intermediate-size satellite telescopes. The results which we had obtained with AFCRL support were very helpful in getting our OAO subcontractors off to a flying start. Our small Princeton group, under the insightful leadership of Jack Rogerson as Executive Director (Co-Principal Investigator after the launch), successfully monitored all this engineering work, and participated actively in the extensive tests. A combination of OAO technical and financial problems, some rather severe, delayed by about six years the launch of our spacecraft (the last of the four OAO's, and the second to yield scientific results). It was for us an unforgettable experience when the Atlas-Centaur rocket carrying our instrument rose on its column of flaming gas, surrounded by the blackness of an early August morning, and accelerated swiftly into bright sunlight above.

Two incidents at and immediately following the launch may be of interest. The evening before launch, as I was checking over my own minutes of meetings with our subcontractors, I noticed an apparent error in computation of the best focus position for the secondary mirror; the focus determined in ground tests had to be corrected for sag of the primary mirror under gravity. Urgent telephone calls to various project engineers, mostly at Cape Kennedy for the launch, soon verified that a serious error had in fact been made in computing this correction. Fortunately, our NASA console at the Cape was still connected by umbilical to the launch rocket. In a final communication with our telescope the secondary mirror was moved to the corrected position. This was most fortunate, since in orbit the mechanism for moving the secondary turned out to be no longer operable! If the telescope had been launched with the focal position uncorrected, only the very brightest stars could have been observed.

A second incident occurred a few days later, shortly after the tense, emotionally charged moment when our equipment was turned on and seemed to be counting photons as planned! The telescope was pointed to the bright star, Zeta Ophiuchi, and as a test we measured the spectrum at the predicted position of one of the strongest $H_2$ lines, a major objective of our research program. We were dismayed and confused when a brief scan showed a weak continuum with no trace of any absorption features! It was only later that we gradually realized that we had scanned back and forth in the wide core of a fully saturated strong line, and had measured only the stray light in the spectrograph. Much wider scans finally revealed in full detail the strong complex structure of $H_2$ (see fig. 4, Paper #12).

The *Copernicus* satellite (as it was christened after the launch) operated effectively, though with diminishing sensitivity, for nearly a decade, until February 1981, when its operations were terminated by NASA. Several features of its design and performance,[1] not emphasized in Paper #9, may be noted here. Partly to make the electronics simpler and more reliable and partly to facilitate programming, a minimum of flexibility was provided, as is evident from the observing routines described in the accompanying paper. This made life simpler for our Guest Observers; in this program, some 200 astronomers from many countries used about half the available observing time during the later *Copernicus* years.

Another item worth noting was the technique which we developed, a half-dozen years after the launch, for reducing sharply the background noise in the high-resolution near-UV phototube (V1). The background noise in these tubes was mostly produced by fluorescence from the $MgF_2$ face plate as a result of excitation by energetic particles. In addition to delayed photon emission during several hours after passage through the South Atlantic Anomaly, a burst of about a hundred counts occurred in succession within $1/8$ second after passage of a cosmic-ray particle; the large number of counts in each burst produced a particularly serious degradation of the signal-noise ratio. Fortunately, the electronics design made possible a real-time rapid readout of phototube counts during ground contacts. Large bursts could then be identified and eliminated. The high signal-noise ratios achieved were useful for special targets.[2]

As we had hoped, a number of *Copernicus* features subsequently provided a helpful model for the Hubble Space Telescope (HST). The division of operational responsibility between Princeton and NASA's Goddard Space Flight Center (GSFC) provides one such example. Princeton was responsible for allocation of observing time and for preparation of observing schedules, expressed in simple language; GSFC translated these schedules into spacecraft language, checked them and controlled the spacecraft generally. Somewhat this same arrangement was adopted for HST, with AURA and its Space Telescope Science Institute replacing Princeton.

The extensive results obtained with *Copernicus* on the interstellar medium are summarized in Paper #12 and other reviews.[3, 4]

### REFERENCES

1. Spitzer, L.—Searching between the Stars, New Haven, CT: Yale Univ. Press, chapter 3 (1982).
2. Barker, E. S., Lugger, P. M., Weiler, E. J., and York, D. G.—*Astroph. J.* **280**, 600 (1984).
3. Spitzer, L. and Jenkins, E. B.—*Annu. Rev. Astron. Astroph.* **13**, 133 (1975).
4. Cowie, L. L. and Songaila, A.—*Annu. Rev. Astron. Astroph.* **24**, 499 (1986).

# *Paper*

### ABSTRACT

The Princeton telescope-spectrometer on the OAO spacecraft *Copernicus* scans stellar spectra with a resolution of about 0.05 Å between 950 and 1450 Å, and twice this in first order between 1650 and 3000 Å. The pointing during several minutes is steady within 0.02″, and the measured photometric precision in the shorter wavelength range is limited only by the statistics of photon counts, with 14-s counts of about $10^3$ on an unreddened B1 star, $m_V = 5.0$, at 1100 Å. In the 1650–3000 Å wavelength range, phototube noise resulting from cosmic rays makes observations difficult on stars fainter than $m_V = 3.0$.

*Subject headings:* instruments—spectra, ultraviolet—spectrophotometry—ultraviolet

## I. INTRODUCTION

The third Orbiting Astronomical Observatory (OAO-3), launched from Cape Kennedy at 5:28 A.M. EST on 1972 August 21, contains a telescope-spectrometer designed for high-resolution studies of stellar spectra in the ultraviolet longward of about 1000 Å, with interstellar absorption lines a primary objective. The satellite, named *Copernicus*, achieved a nearly circular orbit with an eccentricity of 0.00083, an inclination of 35.0° to the equatorial plane, and a semimajor axis of 7123 km. The Princeton equipment is functioning in accordance with design expectations, apart from a relatively high dark count in the phototubes used for the near-ultraviolet, and has been scanning the spectra of some two to three stars per week since the equipment was fully powered on August 26.

This first *Letter* describes the instrumentation and performance of the equipment. Subsequent *Letters* summarize the preliminary scientific results obtained.

## II. INSTRUMENTATION

The *Copernicus* spacecraft, providing power, guidance, communications, and control facilities for the astronomical equipment, is basically the same used in OAO-2 and described by Code et al. (1970). Two significant modifications are (*a*) a light baffle which extends out some 5 feet,

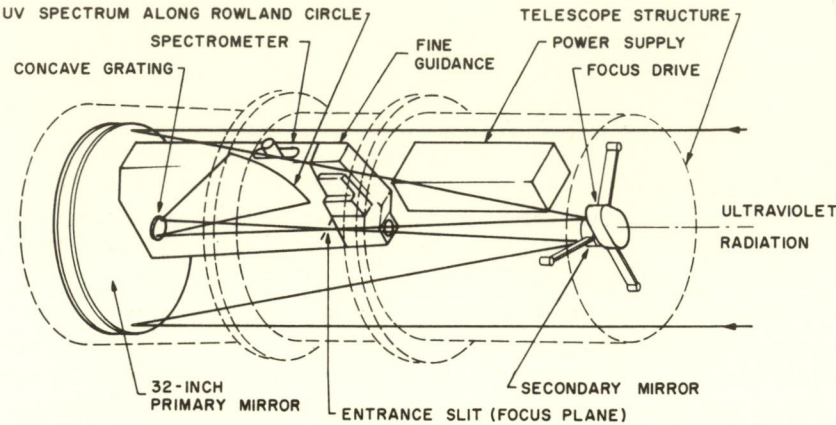

UV SPECTRUM ALONG ROWLAND CIRCLE
SPECTROMETER
CONCAVE GRATING
FINE GUIDANCE
TELESCOPE STRUCTURE
POWER SUPPLY
FOCUS DRIVE
ULTRAVIOLET RADIATION
32-INCH PRIMARY MIRROR
ENTRANCE SLIT (FOCUS PLANE)
SECONDARY MIRROR

FIG. 1. The Princeton telescope-spectrometer on *Copernicus*: *dashed lines*, the telescope structural wall and supporting rings; *solid lines*, the telescope and instrumentation.

and which successfully permits guidance in daylight on a fifth-magnitude star in a field 8 minutes in diameter, provided the optical axis is more than 60° from the Sun and more than about 30° from the sunlit Earth; (*b*) a gyroscopic reference system with a drift of about 2″ per hour, which simplifies and improves the attitude control especially during slewing. Problems of spacecraft operation have been discussed by Code et al. (1970). Data analysis has been facilitated by a data link between the Goddard Space Flight Center and the Princeton University Observatory.

The Princeton equipment, described earlier by Rogerson (1963), uses a Cassegrain telescope, with a fused silica f/3.4 primary of egg-crate construction with a clear aperture of 80 cm. The fused silica has a clear aperture of 7.5 cm and a focal length of 136.7 cm, and in orbit is positioned some 250 cm from the primary by three rods of fused silica. To increase the tolerances on focus and alignment, the primary focal length was made as great as possible and the spectrometer was placed in front of the primary, as shown in Figure 1; the equivalent focal length at the Cassegrain f/20 focus, located at the spectrometer entrance slit, is 1589 cm. Approximately 42 percent of the primary aperture is obscured by the spectrometer, the secondary mirror, and associated structure. Commandable focus positioning of the secondary was set just before launch and has not been used since.

The spectrometer is of Paschen-Runge type, with a concave grating focusing the spectrum on a 1-meter Roland circle, with a dispersion of 4.2 Å mm$^{-1}$ in first order. The entrance slit is 3 mm long and 24.2 $\mu$

TABLE 1
Data Phototubes

| Desig-nation | EMR Type | Photo-cathode | Window Material | Carriage No. | Spectrum Order | Exit Slit ($\mu$) | Wavelength (Å) |
|---|---|---|---|---|---|---|---|
| U1 | FUV | KBr | ......... | 1 | 2 | 23 | 710–1500 |
| V1 | N09 | Bialkali | MgF$_2$ | 1 | 1 | 24 | 1640–3185 |
| U2 | FUV | KBr | ......... | 2 | 2 | 98 | 750–1645 |
| V2 | N09 | Bialkali | MgF$_2$ | 2 | 1 | 96 | 1480–3275 |

wide, corresponding to $39.0'' \times 0.314''$ on the sky. The Bausch and Lomb grating is ruled with 2400 lines per mm, blazed for 2200 Å in first order.

Photons in the spectrum are detected by four EMR phototubes, each with its own exit slit, and movable in pairs along the Rowland circle, two on each carriage—see Table 1. All exit slits are made of three straight segments in a slight curve, to allow for the curvature of the astigmatic image. The U tubes are sensitive only in the second order, except at the extreme ends of the spectrum. The V1 and V2 tubes are restricted to the near-ultraviolet by a fused silica filter and a sapphire prism, respectively. To avoid physical interference along the Rowland circle, carriage 2 receives light from a diagonal mirror placed about 7 cm in front of the focal surface. As shown by Rogerson, Spitzer, and Bahng (1959), image motion at the narrow entrance slit of a spectrophotometer can give appreciable photometric errors unless a monitor tube is provided at an adjacent wavelength, and unless both phototubes have relatively constant responses over their photocathodes. Fabry optics were provided on the V tubes, and detailed tests verified a spatially constant response over the relevant areas of all four photocathodes. However, the small image motion of the spacecraft during exposure makes these features unnecessary. A separate calibration lamp with a Fe-Ne source has been useful for a preliminary wavelength calibration, but is not used for final wavelength standard since it illuminates the grating somewhat differently than does the telescope beam. Ground tests have shown that over the anticipated temperature range of the spectrometer, $-30°$ to $0°$ C, the differential expansion of the fused silica grating and the titanium structure produces a negligible degradation of resolution.

The four optical elements reflecting the light reaching the U tubes—primary, secondary, grating, and carriage-2 diagonal mirror— were given a LiF coating over fresh Al in 1972 January. As pointed out by Hass and his colleagues (Angel et al. 1961), this treatment provides a reflectance $R$ greater than 0.6 at 1025 Å, with less reflectance at somewhat longer wavelengths. Before launch, these optical elements were stored in air of low humidity or in dry nitrogen. Tests indicate that surfaces so treated degrade only slowly, with a drop in $R$ typically from 0.6 to 0.45 in about two years.

The electronics system provides for pulse counting of the signals from all the data tubes, as well as from two monitor tubes, during an integration period with a nominal length of 14 seconds (actually 13.76 s), while the carriages are motionless. The correction of the measured counts for the finite resolution time of the electronics amounts to only about 1.5 percent for a 14-s count of $10^5$ on U1 and U2, and may be ignored. The total counts are stored in the spacecraft memory, and a new integration period starts about 2 seconds later, after the carriages have moved slightly—one-half slit width (about 0.05 Å in first order) toward increasing $\lambda$ in the carriage-1 standard routine, and one full slit width (about 0.4 Å in first order) in the carriage-2 routine. Carriage 1 normally scans over 15 consecutive positions in about 4 minutes and then either scans the same region again (after sufficient backward motion to eliminate backlash) or moves on command to some other region of the spectrum. Carriage 2 normally steps progressively along the spectrum until commanded to stop or reverse. Figure 2 shows typical data obtained with these two standard observing routines.

To avoid arcs produced by the high voltages applied to the photo-tubes (5100 and 3200 volts for the U and V tubes, respectively), a vented design was used, with a void rather than potting relied upon for insulation. To keep ionospheric electrons and ions out of the power supply and the spectrometer, a variety of ion screens charged to $+50$ and $-40$ volts were added. Encoders, geared to the precision screws which move the carriage assemblies, give carriage positions every 16 seconds, and are also used to control the carriage motions.

The guidance-error sensor (see Rogerson 1963 and Gunderson 1964), which gives error signals to the spacecraft guidance system, utilizes light reflected from the two entrance-slit jaws, whose planes intersect at a small angle. An electronically switched four-quadrant EMR photomultiplier tube generates signals from the guidance images.

### III. Performance

The size of the stellar image was determined by using the gyroscopic control to drift a bright star across the slit at a known rate. The counts recorded by the fixed monitor tube at 3428 Å indicated that of the light within a strip 2″ in width, 63 percent passed through the entrance slit, whose 24-$\mu$ width corresponds to 0.3″ on the sky. Measures of the light reaching the guidance tube indicate that the starlight more than 1″ from the slit cannot amount to more than 30 percent of the total,

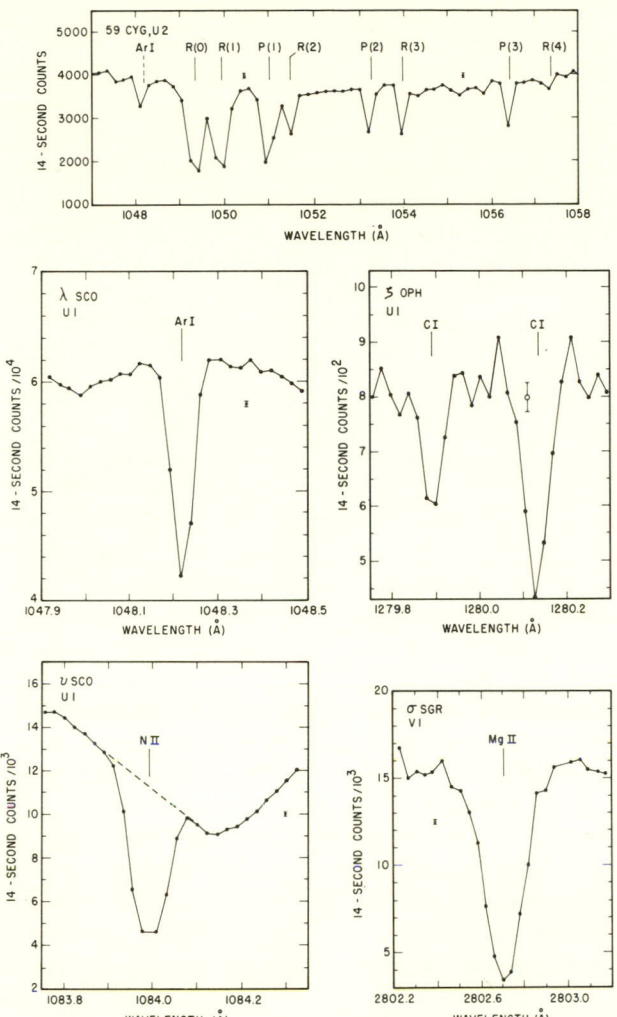

FIG. 2. Sample scans of interstellar lines. Upper diagram shows counts from the intermediate resolution U2 phototube, with carriage-2 in its standard routine; the rotational lines of the 4-0 Lyman band of $H_2$ are shown for a reddened B1.5 V star, $m_V = 4.8$, $E(B - V) = 0.19$. The other diagrams show the shorter, high-resolution scans obtained with carriage-1. The two middle diagrams depict narrow lines of Ar I and C I in second order, with about the instrumental width. The left-hand lower diagram shows the separation of stellar and interstellar N II lines in $v$ Sco ($m_V = 2.8$), while on the right is a V1 scan of a Mg II line in $\sigma$ Sgr ($m_V = 2.1$), showing the much increased noise level in the first-order phototubes. The computed statistical dispersion of counts is shown in each diagram by an error bar. The wavelength scales in the carriage-1 scans have been shifted to correspond to zero relative velocity.

TABLE 2
Photon Counts on a B1 Star, $m_V = 5.0$

| Wavelength (Å) | 4-s Counts, U1 | 14-s Counts, U2 |
|---|---|---|
| 1000 | 870 | 490 |
| 1100 | 1580 | 2600 |
| 1200 | 580 | 1160 |
| 1300 | 220 | 360 |
| 1400 | 50 | 90 |
| Dark | 20–60 | 15–40 |

| Wavelength (Å) | 14-s Counts, V1 | 14-s Count, V2 |
|---|---|---|
| 1650 | 360 | 280 |
| 1900 | 1360 | 1830 |
| 2400 | 1360 | 4700 |
| 2700 | 1230 | 4600 |
| 3000 | 1190 | 4500 |
| Dark | 1500–6000 | 1200–5000 |

showing that at least 44 percent of the light reaching the slit plane enters the spectrometer slit.

The spectrometer resolution can be estimated from the measured widths at half-maximum intensity of calibration spectrum lines and of narrow interstellar absorption lines of C I and Ar I bright stars (see Fig. 2). Each of these methods gives a full width at half-maximum intensity equal to about 0.12 Å in first order (V1) and half this in second order (U1). Allowance for intrinsic line widths gives instrumental widths of 0.10 and 0.05 Å, respectively. For U2 and V2 the corresponding widths are about 4 times as great.

Typical photon counts, corrected for dark counts and for stray light (see below) are given in Table 2 for an unreddened B1 star, based largely on observations of $\lambda$ Sco [B1 V, $E(B - V) = 0.02$]. At 1200 and 2400 Å, near the peak of the grating blaze, the spectrometer counts about 1 percent of the photons passing through the slit. For the U tubes the drop in counts at longer and shorter wavelengths, respectively, is made much more extreme by the reduction in photocathode efficiency and the decreased reflectance of LiF-coated Al.

The mean dark count, averaged over a few minutes, varies slowly over the range indicated in the last column of Table 2, and for the U tubes is about equal to the known flux of primary cosmic rays through the front sections of these tubes, varying in the same way with magnetic latitude. The much larger counts on V1 and V2 depend in part on the length of time since the most recent passage through the South Atlantic Anomaly, where trapped ions produce 14-s counts as great as $10^6$. Moreover, the short-term fluctuation of these counts on the V tubes is typically $\pm 400$, averaging about 8 times the statistical value, indicating that each event produces some 64 counts. As a result, the V1 scan in Figure 2 indicates a scatter very much greater than the expected statistical value, shown by the error bar. This large fluctuating dark count on V1 and V2 in orbit, believed to be a characteristic of conventional phototubes sensitive to visible light, seriously limits the photometric precision obtainable in the near-ultraviolet for all but the brightest stars.

The photometric precision for U1 or U2 during the 4 to 8 minutes required to scan an absorption is normally limited only by the photon statistics (see Fig. 2 with the indicated error bars). When the 14-s U1 count is $3 \times 10^4$, the mean relative deviation of four successive counts when the carriages are stationary averages 0.42 percent, as compared with a theoretical value of 0.40 percent. During a few minutes the pointing of the spacecraft may oscillate with an amplitude usually less than $\pm 0.02''$, with a period of about 6 seconds, but with no obvious effect on the photon counts, each of which covers several jitter cycles. Occasional perturbations of the spacecraft can produce small abrupt shifts in pointing, changing the light entering the slit by as much as 3 percent, as measured on the monitor tube at 3428 Å; this change is taken into account in the photometry. Larger changes are sometimes observed in the U1 and U2 count rates over periods of many minutes, with a resultant nonreproductibility of level sometimes amounting to as much as $\pm 10$ percent; these slow changes in sensitivity, which do not affect the equivalent widths directly, are caused in large part by systematic pointing changes by as much as $\pm 0.05''$ around an orbit. In addition, changes in direction of the Earth's magnetic field can modify the response of the unshielded far-ultraviolet tubes.

As discussed in Paper III below (Rogerson et al. 1973) and in Paper VI (York et al. 1973), stray light reaches the U1 and U2 photocathodes, respectively, from wavelengths longward of the exit slits by about 50 Å for U1 and 20 Å for U2, entering the phototubes through the outgassing holes provided. The resultant background, which is typically some 25 percent of the spectrum seen through the exit slits, can be greatly reduced in U1 by positioning carriage-2 to occult light just longward of the U1 exit slit. The residual background, which then exceeds the dark count by about 10 percent of the adjacent continuum, may be due to scattering by the grating.

The work reported here and in the subsequent five papers was supported by the National Aeronautics and Space Administration under contract NAS5-1810 with Princeton University. Many people have made continuing and central contributions, including, in addition to our Princeton associates: P. Schlueter, R. Kelleher, and D. Drummey of Sylvania Electronics Systems (SES), the principal subcontractor to Princeton for this program; H. Hemstreet, R. Noble, W. Loening, N. Schnog, and L. Medico of Perkin-Elmer Corporation, subcontractor to SES for the telescope and spectrometer hardware; N. Roman and J. Mitchell at NASA Headquarters; J. Purcell, J. Kupperian, L. Koschmeder, R. Schlechter, W. White, and R. White at the Goddard Space Flight Center.

### REFERENCES

Angel, D. W., Hunter, W. R., Tousey, R., and Hass, G., 1961, *J. Opt. Soc. Am.*, **51**, 913.
Code, A. D., Houck, T. E., McNall, J. F., Bless, R. C., and Lillie, C. F., 1970, *Ap. J.*, **161**, 377.
Gunderson, N. A., 1964, *J. Spacecraft and Rockets*, **1**, 91.
Rogerson, J. B., 1963 *Space Sci. Rev.*, **2**, 621.
Rogerson, J. B., Spitzer, L., and Bahng, J. D., 1959, *Ap. J.*, **130**, 991.
Rogerson, J. B., York, D. G., Drake, J. F., Jenkins, E. B., Morton, D. C., and Spitzer, L., 1973, *Ap. J. (Letters)*, **181**, L110 (Paper III).
York, D. G., Drake, J. F., Jenkins, E. B., Morton, D. C., Rogerson, J. B., and Spitzer, L., 1973, *Ap. J. (Letters)*, **182**, (in press).

# SPECTROPHOTOMETRIC RESULTS FROM THE *COPERNICUS* SATELLITE. IV. MOLECULAR HYDROGEN IN INTERSTELLAR SPACE

(WITH J. F. DRAKE et al.)

(ASTROPH. J. LETT. 181, L116, 1973)

## Commentary

THIS paper has been included here because its detailed information on $H_2$ absorption lines is something of a milestone in interstellar studies. Not only is $H_2$ a leading component of the interstellar medium, but its many absorption features provide useful information on several basic properties of the interstellar gas, including the kinetic temperatures within cold diffuse clouds, the density of neutral H in this gas, and the ambient photon flux in the Lyman and Werner bands; such photons dissociate $H_2$ molecules. In addition, the theory of $H_2$ formation and dissociation, based on *Copernicus* observations of diffuse clouds, provides major evidence for the predominance of $H_2$ in the relatively opaque molecular clouds. Infrared emission has been observed in the $H_2$ vibration and rotation bands; transient heating (as in shocks) or ultraviolet radiation can produce the excitation. These various results are described in the survey papers cited in Commentary #9 and in several surveys dealing specifically with $H_2$ in the interstellar gas.[1,2]

During the design of the Hubble Space Telescope there was some discussion of extending the wavelength range down to 1000 Å to permit observations of $H_2$, N II, O VI, and other important species. However, the LiF-coated mirrors and the windowless detectors used on *Copernicus* to provide sensitivity at these short wavelengths were very subject to contamination; their joint response at wavelengths between 1000 and 1200 Å declined by almost two orders of magnitude during the lift of *Copernicus*. So this option was firmly put aside. NASA is now supporting the design of a satellite telescope of intermediate size, designed specifically to observe spectra at these short far-UV wavelengths. A launch date in 1998 is envisaged.

## REFERENCES

1. Spitzer, L.—*Quart, J. Roy, Astr. Soc.* **17**, 97 (1976).
2. Shull, J. M. and Beckwith, S.—*Annu. Rev. Astron. Astroph.* **20**, 163 (1982).

## Paper

ABSTRACT

Strong $H_2$ lines are measured in all 11 reddened stars $[E(B - V) > 0.10]$ observed; the fraction $f$ of hydrogen gas in molecular form exceeds $10^{-1}$. In eight out of nine unreddened stars $[E(B - V) < 0.05]$ there is no trace of $H_2$ absorption, with $f$ less than $10^{-7}$. In two stars of intermediate reddening $f$ is between $10^{-5}$ and $10^{-6}$. The relatively large column densities in higher rotational levels, up to $J = 5$ or 6, correspond to excitation temperatures mostly between 150° and 200°; the ratios of ortho-hydrogen $(J = 1)$ to para-hydrogen $(J = 0)$ correspond to lower temperatures, averaging about 80°. Measures of two HD lines in nine stars indicate a ratio of HD to $H_2$ equal to about $10^{-6}$; correction for the more rapid disruption of HD molecules, in the absence of effective optical shielding by many other such molecules, indicates that one HD molecule is formed and dissociated for about every 200 $H_2$ molecules.

*Subject headings:* abundances—interstellar matter—molecules, interstellar —spectra, ultraviolet—spectrophotometry

## I. INTRODUCTION

The detection and measurement of interstellar $H_2$ absorption lines in the spectra of early-type stars has been a primary objective of the Princeton telescope-spectrometer, which was optimized for performance in the region between 1000 and 1200 Å. Measures of these lines in $\xi$ Per and $\delta$ Sco have been reported recently by Carruthers (1970) and Smith (1973), respectively. This *Letter* reports such measures in 15 stars, with upper limits in eight stars.

## II. ABUNDANCE OF $H_2$

Interstellar $H_2$ molecules are predominantly in the two lowest rotational levels, $J = 0$ (para-hydrogen) and $J = 1$ (ortho-hydrogen). The determination of $N_J$, the column density per $cm^2$ in the line of sight from the Earth to a star, is in principle straightforward for these two $J$-values, since the lines produced are generally on the square-root section of the curve of growth, with wings extending over an angstrom or more. Measures of $R(0)$ and $R(1)$ were made on U2 scans, described in Paper I. The values of $N_J$ were adjusted to give the best fits to the

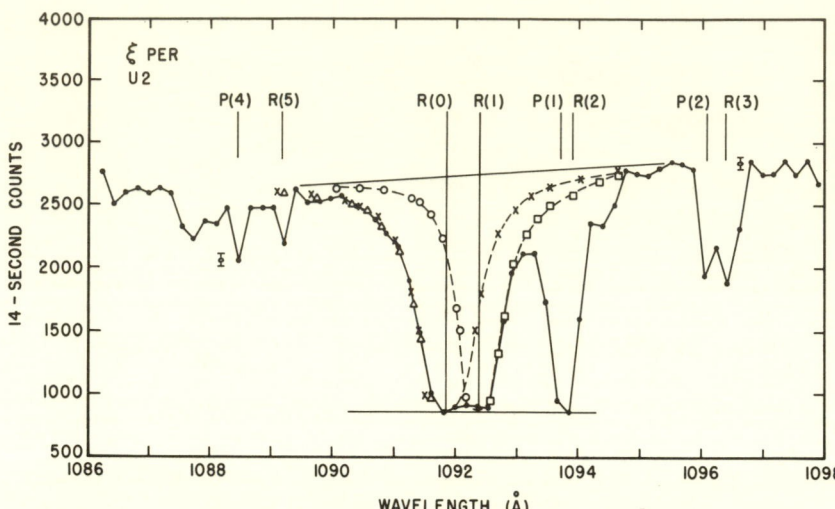

Fig. 1. Profile fitting of $H_2$ lines in U2 scan of $\xi$ Per. The theoretical profile of the $R(0)$ line in the 1−0 vibrational branch is indicated with crosses, obtained with the straight continuum and the flat background level shown. The open circles show a similar profile for $R(1)$. The triangles show the final profile of $R(0)$ when the circles for $R(1)$ are used as a continuum, and the assumed column density is decreased by about 15 percent, while the squares show the analogous profile for $R(1)$. The $P(4)$ and $R(5)$ lines at shorter wavelengths arise from the 2−0 vibrational band. Laboratory wavelengths are less than the indicated values by 0.33 Å.

observed profiles, with the usual damping formula used for $\phi(\Delta\nu)$, the profile of $\tau_\nu$ (Spitzer 1968, eq. [2-43]). In the course of this fit the continuum was also adjusted. Since the wings of $R(0)$, $R(1)$, and $P(1)$ overlap to some extent, at least two of these lines were considered simultaneously for each vibrational band analyzed. The quality of fit obtained is indicated in Figure 1, where the triangles and squares denote the computed profiles for the $R(0)$ and $R(1)$ lines, respectively, in the 1−0 band in $\xi$ Per.

The electronic and vibrational oscillator strengths and damping constants used were taken from Allison and Dalgarno (1970), with the rotational oscillator strengths taken from Spitzer, Dressler, and Upson (1964). Since these lines are much wider than the instrumental width and are fully saturated at the center, the measured count at the center was taken equal to the background level of stray light plus phototube dark count. For the narrower lines the U1 scans were also used in the background determination. The resultant values of $N_0$ and $N_1$ are listed

TABLE 1

$H_2$ Column Densities

| Star | $E(B - V)$ | Log $N_0$ | Log $N_1$ | Log $N_{J,m}$ | $J_m$ | Log $N(H\ I)$ | $f$ |
|------|-----------|-----------|-----------|---------------|-------|---------------|-----|
| $\zeta$ Per | 0.33 | 20.29* | 20.39* | 14.98 | 4 | 20.78 | 0.59 |
| $\xi$ Per | 0.32 | 20.28* | 20.21* | 13.89 | 6 | 20.92 | 0.46 |
| $\epsilon$ Per | 0.10 | 19.59 | 19.56 | 13.40 | 5 | 20.56 | 0.30 |
| $\delta$ Per | 0.03 | 18.91 | 19.18 | 15.06 | 4 | (20.18) | (0.24) |
| HD 21278 | 0.10 | 19.45 | 19.38 | 14.26 | 5 | (20.74) | (0.16) |
| $\alpha$ Cam | 0.32 | 19.81* | 19.94* | 14.58 | 5 | 20.72 | 0.37 |
| 139 Tau | 0.14 | 19.49 | 19.79 | 13.93 | 5 | 20.77 | 0.24 |
| $\zeta$ Ori | 0.09 | 15.2 | ... | ... | 3 | 20.26 | $1.8 \times 10^{-5}$ |
| $\delta$ Ori | 0.07 | 14.1 | ... | 13.20 | 4 | 20.11 | $1.9 \times 10^{-6}$ |
| $\lambda$ Ori A | 0.12 | 19.34 | 19.02 | 13.79 | 5 | 20.72 | 0.11 |
| $\zeta$ Oph | 0.32 | 20.46* | 20.10* | 13.59 | 6 | 20.61 | 0.67 |
| $\gamma$ Ara | 0.09 | 19.08 | 19.24 | ... | ... | 20.58 | 0.13 |
| 67 Oph | 0.12 | 20.35* | 20.00* | ... | ... | (20.74) | (0.54) |
| 59 Cyg | 0.19 | 18.94 | 19.18 | 13.50 | 5 | 20.26 | 0.21 |
| 10 Lac | 0.11 | 19.12 | 19.26 | 13.45 | 5 | 20.84[†] | 0.082 |

* Only one line used for this determination.

[†] $L\alpha$ scan lost in transmission; value taken from Savage and Jenkins (1972).

in Table 1. Except in the few cases indicated, several different vibrational bands were used, and the column densities averaged. The average deviations from the mean ranged up to a maximum of 16 percent for HD 21278, where strong narrow stellar lines made profile fitting difficult, but were generally much less, averaging about 7 percent. For those stars in which only the 1–0 band was used the maximum error is estimated not to exceed 25 percent.

For two of the three Orion stars the $H_2$ lines showed no wings, and the curve of growth was used to determine column densities. Details will be published elsewhere, but results are summarized here. In $\zeta$ Ori measures of five $R(0)$ lines were available, with oscillator strengths ranging over a factor of 13. The equivalent widths did not agree with the theoretical curve computed for a Maxwellian velocity distribution with any value of $b$ (see Paper II); determinations of $b$ from each pair of adjacent lines gave values increasing from 1.6 km s$^{-1}$ for the two weakest lines up to 3.6 km s$^{-1}$ for the strongest pair, with log $N_0$ decreasing from 15.2 to 14.5. As pointed out in Paper II, this is the result anticipated when a number of unequal clouds are present. The column density for this star in Table 1 is somewhat uncertain, and may be regarded as a lower limit. For $\delta$ Ori the lines were weaker ($W_\lambda$ between 2 and 7 mÅ), and the one value of $b$ obtainable was 0.7 km s$^{-1}$. While $R(1)$ and $P(1)$ in these stars appear relatively strong, the data do not permit a determination of $N_1$.

The column densities for atomic H, given in column (7) of Table 1, were taken in part from a detailed analysis (Jenkins 1971) of $L\alpha$ profiles obtained with sounding rockets; in other cases $N(\text{H I})$ was found from approximate measured widths of the $L\alpha$ line at half maximum depth on U2 scans. Values of $f$, the fraction of hydrogen nuclei in molecular form in the interstellar gas, are given in the last column. For stars of spectral type B3 or later, stellar absorption is likely to dominate the $L\alpha$ profile, and for these stars $N_H$, the column density of hydrogen nuclei, was set equal to $5 \times 10^{21} E(B - V)$, following Savage and Jenkins (1972); the corresponding values of $N(\text{H I}) = (1 - f)N_H$ and of $f$ are given in parentheses.

In stars which showed no trace of any of the $H_2$ lines, approximate upper limits on the equivalent widths were obtained from U1 scans, in the manner described in Paper V below (with $M = 4$); the stray light was assumed to equal 0.30 times the observed continuum, an approximate mean value for this wavelength region. The corresponding upper limits on $N_0$, the column density in the $J = 0$ level, are given in Table 2, together with values of $N(\text{H I})$ and $f_{max}$ determined as in Table 1. The spectral types and colors for the unreddened stars $\alpha$ Eri, $\delta$ Per, $\nu$ Sco, and $\lambda$ Sco were provided by Morgan (1972).[1]

Tables 1 and 2 indicate an enormous variation in $N(H_2)$ between different stars, correlated roughly with color excess. All stars with $E(B - V) > 0.10$ show strong $H_2$ absorption, with $f$ exceeding 0.10; except for $\delta$ Per, all stars with $E(B - V) < 0.05$ show no $H_2$ features, with $f$ less than $10^{-7}$. This behavior provides qualitative confirmation of the theory by Hollenbach, Werner, and Salpeter (1971), who find that absorption of the Lyman bands by $H_2$ molecules and grains in the outer layers of a cloud shields the inner molecules from dissociation by ultraviolet light. As a result, $f$ at the cloud center should be a strong function of $\tau_V$, the optical depth to the cloud center in visible light [about half of $3E(B - V)$] through the cloud. They find that $f$ increases from $10^{-7}$ for $\tau_V = 0$ up to 0.9 or more if $\tau_\nu$ exceeds unity. However, the observed concentration of $f$ to very high and very low values seems more extreme, perhaps, than predicted by their theory.

## III. ROTATIONAL EXCITATION OF $H_2$

Equivalent widths were measured for a substantial number of $H_2$ lines from rotationally excited molecules. The maximum value of $J$ for which $R(J)$ or $P(J)$ showed a measurable equivalent width was denoted by $J_m$, and the column density for molecules in this rotational state, by $N_{J,m}$.

---

[1] We are much indebted to Dr. Morgan for these important data.

TABLE 2
Upper Limits on $H_2$ Column Densities

| Star | Spectral Type | $E(B - V)$ | Log $(N_0)_{max}$ | Log $N(\text{H I})$ | $f_{max}$ ($\times 10^8$) |
|------|---------------|------------|-------------------|---------------------|---------------------------|
| $\alpha$ Eri | B3 IV | 0.03 | 12.51 | (20.17) | (4.3) |
| $\gamma$ Peg | B2 IV | 0.01 | 12.81 | 20.20 | 8.0 |
| $\sigma$ Sgr | B2.5 V | 0.03 | 12.63 | 20.58 | 2.3 |
| $\tau$ Sco | B0 V | 0.05 | 12.54 | 20.58 | 1.8 |
| $\nu$ Sco | B2 IV | 0.02 | 12.66 | 20.28 | 4.8 |
| $\lambda$ Sco | B1.5 IV | 0.02 | 12.40 | 19.90 | 6.2 |
| $\alpha$ Pav | B2.5 V | 0.02 | 12.90 | (20.00) | (16.0) |
| $\alpha$ Gru | B7 IV | 0.02 | 12.95 | (20.00) | (18.0) |

The values of $N_{J,m}$ in Table 1 have been computed on the assumption that the lines are formed on the linear portion of the curve of growth, and are thus lower limits. However, these lines are so weak that if the Doppler velocity parameter $b$, defined in Paper II above, is taken to be as low as 3 km s$^{-1}$, $N_{J,m}$ is increased by at most 30 percent over the values given in Table 1, except in the case of $\alpha$ Cam, where the increase of $N_J$ is by a factor 10. Photometric uncertainties are more serious for these relatively weak lines, with equivalent widths mostly between 3 and 5 mÅ, and individual values of $N_{J,m}$ may well be in error by as much as 50 percent.

The excitation energy, $E_J$, of level $J$ is given approximately by

$$E_J = 0.0073\, J(J + 1)\, \text{eV}. \tag{1}$$

Radiative transitions between levels of even and odd $J$ are strictly forbidden, and even collisional transitions occur very rarely, if at all, in the interstellar gas. If it is assumed that the molecules are formed initially with a Boltzmann distribution of $E_J$ at some formation temperature $T_f$, and cascade down to the $J = 0$ or 1 levels, the observed ratio of $N_1$ to $N_0$ can be used to determine $T_f$, using the computations by Spitzer et al. (1964). The values of $T_f$ found from Table 1 range between 55° and 115°, averaging about 80°.

Transitions among levels of even $J$ or of odd $J$ can be produced by collisions, and will occur slowly by spontaneous emission. Thus the rotational temperature, $T_r$, for the higher values of $J$ should depend on the local kinetic temperature and density of the gas. The values of $T_r$ corresponding to the ratio $N_{J,m}/N_0$ (or $N_{J,m}/N_1$ if $J_m$ is odd) in Table 1 are mostly between 150° and 200°, significantly higher than the mean temperature of about 80° found in the 21-cm absorption studies (Hughes, Thompson, and Colvin 1971; Radhakrishnan et al. 1972).

To obtain more detailed information, the curve of growth may be used to give $N_J$ for all the excited rotational levels observed. If a separate curve is plotted for each $J$, using $P(J)$ and $R(J)$ from bands with different oscillator strengths, the horizontal shifts required to fit a pair of curves together gives the difference in the log $N(J)$ values. A preliminary such curve for $\xi$ Per agrees with the theoretical curve for a single cloud with a Maxwellian distribution of velocities and $b = 4.3$ km s$^{-1}$. The resultant values of $N_J/g_J$ drop steeply with $J$ up to $J = 3$, corresponding to a rotational temperature of about 100°; the statistical weight, $g_J$, equals $2J + 1$ and $3(2J + 1)$ for $J$ even and odd, respectively. The drop in $N_J/g_J$ from $J = 3$ to $J = 6$ is much slower, corresponding to a rotational temperature of some 300°, as though a small fraction of the cloud were at this higher temperature. The resultant infrared radiation comes mostly from the $J = 2$ level, with an emissivity of about $10^{-26}$ ergs cm$^{-3}$ s$^{-1}$ per H atom, about equal to the energy input generally assumed for H I regions.

A relatively large density may be required to account for the excitation of the higher rotational levels, whose radiative lifetimes (Spitzer 1949; Dalgarno and Wright 1972) decrease about as $1/J^5$, equaling $10^8$ seconds for $J = 5$. If radiative de-excitation is assumed to be no more rapid than collisional de-excitation, the density required must be at least $10^3$ cm$^{-3}$.

## IV. ABUNDANCE OF HD

Equivalent widths of either or both of the HD lines at 1054.29 and 1066.27 Å, the $R(0)$ lines of the 4–0 and 3–0 bands, respectively, were measured in the more reddened stars, giving values ranging from 6 mÅ in $\epsilon$ Per up to 41 mÅ in $\alpha$ Cam. Column densities obtained for the HD molecules in the $J = 0$ level, using again the $f$-values tabulated by Allison and Dalgarno (1970), and assuming unsaturated lines, are shown in Table 3. No $R(1)$ or $P(1)$ lines from HD were observed, consistent with the short lifetime ($4.0 \times 10^7$ s) for the $J = 1$ level (Dalgarno and Wright 1972). To obtain the overall ratio of HD to H$_2$ molecules, these HD column densities have been divided by the sum of $N_0$ and $N_1$ in Table 1 to give the third column in Table 3. These are minimum estimates, since saturation could increase $N(HD)$ above the value computed from the linear curve of growth; if $b$ is as low as 3 km s$^{-1}$, the value of $N(ND)$ in $\zeta$ Oph and $\zeta$ Per is increased by an order of magnitude, with even larger increases for $\alpha$ Cam and 10 Lac. Evidently the ratio of HD to H$_2$ molecules is about $10^{-6}$ or more.

The detection of the HD lines confirms the presence of deuterium in interstellar space found by radio astronomers, with DCN reported by

TABLE 3

HD Column Densities

| Star | Log $N(HD)$ | Log $\dfrac{N(HD)}{N(H_2)}$ | Log Corr. Factor | $Log\left[\dfrac{N(HD)}{N(H_2)}\right]_{corr}$ |
|---|---|---|---|---|
| $\zeta$ Per | 14.30 | $-6.35$ | $-3.96$ | $-2.39$ |
| $\xi$ Per | 14.15 | $-6.40$ | $-3.97$ | $-2.43$ |
| $\epsilon$ Per | 13.57 | $-6.31$ | $-3.64$ | $-2.67$ |
| HD 21278 | 14.06: | $-5.66$: | $-3.58$ | $-2.08$: |
| $\alpha$ Cam | 14.49 | $-5.68$ | $-3.73$ | $-1.95$ |
| 139 Tau | 13.84 | $-6.12$ | $-3.58$ | $-2.54$ |
| $\zeta$ Oph | 14.23 | $-6.40$ | $-4.06$ | $-2.34$ |
| 59 Cyg | 13.86 | $-5.52$ | $-3.32$ | $-2.20$ |
| 10 Lac | 14.41 | $-5.09$ | $-3.41$ | $-1.68$ |

Jefferts, Penzias, and Wilson (1973) and the 91.6-cm D line possibly detected by Cesarsky, Moffet, and Pasachoff (1973). The ratio of roughly $10^{-6}$ for the abundance of HD relative to $H_2$ shown in Table 3 above is related only indirectly to the relative abundance of deuterons and protons in the interstellar gas, since the probabilities both of molecule formation and of molecule disruption may differ for the two species. The disruption rate by ultraviolet radiation should be much greater for the HD molecule because of the weakness of the HD lines; there is little shielding of the molecule by absorption in the Lyman lines. To correct for this effect we may multiply $N(H_2)$ by a shielding correction factor, equal to the reduction in the rate of ultraviolet disruption of the $H_2$ molecules because of the absorption in the Lyman lines. This factor has been evaluated for the $R(0)$ line of the 10–0 band, using equation (A6) from the Appendix by Hollenbach et al., and taking the optical depth halfway through the cloud; the corrected ratios are listed in the last column of Table 3. The $R(1)$ and $P(1)$ lines, whose inclusion would increase this correction factor by at most 21 percent, have been ignored; higher rotational levels have a much smaller effect. If it is assumed that the HD lines are saturated, then $N(HD)$ must also be multiplied by a corresponding shielding correction factor; however, the resultant decrease is slightly more than offset by the increase of the computed $N(HD)$ if the lines are assumed saturated, and we ignore this effect, which could increase slightly the computed abundance of deuterium.

The corrected values in the last column of Table 3 give the relative numbers of HD and $H_2$ molecules disrupted by ultraviolet photons per unit time, and must equal, of course, the ratio of the corresponding formation rates. If these formation rates were proportional to the relative abundances of D and H atoms, the values in this last column would equal the ratio of $N(D)$ to $N(H)$ in the gas. The average ratio of about $5 \times 10^{-3}$, which is in qualitative agreement with the high DCN abundance found by Jefferts et al., is more than an order of magnitude greater than the ratio $2 \times 10^{-4}$ observed on the Earth. While computation of the shielding correction factor depends on the geometry, it appears unlikely that the computed values could be incorrect by an order of magnitude. However, it seems not unlikely that the rate of formation on grains may be different for HD and $H_2$ and other formation processes may be involved (e.g., exchange interactions between D atoms and $H_2$ molecules).[2] While this particular reaction is unimportant at the low kinetic temperatures of clouds, further analysis is evidently needed to obtain the ratio of deuterons to protons in the interstellar gas.

[2] We are indebted to Dr. R. McCray and Dr. Winifred Morton for pointing out the possible importance of this reaction, and for discussing its rate coefficient.

## REFERENCES

Allison, A. C. and Dalgarno, A. 1970, *Atomic Data*, **1**, 289.

Carruthers, G. 1970, *Ap. J.* (*Letters*), **161**, L81.

Cesarsky, D. A., Moffet, A. T., and Pasachoff, J. M. 1973, *Ap. J.* (*Letters*), **180**, L1.

Dalgarno, A. and Wright, E. L. 1972, *Ap. J.* (*Letters*), **174**, L49.

Hollenbach, D., Werner, M. W., and Salpeter, E. E. 1971, *Ap. J.*, **163**, 165.

Hughes, M. P., Thompson, A. R., and Colvin, R. S. 1971, *Ap. J. Suppl.*, **23**, 323, No. 200.

Jefferts, K. B., Penzias, A. A., and Wilson, R. W. 1973, *Ap. J.* (*Letters*), **179**, L57.

Jenkins, E. B. 1971, *Ap. J.*, **169**, 25.

Morgan, W. W. 1972, informal communication.

Radhakrishnan, V., Murray, J. D., Lockhart, P., and Whittle, R.P.S. 1972, *Ap. J. Suppl.*, **24**, 15, No. 203.

Savage, B. D. and Jenkins, E. B. 1972, *Ap. J.*, **172**, 491.

Smith, A. 1973, *Bull. A.A.S.*, **5**, 32.

Spitzer, L. 1949, *Ap. J.*, **109**, 337.

——. 1968, *Diffuse Matter in Space* (New York, John Wiley & Sons).

Spitzer, L., Dressler, K., and Upson, W. L. 1964, *Pub. A.S.P.*, **76**, 387.

# AVERAGE DENSITY ALONG INTERSTELLAR LINES OF SIGHT

(Astroph. J. Lett. 290, L21, 1985)

## Commentary

WITH space telescopes it became possible to observe interstellar absorption features at wavelengths shortward of 3000 Å, opening up research on roughly a dozen of the more abundant elements, some in several ionization stages. Estimates could be made of the column density for each element along the line of sight to the star used as a light source. An element X was usually found to be mostly in one "dominant" ionization stage, $X^d$, and for many lines of sight hydrogen was mostly in neutral atoms; hence the abundance of an element X relative to H could be set approximately equal to $N(X^d)/N(H^0)$. Often the value of this ratio was below the "cosmic" value (i.e., found in the Sun's atmosphere and in meteorites); hence the ratio of this relative abundance to the corresponding "cosmic" abundance (evaluated for the Sun and meteorites) then provided a measure of the depletion $\delta(X)$ for element X.

By 1984 research with *Copernicus* and the International Ultraviolet Explorer (IUE) had shown for each element X a clear correlation of $\delta(X)$ with $\langle n(H^0)\rangle$, the average density of neutral H in the line of sight, equal to $N(H^0)/R$, where $R$ is the distance of the star being observed. (For reddened stars, with appreciable $H_2$, $2N(H_2)$ is usually added to $N(H^0)$). As $\langle (n(H^0))\rangle$ increased from about 0.1 $cm^{-3}$ to about 5 $cm^{-3}$, the values of $\delta(X)$ measured were observed to decrease by a factor between $1/4$ for Fe to $1/2$ for Mg. While large scatter was clearly present, the correlation seemed real.

Since many different clouds were known to be present along most lines of sight, with a somewhat random distribution, it was not clear how this correlation was related to the actual density within clouds or to any other cloud property. Paper #11 presents an attempt to explain this correlation as the result of an intrinsic difference of $\delta(X)$ between the three dominant types of clouds thought to be present:—warm extended clouds of relatively low density; small, cold, denser clouds; large, cold more opaque clouds. The first two represent the Warm Neutral Medium and the Cold Neutral Medium in the McKee–Ostriker synthesis.[1] The last are the local Giant Molecular Clouds.[2] The observed correlation is then attributed to the statistical distribution of clouds; as shown in fig. 1, the relative numbers of clouds of each type, averaged over many lines of signt, are functions of $\langle n(H^0) \rangle$.

It is important to note that the properties of each cloud type, $i$, which determine the results in fig. 1, are the $H^0$ column density $N_i(H^0)$ (assumed for simplicity to be the same for all clouds of type $i$), together with $n_i$, the average number of such clouds per kiloparsec along the line of sight. Hence, there is no direct connection between the correlation and the actual local density within each cloud. More detailed studies, of the sort described below, are required to show how depletion depends on the properties of each cloud type, such as temperature, density, and evolutionary history.

The theoretical correlation has been used to fit a *Copernicus* survey of depletions for five different elements in some 90 stars, permitting an evaluation of $\delta(X)$ in each cloud type.[3] Somewhat unexpectedly, no differences in depletion appeared between the Cold Neutral Medium and the Giant Molecular Clouds. The intrinsic scatter from the theoretical average curves was 0.10 dex, after removing the comparable effects of known observational errors.

A primary approximation made in this analysis was the neglect of ionized H, which increased the apparent depletion. This effect, which is significant for $\langle n(H^0) \rangle$ less than $0.10$ cm$^{-3}$, has been taken into account in an extended statistical treatment,[4] assuming that each warm neutral cloud is surrounded by a warm ionized shell, where H is largely ionized.

For analyzing the different physical processes that determine the amount of gas locked up in dust grains, these statistical analyses are strictly interim approaches. More fundamentally, the properties of each separate cloud in the line of sight should be measured and analyzed. This function can now be carried out to a large extent with the Goddard High Resolution Spectrograph on the Hubble Space Telescope, which provides a resolving power of about 85,000. This instrument can measure absorption components separated from each other in radial velocity, $v$, by at least a few km s$^{-1}$. Preliminary results[5] show how $\delta(X)$ varies both with $v$ and with temperature $T$. A broad program of further study is needed to establish general results for the interstellar medium.

### REFERENCES

1. McKee, C. F. and Ostriker, J. P.—*Astroph. J.* **218**, 148 (1977).
2. Blitz, L.—In: Giant Molecular Clouds in the Galaxy, eds. P. M. Solomon and M. G. Edmunds, New York: Pergamon 1 (1980).
3. Jenkins, E. B., Savage, B. D., and Spitzer, L.—*Astroph. J.* **301**, 355 (1986).
4. Vladilo, G. and Centurion, M.—*Astron. Astrophys.* **233**, 168 (1990).
5. Spitzer, L. and Fitzpatrick, E. L.—*Astroph. J.* **409**, 299 (1993).

# 11

## Paper

### ABSTRACT

Recent studies indicate that the depletion of certain elements in the interstellar gas is well correlated with $\langle n_H \rangle$, the mean particle density of hydrogen along each line of sight. It is shown that these results can be understood in terms of a simple theoretical model, based on random distributions of the two types of cold clouds which fit the observed distribution of color excess $E_{B-V}$, together with a more uniform warm H I gas of lower density. With increasing $\langle n_H \rangle$ the relative contributions of these three constituents to $N_H$ change, with the low-density warm gas predominant at low $\langle n_H \rangle$ (less than 0.2 cm$^{-3}$), the "standard" diffuse clouds at intermediate $\langle n_H \rangle$ (about 0.7 cm$^{-3}$), and the more absorbing "large" clouds or cloud complexes at the highest $\langle n_H \rangle$ (greater than 3 cm$^{-3}$). Hence, any correlations of physical properties with $\langle n_H \rangle$ may be attributed in part to differences of these properties between such constituents of the interstellar gas.

*Subject headings:* interstellar: abundances—interstellar: matter

## I. INTRODUCTION

Studies of the interstellar medium (ISM) along lines of sight to various stars show how correlations with the average total hydrogen density, $\langle n_H \rangle$, equal to the column density $N_H$ [defined here as $N(H I) + 2N(H_2)$], divided by the stellar distance, $R$. For example, Savage and Bohlin (1979) find that the depletion of Fe correlates rather well with $\langle n_H \rangle$, consistent with an earlier suggestion by Snow (1975); the correlation with $N_H$ is found to be significantly poorer. Similar conclusions for Mg, Cl, Ca, and Ti, respectively, are reached by Murray *et al.* (1984), by Harris and Bromage (1984), by Phillips, Pettini, and Gondhalekar (1984), and by Gondhalekar (1984); see also more general studies by Phillips, Gondhalekar, and Pettini (1982), Harris, Gry, and Bromage (1984), and Jenkins, Savage, and Spitzer (1985). The components produced by individual clouds are not usually resolved; any measured column density represents a sum over all the interstellar gas in each line of sight.

In view of the irregular structure of the ISM, the relationship between $\langle n_H \rangle$ and the physical properties of the intervening gas is not obvious. This *Letter* shows that present ISM models indicate a very direct relationship, on the average, between $\langle n_H \rangle$ and the types of clouds that provide most of the material along the sight. As described below, the irregular distribution of this H I gas may be represented by three different constituents—a warm relatively uniform, low-density gas; and two types of cold clouds: first, the "standard" or diffuse clouds, and second, the more highly clumped, larger, more heavily obscuring clouds. A value of $\langle n_H \rangle$ near the mean value for the low-density gas is possible only if there are no cold clouds of either type contributing much to $N_H$; hence, on lines of sight with such low $\langle n_H \rangle$, this warm gas is predominant. At the other extreme, $\langle n_H \rangle$ can substantially exceed its smoothed value for the galactic disk only if the obscuration by cold clouds along the line of sight much exceeds its expected average value. Such an excess, while intrinsically improbable, is more likely with the highly clumped, large clouds than it is with the more numerous diffuse clouds. Hence, along such lines of sight, the large clouds contribute most of the H I gas. For intermediate $\langle n_H \rangle$, the contribution of the diffuse clouds tends to dominate.

We conclude that lines of sight with different values of $\langle n_H \rangle$ represent properties of these three interstellar constituents in different average proportions.

The interstellar gas model assumed here is idealized but gives a reasonable fit for the observed distribution of color excesses and for other data as well. Hence, the general tendency indicated here should be reasonably well established. In §II, the model is presented, with its various assumptions and parameters. Results of the theoretical calculations based on this model, together with some discussion, are given in §III.

## II. Model for H I Gas

In view of the close correlation between $N_H$ and $E_{B-V}$ (Savage and Mathis 1979), the distribution of neutral hydrogen gas is found from the statistics of $E_{B-V}$. The warm, relatively uniform gas seems to have relatively little effect on observed color excesses; the observed distribution of $E_{B-V}$ may be well fitted (Spitzer 1978) by cold clouds of two different types. The diffuse clouds, with diameters of 2 to 6 pc (Knude 1979), may be seen also at 21 cm. The large clouds, or cloud complexes, with dimensions of some 50 pc, may be seen from their molecular emission.

Table 1 gives for each of these types of cold clouds the values adopted here for $E_i$, the color excess per cloud of type $i$, and for $k_i$, the average number of such clouds along a line of sight in the galactic disk per kiloparsec; these agree within 3% with those given by Spitzer (1978, Table 7.1). The quantity $N_i$ is the H column density, $N_H$, corresponding

TABLE 1
Parameters of Cold H I Clouds

| Number | Type | $E_i$ (mag) | $N_i$ (cm$^{-2}$) | $k_i$ (kpc$^{-1}$) | $n_{si}$ (cm$^{-3}$) |
|--------|------|-------------|-------------------|--------------------|----------------------|
| 1 | Diffuse | 0.060 | $3.5 \times 10^{20}$ | 6.2 | 0.70 |
| 2 | Large | 0.30 | $1.7 \times 10^{21}$ | 0.8 | 0.45 |

to $E_i$, using the relationship (Savage and Mathis (1979)

$$N_H \; (\mathrm{cm}^{-2}) = 5.8 \times 10^{21} E_{B-V}. \qquad (1)$$

The quantity $n_{si}$ in the last column of Table 1 is the smoothed hydrogen particle density, $k_i N_i$.

According to Poisson's law, the probability $P_i(q_i)$ that there will be $q_i$, clouds of type $i$ present along a line of sight of length $R$ is given by

$$P_i(q_i) = \frac{(k_i R)^{q_i} \times e^{-k_i R}}{q_i!}. \qquad (2)$$

The total H column density along the line of sight in clouds, denoted by $N_{H_c}$, is given by

$$N_{H_c} = q_1 N_1 + q_2 N_2. \qquad (3)$$

Since we have assumed that $N_2 = 5N_1$, a given $N_{H_c}$ may be represented by a variety of values of $q_1$ and $q_2$, with $5q_1 + q_2$ held constant. The probability of each configuration equals $P_1(q_1) \times P_2(q_2)$. If a non-integral value of $N_2/N_1$ had been assumed, the analysis would be more complicated, requiring averaging over some range of values of $N_{H_c}$. An integral value for $N_2/N_1$ simplifies the problem and should preserve the essential features.

In addition to these two types of clouds we must consider the low-density H I gas, which is presumably warm. The evidence for this "warm neutral medium," discussed in some detail by McKee and Ostriker (1977) and Spitzer (1978, 1982), comes from several sources. According to 21 cm observations by Payne, Salpeter, and Terzian (1983), the smoothed density, $n_{sw}$, of the "not strongly absorbing" warm gas in the galactic disk may be as great as 0.35 cm$^{-3}$. A lower value, 0.25 cm$^{-3}$, was found by Heiles (1980). Measures of $\langle n_H \rangle$ for unreddened stars ($E_{B-V} \le 0.01$) indicate a wide range of values, with 0.10 cm$^{-3}$ as a rough average (Spitzer 1978). The interstellar matter passing by the Sun has a density of this same order (Weller and Meier 1981).

Evidently these different lines of evidence are somewhat conflicting, probably because of complex structure in the distribution of the warm gas. Here we assume

$$n_{sw} = 0.14 \text{ cm}^{-3} = n_{s1}/5, \tag{4}$$

which may be correct within a factor of 2 in the galactic disk out to about 1000 pc from the Sun.

We do not consider here either the hot gas ($T \approx 10^6$ K), which likely fills an appreciable fraction of the galactic disk (Cox and Smith 1974; McKee and Ostriker 1977), or the familiar H II regions in which the hydrogen is ionized. Along lines of sight which pass mostly through hot ionized gas and no cool clouds, $\langle n_H \rangle$ can be much less than $n_{sw}$. However, any H I absorption lines (O I, Fe II, Mg II, etc.) seen along such lines of sight will be entirely those produced by the warm H I gas. Since this gas occupies perhaps only about one-fourth of the volume of the galactic disk, there must be a small-scale structure which will be apparent for the closest stars. Over distances of more than a few hundred parsecs this structure may well average out, and as a first approximation we assume that this warm gas has a uniform density $n_{sw}$ everywhere in the galactic disk.

We now determine the relative contributions to $N_H$ by each of the three idealized constituents of the neutral interstellar medium—the warm, uniform gas, and the two types of cold clouds. If we let $\alpha$ denote $n_{sw}/n_{s1}$, equal here to 0.2, we obtain for the total $N_H$, expressed as $q$ times $N_1$, the H column density through a diffuse cloud,

$$q \equiv \frac{N_H}{N_1} = \alpha k_1 R + q_1 + 5q_2, \tag{5}$$

$$f_w \equiv \frac{N_w}{N_H} = \frac{\alpha k_1 R}{q}, \tag{6}$$

$$f_1 \equiv \frac{q_1 N_1}{N_H} = \frac{q_1}{q}, \tag{7}$$

$$f_2 \equiv \frac{q_2 N_2}{N_H} = \frac{5q_2}{q}. \tag{8}$$

Evidently $f_w + f_1 + f_2 = 1$. For the relative probability $P_r$ of a particular configuration of $q_1$ and $q_2$, with both $N_H$ and $k_1 R$ fixed, we have

$$P_r = \frac{P_1(q_1)P_2(q_2)}{\sum_{q_2} P_1(q_1)P_2(q_2)}, \tag{9}$$

where $q_1$ is related to $q_2$ by equation (5), with $q$ and $k_1 R$ fixed, and the sum extends from $q_2 = 0$ up to the greatest value of $q_2$ permitted by equation (5). Because of the denominator in equation (9), $P_r$ is the relative probability with $q_1 + 5q_2$, and hence $N_H$, fixed; the absolute probability $P_1 P_2$ takes into account the probability of each particular $N_H$.

### III. RESULTS AND DISCUSSION

These equations have been used to compute $f_n$, $f_1$, and $f_2$ for values of $q/k_1 R$ between 0.2 and 6, with results shown in Figure 1. The average density $\langle n_H \rangle$ along the line of sight equals $q/k_1 R$ multiplied by $n_{s1} = 0.7$ cm$^{-3}$. If $R$ is expressed in pc, $N_H/R$ equals $q/k_1 R$ times $2.2 \times 10^{18}$ cm$^{-2}$ pc$^{-1}$. The solid line in figure 1 represents $f_n$ given by equation (6). The plotted points represent computed values of $f_2$, with different symbols representing different values of $q_1 + 5q_2$, which we denote by $q_c$, equal to the column density $N_c$ in clouds along the line of sight in units of $N_1$. For $q_c \leq 4$ no large cloud can contribute, $f_2 = 0$ and $f_1 = 1 - f_w$. For $q_c \geq 5$, large clouds can also be present, and several different configurations are possible. The computed values of $P_r$ are used to give weighted average values of $f_2$, denoted by $\langle f_2 \rangle$. The weighted root mean square deviations of $f_2$ from the indicated mean values (for fixed $k_1 R$ and $q_c$) are also determined and plotted as $\pm$ error bars. Since $\langle f_1 \rangle = 1 - f_w - \langle f_2 \rangle$ and shows the same dispersion as $f_2$, no individual values for $\langle f_1 \rangle$ are shown.

The dashed curve represents a "best fit" to $\langle f_2 \rangle$. The dashed-dotted curve shows the corresponding "best fit" for $\langle f_1 \rangle$; for $q_c < 5$, this curve gives $f_1$ exactly, as noted above. The indicated dispersions for $f_2$ are appreciable, diminishing as $N_H$ and $q_c$ increase, with increasing numbers of clouds. The deviations from the mean curve are mostly well within the indicated dispersion and show no particular dependence on $q_c$.

The one obvious exception is the point at $q/k_1 R = 5.2$, $\langle f_2 \rangle = 0.95$, higher than any other points. In this unusual case, $q_c = 5$ and $r =$

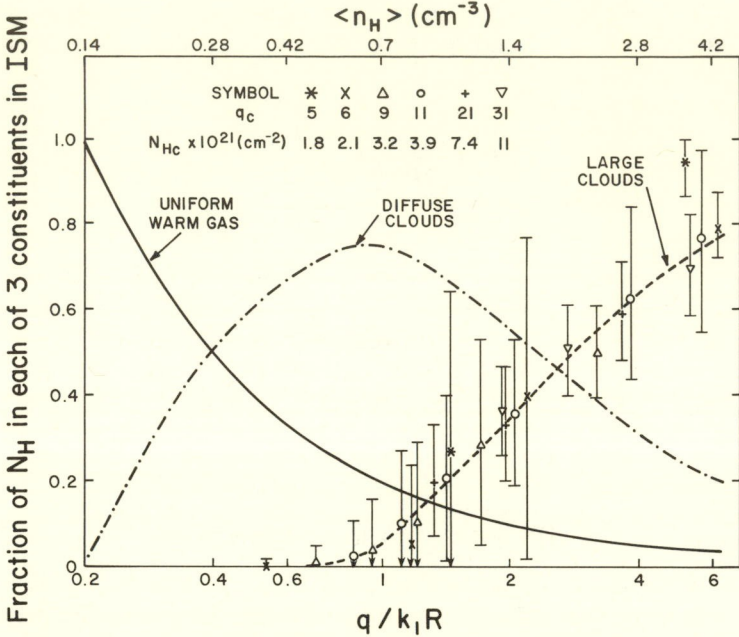

FIG. 1. Fraction of hydrogen column density in different constituents along line of sight, computed from an idealized model of the interstellar gas. The solid curve gives the ratio of $N_n$, the column density of warm H gas along the line of sight, to $N_H$, the total column density for neutral hydrogen, plotted against $q/k_1R$ (see text); the corresponding values of the mean hydrogen particle density, $\langle n_H \rangle = N_H/R$, are shown on the upper horizontal scale. The individual points show the corresponding average ratio for large clouds along the line of sight: the error bars give the dispersion about the indicated average, for a fixed $N_H$ and $R$. The dashed line represents a smooth fit to the points. The dashed-dotted line shows a corresponding fit for diffuse clouds.

$1/k_1 = 170$ pc, with either one large cloud or five diffuse clouds required in this short interval. Both alternatives are improbable, but the second is much less probable, by two orders of magnitude, than the first. Hence, the large cloud strongly predominates on the average.

The mean densities in Figure 1 extend over most of the range of observational interest. Some observed values of $\langle n_H \rangle$ are less than the lower limit of 0.14 cm$^{-3}$ shown in the figure, possibly because of the hot gas present. For some other lines of sight, $\langle n_H \rangle$ exceeds 5 cm$^{-3}$, and domination by large clouds is anticipated. If "domination" by constituent $i$ is interpreted to mean that $f_i$ exceeds $\frac{2}{3}$, then warm clouds dominate if $\langle n_H \rangle$ is less than 0.2 cm$^{-3}$, diffuse clouds, for $\langle n_H \rangle$ between 0.4 and 1.0 cm$^{-3}$, and large clouds, for $\langle n_H \rangle$ exceeding 3 cm$^{-3}$ on the average.

One consequence of this point of view is that for $N_H$ exceeding $N_2$ the various values of $f_i$ show little correlation with $N_H$ for a fixed value of $\langle n_H \rangle$, although an apparent correlation must result from the correlation of $N_H$ with $\langle n_H \rangle$. When $N_H$ is sufficiently small, this situation alters. For $N_H < N_1$, both $f_1$ and $f_2$ are zero. Despite some dispersion in $N_i$, the H column density for a cloud of type $i$, domination of the line of sight by the low-density neutral gas may be reliably indicated by low $N_H$ (or $E_{B-V}$) as well as by low $\langle n_H \rangle$. Thus it seems clear that this warm gas should dominate along the lines of sight to unreddened stars.

We conclude that in most situations the value of $\langle n_H \rangle$ along an interstellar line of sight serves as an indicator of the type of cloud present, with the less frequent clouds of higher column density contributing relatively more material as $\langle n_H \rangle$ increases. In studies of interstellar depletion, this result has two applications. First, determination of column densities from equivalent widths may depend on the type of cloud. For example, large clouds, with high $N_2$, are more likely to possess strong saturated but narrow components which can contribute appreciably to column densities with little effect on total equivalent widths. Second, a variety of physical differences between the constituents, including the ambient interstellar radiation field and evolutionary history as well as local density, may account for observed differences in depletion.

Helpful discussions with B. T. Draine and J. P. Ostriker are gratefully acknowledged.

## REFERENCES

Cox, D. P. and Smith, B. W. 1974, *Ap. J.* (*Letters*), **189**, L105.
Gondhalekar, P. M. 1984, *Ap. J.*, submitted.
Harris, A. W. and Bromage, G. E. 1984, *M.N.R.A.S.*, **208**, 941.
Harris, A. W., Gry, C., and Bromage, G. E. 1984, *Ap. J.*, **284**, 157.
Heiles, C. 1980, *Ap. J.*, **235**, 833.
Jenkins, E. B., Savage, B. D., and Spitzer, L. 1985, in preparation.[†]
Knude, J. 1979, *Astr. Ap. Suppl.*, **38**, 407.
McKee, C. F. and Ostriker, J. P. 1977, *Ap. J.*, **218**, 148.
Murray, M. J., Dufton, P. L., Hibbert, A., and York, D. G. 1984, *Ap. J.*, **282**, 481.

[†]Published 1986, *Ap. J.*, **301**, 355.

Payne, H. E., Salpeter, E. E., and Terzian, Y. 1983, *Ap. J.*, **272**, 540.

Phillips, A. P., Gondhalekar, P. M., and Pettini, M. 1982, *M.N.R.A.S.*, **200**, 687.

Phillips, A. P., Pettini, M., and Gondhalekar, P. M., 1984, *M.N.R.A.S.*, **206**, 337.

Savage, B. D. and Bohlin, R. C. 1979, *Ap. J.*, **229**, 136.

Savage, B. D. and Mathis, J. S. 1979, *Ann. Rev. Astr. Ap.*, **17**, 73.

Spitzer, L. 1978, *Physical Processes in the Interstellar Medium* (New York: Wiley-Interscience).

———. 1982, *Ap. J.*, **262**, 315.

Snow, T. P. 1975, *Ap. J.* (*Letters*), **202**, L87.

Weller, C. S. and Meier, R. R. 1981, *Ap. J.*, **246**, 386.

# CLOUDS BETWEEN THE STARS*

## (PHYSICA SCRIPTA, T11, 5, 1985)

## *Commentary*

EARLY in 1985, when I received word of the Crafoord Prize award, I was invited to come to Stockholm for the award ceremony and to give a lecture on my work,—"if possible, understandable to scientists in general." Since the results from the *Copernicus* satellite were mentioned specifically in the award citation, I decided to center much of my talk on the *Copernicus* program, including also enough historical material for an appropriate background. The published version of my talk is reprinted here as a record of the principal *Copernicus* results, and as a brief survey of the various types of regions found in the interstellar gas.

During the decade since this paper appeared, there have been significant advances in exploring various properties of the different types of regions listed in Table II. Particularly impressive is the knowledge of molecular clouds obtained from radio microwaves and from infrared radiation. The former observational technique has led to active research on chemical processes in giant molecular clouds, while the latter has given important information on the temperature of radiating dust grains. Both have contributed to our information[1] on how gas and dust are distributed in these star-forming regions.

Data on the hot gas at kiloparsec distances from the galactic plane have been extensively obtained with the International Ultraviolet Explorer satellite (see commentary, Paper #7). However, the fraction of the volume actually occupied by this gas in the galactic disk is still controversial,[2] as is the topology of interstellar regions of various types.

The properties of the warm regions have been somewhat elusive, since the atomic absorption by such regions tends to be masked by the stronger absorption in cold, denser clouds. As pointed out in the previous commentary, the high spectral resolution of the Goddard High Resolution Spectrograph ($R \approx 85,000$) now makes it possible to measure separately the absorption produced by "clouds" of these two different types if their radial velocities differ by at least a few km s$^{-1}$.

---

*Crafoord Lecture presented at the Royal Swedish Academy of Sciences, Stockholm, October 3, 1985. Much of this discussion is based on my recent book [Ref. 1 of the paper]. In particular, all the figures, except for the first, were taken from this book.

Early measures[3] are consistent with a 6000 K temperature for the warmer clouds, but regions of lower $T$, between 300 and 3000 K, seem also to be present. More data are required to give a correct statistical picture.

## REFERENCES

1. Scoville, N. Z.—In: Evolution of the Interstellar Medium, ed. L. Blitz, San Francisco: Astron. Soc. Pacific 49 (1990).
2. McKee, C. F.—In: Evolution of the Interstellar Medium, ed. L. Blitz, San Francisco: Astron. Soc. Pacific 3 (1990).
3. Spitzer, L. and Fitzpatrick, E. F.—*Astroph. J.* **409**, 299 (1993) and **445**, 196 (1995).

# 12

## Paper

A primary reason for studying interstellar matter is to understand the evolutionary role of this material. Gas enriched in heavy elements is injected by supernova explosions into interstellar space. This gas becomes mixed with the material already there and then condenses into new stars, which contain some of the recently formed heavy elements. This cosmic cycle is of fundamental importance in galactic evolution, and to analyze how it operates provides a long-range challenge for astronomy.

These evolutionary problems involve a variety of transient dynamical processes, such as supersonic expansion of ionized gas, turbulent velocities in a clumpy medium, and the modification of these flows by ultraviolet photons, cosmic rays and magnetic fields. These effects are enormously complex, and realistic solutions may require much more powerful computers than any now available.

The present paper is devoted to a much simpler problem, in which I have been interested over the years—the structure of the interstellar medium. This topic involves specifying the present properties of the gas and dust between the stars, including density, kinetic temperature, chemical composition, state of ionization, and the distribution functions for particle velocities and photon energies—all of these as a function of position in space. Without some knowledge in this area, attempts to understand the dynamical history of the interstellar gas are futile. Fortunately, the structure of the gas is a sufficiently simple problem so that with a suitable combination of observation and theory, some progress is possible. Indeed the last few decades have revolutionized our knowledge of this topic.

To survey all of this new knowledge here would be quite impossible. Instead, this paper is restricted to the existence of clouds in the interstellar medium—in particular, what the evidence is for a cloudy structure, what different types of regions, or phases, of the interstellar gas are believed to exist, and what are the general properties of each type. The approach adopted is largely chronological. To indicate the physical processes involved some theoretical discussion is included; a few current problems of particular research interest are mentioned.

The first interstellar clouds to be noticed by astronomers were the diffuse nebulae—extended luminous objects. Some of these shine by reflected starlight, but we shall be interested here primarily in emission nebulae, which are energized by a neighboring star with a surface temperature exceeding some 20 000 K (spectral type B0 or earlier); such nebulae show an emission-line spectrum. The brightest of these luminous clouds is the well-known Orion nebula, whose spectrum was photographed as early as 1882. However, it was not until about 1930 that diffuse emission nebulae were identified as clouds of ionized hydrogen, containing appreciable helium and some contamination by heavier elements, such as O, N, and Si.

More specifically, the physical explanation of these clouds in terms of a gas ionized and heated by stellar ultraviolet radiation was proposed within a span of some three years. Eddington showed in 1926 that starlight, despite its very low intensity, contained enough ultraviolet photons to ionize the interstellar gas, and that the electrons ejected by photoelectric absorption would have enough kinetic energy to heat the gas to some 10 000 K. In 1927 appeared Zanstra's classic paper, tracing the cycle of events which follows the photoionization of a hydrogen atom, and showing that the electrons, when they recombine with protons, cascade down to the ground state of the newly formed hydrogen atom, emitting photons in the Balmer and other series of neutral H. In 1928, the paper by Bowen conclusively identified the strong remaining lines from emission nebulae as forbidden lines of the lighter elements; he pointed out correctly that these lines could be excited by electron impact. Some of these ideas are foreshadowed in earlier papers, but these three represent suitable milestones in our understanding of diffuse nebulae. Later extensive work, beginning with Menzel and his collaborators in the late 1930's, elaborated these ideas in a quantitative way. The resultant theory provides impressive agreement with radio and visual observations of ionized gases, ionized and heated by ultraviolet radiation from one or more hot stars [2].

One simple but very important result was established by Strömgren [3], who showed that most interstellar hydrogen was nearly completely ionized or nearly completely neutral. This result is a consequence of the high opacity of neutral H for ultraviolet ionizing radiation. At a density of one neutral H atom per $cm^3$, a typical average value, a photon of wavelength 910 Å, which has just enough energy to ionize hydrogen, has a mean free path before absorption of only 0.05 pc (one parsec is a distance of 3.26 light years), much smaller than the dimensions of a typical cloud. Close to the star the stellar radiation keeps the hydrogen almost fully ionized, and the absorption by H is low. As the level of ionization falls with increasing distance from the star, the absorption grows rapidly and becomes nearly complete at some critical distance, which marks the boundary of the ionized gas. Thus clouds may be divided into regions either of almost completely neutral or of almost fully ionized H, which we denote as H I and H II regions, respectively.

This notation, for which I fear I am responsible [4], is somewhat confusing for physicists, who use the Roman numeral to denote the spectrum of an atom is a particular stage of ionization; of course there is no H II spectrum. In any case we now know that the conspicuous H II regions surrounding early-type stars have temperatures equal to approximately 10 000 K, and have electron densities spanning a wide range, from 10 to $10^3$ per $cm^3$. Denser H II regions exist, usually hidden by obscuring dust clouds and detected by their continuous radio emission. These 10 000-degree H II regions of various densities constitute our first cloud type or phase in the interstellar gas.

Entirely different information about the gas between the stars was obtained by studying the absorption of starlight in this gas instead of emission. This absorption is of two types—first, so-called line absorption by atoms in certain characteristic wavelengths, which are different for different atoms, and second, continuous absorption which extends smoothly over all wavelengths and is produced by small solid dust particles, often called grains. The existence of each type of absorption in interstellar space was first established firmly in about 1930, clearly an important period for the birth of interstellar matter studies. In a systematic extension of earlier work, Plaskett and Pearce [5] made a detailed study of the narrow line of Ca II at 3934 Å, measured in the spectra of some 300 early-type stars; they confirmed that the strength of this line increased with increasing stellar distance, and that the radial velocity varied with stellar distance and galactic longitude just as one would expect for a uniform gas rotating with the Galaxy. Trunpler [6] showed from measured diameters of galactic clusters that the more distant clusters were either much bigger than the closer ones or else that their apparent brightness had been dimmed by a general absorbing medium, presumably containing dust. These and other investigations indicated the presence of both gas and dust between the stars.

Ten years later, Stebbins, Huffer, and Whitford [7] studied the characteristics of this dust absorption. The total absorption is difficult to measure. Instead they measured for about 1000 stars the difference of absorption between two different wavelengths, using the relative strengths of different wide stellar absorption lines (formed in stellar atmospheres) to determine in each star the intrinsic difference of brightness between these two wavelengths of light. This difference of absorption between two wavelengths is called the color excess. Figure 1 shows in one region, extending over 20° of galactic longitude in the galactic plane, the observed values of this color excess, $E$, plotted against the distance determined by comparing the apparent brightness of the stars with their known intrinsic luminosities. The solid line shows the observational limit for a representative hot star (spectral type B0);

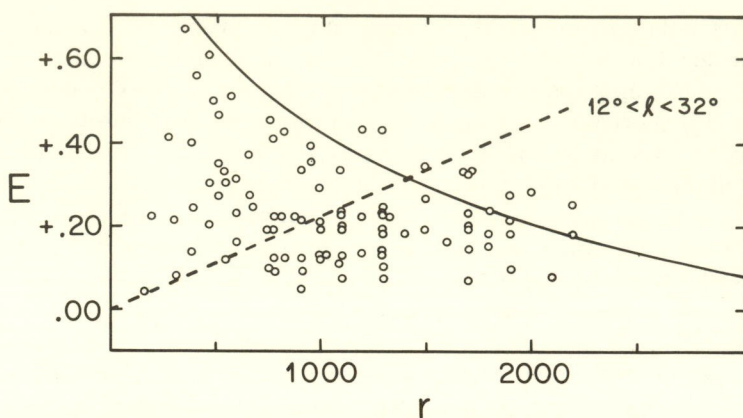

FIG. 1. Distribution of color excesses. Each circle shows the color excess (a measure of the reddening produced by interstellar dust) observed for an early-type star, plotted against the stellar distance in parsecs [7]. The dashed line represents a hypothetical uniform absorbing layer; a B0 star above the solid curve is too faint to observe.

above this line such a star is fainter than 9.0 mag, because of distance and dust absorption, and could not be observed. The tremendous scatter shown in this figure was unexpected and provided a clear proof that the dust was not distributed uniformly.

The nature of this distribution was subsequently analyzed by Schatzman [8] and by Münch [9]. While the concentrations of dust are probably irregular and have all sizes, theory fits the data with a model composed of two types of clouds, with all clouds of each type producing the same color excess. The resultant cloud properties are shown in Table I. The first two rows are the direct result of the statistical analysis, except that the color excesses have been converted to the standard scale, based on the difference of absorption between blue and visual light. The smoothed densities in the third row, based on spreading the cloud material over the volume in the galactic disc, are computed with the gas-to-dust ratio obtained from space observations (see below). The cloud diameters in the next row (on which the values in the last two rows are based) were obtained from separate programs and are somewhat uncertain. The diffuse cloud diameters were determined [10] from the color excesses of A stars; these stars are sufficiently closely spaced in the sky to permit such a measurement. The "large clouds" or cloud complexes can be measured on Milky Way photographs. As we shall see below, these large complexes are also called "giant molecular clouds."

TABLE I
Average Properties of Dust Clouds

|  | Diffuse Clouds | Large Clouds |
|---|---|---|
| Average color excess, $E_{B-V}$, produced by one cloud | 0.061 mag | 0.29 mag |
| No. of clouds per kiloparsec along line of sight | 6.2 | 0.8 |
| Contribution of clouds to the smoothed particle density of hydrogen in the galactic disc | 0.7 cm$^{-3}$ | 0.4 cm$^{-3}$ |
| Cloud diameter | 4 pc | 70 pc |
| Particle density, $n(H)$, of hydrogen within clouds | 40 cm$^{-3}$ | 10 cm$^{-3}$ |
| Mass, if cloud is spherical and uniform | 50 $M_\odot$ | $6 \times 10^4$ $M_\odot$ |

The model of dust clouds in Table I is certainly idealized, but the general concept of many isolated clouds seems well established.

Observations of the interstellar gas showed that just as with the dust, the material has a somewhat clumpy distribution. Detailed measures of interstellar absorption lines are not sufficiently extensive or precise to give the same type of information on clouds as was obtained for the dust. However, line observations by Beals [11], Adams [12] and more recently by Hobbs [13] show that each absorption feature of Ca II or Na I, for example, is composed of many separate components, each with a different velocity in the line of sight, and consequently a different wavelength because of the Doppler effect. This effect is seen clearly in Fig. 2, showing the profile of the weaker of the two Na lines, at 5896 Å, observed at a very high spectral resolution by Hobbs. Six different components of this line are visible, with radial velocities (expressed relative to the local standard of rest for stars and gas) ranging from $-13$ to $+12$ km s$^{-1}$. The number of such components observed somewhat exceeds the 6 per kiloparsec listed in Table I for diffuse clouds. However, many of these components are relatively weak, and 6 per kiloparsec is a not unreasonable value for the stronger ones. One might suspect that the observed dust clouds are also clouds of gas, a speculation which, as we shall see below, was gradually confirmed.

After the classic research on the solar atmosphere by Russell in 1929, the overwhelming abundance of hydrogen in the universe gradually became established. Hence great interest was attached to the observations, starting in 1951, of the 21-cm radiation emitted by neutral H in interstellar space, as predicted by van de Hulst [14]. Some hydrogen was

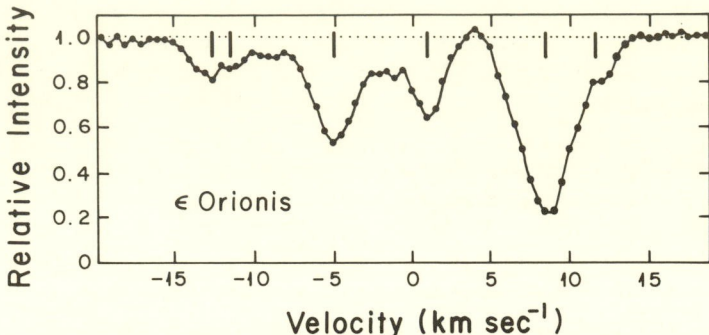

FIG. 2. Profile of interstellar absorption line. The plotted points show the relative intensities in each 0.01-Å measurement band in the region of the Na I absorption feature at 5896 Å [13], plotted against radial velocity in the local standard of rest. The vertical lines above the spectrum indicate six different components, each produced by a group of atoms moving at the indicated velocity.

detected throughout most of our Galaxy, with concentrations in spiral arms. In addition to systematic surveys of the emitted radiation received, absorption of 21-cm radiation was observed in the spectra of extragalactic radio sources.

These absorption measurements, when compared with 21-cm emission in closely adjacent regions, have given important information on interstellar clouds. Typical results [15] are shown in Fig. 3, where the lower diagram shows the intensity of radio waves from the radio source 3C237 (at a galactic latitude of 47°), plotted against the difference of wavelength from the 21 cm line observed in the laboratory; this difference is expressed in terms of the velocity producing the indicated Doppler shift. The velocities of the Earth and the Sun have been subtracted out, leaving the radial velocity of the absorbing gas expressed relative to the "local standard of rest." The upper diagram indicates the 21-cm emission averaged over directions about 4 arcminutes from the source; the contribution from the source has been subtracted out, leaving the emission from the galactic gas.

An important feature of 21-cm emission and absorption by H atoms is that the average rate of emission by such atoms is nearly independent of the mean square random atomic velocity, which is measured by the kinetic temperature, $T$. However, the absorption depends inversely on $T$; collisions among atoms modify somewhat the relative populations of atoms in the two hyperfine levels and affect the absorption coefficient

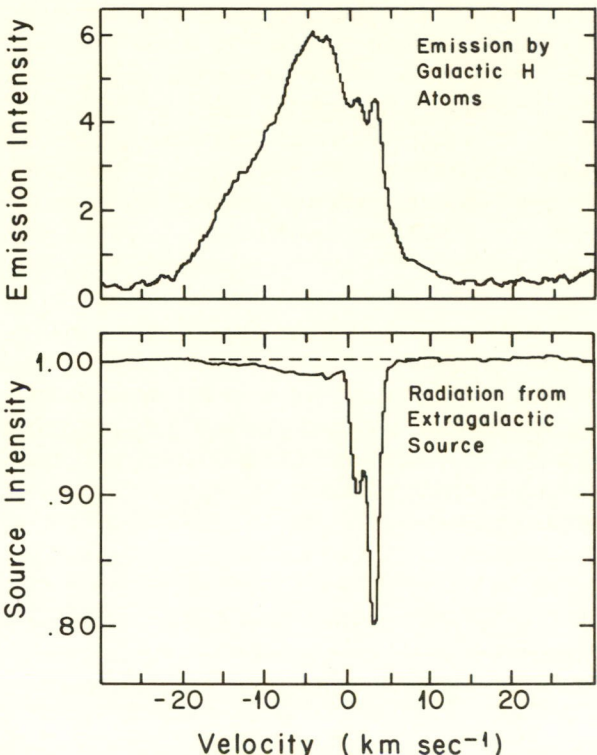

FIG. 3. Comparison of emission and absorption at 21 cm. The lower curve gives the spectrum of the extragalactic radio source 3C237 in the neightborhood of the 21-cm line of neutral H, plotted against radial velocity in the local standard of rest. The upper curve gives the 21-cm emission by galactic H, averaged over regions adjacent to the source [15].

rather strongly. Hence, the ratio of emission to absorption gives us a value for $T$. For the two absorption dips at $v = 1$ and $3$ km s$^{-1}$, the kinetic temperatures computed in this way are 42 and 22 K, respectively; the absorption line widths correspond to temperatures of 45 and 35 K, respectively. Evidently the clouds producing these absorption features are cold. Other 21-cm absorption features are usually not quite this cold, with $T$ ranging from 50 K for the stronger features up to 500 K for the weaker ones. The regions in which H atoms produce emission with very weak absorption, as, for example at $v = -10$ km s$^{-1}$ in Fig. 3, are much warmer, with temperatures exceeding 500 K; we discuss these warm regions later.

If we assume that a gas cloud is also a cloud of dust, these cold clouds seen in 21 cm are essentially the diffuse clouds listed in Table I, and constitute a second type of region of the interstellar gas. These clouds raise several theoretical problems.

The first is how these cold clouds can exist for astronomical times. Self-gravitational attraction does not appear to be quite adequate to hold such a cloud together, except perhaps for the coldest, densest ones. If the clouds are really isolated, with nothing in between, they would be expected to expand at about the sound velocity, which is 0.7 km s$^{-1}$ at 80 K. At this speed a cloud with a radius of 2 pc would double its volume in about $10^6$ years, which is a rather short time astronomically speaking. This expansion can be prevented if the volume between the clouds contains a gas which has the same internal pressure as that within the clouds; then for these perfect gases, the product $nT$ of particle density and temperature should be roughly the same in the cloud and the intercloud medium.

It was pointed out some 30 years ago [16] that such a tendency toward pressure equilibrium should be an important feature of the interstellar gas. Of course pressures far above the interstellar average would be expected in regions where explosions are occurring, such as supernova remnants and in expanding H II gas around early-type stars. Even in the cold clouds some expansion and contraction produced by passing super-novae remnants will give fluctuating pressures. However, the remnants that engulf a cloud at intervals as short as $4 \times 10^6$ years have large radii, about 30 pc, and are relatively weak, with shock temperatures less than $10^6$ K [17]; these will not grossly perturb a cold cloud. Thus the model of interstellar clouds in pressure equilibrium, while clearly idealized, may provide a useful first approximation.

A second theoretical question concerning these clouds is why the gas is so cold. Here theory provides a ready answer, since such low temperatures were predicted [18, 19]. In H II regions, the heating source is the photoionization of H atoms, whenever these are produced by electron-proton recombination. In H I regions, however, there are no ultraviolet photons present capable of ionizing H or He atoms, and atoms of heavier elements, less abundant by factors of $10^{-3}$ or less, provide insufficient heating to warm the gas. On the other hand the cooling which results from excitation of radiating ions by collisions with H atoms is not so very much less than the cooling in H II regions, resulting from ion excitation by electron collisions. As a result, in H I clouds with no dust grains, the resultant temperatures are between about 20 and 50 K. Photoelectric emission by dust grains can under some conditions give higher temperatures; more recent studies [20–22], taking into account the high photoelectric efficiency for ultraviolet radiation, especially with smaller grains, show that temperatures up to 500 K in H I clouds can perhaps be explained by this heating process.

We return now to the warm gas revealed by observations such as those in Fig. 3. The measures usually give only a lower limit on the temperature, partly because of observational errors in measuring very small absorptions and partly because a relatively small amount of cold gas also in the line of sight will give significant absorption and a markedly reduced apparent temperature. The spatial distribution and the density, however, can be measured for the warm gas without knowledge of $T$. This determination can most readily be carried out at high galactic latitudes, where there are relatively few components in the line of sight and less overlapping of different components.

Emission components at 21 cm with no apparent absorption, measured in 28 external radio sources (with $b > 7°$) have been used [23] to determine the thickness of the warm cloud layer. The distance of each component was obtained from its radial velocity, attributing this to galactic rotation. The limited data were best fitted with a Gaussian distribution, with a scale height, $h_w$, of 330 pc, about twice the corresponding value for the cold clouds. The column density $N_{Hw}$ of this non-absorbing gas (the number of such H atoms in a column 1 cm$^2$ in cross-section extending along the line of sight) is about $1.4 \times 10^{20}$ cm$^{-2} \times$ csc b [24], giving for the particle density in the galactic plane,

$$n_{Hw}(z) = 0.16 \, e^{-(z/330 \text{ pc})^2} \text{ cm}^{-3}. \tag{1}$$

Warm hydrogen gas may also be distinguished from cold gas by the relatively large velocity dispersion in each emission component, generally exceeding 7 km s$^{-1}$. This property was used [25] to test the uniformity of the warm gas. While eq. (1) generally gave good average agreement with the observations, for $15° \le |b| \le 25°$, with less irregularity than for the dust distribution, some substantial inhomogeneities were found. In particular, from some regions, tens of degrees in size, at latitudes exceeding about 30° (corresponding to distances from the Sun of less than 300 pc), the observed emission is less than one-fifth of that predicted by eq. (1). Several recent investigations have suggested somewhat different average properties of the warm medium, including an increased $N_{Hw}$ [26], a lower $n_{Hw}(0)$ [27] and an increased scale height, $h_w$ [28]. Because of the irregular distribution of warm gas at close distances, different results are to be expected in different investigations, and in this paper we shall adopt eq. (1).

Evidently the warm gas constitutes a third region or phase of the interstellar gas, and provides a suitable candidate for the intercloud medium discussed above. As we shall see, space astronomy data provide evidence supporting this possibility.

A different type of cold cloud has been revealed by measures of molecular emission lines in the microwave region of the spectrum. In particular, during the past two decades, observations of CO emission at a wavelength of 2.6 mm have revealed the presence of this abundant molecule throughout the Galaxy. A cloud produces such emission only if its particle density exceeds some 100 H atoms per $cm^3$ (mostly in $H_2$ molecules). Many other types of molecules also produce observed emission, mostly from clouds of even higher density. The denser regions within these clouds are too opaque to be explored with visible light and are not conspicuous at 21 cm since the hydrogen is mostly molecular. Within 2000 pc of the Sun a listing of 12 molecular cloud complexes [29] shows objects with a mean size of 90 pc and masses of about $10^5$ $M_\odot$, roughly similar to the large cloud listed in Table I. However, these molecular clouds are irregular in shape, and show a very clumpy distribution of density, with particle densities in the clumps averaging ten times the mean value for the complex as a whole [30]. Nearer the center of our own Galaxy, at galactocentric distances of 4 to 8 kiloparsecs, the "giant molecular clouds" predominate, with about the same linear size but with masses up to $10^6$ $M_\odot$ [31]. Again, much denser cores are present. These opaque molecular clouds are very cold, with $T$ often as low as 10 K, and constitute yet another type of region in the interstellar gas.

Giant molecular clouds provide a birthplace for new star clusters, containing hot very luminous stars. Once such stars have formed, the ionized gas surrounding them can be observed by its radio emission, while infrared astronomers can detect the energy radiated by the surrounding dust. Complex physical processes are involved in the star formation process, including shocks, turbulence, and magnetic forces on any ionized particles. Unravelling all these effects is a challenging task for the future.

While microwave astronomers were finding much cold interstellar gas in giant molecular clouds, X-ray astronomers were finding hot gas in supernova remnants, the exploding clouds produced by supernovae. This discovery had been anticipated theoretically, since the existence of expanding interstellar clouds around former supernovae was well known [32]. The hot gases produced by a supernova have an initial velocity of roughly 10 000 km $s^{-1}$. In the observed supernova remnants these velocities have been reduced by interaction with the interstellar medium, with observed values ranging from a few thousand km/s in the youngest remnants to a few hundred in the oldest. This outward moving material must produce a shock, and the temperature behind such a discontinuity at 1000 km s$^{-1}$ is about $10^7$ K.

Temperatures of this order have indeed been determined from the observed X-ray spectra of supernova remnants [17], with specific values of $T$ ranging from $3 \times 10^6$ K up to $5 \times 10^7$ K. Such a gas constitutes yet another phase of the interstellar medium—very hot but transient. Nearly complete ionization of H, He, and of other light atoms must result from collisions with electrons at such high temperatures.

Another prediction of hot interstellar gas was based [33] on the existence of absorbing clouds at a height, $z$, above the galactic plane equal to about a kiloparsec. Such clouds, absorbing the Ca II line at 3934 Å, had been observed as early as 1955 [34]. As discussed above, an intercloud medium seems required to keep an isolated cloud from expanding. If such a medium is supported at high $z$ by its internal gas pressure, its kinetic temperature must be at least $10^5$ K, the value required to give a density decrease of about $1/2$ in this gas between $z = 0$ and $z = 1000$ pc. The temperature cannot much exceed $3 \times 10^6$ K if the gas is to be confined in the Galaxy. A gas in this temperature range enveloping the Galaxy has been called a "galactic corona"; in such a hot gas the hydrogen and helium would be fully ionized. Some diffuse soft X-rays that might be produced in such an extended gas were observed relatively early [35]; their interpretation was uncertain, partly because unresolved point sources, which are so important for the X-rays of higher energy, seemed to offer a possible interpretation for this lower energy radiation as well.

We pass now to a discussion of the results obtained with the *Copernicus* satellite [36], and how these have modified our views of the interstellar clouds [1, 37]. This satellite, one of a series of NASA Orbiting Astronomical Observatories, was launched in 1972. The ultraviolet spectrometer, with its entrance slit at the focus of an 80-cm primary mirror, was designed primarily for research on interstellar absorption lines. The spectroscopic resolution ($\lambda/\Delta\lambda$, where $\Delta\lambda$ is the instrumental line width a half peak intensity), was 20 000, enough to distinguish interstellar lines from the usually wide stellar features, though insufficient to resolve the individual components apparent in Fig. 2. The photometric accuracy was better than one percent, to permit precise measures on relatively weak features. The wavelength region covered extended down below 1000 Å to permit measures of $H_2$ absorption between 1000 and 1100 Å, the higher lines of the atomic hydrogen Lyman series at 1026, 973, and 950 Å, and C III, N III, and O VI features at 977, 990, and 1032 Å, respectively.

A sample region of the $\zeta$ Ophiuchi spectrum, scanned with *Copernicus*, is shown in Fig. 4. Each point in the scan represents the

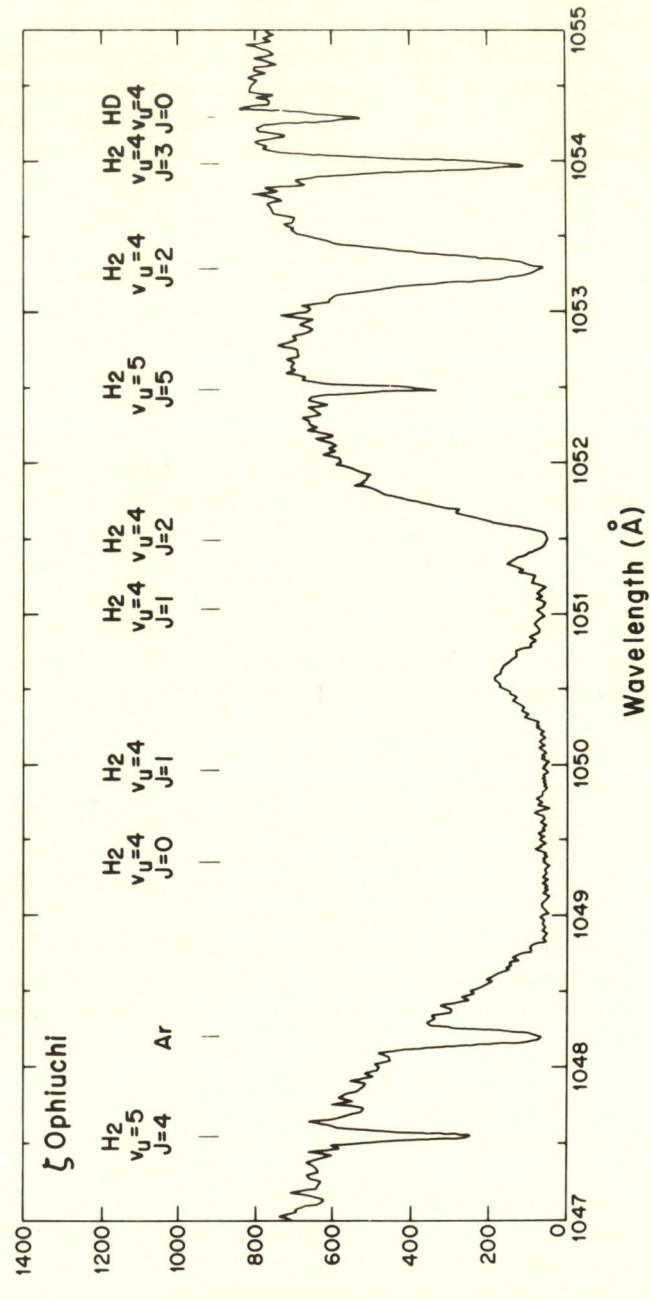

FIG. 4. Interstellar $H_2$ absorption in $\zeta$ Oph. Plot of an 8-Å region in the ultraviolet spectrum of $\zeta$ Oph in the neighborhood of the (4, 0) Lyman band of $H_2$, as indicated by the values of $v_u$, vibrational quantum number of the upper state involved in the transition: $J$ denotes the rotational quantum number of the lower state. One Ar absorption line is also shown, as are two lines absorbed from the $J = 4$ and 5 rotational levels of the (5, 0) $H_2$ band.

number of photons counted during a 14-second period; the exit slit was moved $1/40$ Å (half the slit width) between counting periods. The particular 8-Å region shown is dominated by the $H_2$ absorption features, with the Ar I line at 1048 Å the only atomic feature visible. Such data, obtained for about 100 stars [38], have given accurate values of $N(H_2)$, the $H_2$ column density along each line of sight to the star being observed. Some molecular hydrogen is present in most of the diffuse clouds, with about 10 percent of hydrogen in molecular form in the thicker, more opaque clouds. Observations similar to those in Fig. 4 provided the evidence for the now accepted conclusion that in the opaque molecular clouds most H is present as $H_2$.

More detailed discussion of the $H_2$ results support the theory that $H_2$ molecules form by catalytic action on grain surfaces. The many other types of molecules observed in molecular clouds are thought to form by reaction chains involving the interaction of atomic ions and other molecules with $H_2$.

Transitions between the $J = 0$ and 1 rotational levels of $H_2$ are produced by collisions of protons with $H_2$ molecules: these important reactions involve an exchange of the free proton for one initially bound in the molecule. The measured ratio of column densities in these two levels depends on the kinetic energy of such collisions and hence yields a value for the kinetic temperature. The value of $T$ obtained in this way averages about 80 K, providing important confirmation of the cloud temperatures obtained from 21-cm absorption and emission. Values of the hydrogen particle density, $n(H)$, within clouds are found from the excitation of the higher rotational levels [39]. The values in Table I for $n(H)$ within clouds were obtained partly in this way, partly from the measured cloud diameters.

The strong H I absorption feature, Lyman $\alpha$, at 1216 Å was also measured in the spectra of some 100 stars [40], and values obtained for $N(H)$, the column density of neutral atomic H in each line of sight. Adding $2N(H_2)$ for each star then gives $N(H_{Tot})$, the total column density of neutral H, both atomic and molecular. The values obtained are plotted against $E_{B-V}$, the color excess of each star, in Fig. 5. Since much of the scatter results from observational error, the correlation of dust and gas is very close. There are certainly a few exceptions; the star $\rho$ Ophiuchi, in a dense cloud complex, has a value of $N(H_{tot})/E_{B-V}$ almost three times the average, and falls far off Fig. 5. In addition, a few high-velocity clouds are thought to have lost most of their dust grains, which erode in shocks because of bombardment by energetic atoms. In summary, for typical diffuse clouds in the galactic disc, there is a high linear correlation between the gas and dust present. This result justifies applying to gas clouds the statistical results found for clouds of dust.

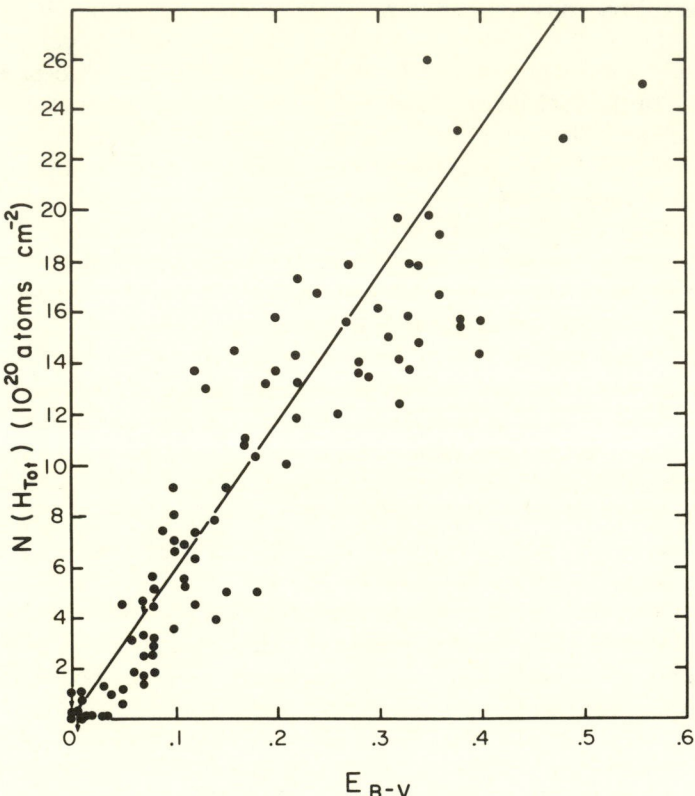

FIG. 5. Correlation of gas and dust. For each star, the column density of hydrogen atoms, including both free neutral atoms and those bound in $H_2$ molecules, is plotted against the color excess, which is proportional to the column density of grains. Two points with arrows represent upper limits. The solid line shows a reasonable fit to the data.

In terms of the mass of material, the mass of hydrogen gas is about 100 times the mass of dust. This ratio is about what one would expect from the condensation into solid particles of most atoms heavier than Ne, and also of some C, N, and O atoms as well.

If the ratio of dust to gas is approximately constant, as indicated by Fig. 5, the warm clouds as well as the cold ones must contain their share of absorbing dust particles. In principle, the presence of warm dust clouds, with a different spatial distribution from the cold ones, should effect the distribution of the observed color excesses. However, a comparison of Table I with eq. (1) indicates that the smoothed particle density of warm gas in the galactic plane (at $z = 0$) is only about 15 percent of the corresponding smoothed density found in the two types of cold clouds together. It is not surprising that such a relatively small contribution has not yet been detected from the color excess statistics.

Another result obtained with *Copernicus* is the kinetic temperature, $T$, in warm clouds of hydrogen. In principle this temperature may be calculated if one knows the velocity dispersion for atoms with different values of the atomic weight, $A$. The rms velocity for a single species is not enough to determine $T$, since this velocity spread may be due in part to a difference of large-scale motions between different regions of the cloud along the line of sight, as well as to an intrinsic dispersion of random velocities within each small volume element, much less than a parsec in size. The large-scale velocities should be about the same for atoms of all atomic weight, $A$, while the true mean square random velocities should vary as $1/A$; equipartition of random kinetic energy between different types of atoms should be quickly established in each small interstellar volume element [4, 19].

A plot of the mean square dispersion of radial velocities against $1/A$ is shown in Fig. 6 for the bright star $\alpha$ Virginis [41]. Since the color excess of this star is only 0.03 mag, no normal diffuse cloud would be expected in the line of sight; according to Table I, the color excess produced by one such cloud averages 0.06 mag. In principle, the velocity dispersion determines the width of the absorption features, unless these features are so strong that even the line wings are heavily absorbed, with resultant widening of the observed lines. In practice, however, the spectroscopic resolution of *Copernicus* was not high enough to measure line widths as narrow as 3 to 8 km s$^{-1}$. Instead, these widths were found indirectly, using for each element the apparent width and depth of several lines of different strengths, and interpreting these with the aid of a theoretical relationship called the "curve of growth." While this indirect approach has it pitfalls, we believe the results are reliable and that the variation of mean square radial velocity with $1/A$ gives the correct kinetic temperature, in this case 7500 K. The contribution of large-scale motions, independent of $A$, appears small. Similar results, with about this same $T$, have been found for some half-dozen other unreddened stars, including $\lambda$ Sco [42].

This evidence is not sufficiently extensive for any firm conclusions, but it encourages one to explore the consequences of a warm gas temperature in the neighborhood of some 7500 K. This is about 100 times the mean temperature of the cold H I clouds. If pressure equilibrium is assumed as a first approximation, $n(\text{H})$ in a warm cloud will then be 0.01 times that in the cold diffuse cloud, or about 0.4 H atoms cm$^{-3}$, according to Table I. The low 21-cm emission from some neighboring

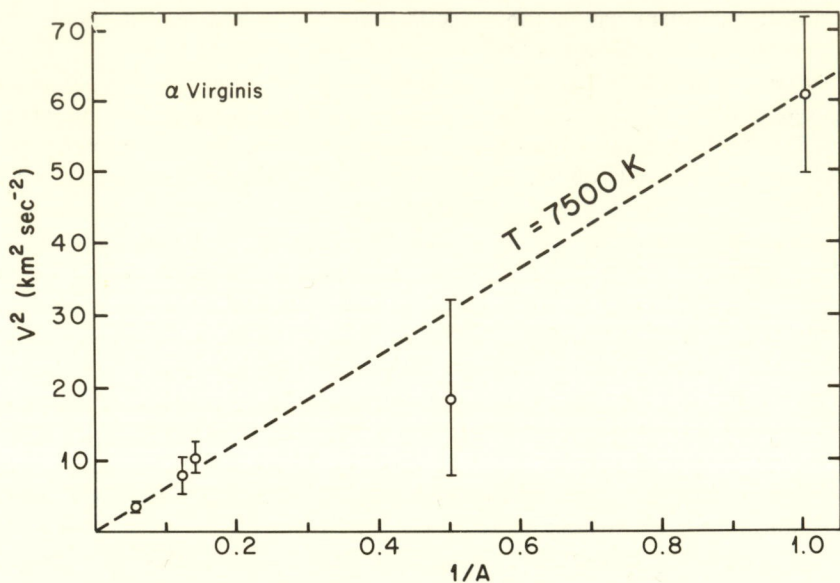

FIG. 6. Velocity dispersion for different elements. The circles show mean square values of $V$ obtained from different interstellar atoms along the line of slight to $\alpha$ Vir, plotted against the reciprocal of the atomic weight, $A$ [41]. The vertical bars show estimated (rms) errors. The dashed line shows the relation expected for equipartition of kinetic energy at $T = 7500$ K.

regions, pointed out above, indicates that the warm gas does not fill all the volume in the galactic disc. If we assume that the warm gas fills 40 percent of this volume, then the smoothed density at $z = 0$ equals 0.16 $cm^{-3}$, in agreement with eq. (1). This model provides a convenient way of remembering the observations and may even constitute a reasonable first approximation to reality.

Physically one wonders how the warm neutral gas is heated to such relatively high temperatures. Photoelectric emission from grains has again been proposed as the effective heating agent [21] and may perhaps be responsible. Another possibility is that this low density gas may be heated by the mechanical energy associated with expanding supernova remnants. The heating energy required is so small that only a small fraction of the energy stored in remnants need be converted. Two specific mechanisms proposed for this conversion involve the storage and subsequent dissipation of mechanical energy in shock-compressed clouds [43] and the dissipation of magneto-acoustic waves, generated when the shocked gas of a remnant impacts a cloud [44]. It is also possible that heating of the warm gas by low-energy cosmic rays (10 to

100 MeV) may be important; since this mechanism was first discussed [45], the theory has been extensively developed [46, 47, and refs. cited]. However, present evidence suggests that the interstellar flux of such cosmic-ray particles is substantially less than the values originally envisaged. Evidently the exact process responsible for heating the warm gas is not yet clear.

A final *Copernicus* result to be mentioned here is the discovery of a widely distributed hot gas, which we regard as yet another region of the interstellar medium. The two O VI absorption features at 1032 and 1038 Å were clearly seen in the spectra of most of the 72 stars examined [48]. While these lines are narrower than the stellar features, they are much wider than the other interstellar lines, as is evident from Fig. 7. The minimum width of these lines corresponds to an rms radial velocity of 10 km/s, and a corresponding kinetic temperature of some $2 \times 10^5$ K. If, as seems likely, the oxygen is ionized by electron collisions, the electron temperature should be between $2 \times 10^5$ and $8 \times 10^5$ K to place at least one percent of the ions in the five-times ionized state. As with the color excess data, the statistical distribution of O VI line strengths is consistent with the concentration of $O^{+5}$ ions in separate regions or clouds, with 6 such regions in the line of sight per kpc.

The observed soft X-rays seem to require gas at higher temperatures [49]; the data can be fitted with some coronal gas at $3 \times 10^6$ K, some at $6 \times 10^5$ K. The $O^{+5}$ ions would all be concentrated in the latter region. While the X-ray emission and the O VI absorption occur in somewhat different regions, there is now no question that a hot component is present in interstellar space. The fraction of the volume in the galactic disc occupied by such gas is uncertain, with estimates ranging from 20 to 80 percent. If the temperature is a hundred times that in the warm gas, pressure equilibrium would require that the particle density (mostly protons and electrons) be correspondingly smaller, with a proton particle density of about $2 \times 10^{-3}$ $cm^{-3}$. The heat source for this corona is presumably the hot gas of supernova remnants, spreading throughout the Galaxy.

In this discussion we have distinguished six different types of regions or phases between the stars, each with gas of different properties. These different regions of the interstellar medium are summarized in Table II. Most of these components can fairly be termed clouds, though No. 3 and No. 6 are probably better described as components of a clumpy intercloud medium.

This classification, like so many, is arbitrary. From the most basic standpoint, a phase of the interstellar medium is perhaps best defined in terms of only two physical quantities—the ionization stage of hydrogen and the temperature. On this basis, phases 1 and 2 would then be

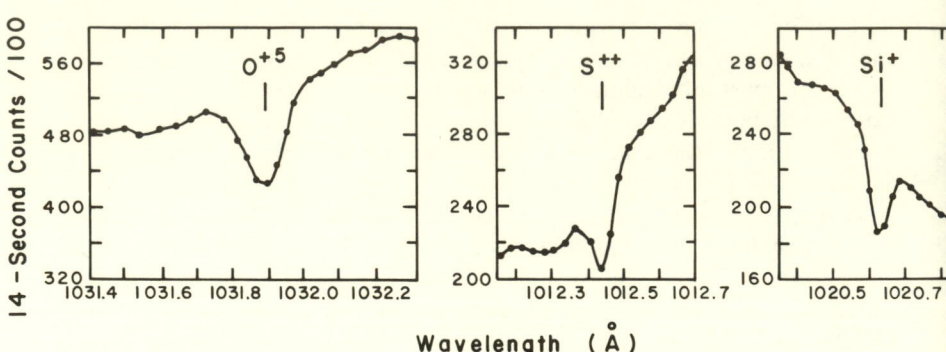

FIG. 7. Profiles of interstellar absorption features. The 14-second counts obtained for α Virginis are plotted against wavelength for the O VI line and for two narrower interstellar features. The vertical lines indicate the central wavelength.

combined, as would 5 and 6. On the other hand, each region in Table II could be divided into many subphases, depending on size and density as well as on previous history. Table II is both idealized and incomplete, but it provides one simple way of organizing a complex mass of data.

No paper on interstellar clouds would be complete without some mention of a recent general theory which ties most of these regions together in a single elegant picture [50]. According to this picture a typical diffuse cloud of cold H is naturally surrounded by envelopes of the other types of regions, as indicated in Fig. 8. The outer envelope is ionized by ultraviolet radiation from O and B stars at various distances and becomes a warm H II regions. The inner envelope is heated by one of the external influences discussed above and becomes warm H I gas. The core, at a higher density, is unaffected by such influences and remains cold and dense. Surrounding the entire assembly is a hot gas. Thus four of the six phases in Table II appear in this model. The spherical form is of course an artifact of the theorist, but the development of the shielding envelopes between the cold core and the hot surrounding gas appears plausible.

One characteristic of this model is that the hot gas close to the outer cloud envelope would be expected to cool by conduction into the warm gas, and thus the temperature of the hot gas nearest to the cloud will be reduced. If most of the hot gas has a temperature of some $10^6$ K, the O VI absorption would then arise in the somewhat cooler layers near the cloud. This point of view is supported by the *Copernicus* observations, which show that the velocities of O VI, Si III and N II absorption features are well correlated with each other [51]. The first of these features is produced by hot gas, and the third by warm gas, presumably

TABLE II
Regions of the Interstellar Gas

| No. | Name | State of Hydrogen Atoms | Temperature | Distinguishing Characteristics |
|---|---|---|---|---|
| 1 | Cold H I clouds | Neutral | Cold (80 K) | $\langle n(H) \rangle \approx 40$ cm$^{-3}$ in diffuse clouds, with higher values in some regions |
| 2 | Molecular clouds | Neutral | Cold (10 K) | Hydrogen molecular, $n(H_2) > 10^2$ cm$^{-3}$ in clumps |
| 3 | Warm H I gas | Neutral | Warm (8000 K) | Low density, $n(H) \approx 0.4$ cm$^{-3}$ |
| 4 | H II regions | Ionized | Warm (8000 K) | Ionized by ultraviolet stellar photons |
| 5 | Supernova remnants | Ionized | Hot ($10^6$ K) | Produced when exploding supernova interacts with interstellar gas |
| 6 | Coronal gas | Ionized | Hot ($10^6$ K) | Maintained by old supernova remnants after high velocities have died out |

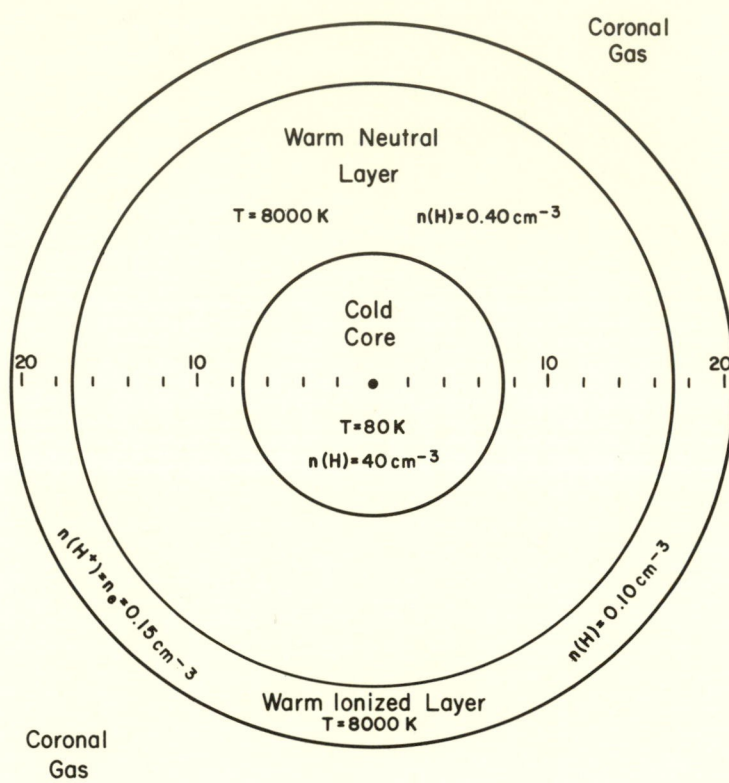

FIG. 8. Structure of a composite cloud. Values of $n$(H), the neutral hydrogen density, and temperature $T$ are indicated for the cold central core and the two warm envelopes; the electron density $n_e$ is also specified for the outer envelope. The horizontal scale shows radii in light years.

the outer H II envelope surrounding the diffuse cloud. Thus the different regions, each with its own temperature, appear to share a common velocity through interstellar space, as would be expected for this theoretical model.

From time to time, at intervals of many million years, an energetic supernova remnant will sweep over a group of clouds, producing the phenomena portrayed in Fig. 9. In this figure no distinction is drawn between the ionized and neutral envelopes, but some low density warm clouds are shown with no cores. The passage of the shock wave at the outer boundary of the supernova remnant produces large perturbations of the clouds. Some envelopes are stripped away; others evaporate as heat is conducted into the warm gas from the hotter layers of the

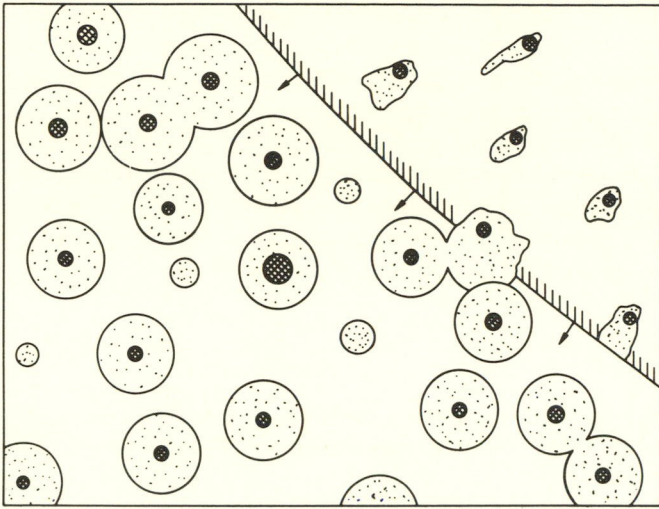

FIG. 9. Clouds in the galactic disc. Each cloud intersecting the galactic plane is represented by its cross-section through the cloud center. The dark central cores represent cold diffuse clouds, while the surrounding dotted circles represent envelopes of warm gas. The hot coronal gas fills the space between the clouds. An expanding supernova remnant advances in the upper right [50].

expanding remnant. Closer to the exploding star some cores will even evaporate entirely, increasing the density of the gas in the remnant. Of course, for the interstellar medium as a whole the destruction of old clouds must be matched by the formation of new ones, partly by radiative cooling and gas compression and partly by gravitational condensation.

This sweeping theory is very appealing, but it does have a number of problems. For one thing, we have seen that the thickness of the warm neutral gas layer in the Galaxy is at least twice as great as that of the cold gas. This result requires that the size of the warm envelope relative to the core, or the fraction of clouds with no cold cores, increases with increasing distance, $z$, from the galactic plane. Until the heating mechanism for this gas is better understood, such effects are not readily analyzed. Also, the detailed mechanism for mass balance—formation of new clouds to replace the old—is not yet explained in quantitative detail. Nevertheless, this theory is most helpful in suggesting new observations and new approaches to old ones. It will be noted that giant molecular clouds and the complexities of star formation are not included in this general synthesis—probably wisely so!

During the last 50 years, which roughly parallel my own scientific career, our knowledge of the interstellar material has grown from a crude beginning to an elaborate structure of observations and theory. To play a part in this exciting enterprise has been a privilege for me, and I have been fascinated to see the many ways in which physical theory has provided a coherent explanation for astronomical observations. A number of the simpler questions have now been partially answered. I am glad that some of the most interesting problems await solutions by younger generations of astronomers.

REFERENCES

1. Spitzer, L., Searching between the Stars, Yale U. Press, New Haven (1982).
2. Osterbrock, D. E., Astrophysics of Gaseous Nebulae, W. H. Freeman, San Francisco (1974).
3. Strömgren, B. *Ap. J.* **89**, 526 (1939).
4. Spitzer, L., *Ap. J.* **93**, 369 (1941).
5. Plaskett, J. S. and Pearce, J. A., Publ. Dominion Ap. Obs. Victoria **5**, 167 (1933).
6. Trumpler, R. J., *Lick Obs. Bull.* **14**, 154 (1930).
7. Stebbins, J., Huffer, C. M., and Whitford, A. E., *Ap. J.* **92**, 193 (1940).
8. Schatzman, E., *Ann. d'Ap.* **13**, 367 (1950).
9. Münch, G., *Ap. J.* **116**, 575 (1952).
10. Knude, J., *Astr. and Ap. Supp.* **38**, 407 (1979).
11. Beals, C. S., *M.N. Roy. Astr. Soc.* **96**, 661 (1936).
12. Adams, W. S., *Ap. J.* **109**, 354 (1949).
13. Hobbs, L. M., *Ap. J.* **157**, 135 (1969).
14. van de Hulst, H. C. and Ned. T., *Natuurk*, **11**, 201 (1945).
15. Dickey, J. M., Salpeter, E. E., and Terzian, Y., *Ap. J. Supp.* **36**, 77 (1978).
16. Spitzer, L., Problems of Cosmical Aerodynamics, IAU-IUTAM Symp. on the Motion of Gaseous Masses of Cosmical Dimensions, Paris, p. 31, Central Air Documents Office, Dayton, Ohio (1951).
17. Gorenstein. P. and Tucker, W. H., *Ann. Reviews Astr. and Ap.* **14**, 373 (1976).
18. Spitzer, L. and Savedoff, M. P., *Ap. J.* **111**, 593 (1950).
19. Spitzer, L., *Ap. J.* **120**, 1 (Russell Lecture) (1954).
20. Watson, W. D., *Ap. J.* **176**, 103 (1972).
21. Jura, M., *Ap. J.* **204**, 12 (1976).
22. Draine, B. T., *Ap. J. Supp.* **36**, 595 (1978).
23. Falgarone, E. and Lequeux, J., *Astr. and Ap.* **25**, 253 (1973).
24. Radhakrishman, V., Murray, J. D., Lockhart, P., and Whittle, R.P.J., *Ap. J. Supp.* **24**, 15 (1972).
25. Heiles, C., *Ap. J.* **235**, 833 (1980).
26. Payne, H. E., Salpeter, E. E., and Terzian, Y., *Ap. J.* **272**, 540 (1983).
27. Jenkins, E. B., Savage, B. D., and Spitzer, L., *Ap. J.* **301**, 355 (1986).

28. Lockman, F. J., Hobbs, L. M., and Shull, J. M., *Ap. J.* **301**, 380 (1986).
29. Blitz, L., Giant Molecular Clouds in the Galaxy (Edited by P. M. Solomon and M. G. Edmunds), p. 1, Pergamon, New York (1980).
30. Blitz, L. and Stark, A. A., *Ap. J. Lett.* **300**, L89 (1986).
31. Solomon, P. M. and Sanders, D. B., Giant Molecular Clouds in the Galaxy (Edited by P. M. Solomon and M. G. Edmunds), p. 41, Pergamon, New York (1980).
32. Minkowski, R., *Ann. Reviews Astr. and Ap.* **2**, 247 (1964).
33. Spitzer, L., *Ap. J.* **124**, 20 (1956).
34. Münch, G. and Zirin, H., *Ap. J.* **133**, 11 (1961).
35. Silk, J., *Ann. Reviews Astr. and Ap.* **11**, 269 (1973).
36. Rogerson, J. B., Spitzer, L., Drake, J. F., Dressler, K., Jenkins, E. B., Morton, D. C., and York, D. G., *Ap. J. Lett.* **181**, L97 (1973).
37. Spitzer, L. and Jenkins, E. B., *Ann. Reviews Astr. and Ap.* **13**, 133 (1975).
38. Savage, B. D., Bohlin, R. C., Drake, J. F., and Budich, W., *Ap. J.* **216**, 291 (1977).
39. Jura, M., *Ap. J.* **197**, 575, 581 (1975).
40. Bohlin, R. C., Savage, B. D., and Drake, J. F., *Ap. J.* **224**, 132 (1978).
41. York, D. G. and Kinahan, B. F., *Ap. J.* **228**, 127 (1979).
42. York, D. G., *Ap. J.* **264**, 172 (1983) (and subsequent unpublished work).
43. Cox, D. P., *Ap. J.* **234**, 863 (1979).
44. Ikeuchi, S. and Spitzer, L., *Ap. J.* **283**, 825 (1984).
45. Hayakawa, S., Nishimura, S., and Takayanagi, K., *Publ. Astr. Soc. Japan* **13**, 184 (1961).
46. Spitzer, L. and Scott, E. H., *Ap. J.* **158**, 161 (1969).
47. Field, G. B., Goldsmith, D. W., and Habing, H. J., *Ap. J. Lett.* **155**, L149 (1969).
48. Jenkins, E. B., *Ap. J.* **220**, 107 (1978).
49. Burstein, P., Borken, R. J., Kraushar, W. L., and Sanders, W. T., *Ap. J.* **213**, 405 (1977).
50. McKee, C. F. and Ostriker, J. P., *Ap. J.* **218**, 148 (1975).
51. Cowie, L. L., Jenkins, E. B., Songaila, A., and York, D. G., *Ap. J.* **232**, 467 (1979).

# PART TWO

# Stellar Dynamics

# THE STABILITY OF ISOLATED CLUSTERS

(MONTHLY NOTICES, ROY. ASTR. SOC., 100, 396, 1940)

## *Commentary*

DURING 1939–1940, when I was a National Research Council Fellow at Harvard, I much enjoyed occasional evening discussions at the home of Bart and Priscilla Bok. At these informal gatherings, attended by a number of younger astronomers, the Boks would bring up for discussion various unsolved and challenging problems of astronomy. On one such occasion, Bart pointed out an apparent paradox in the structure of globular clusters. Random gravitational encounters between stars in the cluster should tend to establish a Maxwell-Boltzmann distribution of velocities, a characteristic of isothermal conditions. However, the equilibrium conditions for an isothermal sphere require an infinite mass. How could these two conditions be reconciled?

This was exactly the type of research problem that appealed to me—involving well-defined physical processes, sufficiently simple so that one could visualize what was happening, and try to predict the likely course of development in advance of any complex mathematics. This visualization of physical phenomena, combined with a comprehending grasp of interactions between components, was a technique that came naturally to me in my first physics course at age fifteen, and despite occasional mistaken insights has been invaluable in my scientific work ever since. So in this case I asked myself, when I was in a relaxed frame of mind, what would an isolated spherical cluster do as its component stars tended to approach a random velocity distribution, of Maxwell-Boltzmann type?

After a little thought it seemed clear that stars which acquired velocities greater than the local escape velocity would escape from the cluster. To compute the actual rate at which stars would escape or "evaporate" from a cluster, I used a very crude analysis, based on the calculations of the "relaxation time" by Jeans and by Karl Schwarzschild some years earlier. The dominant importance of many distant gravitational encounters, relative to the much less frequent close ones, was evident already in these early works.

In addition to the physical arguments presented, Paper #13 includes a straightforward analysis which is still relevant to ongoing research; i.e., the discussion leading to equation (44), giving the rate at which equipartition is established between two groups of particles, each with its own Maxwellian distribution. This rate is important in various plasma physics problems as well as in the dynamics of globular clusters.

Several years later I learned that Ambartsumian had published,[1] two years before Paper #13 appeared, a theoretical discussion very similar to mine. We both considered highly idealized clusters, with uniform density and a square-well gravitational potential (zero outside the cluster and constant with position inside, but varying slowly with time). By coincidence, our rough estimates of relaxation times were also nearly identical.

Within the following decade, several papers discussed the equations needed for more precise computations of collisional relaxation phenomena.[2,3] However, consideration of more realistic cluster models was postponed until high-speed digital computers became available some two decades later. The models considered and the escape rates found for an isolated cluster are reviewed in Paper #20. While an isolated spherical cluster appeals to theorists because it permits a relatively precise solution—a useful reference model—the sizes of actual clusters are generally limited by the tidal field of the Galaxy, with major effects of the escape of stars[4] (see Commentary, Paper #20).

Certainly escape is one of the processes affecting the evolution of a star cluster, though for an isolated cluster the process seems much less important than mass stratification (see Paper #18) or the gravothermal instability (see Paper #20).

## References

1. Ambartsumian, V. A.—Annals Leningrad State Univ. (Astron. Series) **22**, 19 (1938); transl.—In: Dynamics of Star Clusters, IAU Symp. No. 113, eds. J. Goodman and P. Hut, Dordrecht: Reidel 521 (1985).
2. Chandrasekhar, S.—*Rev. Mod. Phys.* **15**, 1 (1943).
3. Cohen, R. S., Spitzer, L., and Routly, P. Mc.R.—*Phys. Rev.* **80**, 230 (1950), Paper #24.
4. Spitzer, L.—Dynamical Evolution of Globular Clusters, Princeton: Univ. Press §3.1, 3.2 (1987).

## *Paper*

### INTRODUCTION

The internal structure of globular clusters presents problems of considerable theoretical interest. Heckmann and Siedentopf* have pointed out that encounters between stars of such a cluster should establish an isothermal condition in times comparable with the short-time scale; it is therefore relevant to investigate the structure of an isothermal gaseous sphere. Unfortunately, both the mass and radius of such a sphere are infinite. To meet this difficulty Heckmann and Siedentopf advanced the assumption, originally made by Martens, of a continuous distribution of galactic field stars, unobservable because of their small masses. These stars would presumably form a local condensation within the cluster, and would provide the extra gravitational attraction necessary to give the rest of the cluster a finite mass and radius.

There is, however, another possible escape from this dilemma. Consider an isolated cluster of known mass and radius. Encounters between stars will gradually set up a Maxwellian velocity distribution; as the cluster approaches isothermal conditions, what will happen to it?

The relation between the velocity of the cluster stars and the velocity of escape from the cluster determines the answer to this question. From the virial theorem it follows that the kinetic energy of the cluster stars is one-half the total gravitational energy, measuring this in the negative sense. This assumes, of course, that the cluster is in a statistically steady state. It is shown in section 3 that one may neglect in this connection the deviations from the steady state produced by the phenomenon of evaporation considered here. The mean kinetic energy of a cluster star is therefore one-half the potential energy of the cluster per star, and one-fourth[†] the mean potential energy of the cluster stars. It is evident that if the kinetic energy of any star exceeds its potential energy the star will escape. Hence if the kinetic energy of any star exceeds four times the mean kinetic energy of all the stars, that star will leave the cluster, provided that its potential energy is not greater than the mean potential

---

* O. Heckmann and H. Siedentopf, *Zs. f. Ap.*, **1**, 43, 1930.

[†] The additional factor of one-half may be derived as follows. Consider a cluster of mass $M$ with some arbitrary density distribution. Let the energy required to remove a single gram of matter to infinity be equal, on the average, to $\psi GM$, where $\psi$ is some function of the size and density distribution of the cluster, and $G$ is the usual gravitational constant. Then the potential energy of the cluster as a whole will be the integral over $dM$ of $\psi GM\,dM$, or simply $\frac{1}{2}\psi GM^2$. Hence the potential energy of the cluster per unit mass is $\frac{1}{2}\psi GM$, or one-half of the mean potential energy of a unit mass.

energy of the cluster stars. In other words, the mean square velocity of the cluster stars is one-fourth the mean square velocity of escape.

If the kinetic energy of the cluster is divided among the stars by the usual Maxwell–Boltzmann formula, some stars will clearly have more than enough energy to leave the cluster entirely. The proportion of stars with such velocities will at any time be small, but as long as the members of the cluster behave as mass points with a random isothermal distribution of velocities, the cluster will lose stars by this process of "evaporation" and will continually contract. This is the dynamical explanation of the fact that an isothermal gaseous sphere in a steady state and in an infinite space must have an infinite mass and an infinite radius.

The problem of relevance to globular clusters is clearly the determination of the length of time involved in such a process. To calculate this we must know the time of relaxation for an enclosed system of particles —the time required to establish a Maxwellian velocity distribution. Let this quantity be denoted by $\tau$, a function, in general, of the density and mean square velocity of the particles. Let $K$ denote the fraction of stars whose velocities would be greater than the velocity of escape if the velocity distribution were Maxwellian; this will vary only with the ratio of the velocity of escape to the root mean square velocity. At any epoch $t_0$ consider all those stars within some region of the cluster which have velocities less than the velocity of escape from that region. If the region were enclosed, then at an epoch $t_0 + \tau$ a fraction $K$ of these stars would have attained velocities exceeding the velocity of escape. Since the system is not enclosed, however, and since the mean free path of a cluster star is greater than the radius of the cluster, we assume that this fraction $K$ escapes during the interval $\tau$; hence the probability per unit time that any particular star escapes from the cluster will be $K/\tau$. Since $K$ is small this procedure should give a valid estimate of the rate of escape.

To calculate $\Lambda$, the actual rate of evaporation, we should average $K/\tau$ over all regions of the cluster. The true density distribution of clusters is not well known, however, and to obtain an order-of-magnitude result we shall consider an idealized isothermal cluster of uniform density. The time of relaxation will be constant throughout such a cluster and only $K$ need be averaged. The results of the calculations may then be applied to an approximate discussion of actual clusters.

In the first of the following six sections the value of $K$ is calculated for this simplified model. Section 2 is devoted to a rough determination of the best value of $\tau$. The resultant value of $\Lambda$ is given in section 3 together with an evaluation of $F(r)$, the number of escaping stars within a radius $r$ of the cluster center; a determination is also given of the deviation from a steady state introduced by the escape of stars from the cluster. Section 4 extends the analysis to centrally concentrated clusters; section 5 examines the effect of a dispersion in stellar masses. In section 6 the relevance of these results to the age and probable fate of globular and galactic clusters is briefly examined.

1. The idealized isothermal cluster may be assumed to have a radius $R$ and a total mass $M$, while the individual stars may be assumed each to have the same mass $m$. All quantities will be given in terms of a macroscopic system of units with the parsec, the solar mass, and the sidereal year replacing the centimeter, the gram, and the second, respectively. In these units the gravitational constant $G$ has the value $4.49 \times 10^{-15}$.

If $-\Omega$ represents the gravitational energy of the cluster and $T$ the total kinetic energy of all the cluster stars, then we have by the virial theorem*

$$T = \tfrac{1}{2}\Omega + \frac{1}{4}\frac{d^2 I}{dt^2}, \tag{1}$$

where $I$ is the moment of inertia of the cluster about its origin. The last term will be neglected here since, as shown in section 3, its effect on $T$ is small. For a uniform sphere we have

$$\Omega = \frac{3GM^2}{5R}, \tag{2}$$

where $G$ is the gravitational constant; if we define $w^2$ as the mean square velocity of the cluster stars, the kinetic energy is given by

$$T = \tfrac{1}{2}Mw^2. \tag{3}$$

From (1), (2) and (3) we find

$$w^2 = 0.6\frac{GM}{R}. \tag{4}$$

Let $P(v)\,dv$ be the probability that the velocity of any particular cluster star lies between $v$ and $v + dv$. Then, if the mean square

---

* H. Poincaré, *Leçons sur les hypothèses cosmogoniques*, p. 94; A. S. Eddington, *M.N.*, **76**, 525, 1916.

velocity is $w^2$, the Maxwellian distribution is given by

$$P(v) = 3\left(\frac{6}{\pi}\right)^{1/2}\frac{v^2}{w^3}\exp\left(-\frac{3v^2}{2w^2}\right).$$   (5)

For any value of the escape velocity $v_\infty$, $K$ is simply

$$K = \int_{v_\infty}^\infty P(v)\,dv;$$   (6)

the introduction of (5) into (6) and the use of the substitution

$$z = \frac{3v_\infty^2}{2w^2}$$   (7)

yields

$$K = 2\left(\frac{z}{\pi}\right)^{1/2}e^{-z} + 1 - \operatorname{erf} z^{1/2}.$$   (8)

The function erf $x$ is the usual error function of $x$.

Since at the surface $v_\infty^2 = 2GM/R$, it follows that at the boundary of a cluster of uniform density $z$ equals five; from (8) we find that $K$ in this case equals $1.86 \times 10^{-2}$. When the mean square velocity of escape, which was shown in the introduction to equal $4w^2$, is used in (7), $z$ equals 6 and $K$ becomes $7.4 \times 10^{-3}$. In averaging $K$ over a uniform sphere we must take the usual polytropic function* for the polytropic index zero. In the usual notation we find for the velocity of escape as a function of distance from the center of the cluster

$$v_\infty^2 = \frac{2GM}{R}(1 + \tfrac{1}{2}\theta_0);$$   (9)

with the use of (4) this yields, in (7),

$$z = 5(1 + \tfrac{1}{2}\theta_0).$$   (10)

The substitution of (10) into (8) and integration over the mass of the cluster is tedious but straightforward, involving integration by a series expansion. The final average value of $K$ is found to equal $8.8 \times 10^{-3}$.

* S. Chandrasekhar, *Stellar Structure* (University of Chicago Press, 1939), 91.

2. The time of relaxation $\tau$ is a rough measure of the time required for a gaseous assembly to attain an approximately Maxwellian distribution of velocities. A precise definition would be of little use here since the calculation of $\tau$ is not accurate to much better than half an order of magnitude. A rigorous analysis has been carried through only for the case of inverse fifth-power forces between the particles of an assembly, and even then only in so far as the actual distribution of velocities can be neglected. In such a special case, $\tau$, defined as the time in which deviations from a Maxwellian distribution fall to $1/e$ of their original value, is comparable with the mean time between collisions. This relationship should hold approximately for other laws of force as well. The point is essentially that by the time each particle has lost its original velocity, on the average, the Maxwellian distribution will have prevailed.

In the present instance it is difficult to know precisely what a collision, or an encounter, is, since in the case of inverse-square attraction between stars the deflection produced by a single encounter decreases so slowly with increasing distance of closest approach between the two stars that the cumulative effect of many distant encounters per unit time is actually greater than the effect of a few close encounters. One may take for $\tau$ the time required on the average for a single star to be deflected through an angle of $\pi/2$. The analysis for this case has been given by Jeans[†] and by Smart.[‡]

We let $\tau_s$ denote the value of $\tau$ found from a consideration of single deflections alone. If $\nu$ is the number of stars per cubic parsec, and $V$ is their relative velocity, we have Smart's formula

$$\tau_s = \frac{V^3}{4\pi G^2 m^2 \nu}. \tag{11}$$

The average value of $V^3$ may be found from the previous section. If we assume again that all stars have the same mass $m$, then the mean square of the relative velocity $V$ equals twice $w^2$, the mean square of

---

[†] J. H. Jeans, *Astronomy and Cosmogony* (Cambridge University Press, 1929), 318. The formulæ as given by Jeans are not in the most convenient form, since they refer to velocities and distances relative to the center of mass of the two stars. In these units, furthermore, his use of the formula $\pi\nu V p^2$ for the frequency of collisions is apparently incorrect and leads to a value of $\tau_s$ eight times too great.

[‡] W. M. Smart, *Stellar Dynamics* (Cambridge University Press, 1938), 318.

the individual velocity, and we have from (4)

$$\overline{V^2} = 2w^2 = 1.2\frac{GM}{R}.$$ (12)

From the Maxwellian distribution law it may be shown that

$$\overline{V^3} = \frac{2^{7/2}\left(\overline{V^2}\right)^{3/2}}{3^{3/2}\pi^{1/2}} = 3.47w^3.$$ (13)

Since $\nu$ equals $3N/4\pi R^3$, where $N$, the total number of stars in the cluster, is equal to $M/m$, we have finally from (11), (12), and (13)

$$\tau_s = 0.538N^{1/2}R^{3/2}/m^{1/2}G^{1/2}.$$ (14)

The consideration of single deflections alone gives rather a poor approximation, since the cumulative effect of the more distant encounters is actually more important. Jeans* has shown that such encounters will on the average produce a cumulative deflection of $\pi/2$ in a time[†] which may be expressed in the form

$$\tau = \frac{\pi^2}{32\ln(\pi/2\delta)}\tau_s,$$ (15a)

where ln denotes the natural logarithm; $\delta$ is the deflection produced by an encounter with a star at the interstellar distance $\nu^{-1/3}$. Expressing $\delta$ in terms of $G$, $m$, $\nu$, and $V^2$, and determining $\overline{V^2}$ from (12), we find

$$\tau = \frac{0.20}{\log N - 0.18}\tau_s,$$ (15b)

where $\log N$ is the common logarithm of $N$.

The time $\tau$ has also been calculated by Schwarzschild[‡] from rather different considerations. This other approach rests on the assumption that $\tau$ is equal to the time in which the average interchange of energy between the stars is just equal to the average value of the kinetic energy

* J. H. Jeans, loc. cit.
[†] W. M. Smart, *Stellar Dynamics* (Cambridge University Press, 1938), 320.
[‡] K. Schwarzschild, *Seeliger Festschrift*, 94, 1924.

per star. The method leads to the formula

$$\tau = \frac{0.044}{\log N - 1.6} \, \tau_s,$$ (16)

provided that $\log N$ is considerably greater than unity, and that $\tau_s$ is taken as a parameter whose value is defined by (14).

These two results (15b) and (16) differ by a factor of 4.5, which is largely accounted for by the different average values of $V$ which are used in the two formulæ. In Schwarzschild's analysis $V^{-2}$ and $V^2$ are averaged over a Maxwellian velocity distribution, whereas in the derivation of (15a) from Jean's formula, the average value of $V^8$ has been used, increasing $\tau$ by a factor of 3.7. The encounters with low relative velocity are certainly not so effective as the Schwarzschild treatment assumes, since other stars will usually intervene before a very slow encounter has progressed very far. On the other hand, in computing (14) one should average the square of the total deflection per unit time over all velocities, rather than the time necessary to produce a given deflection. This would decrease the value of $V$ used in (14) and would lead to a smaller $\tau$. Unless some method were used, however, to eliminate the effect of the lowest-velocity encounters, this process would yield the average value of $V^{-3}$, which is infinite. This approach is hence not a convenient one. To obtain an order of magnitude result we may take a rough average of the two formulæ, and set

$$\tau = \frac{0.10}{\log N - 0.5} \, \tau_s,$$ (17a)

with an uncertainty of somewhat less than half an order of magnitude. Substituting from (14) for $\tau_s$ we see that (17a) becomes

$$\tau = 8.0 \times 10^5 \, \frac{N^{1/2}R^{3/2}}{m^{1/2}(\log N - 0.5)} \text{ years,}$$ (17b)

where $R$ and $m$ again denote the radius of the cluster in parsecs and the mass of a cluster star in units of the solar mass.

The values of $\tau$ found from (17b), with $m$ set equal to unity, are shown in Table 1 below for various values of $N$ and $R$. These are maximum values for any isothermal cluster with a finite radius $R$, and an average mass $m$ equal to unity, since any change in the density distribution will increase $\nu$ in (11) more than it will increase $V^3$. If one considers a polytrope with $n$ equal to 3, for instance, the average value

TABLE 1
Time of Relaxation in a Globular Cluster of Uniform Density

| $N$ | $R = 1$ | $R = 3$ | $R = 10$ | $R = 30$ | $R = 100$ |
|------|---------|---------|----------|----------|-----------|
| $10^2$ | $5.3 \cdot 10^6$ | $2.8 \cdot 10^7$ | $1.7 \cdot 10^8$ | $8.7 \cdot 10^8$ | $5.3 \cdot 10^9$ |
| $10^4$ | $2.3 \cdot 10^7$ | $1.2 \cdot 10^8$ | $7.2 \cdot 10^8$ | $3.8 \cdot 10^9$ | $2.3 \cdot 10^{10}$ |
| $10^6$ | $1.5 \cdot 10^8$ | $7.6 \cdot 10^8$ | $4.6 \cdot 10^9$ | $2.4 \cdot 10^{10}$ | $1.5 \cdot 10^{11}$ |
| $10^8$ | $1.1 \cdot 10^9$ | $5.6 \cdot 10^9$ | $3.4 \cdot 10^{10}$ | $1.7 \cdot 10^{11}$ | $1.1 \cdot 10^{12}$ |
| $10^{10}$ | $8.4 \cdot 10^9$ | $4.4 \cdot 10^{10}$ | $2.7 \cdot 10^{11}$ | $1.4 \cdot 10^{12}$ | $8.4 \cdot 10^{12}$ |

The values of the radius $R$ are given in parsecs; $N$ is the number of stars in the cluster. The values of $\tau$ give the number of years required to establish a Maxwellian velocity distribution among stars of solar mass. For a centrally concentrated cluster, $R$ is the radius containing half the mass.

of $V^3$ will be 4.0 times as great as in the case of a homogeneous cluster with the same mass and radius; the value of $\nu$ averaged over the mass of the cluster, however, will increase by a factor of about 10.

The values in the table are somewhat smaller than those usually given. Heckmann and Siedentopf, for instance, give $1.6 \times 10^{10}$ years for the time of relaxation in a cluster with $N$ equal to $10^6$ and $R$ equal to 10 parsecs. The analysis by Rosseland*, from which their value is derived, is based on the same physical principles as that by Schwarzschild. Rosseland's analysis, however, assumes that one of the stars in each encounter is at rest, and his results give an upper limit for $\tau$.

3. As we have seen in the introduction, we may divide $K$ by $\tau$ to find $\Lambda$, the probability of escaping from the cluster per unit time per star. This procedure assumes that in the time $\tau$ a Maxwellian distribution is established for energies four or five times the average. If the initial distribution were one in which all velocities were equal, this assumption would certainly be incorrect, since several encounters would be necessary to give some stars the relevant energies. The situation contemplated, however, is one in which a cut-off Maxwellian velocity distribution already exists, the cut-off coming at the escape velocity. In this case $K/\tau$ should give a fairly close approximation to $\Lambda$.

Combining (17$b$), then, with the value of $K$ found at the end of section 1, we have

$$\Lambda = 1.1 \times 1.0^{-8} m^{1/2} (\log N - 0.5)/N^{1/2} R^{3/2}. \tag{18}$$

The stipulation that the cluster does not lose an appreciable fraction of its mass during the short-time scale of $2 \times 10^9$ years leads to the

* S. Rosseland, *M.N.*, **88**, 208, 1928.

condition that $2 \times 10^9 \Lambda$ is less than unity, or that

$$\frac{NR^3}{(\log N - 0.5)^2} > 4.8 \times 10^2 m. \tag{19}$$

The number of stars evaporating from the cluster per unit time is simply $\Lambda N$. The root mean square velocity of the escaping stars will be roughly equal to that of the stars in the cluster. Hence the star density $\nu(r)$ will be given by

$$4\pi r^2 \nu(r)w = \Lambda N. \tag{20}$$

This expression is valid when $r$ is large compared to four times $R$, the cluster radius. A consideration of the distribution of velocities reduces the effective velocity in (20) by a factor of $\pi^{1/2}$. If we substitute from (18) for $\Lambda$, a simple integration yields for $F(r)$, the number of such stars within a sphere of radius $r$,

$$F(r) = 0.38(\log N - 0.5)r/R, \tag{21}$$

since, for a uniform sphere, $w^2$ is $0.6GM/R$. If we let $G(r)$ equal the number of stars within a cylinder of radius $r$ (the axis of which passes through the center of the cluster), then $G(r)$, as found from the usual integral equation, is also given by (21), with 0.59 replacing 0.38.

As the evaporation of stars from the cluster proceeds, each escaping star will carry away on the average a positive energy $\frac{1}{2}mw^2$ [†], and the energy of the cluster will accordingly diminish by the same amount. But from the virial theorem it follows that the total energy $U = T - \Omega$ is equal to $-T$, or $-\frac{1}{2}Mw^2$. Hence we have for the rate of loss of energy by the cluster the two equations

$$\frac{dU}{dt} = \frac{1}{2}w^2 \frac{dM}{dt}, \tag{22}$$

$$\frac{dU}{dt} = \frac{d}{dt}\left(-\frac{1}{2}Mw^2\right). \tag{23}$$

[†] This is an overestimate and gives an upper limit for $\gamma$ in (30).

Combining these two equations we have

$$w^2 \frac{dM}{dt} = -w^2 \frac{dM}{dt} - M \frac{dw^2}{dt}, \tag{24}$$

$$\frac{2}{M} \frac{dM}{dt} = -\frac{1}{w^2} \frac{dw^2}{dt}, \tag{25}$$

$$w \propto \frac{1}{M}. \tag{26}$$

Since $w$ is the proportional to $1/M$, the kinetic energy $T$ will also vary as $1/M$. Since by the virial theorem $T$ and $\Omega$ are proportional, we have

$$\Omega \propto \frac{GM^2}{R} \propto \frac{1}{M}, \tag{27}$$

from which it follows that

$$R \propto M^3, \tag{28}$$

$$I \propto MR^2 \propto M^7. \tag{29}$$

From this proportionality we may readily calculate $d^2I/dt^2$ in (1). We find from (18), (28) and (29) that (1) then becomes approximately

$$T = \tfrac{1}{2}\Omega \left\{ 1 + \gamma \left( \frac{\log N - 0.5}{N} \right)^2 \right\}, \tag{30}$$

where $\gamma$ is a numerical constant depending on the distribution of stars within the cluster; if the star density is assumed to be uniform, $\gamma$ equals 0.22. It is evident that if $N$ is greater than 20, this correction factor is less than one-tenth of 1 percent, and quite negligible. Hence the use of the virial theorem leads to consistent results, and the relationship between $T$ and $\Omega$ is essentially unchanged by the escape of stars.

4. An actual cluster may differ from the ideal configuration analyzed here in at least three important respects. In the first place the motions of the stars may not be at random. This will be the case, in part at least, for any rotating cluster. In the second place the stars may be much more concentrated towards the center of the cluster than they would be in the homogeneous model discussed here. Finally the different stars

may have different masses. Each of these possibilities will be discussed in turn in this section and the next.

In connection with the first point, it may be noted that if the kinetic energy of the cluster stars appeared largely in the form of a rapid rotation of the cluster about some fixed axis, the effective velocity temperature would be very much lower than would otherwise be the case, since this temperature refers to the relative motion of neighboring stars. Hence the random kinetic energy per star would be very much less than the energy of escape, and practically no stars would ever leave the cluster through this evaporation process. In such a case viscosity would act to accelerate the outer layers and eventually the cluster would eject stars, but at a very slow rate. Since such configurations would be very oblate, however, it is evident that they do not correspond to the observed clusters, as these are nearly spherical.

The case of a cluster with a considerable concentration of stars toward its center is a much more important one. To extend the analysis to such a case we may apply (19) to the inner core of a cluster, taking this core to include half the total mass $M$ of the cluster: the radius of the core will be denoted by $R_{1/2}$. One may expect that this core will not exhibit a marked density concentration toward its center. Even for the limiting case of the polytrope with $n$ equal to five, the ratio of the central density to the mean density in a sphere with half the total mass is only 4.4. Such a polytrope, it may be noted, corresponds to Schuster's law of density distribution for globular clusters, and gives apparently better agreement with the available observations than any other simple law.*

Because of this relatively slight density concentration, such an inner core will have properties not unlike those of a uniform sphere. At the boundary of this core the value of $v_\infty^2$ should be roughly equal to its average value for the cluster as a whole, which, as we have already seen, is four times the mean square velocity $w^2$. This expectation is verified by simple calculations based on the Emden function for the polytrope $n$ equals 5.† It is readily shown that when $M(r)$ equals $\frac{1}{2}M$ in this polytrope, $v_\infty^2$ equals 1.03 times its average value for the entire polytrope. Hence the value of $K$ for this inner core should not exceed $10^{-2}$.

The average value of $V^2$ given by (12) should be approximately correct even for this more general case, provided that $\frac{1}{2}M$ and the corresponding radius $R_{1/2}$ are used instead of $M$ and $R$. We may again make a comparison with the polytrope for $n$ equal to 5, for which we find by use of the virial theorem that the kinetic energy of the

* P. ten Bruggencate, *Sternhaufen* (J. Springer, 1927), 38 *et seq.*
† S. Chandrasekhar, loc. cit., 93.

entire polytrope per unit mass is $3\pi(2^{2/3} - 1)^{-1/2}/32 \times G(\tfrac{1}{2}M)/R_{1/2}$ or $0.384G(\tfrac{1}{2}M)/R_{1/2}$, which is only 28 percent greater than the corresponding quantity for an isothermal sphere of mass $\tfrac{1}{2}M$ and radius $R_{1/2}$.

If, then, we apply to the core of a cluster the formulæ developed for a uniform sphere, $K$ will be substantially correct, the value of $\nu$ used will be too small by a factor of not more than 2 or 3, and the average value of $V^3$ will be too small by some 50 percent; the resultant value of $\Lambda$ calculated from (18) will accordingly be too small by a factor of not more than 2, which is less than the inaccuracy inherent in the analysis in any case. Unless the inner half of a cluster is more centrally concentrated than the corresponding inner half of the polytrope for which $n$ equals 5, we may conclude that the general order of magnitude given by (18) is substantially correct.

5. The assumption that all stars have the same mass is obviously not very realistic. It is clear that stars whose mass is less than one-fourth of $m$, the average mass, will tend to leave the cluster, since the root mean square equilibrium velocity for such stars will be greater than the average velocity of escape. To find the rate of escape, however, one must investigate more closely the exchange of energy between stars.

If we consider a single encounter between star $A$ and star $B$, of mass $m_A$ and $m_B$, respectively, then the gain of energy of star $A$ is given by

$$\Delta E_A = \tfrac{1}{2}m_A(\mathbf{V}_g + \mathbf{U}_{A2})^2 - \tfrac{1}{2}m_A(\mathbf{V}_g + \mathbf{U}_{A1})^2, \qquad (31a)$$

where $\mathbf{V}_g$ is the velocity of the center of mass of the two stars; $\mathbf{U}_{A1}$ and $\mathbf{U}_{A2}$ are the initial and final velocities of star $A$ relative to the center of mass. The orbit of star $A$ relative to the center of mass is a simple hyperbola; $U_{A1}$, the scalar value of the vector $\mathbf{U}_{A1}$, is clearly equal to the corresponding $U_{A2}$.

If we introduce $\psi_1$, the angle between $\mathbf{U}_{A1}$ and $\mathbf{V}_g$, and $\psi_2$, that between $\mathbf{U}_{A2}$ and $\mathbf{V}_g$, we have from (31a),

$$\Delta E_A = m_A U_{A1} V_g(\cos \psi_2 - \cos \psi_1). \qquad (31b)$$

From Fig. 1 it is evident that $\psi_1 - \psi_2$ equals $\pi - 2\alpha$, where $2\alpha$ is the angle between the asymptotes of the relative orbit.

To determine $\alpha$ we have the relationship derived* from simple geometry and the conservation of energy and angular momentum

$$\cot \alpha = \frac{G(m_A + m_B)}{pV^2}, \qquad (32)$$

* Smart, loc. cit., 317.

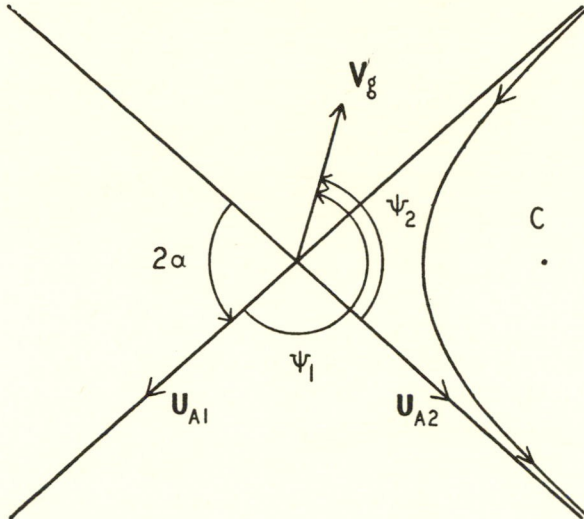

FIG. 1. The center of mass of the two stars is actually at C, but to clarify the meaning of the angles involved, its vector velocity $\mathbf{V}_g$ is shown by an arrow drawn at the intersection of the asymptotes of the orbit of one of the stars. The other star is not shown, but its orbit relative to C is a hyperbola of the same shape.

where $p$ is the distance at which the stars would have passed had there been no gravitational attraction. Hence we have from (31b), eliminating $\psi_2$, and using (32) to evaluate $\sin 2\alpha$ and $\cos 2\alpha$,

$$\Delta E_A = 2m_A U_{A1} V_g \left\{ \frac{Y \sin \psi_1 - \cos \psi_1}{1 + Y^2} \right\}, \qquad (33)$$

where

$$Y = \tan \alpha. \qquad (34)$$

To find $\tau$ by Schwarzschild's method,[*] one computes the average value of $(\Delta E_A)^2$. In the present case, however, the average value of $\Delta E_A$ is not zero. Although the average value of $\sin \psi_1$ obviously vanishes, since positive and negative values of $\psi_1$ are equally probable, $\cos \psi_1$ may have a non-zero average. From the definition of a scalar

---

[*] K. Schwarzschild, *Seeliger Festschrift*, 94, 1924.

product we see that

$$U_{A1}V_g \cos \psi_1 = \mathbf{U}_{A1} \cdot \mathbf{V}_g = (\mathbf{v}_{A1} - \mathbf{V}_g) \cdot \mathbf{V}_g$$

$$= \frac{m_B}{(m_A + m_B)^2} \{m_A v_{A1}^2 - m_B v_{B1}^2 + (m_B - m_A)\mathbf{v}_{A1} \cdot \mathbf{v}_{B1}\},$$

$$(35)$$

where $\mathbf{v}_{A1}$ and $\mathbf{v}_{B1}$ are the original velocities of the two stars in the inertial frame of the cluster. In the derivation of (35) use has been made of the usual formula for $\mathbf{V}_g$,

$$\mathbf{V}_g = \frac{m_A}{(m_A + m_B)} \mathbf{v}_{A1} + \frac{m_B}{(m_A + m_B)} \mathbf{v}_{B1}. \qquad (36)$$

To find the total energy gained per second by star $A$ we must first multiply (33) by $2\pi V p \, dp$, the number of times per year that star $A$ encounters a star of relative velocity $V$ in such a way that $p$ lies between $p$ and $p + dp$; $\nu$ is the number of stars of mass $m_B$ per unit volume. Then we must integrate over all relevant values of $p$. Since the infinite integral diverges, we may take $\nu^{-1/3}$ as the upper limit of integration over $p$, since in general encounters such that $p$ is greater than this limit will be screened by other intervening stars. This somewhat crude procedure is apparently the only way by which an analysis of single encounters can be made to apply to the present problem. Finally the resultant expression must be averaged over all values of $\theta$, the angle between $\mathbf{v}_{A1}$ and $\mathbf{v}_{B1}$, and averaged also over $v_{B1}$.

The integration over $p$ yields

$$2\pi \nu V \int_0^{\nu^{-1/3}} \Delta E_A \, p \, dp = -4\pi \nu V m_A U_{A1} V_g \cos \psi_1$$

$$\times \frac{1}{2} \frac{G^2(m_A + m_B)^2}{V^4} \ln\left(1 + \frac{V^4}{G^2(m_A + m_B)^2 \nu^{2/3}}\right); \quad (37)$$

if we substitute for $\cos \psi_1$ from (35) and drop the subscript 1 from $v_{A1}$

and $v_{B1}$, we find that (37) becomes

$$2\pi v V \int_0^{v^{-1/3}} \Delta E_A p \, dp = 2\pi v V m_A m_B G^2 \ln\left(1 + \frac{V^4}{G^2(m_A + m_B)^2 v^{2/3}}\right)$$

$$\times \left\{\frac{m_B v_B^2 - m_A v_A^2 + (m_A - m_B)\mathbf{v}_A \cdot \mathbf{v}_B}{V^3}\right\}. \quad (38)$$

Let $\theta$ equal the angle between $\mathbf{v}_A$ and $\mathbf{v}_B$. Then we have

$$V^2 = v_A^2 + v_B^2 - 2\mathbf{v}_A \cdot \mathbf{v}_B, \quad (39a)$$

$$\mathbf{v}_A \cdot \mathbf{v}_B = v_A v_B \cos \theta. \quad (39b)$$

To average over $\theta$ we multiply (38) by $\frac{1}{2}\sin\theta \, d\theta$ and integrate from zero to $\pi$. Unless $v_A$ is nearly equal to $v_B$ the logarithmic term will not change appreciably in the course of the integration, and we may take it outside the integral with its value for $\cos\theta$ equal to unity. Since the relative change in the logarithm with $\theta$ becomes less as $v$ decreases, this procedure becomes asymptotically correct with decreasing $v$. If we let $dE_A/dt$ denote the average of (38) over $\theta$, and let $x$ equal $\cos\theta$, then we find

$$\frac{dE_A}{dt} = \frac{1}{2}Q \int_{-1}^{+1} \frac{m_B v_B^2 - m_A v_A^2 + (m_A - m_B)v_A v_B x}{(v_A^2 + v_B^2 - 2v_A v_B x)^{3/2}} \, dx, \quad (40)$$

where

$$Q = 2\pi v m_A m_B G^2 \ln\left(1 + \frac{(v_A - v_B)^4}{G^2(m_A + m_B)^2 v^{2/3}}\right). \quad (41)$$

The integration is elementary and yields

$$\frac{dE_A}{dt} = \frac{1}{2}Q\left\{\frac{m_B v_B^2 - m_A v_A^2}{v_A v_B}\left(\frac{1}{|v_A - v_B|} - \frac{1}{v_A + v_B}\right)\right.$$

$$\left. + \frac{m_A - m_B}{v_A v_B}\left(\frac{v_A^2 + v_B^2 - v_A v_B}{|v_A - v_B|} - \frac{v_A^2 + v_B^2 + v_A v_B}{v_A + v_B}\right)\right\}, \quad (42)$$

where $|v|$ denotes the absolute value of $v$. If $v_A$ is greater or less than

$v_B$ respectively, (42) becomes

$$\frac{dE_A}{dt} = \begin{cases} -Qm_A/v_A & v_A > v_B, \\ Qm_B/v_B & v_A < v_B. \end{cases} \tag{43}$$

If (43) is integrated over a Maxwellian distribution for $v_B$, one obtains the average rate of increase in energy for stars of mass $m_A$ and velocity $v_A$. This rate is a quantity of not much significance, however, since it bears little relation to the rate of increase in the number of stars with velocities greater than $v_A$. It is rather the average of the rate of increase of $E_A$ over all values of $v_A$ as well that gives the true rate of approach toward equipartition. The velocity distribution, over which this average should be taken, is unknown. To carry through the integration we shall assume a Maxwellian distribution with, of course, a different mean square velocity from that which equipartition of energy with the more massive stars would establish. It is readily shown that other reasonable distributions do not change the results appreciably.

Let $\bar{E}_A$ and $\bar{E}_B$ denote the average values of $E_A$ and $E_B$, respectively; from (43) we find by straightforward integration,

$$\frac{d\bar{E}_A}{dt} = \left(\frac{3}{\pi}\right)^{1/2} Q \frac{\bar{E}_B - \bar{E}_A}{\left(\bar{E}_A/m_A + \bar{E}_B/m_B\right)^{3/2}}. \tag{44}$$

We may assume that $v$, $\overline{v_B^2}$ and $m_B$ are equal to the number of all stars per cubic parsec, the mean square velocity $w^2$ and the average mass $m$ (or $M/N$) respectively. Then if we substitute from (41) for $Q$, and recall formulae (11), (13) and (17a) for $\tau$, (44) becomes, on integration,

$$t = 0.68\tau \int \frac{(1+y)^{3/2}}{1 - m_A y/m_B} \, dy, \tag{45}$$

where we have introduced the new variable

$$y = \frac{m_B \bar{E}_A}{m_A \bar{E}_B} = \frac{\overline{v_A^2}}{\overline{v_B^2}}. \tag{46}$$

In deriving (45) from (44) one should replace the constant $-0.5$ in the denominator of (17b) by $-0.7$; this change, however, is negligible. Since (44) is much more accurate than any of the formulae derived for the time of relaxation $\tau$, (45) may be regarded simply as a parameter whose value is defined by (11) and (17a).

When $m_A$ is greater than $m_B$, and $y$ is less than unity, the factor $(1 + y)^{3/2}$ in the integrand of (45) may be neglected, and we have

$$\frac{m_A}{m_B} y = \frac{\overline{E}_A}{\overline{E}_B} = 1 - C \exp\left(-1.47 \frac{m_A t}{m_B \tau}\right), \qquad (47)$$

where $C$ is a constant of integration. Hence equipartition of energy for stars more massive than the average will be attained in the time $\tau$ or less. It follows that stars of relatively high mass will rarely be ejected from the cluster. A solution similar to (47) may be found when $y$ is large, provided that $m_A y/m_B$ is nearly unity; in this case $(m_A/m_B)^{5/2}$ replaces $m_A/m_B$ in the exponent.

When $m_A y/m_B$ is small we may expand the denominator in (45). In this way we find

$$t = 0.27\tau(1 + y)^{5/2}$$

$$\times \left\{1 + \frac{1}{7} \frac{m_A}{m_B}(5y - 2) + \frac{1}{63}\left(\frac{m_A}{m_B}\right)^2 (35y^2 - 20y + 8) \cdots \right\}.$$

$$(48)$$

It is evident from (48) that within a time $\tau$ stars of small mass will increase their average energy by about 50 percent, provided that all stars had initially the same velocity. This would account for a considerable stratification of stars of different mass in the extended atmospheres of globular clusters.

Most stars of small mass will escape from the cluster if $y$ is about 4, since, as we have seen in the introduction, the root mean square velocity of the average cluster star is one-half the average velocity of escape. The increased rate of evaporation with increasing velocity, however, will eliminate most of these less massive stars before $y$ has reached so high a value. More specifically, a star of zero mass and originally of average velocity will require a time equal to $14\tau$ to double its velocity; on the other hand, a time of $2.7\tau$ will suffice to increase $y$ from 1 to 2, increasing $K$—the probability of evaporation of a single star during a time equal to the time of relaxation $\tau$—by approximately 10. Since the average value of $K$ is about 0.01, as we have seen in section 1, it is evident that at this increased rate a time of $10\tau$ will suffice to eliminate most of the less massive stars. While this effect will be diminished by the tendency of the less massive stars to rise further from the center of the cluster into regions of low star density, one may nevertheless conclude that by the time 20 percent of the average stars have been lost, most of the stars less massive than one-fourth the average will have left the cluster. If the cluster is sufficiently old, this effect should produce a sharp cut-off at the lower end of the mass distribution function.

6. These results may be applied to actual clusters, both globular and galactic. The values of $N$ and $R_{1/2}$ for globular clusters are not well known. A lower limit of $10^5$ may safely be adopted for $N^*$; since $R_{1/2}$, the radius containing half the mass of the cluster, is scarcely less than 5 parsecs, and $m$ is not much greater than unity, (19) is obviously satisfied by a wide margin for all globular clusters. Similarly, it is evident from (18) that for such clusters $\Lambda$ is less than $1.5 \times 10^{-11}$ years$^{-1}$; hence a globular cluster will lose not more than roughly half its mass in $10^{11}$ years.

Even for the long-time scale of $2 \times 10^{12}$ years the loss of stars by evaporation is not necessarily serious. In this case, the right-hand side of (19) is increased by a factor of $10^6$. Even with the extreme value of $R$ above, a cluster with $m$ equal to 0.2 need contain only some $10^8$ stars to be stable during so long an interval. The loss of the less massive stars, however, which is just on the verge of becoming important in $10^9$ years, would presumably be complete in this latter case.

It is evident from (21) that the number of escaping stars in the neighborhood of a cluster depends only logarithmically on the total population of the cluster. If log $N$ is 6.5 and $R_{1/2}$ is 6 parsecs, $G(r)$, the number of stars on the photographic plate within a radius $r$ parsecs from the center of the cluster, should be equal to $0.6r$; thus there should be 120 stars moving outwards from the cluster within a radius of 200 parsecs. This quantity depends on the determination of $\tau$, and is therefore uncertain to within half an order of magnitude. Any dispersion in the masses of cluster stars will increase $G(r)$; the results here give a lower limit for this quantity.

It has been known for some time that a few cluster-type Cepheids may be observed far from the center of globular clusters. Recent investigations by Kopal[†] indicate that the above order of magnitude is at least a lower limit for the distribution of these short-period variables. Star counts in the vicinity of globular clusters would be valuable in determining the total number and distribution of the stars associated with these clusters and in providing a test for the theory.

---

* H. Shapley, *Mt. Wilson Contr.*, **152**, 27, 1918.
† Z. Kopal, unpublished work, to appear shortly as a *Harvard Circular*.

It is of interest to note that a mass of $10^9$ suns is an upper limit for the observed clusters, since otherwise we should frequently observe direct collisions between cluster stars. If two stars, each of radius $r_0$, pass within a periastron distance $q$ of each other, where $q$ equals $0.29r_0$, corresponding to an $M(r)$ equal to $\frac{1}{2}M(r_0)$ on the polytropic model $n$ equal to 3, then an enormous liberation of energy will certainly take place. As Whipple has suggested*, this is perhaps the mechanism responsible for supernovæ. At any rate, such a cataclysm would lead to a much greater increase in the rate of energy radiation than is observed for ordinary novæ; one may assume that if more than one such collision occurred every twenty years in the globular clusters of our own galaxy, at least one such would have been observed.

Following Whipple, we have for $\Gamma$ the total number of collisions per year in a homogeneous cluster of uniform density

$$\Gamma = \frac{2(3\pi)^{1/2}GmqSv^2}{w}, \tag{49}$$

where $m$ is the mass and $w$ is the root mean square velocity of a cluster star; $S$ is the volume of the cluster and $v$ is again the number of stars per cubic parsec. As in section 2, we may determine $w$ from formula (4), and if $R$ is again the radius of the cluster, we have

$$\Gamma = \frac{3}{2}\left(\frac{5}{\pi}\right)^{1/2}\frac{G^{1/2}M^{3/2}q}{R^{5/2}m}. \tag{50}$$

Let $t_c$ be defined as the average interval in years between such collisions in each cluster; then we have simply

$$t_c = 1/\Gamma = 3.5 \times 10^{14}R^{5/2}m/M^{3/2}q', \tag{51}$$

where $q'$ denotes the value of $q/R_\odot$.

If we set $q$ equal to $0.29r_0$, and assume that the stars under consideration have the same mean density as the Sun, then we find for $M$,

$$M = 1.1 \times 10^{10}R^{5/3}m^{4/9}t_c^{-2/3}. \tag{52}$$

If seventy globular clusters produce one such encounter every twenty years, $t_c$ equals 1400 years. The integrated magnitudes of globular clusters indicate that if $M$ is as great as $10^8$, $m$ must be considerably less than unity; we may accordingly set $m$ equal to 0.1. As in section 3,

---

* F. L. Whipple, *Proc. Nat. Acad. Sciences*, **25**, 118, 1939.

these results may be applied to the inner core containing half the mass of the cluster. Since the radius of such a core will not exceed 10 parsecs in general, (52) gives for $M$ the upper limit

$$M < 1.5 \times 10^9. \tag{53}$$

Any rotation of the cluster as a whole will decrease the relative velocity and will decrease $M$ in (53). Since in addition the observations probably indicate a considerably greater value of $t_c$ than 1400 years, one may infer that (53) is a generous upper limit for $M$.

The galactic clusters present an essentially different picture. As Bok[*] has recently shown, a galactic cluster will gain energy by encounters with field stars; this will lead to an expansion of the cluster until it can no longer hold together under the disruptive influence of galactic rotation. The effect is greatest for extended, diffuse configurations. The process of evaporation analyzed here, however, leads to a contraction of the cluster and is greatest for the densest clusters. Hence galactic clusters may be divided into two classes: the loose clusters, which are expanding and will gradually break up, and the dense clusters, which are contracting and ejecting stars.

Such clusters as the Pleiades, the Hyades, Praesepe, and Messier 37 are clearly loose clusters in this sense. Their radius is great enough to satisfy (19) by a considerable margin, while the galactic effects will, as shown by Bok, disrupt these clusters in about $2 \cdot 10^9$ years.

For some of the smaller aggregations, however, the galactic effects are negligible and such clusters are presumably contracting. Cuffey[†] has recently determined color indices and hence parallaxes and true diameters for several of these objects. The smallest cluster on his programme, N.G.C. 2129, contains 40 stars, and is found to have a diameter of 0.8 parsec. The star density in this cluster is about 150 stars per cubic parsec, more than a thousand times the density in the neighborhood of the Sun. So dense a cluster will not be disrupted by any of the galactic effects in a time comparable with $10^9$ years. While the analysis leading to (15) and (16) breaks down when $\log N$ is of order unity, (14) still gives an upper limit for $\tau$; setting $R_{1/2}$ equal to 0.3 and $m$ equal to unity, we see that the time of relaxation is less than $8 \cdot 10^6$ years. Hence the cluster should lose 1 percent of its mass every $10^7$ years, or one star every $2 \cdot 10^7$ years. If these views are correct, either the cluster must have been formed more recently than $2 \cdot 10^9$ years ago, or else we now observe a cluster which has lost at least half its original members. From

[*] B. J. Bok, *Harvard Circular* No. 384, 1934.
[†] J. Cuffey, *Harvard Annals*, **106**, No. 2, 1937.

the previous section it follows that this cluster should contain very few stars of mass much less than the average.

It is of some theoretical interest to investigate in this connection the eventual fate of an isolated cluster.[‡] It is evident from section 3 that the loss of stars from a cluster proceeds at a continually accelerating rate as the cluster contracts. On the other hand, it is obvious that the cluster has not sufficient energy to eject all its stars to infinity. It may be readily shown that the mean free path of a cluster star is always greater than the radius of the cluster, provided that the cluster has more than ten stars. Hence the evaporation of stars would presumably proceed until one of two alternatives occurred. Collisions between stars could become important. Or, as the cluster continued to lose stars, the remaining ones might conceivably find themselves in periodic orbits, with no encounters to upset things. If, for instance, the cluster possessed angular momentum, it would presumably become a highly oblate ellipsoid rotating with an angular velocity varying along the radius. If the stars in such a disc had very little random velocity, this would be very nearly a permanently stable configuration.

## SUMMARY

The fact that an isothermal gaseous sphere has an infinite mass and an infinite radius is shown to correspond to the dynamical fact that a finite, nearly isothermal sphere will eject matter to infinity. The fraction of stars which have velocities greater than the escape velocity is calculated for a simplified ideal cluster with uniform density by the use of the virial theorem and the Maxwell-Boltzmann distribution of velocities. A deter-

---

[‡] Evaporation of stars from a cluster and the resultant contraction and evolution of the cluster have been qualitatively discussed by R. H. Dicke (*Astr. J.*, **48**, 108, 1939) in a paper which appeared after this work had been completed.

The recent investigation by H. Mineur (*Annales d'Astrophysique*, **2**, 167, 1939) refers not to the loss of stars by evaporation but to the shearing effect of galactic rotation. His result that the rate of loss from this source alone becomes infinite as the cluster becomes more and more compact seems to contradict the present analysis, but actually results from the use of the steady-state approximation. As in the case of a nearly isothermal sphere, this approximation is not valid far out from the effective cluster radius. The radius of the critical equipotential surface $S_1$ discussed by Mineur (all stars passing through $S_1$ are assumed to leave the cluster) will clearly be independent of the cluster radius for a spherical cluster, and will in fact vary only with the cluster mass. Hence if the density of the cluster is high, the radius of the cluster will be very much less than the radius of $S_1$, and the steady-state approximation cannot be used to find $\rho_1$, the density at $S_1$, or $\epsilon$, the rate at which stars pass through $S_1$. In the case of a compact cluster, most of the stars at $S_1$ will be those which have already left the cluster through the process of evaporation discussed here, and which are moving outwards with velocities greater than the velocity of escape. The rate of loss in such a case is given by (18), not by Mineur's formulae.

mination of the time of relaxation gives the rate at which an equilibrium velocity distribution is established, and hence determines the rate at which stars leave the cluster through this process of "evaporation."

The validity of the formulae is extended to include the case in which the density increases toward the center of the cluster; computations are made for the polytrope $n$ equal to 5, corresponding to the Schuster density law. A determination is made of the rate at which energy becomes equalized between stars of unequal masses; it is shown that a cluster which has lost 20 percent of its total mass through evaporation will contain very few stars of mass less than one-fourth the average.

The calculations indicate that even in $10^{11}$ years, however, a globular cluster will not lose more than half of its stars through this process of evaporation. For the short-time scale the effect on the evolution of such clusters is therefore quite negligible. The number of escaping stars within a radius $r$ of the cluster should vary directly with $r$ and should depend only logarithmically on the population of the cluster; if half the mass of the cluster is concentrated within a sphere of radius 6 parsecs, 120 such stars should be observed within 200 parsecs of the center. The fact that no stellar collisions have been observed in globular clusters leads to an upper limit of $10^9$ suns for the mass of an average cluster.

In the galactic clusters the effect may be more important because of the lower average stellar velocities. Such clusters may be divided into two classes: the loose extended configurations expanding and dissipating under galactic influences, the denser ones ejecting stars and contracting. The cluster N.G.C. 2129, for instance, with 40 stars within a radius of 0.4 parsec, should eject one star every $2 \times 10^7$ years; the rate of loss, moreover, should increase with the time. The possible fate of such systems is briefly discussed.

The author is very grateful to Dr. Shapley for the use of the (Harvard) observatory facilities, to Dr. Schwarzschild and Dr. Kopal for several suggestions, to Dr. Bok for many valuable discussions, and to the National Research Council for a fellowship during the past year.

*Sloane Physics Laboratory*
*Yale University*
*January* 1940

# STELLAR POPULATIONS AND COLLISIONS
# OF GALAXIES

(WITH W. BAADE*)

(ASTROPH. J., 113, 413, 1951)

## *Commentary*

WHEN Martin Schwarzschild and I moved to Princeton in 1947, we started a program of Visiting Professors, inviting active and stimulating astronomers to spend a month or more in Princeton and to give a series of lectures for our small group. Chandrasekhar came in the first year of the program, Walter Baade and Bengt Strömgren in the second. In the course of Baade's fascinating lectures on stellar populations, he discussed S0 galaxies, which have about the same shape as spirals—disk plus bulge—but lack spiral arms and evident dust lanes. These systems are relatively much more frequent in rich clusters of galaxies than in the general field.

Perhaps because of a passion for science fiction in my youth, I was fascinated by spectacular physical phenomena. It occurred to me that S0's might result from collisions between two spirals. I enjoyed trying to visualize what would happen in such a collision. It soon became apparent that at the high relative speeds characteristic of dense clusters a collision between two galaxies would have very little effect on the stars themselves, and the two stellar systems would simply interpenetrate and then separate with little change. The gas, however, would behave quite differently because of the short mean free path for the atoms involved. While the detailed behavior would be difficult to predict, relatively little gas would likely remain in either of the two interacting galaxies. These conclusions were, of course, very tentative in view of limited knowledge, but we felt that the suggested mechanism was a promising one.

During the decades since Paper #14 was published, the observational situation has, of course, changed tremendously. The distance scale we assumed was based on a Hubble constant, $H_0$, of 530 km s$^{-1}$ Mpc$^{-1}$, as compared with the present range of values[1] between 50 and 90 km s$^{-1}$

---

* Visiting professor at Princeton University during March 1950, when most of this paper was written.

Mpc$^{-1}$. If we take a rough value of 70 km s$^{-1}$ Mpc$^{-1}$ for this quantity, our distances and times must all be increased by a factor 7.6. The galactic radii we assumed in 1950 may have been estimated more from the size of our own Galaxy than from the angular size of distant systems, but the extended radii found from the stronger absorption-line systems in quasars (see Commentary, Paper #8) suggest that the relevant galactic radii should be increased by the same factor used for galactic distances. This range of radii then becomes 15–38 kpc.

With all distances increased in proportion to the Hubble time $1/H_0$, the number of collisions per galaxy in Table 1 remains valid, provided that the time interval is also increased in proportion to $1/H_0$. It follows that with $H_0 = 70$ km s$^{-1}$ Mpc$^{-1}$, Table 1 then refers to a time of $23 \times 10^9$ yrs, not much longer than the estimated globular cluster age of some $17 \times 10^9$ yrs. If the assumed radii are realistic, one may conclude that galaxies whose orbits pass through the central regions of a rich cluster have probably experienced an appreciable number of interpenetrating collisions with other galaxies.

Another result of modern observations is that most S0 galaxies do not look like spirals whose gas has been stripped. S0 bulges tend to be larger and brighter than those in spirals, with a much wider distribution of central velocity dispersions than for any other galactic type;[2] in the S0 disks the stellar surface brightness decreases outward more steeply than does the brightness of spiral disks.[2] S0 colors indicate that the stars in their disks are systematically younger, by 3 to 5 Gyrs, than the bulge stars.[3] To fit the detailed spectra, some A to F stars, as well as occasionally some emitting H and O$^+$ gas, are required.[4]

These results on the stellar content of S0's suggest that if the cluster environment has led to gas stripping of former spirals, the same environment has also produced changes in the stellar populations, increasing in particular the size and luminosity of the central bulges. It is not obvious whether gas stripping can produce such a result, but other processes occurring in dense systems, such as the merger of a spiral with a small elliptical galaxy, might do so. It is also possible, as suggested several times, that the observed correlation between rich clusters and a relatively high frequency of S0 galaxies might result from the specific processes by which a cluster of galaxies forms within a primordial condensation.

While the origin and evolution of S0 systems is still obscure, some gas stripping must certainly occur, especially in rich clusters. Indeed, images of several galaxies and their neighboring gas clouds have been interpreted[5] in terms of such a process. Hence more detailed analyses of gas stripping, which have appeared frequently in the last two decades, are of continuing interest. One of these studies[6] applies the collision process in Paper #14 to a galaxy with a cloudy interstellar medium, and concludes that cold diffuse H I clouds will be swept out in a single collision, while giant molecular clouds, assumed to have a greater mass per unit area, will not; the weakness of H I compared to $H_2$ in spiral galaxies of the Virgo and Coma clusters has been attributed to such an effect.[6,7]

Stripping by ram pressure as a galaxy moves through the hot gas of a rich cluster has also been proposed.[8] The pressure produced by such motion is certainly high, but the hot gas is of low density, and its push against the denser interstellar gas in the disk of a spiral galaxy will tend to produce Rayleigh-Taylor instabilities. While a suitably specified magnetic field could, perhaps, produce stability, this process seems much more likely to churn up the interstellar gas in the disk rather than to push it out of the galaxy. The Kelvin-Helmholtz instability produced by shear flow can lead to similar effects; the resultant turbulent convection has been cited[9] as the basis for increased ablation. Fluid dynamical simulations[10] indicate that ram pressure can strip gas of very low density from an elliptical galaxy in a cluster. Similarly, this process could also be effective for halo gas in spiral systems and might lead to a gradual loss of mass from such a galaxy as a whole.[11] Evaporation of gas by thermal conduction from the hot cluster gas has also been suggested[12] as a gas stripping process. It is not yet evident which of these and other processes will give the most effective stripping of galaxies under various conditions.

### REFERENCES

1. Jacoby, G. H. et al.,—*Publ. Astr. Soc. Pacific* **104**, 599 (1992); Saha, A. et al. —*Astroph. J.* **425**, 14 (1994).
2. Dressler, A. and Sandage, A.—*Astroph. J.* **265**, 664 (1983).
3. Bothun, G. D. and Gregg, M. D.—*Astroph. J.* **350**, 73 (1990).
4. Gregg, M. D.—*Astroph. J.* **337**, 45 (1989).
5. Rood, H. J. and Williams, B. A.—*Astroph. J.* **288**, 535 (1985); van Driel, W. and van Woerden, H.—*Astron. Astroph.* **243**, 71 (1991).
6. Valluri, M. and Jog, C. J.—*Astroph. J.* **357**, 367 (1990).
7. Casoli, F., Boissé, P., Combes, F., and Dupraz, C.—*Astron. Astroph.* **249**, 359 (1991).
8. Gunn, J. E. and Gott, J. R.—*Astroph. J.* **176**, 1 (1972).
9. Nulsen, P.E.J.—Monthly Notices, *Roy. Astr. Soc.* **198**, 1007 (1982).
10. Balsara, D., Livio, M., and O'Dea, C. P.—*Astroph. J.* **437**, 83 (1994).
11. Ostriker, J. P.—Informal communication (1994).
12. Cowie, L. L. and Songaila, A.—*Nature* **266**, 501 (1977).

## *Paper*

### Abstract

Dense clusters of galaxies, such as the Coma and Corona clusters, contain large numbers of S0 galaxies, which are highly flattened but show no obscuration or spiral structure and presumably contain stars only of population type II. It is suggested that collisions between galaxies sweep any interstellar matter out of the galaxies in such clusters and thereby prevent the appearance of any type I systems. In the Coma cluster each galaxy will collide with at least twenty other galaxies during $3 \times 10^9$ years if the galactic motions within the cluster are largely radial. An analysis of galactic collisions shows that two galaxies will interpenetrate freely with relatively little effect on the velocities and positions of the stars in each galaxy but that interstellar matter will be swept completely out of the two galaxies and left in intergalactic space, provided that the initial density is not much lower than 0.1 H atoms per cubic centimeter. Thus highly flattened systems, which might otherwise retain interstellar matter and develop into normal spirals, become pure type II systems.

## I. Introduction

The existence of broad differences between stars of population types I and II[1] seems fairly well established. In particular, the highly luminous stars of population I, such as main-sequence O and B stars, supergiant stars of types F–M, and classical cepheids, are found almost entirely in spiral arms of late-type spiral galaxies and in irregular galaxies, such as the Magellanic Clouds. In elliptical galaxies and outside the spiral arms in spiral galaxies only stars of population II are found, with a sharp cutoff in the luminosity function at $M_v$ equal to about $-2.5$ mag.

Theory indicates and observation confirms that the presence of dense interstellar matter is both a necessary and a sufficient condition for the appearance of population type I. The stars as bright as absolute magnitude $-5$, which are so characteristic of type I, cannot have been shining at this present rate for anything like $3 \times 10^9$ years, and theories have been advanced[2] which may explain the formation of supergiant stars

---

[1] Baade, W., *Ap. J.*, **100**, 137, 1944.

[2] Hoyle, F. and Lyttleton, R. A., *Proc. Cambridge Phil. Soc.*, **35**, 405, 1939, and **36**, 424, 1940; Whipple, R. L., *Ap. J.*, **104**, 1, 1946; Spitzer, L. Jr., *Centennial Symposia* ("Harvard Obs. Monographs," No. 7 [1948]).

from the interstellar matter. The cutoff in the luminosity function in type II systems is consistent[3] with the belief that all these stars were formed some $3 \times 10^9$ years ago; and the observed absence of much interstellar matter in these systems, in accord with theoretical expectations,[4] accounts naturally for the absence of young stars.

While the correlation between dense interstellar matter and type I stars is extremely good, the correlation between the degree of flattening and the presence of strong obscuration produced by interstellar grains is broken by one striking exception. It is well known that in the very dense clusters of galaxies, such as the Coma and Corona clusters, most of the members seem to belong to the early galactic classes, and Hubble assigns[5] the majority of the members now to the class S0. This class, which in a way is a continuation of the E galaxies, where the intensity distribution from the center outward is smooth, is characterized by the segregation of a central lens of high density from an outer disk of much lower density, no spiral structure appearing. A closer study of the S0 members of the Coma and Corona clusters shows the following interesting features:

1. As far as flattening (angular momentum) is concerned, the S0 galaxies exhibit the full range from relatively small to very high flattening (see fig. 1).

2. With increasing flattening, the central lens shrinks in dimensions, a feature analogous of the shrinking of the central lens in the series of spirals Sa, Sb, Sc (see fig. 1).

3. None of the S0 galaxies shows spiral structure or exhibits any signs of obscuration, although quite a number of the galaxies are seen edge-on. This absence of obscuration is particularly striking in highly flattened systems seen edgewise, because, without exception, the small central nuclei are visible (see fig. 1).

These three features strongly suggest that the galaxies classes together as S0 by Hubble represent actually a series of forms paralleling the series of normal spirals, Sa, Sb, Sc. But the galaxies of the S0 series contain no obscuring matter and presumably are therefore unable to develop spiral structure. That their stellar composition is the same as that of the E galaxies (pure type II) follows from the fact that, according to unpublished measures of Stebbins and Whitford, E and S0 galaxies have identical color indices.

[3] Russell, H. N., *Pub. A.S.P.*, **60**, 202, 1948.
[4] Spitzer, L. Jr., *Ap. J.*, **95**, 329, 1942.
[5] Informal communication to W. Baade.

FIG. 1. Central region of the Corona Borealis cluster, photographed with the Hale 200-inch telescope.

It is the purpose of the present paper to point out that collisions between galaxies may explain the lack of type I systems in the denser clusters. In less massive or less dense systems, such as the Virgo and Hercules clusters, collisions are less frequent, corresponding with the observed fact that many spirals are observed in these aggregations. The relatively high frequency of collisions between galaxies within a dense cluster has already been pointed out,[6] but the detailed processes occurring in such collisions do not seem to have been considered. In Section II below, the collisional frequency is reinvestigated, and it is shown that a typical galaxy in the Coma cluster will have passed through some twenty to five hundred galaxies during $3 \times 10^9$ years, if it has been moving nearly radially in the cluster. Section III investigates a typical collision. Encounters between stars are negligible, but each stellar system will be distorted somewhat by the gravitational field of the other. The chief effect is that, if gas is present in the two galaxies, it will be swept out of both by the collision and left behind in intergalactic space. While further details of this process remain to be worked out, one may apparently conclude that interstellar matter cannot be present in any appreciable amount in many of the galaxies in the Coma cluster. As a tentative hypothesis, one may infer that this lack of interstellar matter is the reason why the highly flattened S0 galaxies show no obscuration or spiral structure and, presumably, represent pure population type II.

## II. Frequency of Collisions

If a galaxy is assumed to move in a straight line through the cluster at a very great velocity, passing at a distance $r$ from the cluster center, the average number of collisions that this galaxy will suffer with other galaxies during a single passage through the cluster will equal simply $\sigma N(r)$, where $\sigma$ is the collisional cross-section in square parsecs and $N(r)$ is the number of galaxies in a cylinder 1 square parsec in cross-section passing $r$ parsecs from the cluster center. For an actual galaxy, moving at a velocity about the average, the actual number of collisions per passage through the cluster will exceed $\sigma N(r)$ by $2^{1/2}$ because of the velocity of the other galaxies. The curvature of the galaxy's path will also modify the actual number of collisions somewhat, but the difference will not be large, provided that the orbit is not nearly circular and provided that $r$ is set equal to the actual distance of closest approach.

The appropriate value of $\sigma$ is somewhat uncertain. The interstellar matter and spiral arms in a flattened galaxy may lie anywhere from 1000

---

[6] Zwicky, F., *Ap. J.*, **86**, 217, 1937.

out to 20,000 parsecs from the galactic center. We shall here take 2000 and 5000 parsecs as typical distances. With these values, a "collision" will occur if the centers of two galaxies pass within 4000 or 10,000 parsecs, respectively, from each other. The effective collisional cross-section for the average galaxy should lie somewhere between the two limiting values computed in this way.

Values of $N(r)$ may be obtained from nebular counts by Zwicky[6] and by Omer, Page, and Wilson.[7] We shall here use the latter counts. These are based on blue and red plates obtained with the 48-inch Schmidt telescope and consequently have a lower limiting magnitude (19.2) than the earlier counts. These later counts yield about 790 galaxies in the Coma cluster, with about half within a cylinder 40' from the center. To obtain $N(r)$ from the counts of galaxies per square degree, the distance of the cluster must be known. According to unpublished velocity measures by Humason,[8] the mean radial velocity of the Coma cluster is 6570 km/sec, corresponding to a distance of $1.25 \times 10^7$ parsecs. This adopted value of the distance has been used to obtain the values of $N(r)$ given in Table 1.

To find the total number of collisions within $3 \times 10^9$ years, we must compute the number of times that a typical galaxy traverses the cluster. The radius of 40', containing about half the galaxies (in projection), corresponds to a distance of some 150 kiloparsecs. If the root-mean-square (r.m.s.) velocity is taken to be 1700 km/sec, corresponding to the observed r.m.s. radial velocity of 1000 km/sec,[8] the time for a single passage through a diameter of 280 kiloparsecs is $1.7 \times 10^8$ years. Hence the values of $\sigma N(r)$ must be multiplied by about 18 times $2^{1/2}$ to find the total number of collisions per galaxy within the presently accepted "age of the universe." The numbers of collisions computed in this way are given in Table 1 for the two values of the collision distance.

It should be emphasized that these computations are very preliminary and are designed to give only the order of magnitude. In particular, the values obtained for large distances of closest approach are not very reliable, since, for these, one should presumably consider nearly circular orbits about the center. More precise computations should also take into account a number of other effects neglected here.

One may infer from Table 1 that, if the galaxies move nearly radially, passing within 10' of the cluster center, the number of collisions during $3 \times 10^9$ years is at least 20 and probably nearer to 150. Even if only 10 percent of the galaxies had interstellar matter originally, these would be mostly swept clean by collisions in $3 \times 10^9$ years. For galaxies moving in

[7] Unpublished.
[8] Private communication to W. Baade.

TABLE 1
Number of Collisions per Galaxy in $3 \times 10^9$ Years

| Collision Distance (Parsecs) | Distance of Closest Approach to Center of Coma Cluster | | | | | | |
|---|---|---|---|---|---|---|---|
| | 2' | 5' | 10' | 20' | 40' | 60' | 80' |
| 10,000 | 490 | 260 | 140 | 51 | 27 | 13 | 5 |
| 4,000 | 79 | 42 | 22 | 8.2 | 4.3 | 2.1 | 0.8 |
| $10^8 \times N(r) \, (\mathrm{pc}^{-2})$ | 5.7 | 3.0 | 1.6 | 0.6 | 0.3 | 0.15 | 0.06 |

nearly circular orbits far out in the cluster, the number of collisions is much smaller, and it is doubtful whether the interstellar matter would be swept out of many such spirals in the Coma cluster. It is difficult to see how a cluster of galaxies could be formed in such a way as to leave galaxies moving in these large, nearly circular orbits.[9]

### III. DYNAMICS OF GALACTIC COLLISIONS

While the details of a collision between two galaxies are naturally very complicated, several results can be established immediately. In the first place, encounters between individual stars are almost entirely negligible. In general, the velocity with which two galaxies collide must be at least several hundred kilometers per second, about equal to the escape velocity from either galaxy. At such high velocities two stars must pass within several stellar radii to deflect each other appreciably. In the Coma cluster, the r.m.s. relative velocity of the galaxies is 2400 km/sec (equal to $2^{1/2}$ times the r.m.s. individual galactic velocity of 1700 km/sec). At this very great velocity, two stars in different galaxies would have to collide physically to be deflected several degrees, and even one such collision is unlikely when one galaxy passes through another. Hence the only force which must be considered on a star during the collision is the smoothed gravitational force produced by both galaxies.

In the second place, it is evident that, while stellar systems may interpenetrate one another with relatively little mutual disturbance, the effect on any interstellar matter present will be catastrophic. Even if the interstellar density is as low as 0.1 H atom/cm$^3$, a single neutral H atom will move less than 0.1 parsec, on the average, before it collides with a similar atom. At the relative velocity of 2400 km/sec under

[9] This point has been discussed by several authors, including E. von der Pahlen (*Zs. f. Ap.*, **24**, 68, 1947) and G. Gamow and G. Keller (informal communication), in connection with the motion of stars in globular clusters and spherical galaxies, respectively.

consideration, corresponding to about 7600 electron volts (e.v.) of energy for each H atom, relative to the center of gravity of the two systems, both atoms are likely to be ionized, and the gas, even if neutral originally, will rapidly become ionized. The high proton velocities will then reduce the interaction cross-section somewhat; according to the equations previously given,[10] the time of relaxation of the protons, for a kinetic energy of 7600 e.v. and a proton density of 0.1 cm$^{-3}$, is about $7.5 \times 10^4$ years. During this time the protons will travel about 180 parsecs relative to one another, somewhat less than the thickness of the interstellar gas in a typical galaxy. Heavier atoms will be more highly charged and will be slowed down more rapidly. These estimates consider only the interactions with protons. While there can be no large-scale separation of electric charges, it is readily shown that the electrons will be stopped almost at once. Interaction with electrons will then tend to slow down the protons and other positive ions.

A small grain, $3 \times 10^{-5}$ cm in radius, will lose about half its initial energy in a distance of 30 parsecs, if the proton density is 0.1 cm$^{-3}$. This estimate neglects the electrical charge on the grain, which may considerably reduce the distance traveled. Evidently, densities as low as $10^{-2}$ H atoms/cm$^3$ and probably less than this value would be required for two gaseous masses several hundred parsecs thick to pass through each other.

The subsequent motions of the gas will depend on the temperature. If the gas retained the temperature of 60,000,000° produced by the collision, the resultant expansion would be so vigorous that little gas could be left in either galaxy. However, cooling may be appreciable in the $10^5$–$10^7$ years required for the galaxies to pass through each other. In this case the gas would have a low velocity relative to the center of gravity of the two systems, while the galaxies will be moving away in opposite directions at velocities greater than the escape velocity from either galaxy separately. It is possible that some gas might fall back into each galaxy, but it seems likely that most of it will be left in intergalactic space. We conclude that any interstellar gas will be swept out of two colliding galaxies (or, more accurately, out of those sections of the galaxies which interpenetrate one another), provided that the gas densities are at all comparable to those in the solar neighborhood.

The ultimate fate of the gas swept out is difficult to predict. When two field galaxies collide, some of the gas might condense to form an irregular galaxy. This suggestion is consistent with the fact that irregular galaxies have mostly pure type I population. In a dense cluster of galaxies other possibilities are present. Some gas might escape from the

[10] Spitzer, L. Jr., *Ap. J.*, **93**, 369, 1941.

cluster, though much loss of matter in this way seems unlikely in view of the velocities in excess of 2400 km/sec that would be required.

Some of the gas swept out might be expected to form new galaxies. If such a formation had occurred long ago and if the new galaxies had lost their interstellar matter in subsequent collisions, such systems would probably be indistinguishable from the original galaxies. Finally, the gases swept out of different galaxies might collide, cool off, and collect at the center of the cluster. The passage of galaxies through the cluster center might then sweep out any remaining gas in such systems. Some interstellar matter might still remain at the center of the cluster. It seems more likely that any central core of gas, with no net angular momentum to impede contraction and condensation, would have condensed into stars long ago.

It remains to consider the effect of collisions on the stars themselves. Zwicky has suggested that the collisions would sweep individual stars out of a galaxy, as well as interstellar matter, but, in view of the negligible interaction between specific stars, this seems unlikely. There will be, as he suggests, a tendency for the internal energy of the galaxies to increase at the expense of the kinetic energy of the galaxies as a whole. While this effect may be appreciable for collisions between field galaxies, whose relative velocities are not very large, it is certainly less important for collisions between cluster galaxies, whose relative velocities are great.

Let us consider in more detail a collision between two galaxies moving with a relative velocity $V$ of 2400 km/sec. This velocity is so much greater than the velocity of an individual star relative to the center of its own galaxy that we may compute the perturbations of each galaxy on the other as if the stars in each system were at rest relative to one another. Let us consider a particular star in galaxy 1, with a distance $a_1$ from the center of galaxy 1; let $a_2$ be the distance of closest approach of galaxy 2 to the star in question. We assume that the masses $M_1$ and $M_2$ of each galaxy are effectively concentrated at the galactic centers. Then it is readily shown that $\Delta v$, the change of velocity of the particular star under consideration, is given by

$$\Delta v = \frac{2GM_2}{a_2 V} = \frac{2v_{c2}^2(a_2)}{V}, \tag{1}$$

where $v_{c2}(a_2)$ is the circular velocity around $M_2$ at the distance $a_2$. The velocity change, $\Delta v$, will be directed toward the point of closest approach of $M_2$. Values of $v_{c2}$ are observed[11] to range up to about 100 km/sec for a dwarf galaxy, such as M 33, and up to about 300 km/sec for a giant galaxy, such as M 31. We shall use here an intermediate value of 200 km/sec. The value of $\Delta v$ is then 33 km/sec.

Other parts of galaxy 1 will also be accelerated. On these assumptions, the velocity changes will all be parallel and will decrease proportionally to $1/a_2$. Let $\Delta v_R$ be the velocity change of a star relative to the galactic center, i.e., $\Delta v_R$ equals $\Delta v$ for the star under consideration minus $\Delta v$ for the center of galaxy 1. We shall compute the r.m.s. $\Delta v_R$ for stars moving in the same plane with the same value of $a_1$. Two simplifying assumptions will be made: first, that the relative motion of the two galaxies is perpendicular to the orbits of the star under consideration and, second, that the distance of closest approach of the two galactic centers is $2a_1$. Thus the distance $a_2$ of closest approach of the center of galaxy 2 varies from $a_1$ for a star on one side of galaxy 1 to $3a_1$ for a star on the other side. Then $\Delta v_R$ will vary from 17 km/sec on one side to 6 km/sec on the other side. Most of the stars will have intermediate values of $\Delta v_R$, and it may be shown that the r.m.s. value of $\Delta v_R$ is in the neighborhood of 8 km/sec.

It is evident that these computations are very crude; they give too high a value for the r.m.s. $\Delta v_R$. First, the mass of a galaxy is not concentrated at its center. Let us denote by $M_2(a_2)$ the mass of galaxy 2 lying within a sphere of radius $a_2$ about the galactic center. Then the increase of $M_2(a_2)$ with increasing $a_2$ will prevent $\Delta v$ from falling off as rapidly with increasing distance as equation (1) would predict. As a result, the differential change of velocity between different regions of galaxy 1 will be less than was found in the previous simplified computations. Second, a change in the values of $a_1$ or $a_2$ is likely to decrease the r.m.s. $\Delta v_R$; for example, with a fixed $a_1$, an increase of $a_2$ will evidently decrease $\Delta v_R$, while a decrease of $a_2$ will decrease $M_2(a_2)$ and may also decrease the r.m.s. $\Delta v_R$. Finally, if the direction of relative motion of the two galaxies is not perpendicular to the orbital plane of the stars under consideration, $\Delta v_R$ will again be decreased. While detailed computations are desirable for more definite results, it would appear that the r.m.s. $\Delta v_R$ computed above is too great by at least a factor of 2 and possibly much more. We shall tentatively take 3 km/sec as a final average value.

The effect of such velocity increments is apparently small. In successive collisions the mean-square velocities will add, but 100 collisions, each with a value of $\Delta v_R$ of 3 km/sec, will produce an increase of only

[11] Babcock, H. W., *Lick Obs. Bull.*, **19**, 41, 1939 (No. 498); Mayall, N. U. and Aller, L. H., *Ap. J.*, **95**, 24, 1942.

900 km$^2$/sec$^2$ in the mean-square velocities of the stars in the galaxy. In an elliptical galaxy, where random velocities are already some 200 km/sec, the internal energy will increase by about 2 percent. Such an increase would have a completely negligible effect on the size and shape of the galaxy. In a highly flattened system each collision will yield velocity increments of about 1.7 km/sec perpendicular to the galactic plane, and 100 such collisions would increase the mean-square velocities in this direction by some 300 km$^2$/sec$^2$. This change, also, is probably below the limit of detection.

We may conclude that collisions between galaxies in a dense cluster, where the random galactic velocities are as great as 1700 km/sec, have a relatively small effect on the distribution of the stars already present in the galaxies.

# THE POSSIBLE INFLUENCE OF INTERSTELLAR CLOUDS ON STELLAR VELOCITIES*

(WITH M. SCHWARZSCHILD)

(ASTROPH. J. 114, 385, 1951)

## *Commentary*

MARTIN Schwarzschild has reminded me of the circumstances which led to this joint paper. He had begun to teach a course on stellar systems, and for this activity had gathered available data on stellar velocity dispersions for stars of different ages. By this time, theoretical calculations had given the maximum ages for main-sequence stars of each spectral type; the average age for such stars may be approximated by half this maximum age. Martin was impressed by the increase of these dispersions with increasing age, and wondered whether random gravitational encounters might account for such an effect. Because of my own research in collisional relaxation processes, he discussed this topic with me, and we soon realized that perturbations by aggregations much more massive than any star would be required. Paper #24, with its analysis of electron-electron collisions by means of the Fokker-Planck equation, had been completed the year before, and I was enthusiastic about applying this same technique to Martin's problem of how stars in the Galaxy might be gradually accelerated.

A primary result of Paper #15 is that the velocity distribution, $f$, computed for the idealized isotropic model, does not differ greatly from the Maxwellian distribution. If this result could have been assumed at the outset, the analysis would have been much shortened, since equation (44) of Paper #13 could have been used to compute the change in $\sigma$, the velocity dispersion. If the stellar kinetic energy is represented by $E_A$ (assumed much smaller than $E_B$, which we regard as constant), and if $\langle E_A \rangle / m_A$ exceeds $\langle E_B \rangle / m_B$ (stellar velocity much exceeding cloud velocity), then this equation can be integrated to give $E_A$ proportional to $t^{2/5}$. Thus, as in the approximate relation (18) below, $\sigma$ varies as $t^\alpha$,

*This work was supported in part by funds of the Eugene Higgins Trust allocated to Princeton University.

with $\alpha = 0.2$. If, on the other hand, the cloud velocity much exceeds the stellar velocity, then the right-hand side of equation (44) is essentially constant, and $\alpha$ equals 0.5. As we shall see below, the value of $\alpha$ plays an important role in evaluating different theories for accelerating stars in the Galaxy.

To explain the observed velocity dispersion, $\sigma$, of old stars, the theory of Paper #15 required cloud masses of $10^6 M_\odot$, with random velocities of some 10 km s$^{-1}$. While cloud complexes with masses of order $10^5 M_\odot$ seemed to exist, with larger masses not excluded, it was difficult to believe that the centers of gravity of these extended structures had such large random velocities. This particular difficulty with the theory was eliminated in a second paper,[1] showing that the differential galactic rotation removed any necessity for such random cloud motions. As a star moves around its epicycle, in the course of galactic rotation, its velocity varies with galactocentric distance; deflection by a massive cloud will shift the star into a different epicycle, with an increased amplitude of motion on the average. If the velocity $v_Z$ perpendicular to the galactic plane is ignored, the velocity dispersion varies again as $t^\alpha$, with now $\alpha = 1/3$. More detailed and realistic analyses[2] of this process indicate that for the dispersion $\sigma_R$ of velocities in the galactic center direction (or for the azimuthal $\sigma_\Phi$) $\alpha$ should equal 0.25.

Since Paper #15 was written, the observational database has been expanded with age determinations of older stars (mostly K and M dwarfs) from the intensities of the Ca II H and K emission features in their spectra.[3] The resultant plot of $\sigma_R$ and $\sigma_\Phi$ as a function of age has become a "standard pattern," used to test various theories of star acceleration in the Galaxy. A more precise test would require that these data be refined in two ways. First, the calibration of Ca II emission intensity with age,[3] then "very crude," should be improved. Second, allowance should be made for possible contamination of the data by Population Type II stars, or other late-type old populations with a high initial velocity dispersion (an effect discussed in Paper #15). As a first approximation we may disregard these uncertainties, and compare the observational standard pattern with the theory of cloud-star encounters. If cloud properties are constant over time and cloud masses are of order $10^6 M_\odot$ (see Commentary #16), this process can explain increases of $\sigma_R$ up to some 30 km s$^{-1}$, but not the even higher 60 km s$^{-1}$ for stars about $10^{10}$ years in age.[3] Apparently a value of $\alpha$ about equal to 0.5 instead of 0.25 is required to fit the standard pattern.

However, there seems to be no strong reason to assume that cloud properties have been constant with time. The star formation rate may have varied by as much as half an order of magnitude during the last $10^{10}$ years.[4] If acceleration of stars by clouds has increased by the same amount, the standard pattern of $\sigma_R$ vs age can be reproduced by cloud-star encounters. In view of the uncertainties in this pattern and in possible changes within our Galaxy, one can conclude only that such encounters might account for most of the increase of $\sigma_R$ with age.

Other explanations for the observed increase of stellar velocity dispersions with age have been considered. These are somewhat similar to that in Paper #15 in that they all invoke the stochastic acceleration of stars by irregular variations in the gravitational potential. One of these theories makes use of large-scale transient fluctuations of the mass distribution in our Galaxy. In particular, if spiral arms are transient phenomena, forming then disrupting, the fluctuations of potential could accelerate stars, perhaps intermittently.[2] Another theory is based on the assumption that massive black holes, with $M \approx 10^6 M_\odot$, populate the galactic halo,[5] passing though the galactic disk with velocities of order 100 km s$^{-1}$. If the dark matter in the halo is attributed to such $10^6 M_\odot$ black holes, the resultant number density of these objects provides about the right amount of heating to explain the higher velocity dispersions of the older stars. Moreover, since the black holes are moving faster than the stars, the latter must gain energy at a constant rate, and $\alpha$ equals 0.5, in apparent agreement with the observations. Presumably such black holes, if they are indeed present, can be detected in due course by their excitation of the interstellar clouds through which they must occasionally pass. Either of these two processes could be important for accelerating stars during $10^9$ to $10^{10}$ years.

REFERENCES

1. Spitzer, L. and Schwarzschild, M.—*Astroph. J.* **118**, 106 (1953).
2. Binney, J. and Tremaine, S.—Galactic Dynamics, Princeton: Univ. Press §7.5 (1987).
3. Wielen, R.—*Astron. Astroph.* **60**, 263 (1977).
4. Soderblom, D. R., Duncan, D. K., and Johnson, R. H.—*Astroph. J.* **375**, 722 (1991).
5. Lacey, C. G. and Ostriker, J. P.—*Astroph. J.* **299**, 633 (1985).

# *Paper*

## Abstract

Gravitational encounters between stars and interstellar clouds produce a much shorter relaxation time of the galaxy in the solar neighborhood than do star-star encounters. This result is caused by the much larger masses of the interstellar clouds as compared with stars. In the extreme case that the largest cloud complexes acting as gravitational units should have masses of the order of a million solar masses, it is found that low-velocity stars have been speeded up appreciably by star-cloud encounters during $3 \times 10^9$ years. This speedup of the stars, which is the same for stars of all masses, arises from the tendency of the encounters to act toward equipartition of energy between clouds and stars, though at the present time equipartition must be far from reached. If the masses of the large cloud complexes are, in fact, high enough to make the star-cloud encounters sufficiently effective, one may suppose that all low-velocity population I stars have been formed from interstellar clouds with initial average velocities equal to those of the present clouds and that the present differences in the velocity dispersions of population I stars have been caused entirely by star-cloud encounters. Under this assumption, the encounters would have increased the average velocity of older groups (late dwarfs and red giants) by about a factor of 2, while they would not have had time to affect the velocities of the younger stars (early main sequence).

Even under extreme assumptions the star-cloud encounters are found to be incapable of changing noticeably the velocities of the fast population II stars. This may indicate that all population II stars were formed from the interstellar matter at an early stage, when the velocities of the primeval clouds were still high.

## I. Introduction

If one tries to interpret the observed motions of the stars in our galaxy, one encounters the question whether every star has been following its orbit in the general gravitational field of the galaxy without perturbation for the last three billion years or whether perturbing processes have been effective in changing the stellar orbits. In the first alternative the velocity dispersions now observed would be completely determined by events at the time the galaxy settled to its present state. In the second alternative, at least some of the present velocity characteristics might be determined by perturbation processes which might occur continuously through the life of the stars.

The type of perturbation process that has been fully investigated previously consists of encounters between stars.[1] It has been found that, for regions in our galaxy similar to the neighborhood of the sun, encounters between stars are so ineffective that their results would become appreciable only after a relaxation time of about $10^{14}$ years; accordingly, the effects of encounters between stars over a time interval of $3 \times 10^9$ years should be negligible.

No other perturbation processes seemed apparent as long as it was believed that the interstellar matter, which contains about half the mass in the solar neighborhood, was distributed fairly evenly, so that it would contribute only to the general gravitational field of the galaxy but not to the fluctuating gravitational field. In recent years, however, evidence has been accumulating which shows that the interstellar matter is far from evenly distributed but is rather concentrated into clouds. The uneven distribution of the interstellar matter will produce perturbations on the stellar orbits. Hence the question arises whether the perturbations by interstellar clouds will affect the stellar velocities noticeably in $3 \times 10^9$ years.

To obtain a first estimate of the effectiveness of gravitational perturbations by interstellar clouds, one may approximate the interactions by simple two-body encounters between stars and clouds. For such encounters the relaxation time $T$—i.e., the time in which the accumulated velocity differences reach the order of magnitude of the original velocities—depends on the average mass of the perturbing clouds, $m_c$; on the average number of clouds per unit volume, $n$; and on the average relative velocity, $V$, according to the relation[2]

$$T \propto \frac{V^3}{m_c^2 n_c}. \tag{1}$$

Since $mn$ is the average density, which is of the same order of magnitude for clouds and stars, the essential change arises from the remaining factor, $m_c$, which is much larger for clouds than for stars. Since $m_c$ might be 100 solar masses for individual small clouds, the relaxation time would be reduced to $10^{12}$ years, which is still long compared with

[1] S. Chandrasekhar, *Principles of Stellar Dynamics* (Chicago: University of Chicago Press, 1942), chap. ii.

[2] Ibid., eq. (2.379).

the probable age of the galaxy. However, if the large cloud complexes are considered, masses of the order of $10^5$ solar masses may be possible. This would bring the relaxation time down to the order of $10^9$ years.

What would be the consequences if the relaxation time for star-cloud encounters were actually fairly low? If the relaxation time were much shorter than the age of the galaxy, equipartition of energy should exist between the cloud complexes and the stars. That this is not the case is shown by the velocity measurements for clouds and stars. Nor does it seem likely that the masses of the cloud complexes are great enough to produce so low a relaxation time. If, then, the relaxation time is of the same order as the age of the galaxy, equipartition of energy will not have been established; nevertheless, the star-cloud encounters may already have produced noticeable effects on the stellar velocities. The main effect will have been a steady increase in the stellar velocities, since the encounters work toward equipartition of energy even if the final state is far from being reached.

Under the hypothesis that all stars of population I are formed from interstellar clouds and had, whenever they were formed, an average velocity equal to that of the present clouds, one would conclude that, as a result of star-cloud encounters, the average velocity of any group of population I stars is higher, the older the group. This seems in qualitative agreement with observations: the late-type giants and faint dwarfs, which are expected to be older stars, show fairly big velocity dispersions, whereas the younger early main-sequence stars have low velocities.

The form and amount of this continuous increase in velocity by star-cloud encounters are treated theoretically in the second section. In Section III the theoretically derived effects are related to the observed velocity dispersions.

## II. Statistics of Star-Cloud Encounters

It is well known[3] that, when particles attract one another according to inverse-square forces, the cumulative effect of many small collisions, produced by relatively distant encounters, outweighs the large deflections produced by a few relatively close encounters. Under these conditions the change in velocity of a particle is formally similar to the change of position in Brownian motion, an analogy first pointed out by Chandrasekhar.[4] The change of velocity of a star, interacting with other gravitational centers, is therefore given by a diffusion equation in

---

[3] J. H. Jeans, *Astronomy and Cosmogony* (Cambridge: Harvard University Press, 1929), p. 319; Chandrasekhar, op. cit., chap. ii.

[4] *Ap. J.*, **97**, 255, 1943; see also *Rev. Mod. Phys.*, **15**, 1, 1943.

velocity space. The form of this equation has been derived in a paper by Cohen, Spitzer, and Routly,[5] subsequently referred to as "CSR," for the interaction between electrically charged particles. With obvious modifications, these equations may be utilized for the interaction between gravitating centers.

Let $f$ be the velocity distribution function for some particular type of star; i.e., let $f dv_x dv_y dv_z$ be the number of stars for unit volume in the velocity range $dv_x dv_y dv_z$. We neglect the smoothed gravitational field of the galaxy. This produces galactic rotation and the concentration of the stars to the galactic plane. It is the fluctuations of the field that produce the velocity changes in which we are interested. In this situation, the Boltzmann equation becomes

$$\frac{\partial f}{\partial t} = -K(ff_c), \qquad (2)$$

where $f_c$ is the velocity distribution function for the clouds, and $K(ff_c)$ is defined in equation (7) of CSR.

The computation of $K(ff_c)$ for interactions between stars and clouds is much complicated by two considerations. First, the velocity distributions $f$ and $f_c$ are not isotropic. While this fact will alter the detailed variation of $f$, the resultant change of stellar energy with time should be much the same as for spherical velocity distributions, and we shall accordingly assume that $f$ and $f_c$ are functions only of the total scalar velocities $v$ and $v_c$ and are independent of direction. Second, the velocities of the interstellar clouds are apparently non-Maxwellian.[6] This fact should also have only a secondary influence on the increase of velocity of the stars, especially since the more rapidly moving clouds have apparently less mass than the typical clouds, whose root-mean-square radial velocity is apparently[6,7] between 7 and 9 km/sec. We shall therefore consider a set of clouds with a Maxwellian velocity distribution and use, accordingly,

$$f_c(v_c) = \frac{n_c}{V_c^3} \frac{1}{(2\pi)^{3/2}} e^{-v_c^2/2V_c^2}, \qquad (3)$$

where $n_c$ is the number of clouds per unit volume and $V_c$ denotes the root mean square of one component of the cloud velocities.

[5] *Phys. Rev.*, **80**, 230, 1950.

[6] F. L. Whipple, *Centennial Symposia* (Cambridge, Mass.: Harvard College Observatory, 1948).

[7] L. Spitzer, Jr., *Ap. J.*, **108**, 276, 1948.

On these assumptions, equation (23) of CSR becomes

$$K(ff_c) = \frac{1}{v^2} \frac{\partial}{\partial v} \left[ v^2 f \left\{ r_0 + \frac{q_0}{v} - \frac{1}{2v^2} \frac{\partial}{\partial v}(v^2 p_0) \right\} \right]$$
$$- \frac{1}{2v^2} \frac{\partial}{\partial v} \left[ v^2 p_0 \frac{\partial f}{\partial v} \right], \tag{4}$$

where $r_0$, $q_0$, and $p_0$ are defined in equation (31) of CSR. From the functional forms of these quantities, given in equation (32) of CSR, we find

$$r_0 + \frac{q_0}{v} - \frac{1}{2v^2} \frac{\partial}{\partial v}(v^2 p_0) = \frac{-3L}{4V_c^2 x^2} \frac{m}{m_c}(\phi - x\phi') \tag{5}$$

and

$$p_0 = \frac{3L}{2^{3/2} V_c x^3}(\phi - x\phi'), \tag{6}$$

where $m$ and $m_c$ are the masses of star and cloud, respectively, and, by definition,

$$x = \frac{v}{\sqrt{2} V_c} \tag{7}$$

and $\phi(x)$ is the familiar error function,

$$\phi(x) = \frac{2}{\pi^{1/2}} \int_0^x e^{-u^2} du. \tag{8}$$

The quantity $L$, defined in the electrostatic case in equation (29) of CSR, here becomes

$$L_c = \frac{8\pi G^2 m_c^2 n_c \ln \alpha}{3}; \tag{9}$$

the value of $\alpha$ is discussed below.

If these various expressions are combined, equation (2) becomes

$$\frac{\partial f}{\partial t} = \frac{3}{4} \frac{L_c}{2^{3/2} V_c^3} \frac{1}{x^2} \frac{\partial}{\partial x} \left\{ \frac{(\phi - x\phi')}{x} \left( \frac{\partial f}{\partial x} + \frac{2xmf}{m_c} \right) \right\}. \tag{10}$$

As equipartition is approached, $\partial f/\partial x$ approaches $-2x fm/m_c$. However, the situation of interest here is far from equipartition; $m_c$ is so large that we may neglect the term in $m/m_c$ in equation (10). Thus the mass of the stars considered does not enter the following derivations. If we define a dimensionless time by the relationship

$$\theta = t \cdot \frac{3}{2^{7/2}} \frac{L_c}{V_c^3} = t \cdot \frac{\pi}{2^{1/2}} G^2 \frac{m_c^2 n_c}{V_c^3} \ln \alpha, \tag{11}$$

we have, finally,

$$\frac{\partial f}{\partial \theta} = \frac{1}{x^2} \frac{\partial}{\partial x} \left( \frac{\phi - x\phi'}{x} \frac{\partial f}{\partial x} \right). \tag{12}$$

The value of $\alpha$ in equation (11) still remains to be determined. For interactions between stars, $\alpha$ is a very large quantity, essentially the ratio of the kinetic energy of a star to the gravitational potential energy of two stars at their maximum separation. In the present instance, where interstellar clouds are responsible for the perturbations, $\alpha$ is reduced by the fact that the cloud is diffuse and its gravitational attraction is less than is assumed if the star passes through the cloud. A more detailed consideration of this situation shows that $\alpha$ is essentially the ratio of $b_{max}$, the largest collision parameter (distance of closest approach), to the radius of the cloud. Since the large clouds considered below have a radius comparable with the thickness of the galaxy, though much smaller than the galactic radius, we shall here assume that $\alpha$ is about 10.

Equation (12) determines a unique solution if the initial velocity distribution, $f$ at $\theta = 0$ for all $x$, is given. Since equation (12) is to be applied here to population I stars, which are supposedly formed from the clouds, the initial star velocities must be equal to the cloud velocities. Hence, in the present notation,

$$f = e^{-x^2} \quad \text{at} \quad \theta = 0, \tag{13}$$

where, for convenience, $f$ is normalized to 1 at $x = \theta = 0$.

The particular solution of equation (12) corresponding to the initial condition (13) has been derived by numerical integration. To bridge the singularity at $x = 0$ in equation (12), the solution for small $x$ values was represented by an error-curve, and a varying dispersion for increasing $\theta$. Near the origin, $x = \theta = 0$, the step values used were $\Delta x = 0.1$ and $\Delta \theta = 0.01$. For larger $x$ and $\theta$ values, 0.2 was used for $\Delta x$, while $\Delta \theta$

TABLE 1
Velocity Distribution Resulting from Star-Cloud Encounters*

| $\theta$ | $x$ | | | | | | | | | | |
|---|---|---|---|---|---|---|---|---|---|---|---|
| | 0.0 | 0.2 | 0.4 | 0.6 | 0.8 | 1.0 | 1.2 | 1.4 | 1.6 | 1.8 | 2.0 |
| 0.00 | 1.000 | 0.961 | 0.852 | 0.698 | 0.527 | 0.368 | 0.237 | 0.141 | 0.077 | 0.039 | 0.018 |
| 0.04 | 0.853 | .826 | .752 | .639 | .504 | .366 | .244 | .148 | .083 | .042 | .020 |
| 0.08 | 0.750 | .731 | .676 | .589 | .479 | .361 | .248 | .155 | .088 | .045 | .021 |
| 0.12 | 0.674 | .658 | .616 | .547 | .456 | .353 | .251 | .161 | .093 | .048 | .023 |
| 0.16 | 0.613 | .601 | .568 | .511 | .435 | .345 | .251 | .166 | .098 | .052 | .025 |
| 0.20 | 0.565 | .554 | .527 | .480 | .415 | .336 | .250 | .169 | .102 | .055 | .026 |
| 0.24 | 0.524 | .516 | .493 | .453 | .396 | .326 | .249 | .172 | .106 | .058 | .028 |
| 0.28 | 0.490 | .482 | .464 | .429 | .380 | .317 | .246 | .174 | .110 | .061 | .030 |
| 0.32 | 0.461 | .454 | .438 | .408 | .364 | .308 | .243 | .175 | .113 | .064 | .032 |
| 0.36 | 0.436 | .429 | .415 | .389 | .350 | .300 | .240 | .176 | .116 | .067 | .034 |
| 0.40 | 0.414 | .408 | .395 | .372 | .337 | .292 | .237 | .176 | .118 | .070 | .036 |
| 0.50 | 0.368 | .364 | .354 | .336 | .309 | .273 | .227 | .176 | .122 | .076 | .040 |
| 0.60 | 0.333 | .330 | .322 | .308 | .286 | .256 | .218 | .173 | .125 | .080 | .045 |
| 0.70 | 0.305 | .303 | .296 | .284 | .266 | .242 | .209 | .170 | .127 | .084 | .049 |
| 0.80 | 0.282 | .280 | .275 | .265 | .250 | .229 | .201 | .166 | .127 | .087 | .052 |
| 0.90 | 0.263 | .261 | .257 | .248 | .235 | .217 | .193 | .162 | .127 | .090 | .056 |
| 1.00 | 0.246 | .245 | .241 | .234 | .223 | .207 | .186 | .158 | .126 | .091 | .058 |
| 1.10 | 0.232 | .231 | .228 | .222 | .212 | .198 | .179 | .155 | .125 | .092 | .061 |
| 1.20 | 0.220 | .219 | .216 | .211 | .202 | .190 | .173 | .151 | .124 | .093 | .063 |
| 1.30 | 0.209 | .209 | .206 | .201 | .193 | .182 | .167 | .147 | .122 | .094 | .064 |
| 1.40 | 0.200 | .199 | .197 | .192 | .185 | .176 | .162 | .143 | .120 | .094 | .066 |
| 1.60 | 0.182 | .181 | .180 | .177 | .172 | .164 | .152 | .136 | .117 | .094 | .068 |
| 1.80 | 0.169 | .168 | .166 | .164 | .160 | .153 | .143 | .130 | .114 | .093 | .070 |

*Continued on next page*

TABLE 1—*Continued*

| θ | x | | | | | | | | | | |
|---|---|---|---|---|---|---|---|---|---|---|---|
| | 0.0 | 0.2 | 0.4 | 0.6 | 0.8 | 1.0 | 1.2 | 1.4 | 1.6 | 1.8 | 2.0 |
| 2.00 | 0.157 | .156 | .154 | .153 | .149 | .143 | .135 | .124 | .109 | .091 | .070 |
| 2.20 | 0.146 | .145 | .143 | .142 | .140 | .136 | .129 | .119 | .106 | .089 | .070 |
| 2.40 | 0.138 | .137 | .136 | .135 | .133 | .129 | .123 | .114 | .103 | .087 | .070 |
| 2.60 | 0.131 | .130 | .129 | .128 | .126 | .123 | .118 | .110 | .099 | .086 | .070 |
| 2.80 | 0.125 | .124 | .123 | .122 | .120 | .118 | .113 | .106 | .096 | .084 | .069 |
| 3.00 | 0.118 | .118 | .117 | .115 | .113 | .112 | .109 | .103 | .095 | .084 | .069 |
| 3.50 | 0.105 | .105 | .104 | .102 | .101 | .100 | .098 | .094 | .088 | .080 | .068 |
| 4.00 | 0.096 | .096 | .096 | .095 | .094 | .093 | .091 | .088 | .083 | .076 | .066 |
| 5.00 | 0.084 | .084 | .084 | .083 | .083 | .083 | .082 | .080 | .076 | .070 | .062 |
| 6.00 | 0.074 | .074 | .074 | .073 | .073 | .072 | .072 | .071 | .068 | .064 | .058 |
| 7.00 | 0.068 | .068 | .068 | .067 | .067 | .066 | .066 | .065 | .063 | .060 | .055 |
| 8.00 | 0.063 | .063 | .063 | .062 | .062 | .061 | .061 | .060 | .058 | .055 | .052 |
| 9.00 | 0.058 | .058 | .058 | .057 | .057 | .056 | .056 | .055 | .054 | .052 | .049 |
| 10.00 | 0.055 | 0.055 | 0.055 | 0.054 | 0.054 | 0.054 | 0.054 | 0.053 | 0.052 | 0.050 | 0.047 |
| 1.00 | 0.431 | 0.424 | 0.405 | 0.374 | 0.331 | 0.280 | 0.224 | 0.168 | 0.115 | 0.072 | 0.040 |
| 2.00 | 0.312 | .309 | .298 | .280 | .256 | .226 | .191 | .154 | .116 | .082 | .053 |
| 3.00 | 0.254 | .252 | .243 | .231 | .214 | .193 | .168 | .140 | .111 | .084 | .058 |
| 4.00 | 0.217 | .215 | .208 | .199 | .186 | .170 | .150 | .129 | .106 | .083 | .061 |
| 5.00 | 0.192 | .190 | .185 | .177 | .166 | .154 | .138 | .120 | .100 | .081 | .061 |
| 6.00 | 0.173 | .172 | .167 | .160 | .152 | .141 | .128 | .112 | .096 | .078 | .061 |
| 7.00 | 0.158 | .157 | .153 | .148 | .140 | .131 | .119 | .106 | .091 | .076 | .060 |
| 8.00 | 0.147 | .146 | .142 | .137 | .131 | .122 | .112 | .101 | .088 | .074 | .060 |
| 9.00 | 0.137 | .136 | .133 | .129 | .123 | .115 | .106 | .096 | .084 | .072 | .059 |
| 10.00 | 0.129 | 0.128 | 0.126 | 0.121 | 0.116 | 0.109 | 0.101 | 0.092 | 0.081 | 0.070 | 0.058 |

TABLE 1—*Continued*

| θ | x | | | | | | | | | |
| --- | --- | --- | --- | --- | --- | --- | --- | --- | --- | --- |
| | 2.2 | 2.4 | 2.6 | 2.8 | 3.0 | 3.2 | 3.4 | 3.6 | 3.8 | 4.0 |
| 0.00 | 0.008 | 0.003 | 0.001 | 0.0004 | 0.0001 | 0.0000 | 0.0000 | 0.0000 | 0.0000 | 0.0000 |
| 0.04 | .009 | .003 | .001 | .0004 | .0001 | .0000 | .0000 | .0000 | .0000 | .0000 |
| 0.08 | .009 | .004 | .001 | .0004 | .0001 | .0000 | .0000 | .0000 | .0000 | .0000 |
| 0.12 | .010 | .004 | .001 | .0005 | .0001 | .0000 | .0000 | .0000 | .0000 | .0000 |
| 0.16 | .011 | .004 | .002 | .0005 | .0002 | .0000 | .0000 | .0000 | .0000 | .0000 |
| 0.20 | .011 | .004 | .002 | .0005 | .0002 | .0000 | .0000 | .0000 | .0000 | .0000 |
| 0.24 | .012 | .005 | .002 | .0006 | .0002 | .0001 | .0000 | .0000 | .0000 | .0000 |
| 0.28 | .013 | .005 | .002 | .0006 | .0002 | .0001 | .0000 | .0000 | .0000 | .0000 |
| 0.32 | .014 | .006 | .002 | .0006 | .0002 | .0001 | .0000 | .0000 | .0000 | .0000 |
| 0.36 | .015 | .006 | .002 | .0007 | .0002 | .0001 | .0000 | .0000 | .0000 | .0000 |
| 0.40 | .016 | .006 | .002 | .0007 | .0002 | .0001 | .0000 | .0000 | .0000 | .0000 |
| 0.50 | .019 | .007 | .003 | .0009 | .0003 | .0001 | .0000 | .0000 | .0000 | .0000 |
| 0.60 | .021 | .009 | .003 | .0010 | .0003 | .0001 | .0000 | .0000 | .0000 | .0000 |
| 0.70 | .024 | .010 | .004 | .0012 | .0004 | .0001 | .0000 | .0000 | .0000 | .0000 |
| 0.80 | .027 | .012 | .004 | .0014 | .0004 | .0001 | .0000 | .0000 | .0000 | .0000 |
| 0.90 | .030 | .013 | .005 | .0017 | .0005 | .0001 | .0000 | .0000 | .0000 | .0000 |
| 1.00 | .032 | .015 | .006 | .0020 | .0006 | .0002 | .0000 | .0000 | .0000 | .0000 |
| 1.10 | .034 | .017 | .007 | .0023 | .0007 | .0002 | .0000 | .0000 | .0000 | .0000 |
| 1.20 | .037 | .018 | .008 | .0027 | .0008 | .0002 | .0001 | .0000 | .0000 | .0000 |
| 1.30 | .039 | .020 | .008 | .0031 | .0009 | .0003 | .0001 | .0000 | .0000 | .0000 |
| 1.40 | .041 | .021 | .009 | .0035 | .0011 | .0003 | .0001 | .0000 | .0000 | .0000 |
| 1.60 | .044 | .024 | .011 | .0044 | .0014 | .0004 | .0001 | .0000 | .0000 | .0000 |

*Continued on next page*

TABLE 1—*Continued*

| $\theta$ | 2.2 | 2.4 | 2.6 | 2.8 | 3.0 | 3.2 | 3.4 | 3.6 | 3.8 | 4.0 |
|---|---|---|---|---|---|---|---|---|---|---|
| 1.80 | .046 | .027 | .013 | .0054 | .0018 | .0005 | .0001 | .0000 | .0000 | .0000 |
| 2.00 | .049 | .029 | .015 | .0065 | .0023 | .0007 | .0002 | .0000 | .0000 | .0000 |
| 2.20 | .050 | .032 | .017 | .0076 | .0028 | .0009 | .0002 | .0001 | .0000 | .0000 |
| 2.40 | .051 | .033 | .019 | .0088 | .0034 | .0011 | .0003 | .0001 | .0000 | .0000 |
| 2.60 | .052 | .035 | .020 | .0100 | .0041 | .0014 | .0004 | .0001 | .0000 | .0000 |
| 2.80 | .053 | .036 | .022 | .0111 | .0047 | .0016 | .0005 | .0001 | .0000 | .0000 |
| 3.00 | .053 | .037 | .023 | .0122 | .0054 | .0020 | .0006 | .0001 | .0000 | .0000 |
| 3.50 | .054 | .040 | .026 | .0148 | .0071 | .0029 | .0009 | .0003 | .0000 | .0000 |
| 4.00 | .054 | .041 | .028 | .0170 | .0088 | .0039 | .0014 | .0004 | .0001 | .0000 |
| 5.00 | .053 | .042 | .031 | .0206 | .0120 | .0060 | .0025 | .0009 | .0002 | .0001 |
| 6.00 | .051 | .042 | .033 | .0231 | .0146 | .0081 | .0038 | .0015 | .0005 | .0001 |
| 7.00 | .049 | .042 | .034 | .0248 | .0167 | .0100 | .0052 | .0023 | .0009 | .0003 |
| 8.00 | .047 | .041 | .034 | .0259 | .0183 | .0116 | .0065 | .0032 | .0013 | .0005 |
| 9.00 | .045 | .040 | .034 | .0265 | .0195 | .0130 | .0078 | .0041 | .0019 | .0009 |
| 10.00 | 0.043 | 0.039 | 0.033 | 0.0269 | 0.0204 | 0.0141 | 0.0089 | 0.0050 | 0.0025 | 0.0013 |
| 1.00 | 0.019 | 0.008 | 0.003 | 0.0010 | 0.0003 | 0.0001 | 0.0000 | 0.0000 | 0.0000 | 0.0000 |
| 2.00 | 0.030 | .015 | .007 | .0025 | .0008 | .0002 | .0001 | .0000 | .0000 | .0000 |
| 3.00 | .037 | .022 | .011 | .0048 | .0018 | .0006 | .0002 | .0000 | .0000 | .0000 |
| 4.00 | .042 | .026 | .015 | .0073 | .0031 | .0012 | .0004 | .0001 | .0000 | .0000 |
| 5.00 | .045 | .029 | .018 | .0096 | .0046 | .0019 | .0007 | .0002 | .0000 | .0000 |
| 6.00 | .045 | .031 | .020 | .0116 | .0061 | .0028 | .0011 | .0004 | .0001 | .0000 |
| 7.00 | .046 | .033 | .022 | .0134 | .0074 | .0037 | .0016 | .0006 | .0002 | .0000 |
| 8.00 | .046 | .034 | .023 | .0149 | .0087 | .0046 | .0021 | .0009 | .0003 | .0001 |
| 9.00 | .046 | .035 | .024 | .0162 | .0098 | .0054 | .0027 | .0012 | .0004 | .0002 |
| 10.00 | 0.046 | 0.035 | 0.025 | 0.0172 | 0.0108 | 0.0062 | 0.0032 | 0.0015 | 0.0006 | 0.0002 |

\* In the upper section, $f$ is tabulated as a function of the normalized velocity, $x$, and of the normalized time, $\theta$. The lower section gives $f_a$.

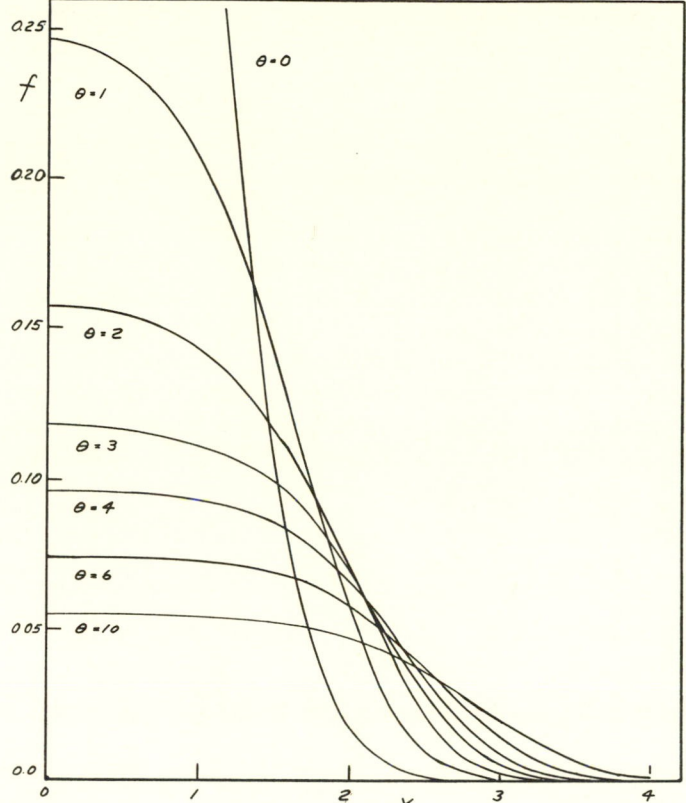

FIG. 1. Velocity distribution, $f(x)$, resulting from star-cloud encounters, as a function of time, $\theta$.

was several times increased until the value 0.2 was used from $\theta = 1.4$ on. Since these $\Delta\theta$ values were close to the stability limit, the results had to be slightly smoothed after every few steps, to avoid the appearance of spurious oscillations. The solution was carried to 66 steps until $\theta = 10$. The results are shown in Table 1 and Figure 1. The last digits given in Table 1 are not secure for the higher $\theta$ values.

Any one line in Table 1 gives the velocity distribution for a group of stars, all of the same age, $\theta$. On the other hand, if type I stars have been forming continuously, there will be an appreciable spread of ages in any group of such stars. For example, in a group of low-velocity G dwarfs some stars may be as old as $3 \times 10^9$ years, while others may have been formed only very recently. Such an age distribution can be taken into account if one assumes that the rate of star formation has been fairly constant so that the age distribution is uniform. The corresponding average velocity distribution, $f_a$, can then be computed according to the relation

$$f_a = \frac{1}{\theta} \int_0^\theta f \, d\theta. \tag{14}$$

The results of this computation are added at the foot of Table 1.

By definition, $x^2 f(x)$ or $x^2 f_a(x)$ gives the distribution of total velocities for a group of stars at a given time. To compare these theoretical results directly with the observed total velocities would seem of uncertain value, since the ellipsoidal distribution of actual velocities differs appreciably from the spherical distribution assumed in the above derivations. Therefore, it seems safer to compare the theoretical results with observations only in terms of single velocity components. If $v_y$ is a particular velocity component, it may be normalized in the same way as in equation (7) by

$$y = \frac{v_y}{2^{1/2} V_c}. \tag{15}$$

Again, by definition, $f(x)$ or $f_a(x)$ gives the distribution of the single component $y$ for those stars for which the other two velocity components are negligible, since, under this selection, $y = x$. But here again direct comparison with observations does not seem practical, since the restriction to stars with two negligible—or at least small—velocity components would limit the number of available stars so much that significant distribution functions could probably not be derived from the present observational data. Thus, for the comparison with observation, one is limited at present to the frequency distribution of a single velocity component for all stars of a given group, irrespective of the values of the other two velocity components.

To derive the distribution function of a velocity component for all stars in a group, $f^*(y)$ or $f_a^*(y)$, one has to integrate $f$ or $f_a$ over the other two velocity components. Because of the spherical symmetry assumed above, the double integral reduces to the following single integral

$$f^*(y) = \int_{|y|}^\infty f(x) 2\pi x \, dx, \qquad f_a^*(y) = \int_{|y|}^\infty f_a(x) 2\pi x \, dx. \tag{16}$$

This integration has been performed for one time, $\theta = 10$. The results are given in Table 2 and shown in Figure 2. For comparison, error-curves

TABLE 2
Distribution of One Velocity Component at Time $\theta = 10$*

| $y$ | $f^*$ | $f_a^*$ | $y$ | $f^*$ | $f$ |
|---|---|---|---|---|---|
| 0.0 | 1.37 | 1.69 | 2.0 | 0.71 | 0.54 |
| 0.2 | 1.36 | 1.67 | 2.2 | .59 | .40 |
| 0.4 | 1.34 | 1.62 | 2.4 | .47 | .28 |
| 0.6 | 1.30 | 1.55 | 2.6 | .36 | .19 |
| 0.8 | 1.26 | 1.44 | 2.8 | .26 | .12 |
| 1.0 | 1.19 | 1.32 | 3.0 | .17 | .07 |
| 1.2 | 1.12 | 1.17 | 3.2 | .10 | .03 |
| 1.4 | 1.03 | 1.02 | 3.4 | .06 | .02 |
| 1.6 | 0.93 | 0.85 | 3.6 | .03 | .01 |
| 1.8 | 0.82 | 0.69 | 3.0 | 0.01 | 0.00 |

* The values of $f^*$ refer to a group of stars of the same age, $\theta$. The values of $f_a^*$ apply to a group of stars with a uniform distribution of ages from 0 to $\theta$.

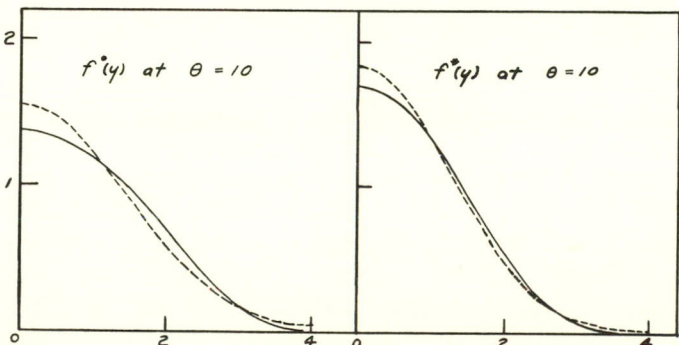

FIG. 2. Distribution of one velocity component as produced by star-cloud encounters. For comparison, the dashed curves represent normal distributions with the same area and dispersion. The velocities are given in units of $2^{1/2}Vc$ (root-mean-square cloud velocity component). The left figure refers to a group of stars all of the same age, $\theta = 10$. The right figure applies to a group of stars with a uniform distribution of ages, from $\theta = 0$ to $\theta = 10$.

corresponding to the same number of stars and to the same root-mean-square velocity have been added in Figure 2. As shown by this figure, the distribution-curves computed from equation (16) differ only slightly from error-curves and could probably not be distinguished from the latter on the basis of the observational material presently available.

TABLE 3

Root-Mean-Square Star Velocity as a Function of Time
in Units of the Root-Mean-Square Cloud Velocity*

| $\theta$ | $V_s / V_c$ | $V_{sa} / V_c$ | $\theta$ | $V_s / V_c$ | $V_{sa} / V_c$ |
|---|---|---|---|---|---|
| 0 | 1.00 | 1.00 | 6 | 1.82 | 1.56 |
| 1 | 1.32 | 1.18 | 7 | 1.88 | 1.61 |
| 2 | 1.48 | 1.30 | 8 | 1.93 | 1.65 |
| 3 | 1.59 | 1.38 | 9 | 1.97 | 1.68 |
| 4 | 1.68 | 1.45 | 10 | 2.01 | 1.72 |
| 5 | 1.76 | 1.51 | | | |

*The second column refers to a group of stars all of the same age, $\theta$. The third
column applies to a group of stars with a uniform distribution of ages from 0 to $\theta$.

Hence the root mean square of a velocity component for a group of
stars, $V_s$, appears to be the only item available at present for comparison with observations. In preparation for such a comparison, the variation of $V_s$ with time, to be expected from the star-cloud encounters, has
been computed from the above $f$ functions with the help of the
following relation:

$$\frac{V_s^2}{V_c^2} = 2\overline{y^2} = 2\frac{\int_{-\infty}^{+\infty} y^2 f^*(y)\, dy}{\int_{-\infty}^{+\infty} f^*(y)\, dy} = \frac{2}{3}\frac{\int_0^\infty f(x) x^4\, dx}{\int_0^\infty f(x) x^2\, dx} = \frac{2}{3}\overline{x^2}. \quad (17)$$

A corresponding relation holds for the quantities with subscripts $a$. The
results are listed in Table 3. These numerical values can be represented
with fair accuracy by the following interpolative relations:

$$V_s \approx V_c \cdot (1 + 3.2\theta)^{1/5}, \qquad V_{sa} \approx V_c \cdot (1 + 1.4\theta)^{1/5}. \quad (18)$$

An equation of this general functional form may be derived theoretically[8] if it is assumed that the distribution of stellar velocities remains
Maxwellian at all times.

Table 3 and equation (18) show that a group of stars reaches twice
the average cloud velocity at a time $\theta = 10$ if the stars were all formed
simultaneously, and at a time $\theta = 22$ is stars were formed continuously
at a steady rate.

In comparing the second and third columns of Table 3, one finds that
any entry in the third column for a given $\theta$ shows nearly the same value
as an entry in the second column for $\frac{1}{2}\theta$. On the other hand, any entry
in the third column corresponds to a group of stars with a uniform age

---

[8] L. Spitzer, Jr., *M.N.*, **100**, 396, 1940.

distribution from 0 to $\theta$, i.e., an average age of $\frac{1}{2}\theta$. Thus one may conclude that the dispersion shown by a group of stars as a result of star-cloud encounters is a function of the average age of the stars in the group, irrespective of the age distribution in the group.

### III. Application to Observed Velocities

The observed velocity dispersions in relation to the effects of the star-cloud encounters will be discussed in this section consecutively for four types of stars: the early main-sequence stars, the late main-sequence stars, the red giants, and the high-velocity stars. The velocities perpendicular to the galactic plane have not been used in this discussion, since there is only limited information available on the velocities of interstellar clouds and early-type stars in this direction. Correspondingly, the values of $V_c$ and $V_s$ here used correspond to root mean squares of one velocity component, the mean being taken over all directions in the galactic plane. The observational data are summarized in Table 4.

To start with the earliest main-sequence stars, the O and the early B stars should have an average age short compared with that of the galaxy because of their high rate of energy output. Consequently, one should expect the star-cloud encounters to have had insufficient time to affect the velocities of these stars. Thus the velocities of the O and early B stars should represent the velocities of the clouds from which they were formed. As Table 4 shows, the velocity dispersion of the earliest main-sequence stars is in good agreement with the velocities of the large clouds. The small clouds, with their relatively larger velocities, may have given rise to the formation of rather few stars.

Following down the main sequence, the average age should be increasing because of the slower and slower rate of expenditure of nuclear energy. Under the hypothesis that all main-sequence stars of population I have been formed from interstellar clouds, they all must have had a velocity dispersion equal to that of the clouds at the time they were formed. Consequently, the higher the average age, the greater should be the average velocity as a consequence of the star-cloud encounters, in qualitative agreement with observations.

For early F stars the energy output is sufficiently low so that the oldest stars of this type may have an age equal to that of the galaxy. If one assumes that the normal main-sequence F stars have been continuously formed from interstellar clouds at a fairly constant rate, the average age of early F stars must be half the age of the galaxy, or approximately $1.5 \times 10^9$ years. On the other hand, the average velocity of early F stars is about twice that of the larger clouds. Hence,

TABLE 4
Observed Root-Mean-Square Velocities*

| Objects | Vc or Vs (Km / Sec) | Ref.[†] |
|---|---|---|
| *Interstellar clouds:* | | |
| Large clouds .................. | 8 | 1, 2 |
| Small clouds .................. | 25 | 1 |
| *Early main-sequence stars:* | | |
| Oe5–B5....................... | 10 | 3 |
| B8–B9 ....................... | 12 | 3 |
| A0–A9....................... | 15 | 4 |
| F0–F9 ....................... | 20 | 4 |
| *Late main-sequence stars:* | | |
| F5–G0....................... | 23 | 4 |
| G0–K6....................... | 25 | 4 |
| K8–M5 ...................... | 32 | 4 |
| *Red giant stars:* | | |
| K0–K2....................... | 25 | 5 |
| K0–K9....................... | 21 | 4 |
| M0–M9 ...................... | 23 | 4 |
| *High-velocity stars:* | | |
| General high-velocity stars .... | 50 | 6 |
| RR Lyrae variables ........... | 120 | 7 |
| Subdwarfs ................... | 150 | 8 |

* Mean of two components in galactic plane.

[†] The following references are cited:

1. F. L. Whipple, *Centennial Symposia* (Cambridge, Mass., Harvard College Observatory, 1948), p. 118.

2. L. Spitzer, Jr., *Ap. J.*, **108**, 276, 1948.

3. H. Nordstrom, *Lund. Obs. Medd.*, Ser. II, No. 79, p. 161, 1936: $[(\frac{1}{2}\sigma_1^2 + \frac{1}{2}\sigma_2^2)^{1/2}]$.

4. E.T.R. Williams and A. N. Vyssotsky, *A. J.*, **53**, 92, 1947 (particularly Table 10.1): $[(\frac{1}{2}\sigma_a^2 + \frac{1}{2}\sigma_b^2)^{1/2}]$.

5. A. N. Vyssotsky, *A. J.*, **56**, 62, 1951 (particularly Table 12.1).

6. J. Oort, *Groningen Pub.*, **40**, 45, 1926: $[2^{-1/2}/0.015]$; also G. Miczaika, *A.N.*, **270**, 258, 1940.

7. J. Oort, *B.A.N.*, **8**, 337, 1936: $[94(\pi/2)^{1/2}]$.

8. A. N. Joy, *Ap. J.*, **105**, 102, 1947: $[121(\pi/2)^{1/2}]$.

according to the second column of Table 3, $\theta \approx 10$ should correspond to the average age of the early F stars. These values of $t$ and $\theta$ may be introduced into equation (11) to test whether the effects of star-cloud encounters can be brought into quantitative agreement with observations. The most uncertain quantity in equation (11) is probably $m_c$. If one correspondingly solves equation (11) for $m_c$ and uses $V_c = 10$ km/sec, $\log \alpha = 3$, and $m_c n_c = \bar{\rho} = 3 \times 10^{-24}$, one obtains $m_c = 10^6 m_\odot$ by order of magnitude. It is clear that this very large mass cannot be represented by an individual interstellar cloud, but, at best, by a large cloud complex. If one assumes for such a cloud complex ten times the average density of interstellar space, one obtains about 100 parsecs for the diameter of such a cloud complex. This diameter size falls well within the range of the observed diameters of cloud complexes.[9] Bok[10] has deduced a mass of $300 m_\odot$ for the solid particles within a typical large cloud, with an assumed diameter of 40 parsecs. If the mass is a hundred times that of the obscuring grains, the total mass of such a cloud becomes $3 \times 10^4 m_\odot$. Whether aggregations with a mass as great as $10^6 m_\odot$ can be moving as a unit through interstellar space with velocities of 5 to 10 km/sec is uncertain, and, as a result, the entire present discussion is necessarily hypothetical. Since no decision in this question seems possible at present, the discussion will here be followed through under the assumption that the masses of the cloud complexes are large enough and their mean velocities high enough to make the star-cloud encounters effective.

Turning, next, to the late main-sequence stars, one finds that the age of a dwarf with spectral type later than F5 may be anywhere between zero and the age of the galaxy. Therefore, if one again assumes that these stars have been formed from interstellar clouds at an essentially constant rate, the average age of any group of dwarfs later than F5 should be equal to half the age of the galaxy, irrespective of spectral type. With the same average age and the same initial velocities, one should expect all the dwarfs later than F5 to have the same velocity dispersions at the present time. Furthermore, this expectation is not affected by the circumstance that the stellar masses decrease with advancing spectral type, since the encounter effects are independent of the stellar masses, as long as the state is far from approaching equipartition of energy with the interstellar clouds.

The above deduction is in disagreement with the observations, which, according to Table 4, show a further increase of the velocity dispersion for the latest main-sequence stars. However, the later main sequence

[9] J. L. Greenstein, *Harvard Ann.*, **105**, 359, 1937.
[10] *Centennial Symposia* (Cambridge, Mass.: Harvard College Observatory, 1948).

does not consist purely of population I stars, like the early main sequence, but contains an appreciable fraction of population II stars. Thus the question arises whether the apparent increase in velocity dispersion for the later main-sequence stars may arise solely from an increase in the percentage of population II stars for the later spectral types. To test this possibility, the available velocities of M dwarfs were analyzed.[11] Only the velocity component in the center-anticenter direction was used, to avoid difficulties with asymmetry. It has previously been pointed out that the distribution-curve for this velocity component cannot be represented by one error-curve.[11] Here it was attempted to represent the data by the sum of two error-curves, and it was found that the observations could be fitted if two-thirds of the stars were assumed to belong to an error distribution with a velocity dispersion of 23 km/sec and the remaining third of the stars to an error distribution with a dispersion of 67 km/sec. Such a representation by the sum of two error-curves is, of course, not the only possible representation, but seems to be a possible one. If one assumes the normal ratio for the axes of the velocity ellipsoid in the galactic plane, one can transform the above values for the dispersions in the center-anticenter direction into values of $V_s$, for which one obtains 20 and 56 km/sec, respectively. Since the latter value is close to that of the high-velocity population II stars in general, one may well assign one-third of the M dwarfs to population II. The remaining two-thirds of the M dwarfs have, then, according to the first of the above $V_s$ values, an average velocity equal to that of the F dwarfs. One may conclude that the later main-sequence stars of pure population I may have a constant average velocity, irrespective of spectral type, in agreement with what one would expect from the effects of star-cloud encounters.

Turning, next, to the red giants, one sees from Table 4 that the average velocities for these stars are somewhat higher than for main-sequence stars of the same brightness. This velocity excess seems to remain even after the small fraction of high-velocity population II giants has been taken account of. (From his material on early K giants Vyssotsky has separated those stars with small velocities perpendicular to the galactic plane; for this group of stars, which should contain hardly any population II stars, he obtained $V_s = 21$ km/sec, a value still higher than the corresponding value for main-sequence A stars.) According to the present hypothesis, all population I stars started out with the low-velocity dispersion of the large interstellar clouds and increased their velocity dispersions by star-cloud encounters according to their age. Because of their higher velocity dispersions, one might conclude,

[11] A. N. Vyssotsky, *Ap. J.*, **104**, 239, 1946.

therefore, that the red giants must be somewhat older, on the average, than the bright main-sequence stars—a conclusion which fits well with the current speculations regarding the interior of the red giants.

There remain, finally, the high-velocity population II stars to be discussed. As shown by the bottom part of Table 4, even the general high-velocity stars—not to include the extreme groups like the RR Lyrae variables and the subdwarfs—have velocity dispersions about two and a half times larger than the older stars of pure population I. It seems impossible to explain these high velocities by the effects of star-cloud encounters. On the one hand, since the increase in velocity by encounters rapidly diminishes with increasing velocity (as shown by eq. (1) or more precisely by eq. (18)), the necessary masses for the cloud complexes would have to be even larger than those discussed above. On the other hand, if contrary to expectation, the relaxation time through star-cloud encounters should be so short that these encounters could produce $V_s = 50$ km/sec, the velocity dispersions of the older main-sequence stars should also have reached high values; but this consequence would disagree with observation.

Thus it seems necessary to conclude that the velocity dispersions of population II stars have not been altered essentially by any encounters and must represent the initial average velocities of these stars at their formation. Other evidence has already indicated that population II stars may all be of about the same age as the galaxy itself. Hence it seems reasonable to surmise that at the early stage of the development of the galaxy, when presumably the population II stars were formed from interstellar material, the large clouds had average velocities not of 10 km/sec as now, but rather 50 km/sec—and in part even higher velocities to account for the present extreme high-velocity stars.

## IV. CONCLUSIONS

The results of the above discussion may be summarized in the following speculative working hypothesis.

At an early stage of the galaxy the interstellar matter had large random velocities, possibly corresponding to violent turbulence with great density fluctuations. At this stage all population II stars were formed out of the interstellar matter.

Subsequently, the velocities of interstellar matter soon decayed to their present average values. Ever since this settling-down process, the population I stars have been forming out of the interstellar clouds at a more or less constant rate.

Only after the settling-down process, was there enough time for interstellar grains to form abundantly. Hence grains did not play a role in the formation of population II stars but did in the formation of population I stars. This accounts for the difference in chemical composition between the two populations.[12]

All stars are continually having encounters with massive cloud complexes. The relaxation time for high-velocity stars, in encounters with large clouds, is too long to have affected appreciably the velocity dispersions of population II stars. This relaxation time is sufficiently short for low-velocity stars, however, so that the star-cloud encounters have speeded up the older groups of population I stars markedly. This accounts for the increased velocity dispersion of the later dwarfs and giants of population I as compared with the early main-sequence stars.

The above statements are purely hypothetical at present. They seem, however, not to be contradicted by the observational evidence, and they contain a possible, though far from proved, explanation of the velocity dispersions of population I stars. If this working hypothesis should gain substance by future evidence, it would be necessary to expand the theoretical discussion of Section II to take account of the nonsphericity of the actual velocity ellipsoids.

It is a pleasure to acknowledge the carefulness and efficiency with which Mr. Härm carried through the extensive numerical integration described in Section II. To Dr. A. N. Vyssotsky we are most grateful for stimulating discussions regarding the observational basis for this investigation.

[12] Schwarzschild, Spitzer, and Wildt, *Ap. J.*, **114**, 398, 1951.

# DISRUPTION OF GALACTIC CLUSTERS

(ASTROPH. J. 127, 17, 1958)

## *Commentary*

AT A small conference in Rome, in May 1957, Jan Oort raised one of his typically challenging questions.[1] "What has happened," he asked, "to the galactic clusters that were presumably formed more than $5 \times 10^8$ years ago? There are relatively few of such old systems around."

Back in Princeton I thought about this problem, wondering if the mechanisms discussed at the conference, including the familiar evaporation of stars from a cluster, might suffice. A few years earlier, Martin Schwarzschild and I had shown that gravitational encounters with interstellar clouds might explain the higher random velocities of the older stars (see Paper #15). It occurred to me that perhaps these same clouds might, through their tidal effects, increase the internal energy of a stellar cluster in the galactic disk, leading to expansion and complete disruption of the cluster. Within a few months, a brief analysis led to the paper above.

This work is referred to occasionally for its introduction of the "impulsive approximation." In fact, this simplified approach to dynamical problems was used in 1913 by Born in his classical theory[2] of the stopping power of matter for fast electrons and positive ions; in this approximation, the motion of atomic electrons is ignored, and the energy gained by each of these electrons is computed from the momentum impulse imparted. For computations in stellar dynamics the impulsive approximation is still useful in preliminary analyses, but rapid digital computers increasingly provide the basis for most contemporary research.

During the years since Paper #16 was published, more detailed computations have been made[3] on the disruption of galactic clusters. One modification has been to place increasing emphasis on the role of giant molecular clouds (GMC's). These structures were to some extent included in my paper above, since the numerical computations took into account the effect produced by the "large clouds" revealed by the statistics of color excesses. The reality of these extended, irregular, features could scarcely be questioned, in view of the large-scale distribution of dust in the Milky Way.[4] Originally, the masses of these clouds were highly uncertain, but recent research has shown that these large clouds have individual masses of about $10^5 M_\odot$, and that they constitute the GMC's in the solar neighborhood (see Table 1 and ref. 29 of Paper #12), contributing about two-thirds of the tidal heating produced by all interstellar clouds near the Sun. Clouds of this mass can disrupt a galactic cluster in the improbable event of a direct central collision,[5] but like the cold diffuse clouds, have usually a somewhat minor effect compared to all the other perturbations experienced by a galactic cluster.[3, 5]

In other regions of the Galaxy, especially somewhat closer to the galactic center, the masses of the GMC's can be as great as $10^6 M_\odot$. Since the tidal heating varies as the cloud mass, for the same smoothed interstellar density, such clouds produce at least one order of magnitude more tidal heating than do the less massive clouds nearer to the Sun. Clusters with a relatively low density tend to be disrupted in a single encounter with such a massive cloud,[3] while denser clusters dissipate gradually, as envisaged in Paper #16. Analyses indicates and numerical simulations confirm that GMC's of this mass can apparently account[3] for the observed relative scarcity of old galactic clusters.

Another recent addition to the theoretical study of tidal heating has been the suggestion that a population of massive black holes in the Galaxy might play an important part. Such a black-hole population has been invoked to explain the acceleration of older stars to higher random velocities (see Commentary, Paper #15). With masses of some $10^6 M_\odot$, a velocity dispersion of some hundred km s$^{-1}$, and a number density consistent with the dark matter deduced for the halo of our Galaxy, such massive objects, disrupting a cluster in a single encounter, could apparently also explain[3] the relative scarcity of the older clusters.

An important theoretical question in connection with all the studies of tidal heating is the extent to which such heating can be produced by slow encounters, with a duration of several orbital periods for many of the cluster stars. In this situation, the impulsive approximation is invalid, and in the limiting case of sufficiently low impact velocities the orbits of cluster stars change "adiabatically" in the course of the encounter. For such adiabatic perturbations the final orbit is nearly identical with the initial one, apart from a likely change of phase.

The usual assumption for slow encounters, valid in simple cases, is that the change in orbital energy, for example, decreases exponentially with increasing "slowness parameter," $\beta$, proportional to the ratio of the impact time, $p/V$, to the orbital period $2\pi/\omega$ for the cluster stars (see equation (34) in Paper #16). Recently it has been shown[6] that in the case of actual clusters, where several comparable frequencies may be involved (such as radial and azimuthal periods), this simple result may no longer hold. Instead, a slow perturbation can produce a heating effect which varies asymptotically as $1/\beta^2$, not as $\exp(-2\beta)$. This additional heating increases somewhat the disruption rate both of globular and of open clusters.

Another effect which can increase the disruption rate of clusters is the accelerated diffusion of stars in velocity space as a result of transient perturbations induced by external gravitational fields (see commentary, Paper #20). This process will probably not decrease the lifetime of galactic clusters by an order of magnitude, but should certainly be included in future theoretical models.

## REFERENCES

1. Oort, J. H.—In: Stellar Populations, Scripta Varia No. 16, Pontifical Academy of Sciences, ed. D.J.K. O'Connell, Amsterdam: North Holland Publ. p. 63, (1958).
2. Bohr, N.—Phil. Mag. **25**, 10 (1913) and **30**, 581 (1915).
3. Wielen, R.—In: Dynamics of Star Clusters, IAU Symp. No. 113, eds. J. Goodman and P. Hut, Dordrecht: Reidel 449 (1985); In: The Formation and Evolution of Star Clusters, ed. K. Janes, *Astron. Soc. Pacific Conference Series*, **13**, 343 (1991).
4. Greenstein, J. L.—*Harvard Ann.* **105**, 359 (1937), see Table II.
5. Terlevich, E.—*Monthly Notices Roy. Astron. Soc.*, **224**, 193 (1987).
6. Weinberg, M. D.—*Astron. J.*, **108**, 1403 and 1414 (1994).

*Paper*

### ABSTRACT

The tidal force of passing interstellar clouds accelerates the stars in a galactic cluster, increases the total cluster energy, and leads to the expansion and ultimate disruption of the cluster. The magnitude of this effect is computed on the impulsive approximation, i.e., on the assumption that the stars do not move appreciably as a cloud passes by. Solution of the more general problem, on the simplifying assumption that the gravitational potential of the cluster varies as the square of the distance from the center, shows that the impulsive approximation should usually be adequate. The computed rate of increase of internal cluster energy, resulting from encounters with clouds, is about thirty times the corresponding rate for encounters with field stars, if a cluster radius of 5 pc is assumed. The "disruption time" required for a cluster to dissociate into separate stars varies directly with the cluster density and is about $2 \times 10^8$ years for a mean density of $1 M_\odot / \mathrm{pc}^3$. The stars of relatively low mass in a cluster will be lost more rapidly, partly by evaporation and partly by enhanced disruption of the extended aura which these stars presumably form about the cluster. This mechanism may account for the scarcity of galactic clusters older than $10^9$ years.

### I. INTRODUCTION

The evolution of a galactic cluster is dominated by two major effects. The first of these, pointed out several decades ago by Bok (1934), is the increase in mean energy as a result of encounters with passing field stars. Since this effect leads to a gradual expansion of the cluster, we shall refer to it here as "disruption." This process has been analyzed in detail by Mineur (1939) and Chandrasekhar (1942). The second effect, pointed out independently by Ambarzumian (1938) and Spitzer (1940), is the "evaporation" of cluster stars as a result of gravitational encounters between stars in the cluster. In this process, an individual star acquires positive energy and departs from the cluster, leaving the remaining stars with a more negative mean energy than before. As a result, the cluster contracts. This effect has been analyzed by Chandrasekhar (1942, 1943*a, b*).

Previous analyses of disruption have referred to the gravitational encounters with field stars only. It has been pointed out by Spitzer and

Schwarzschild (1950) that the velocity distribution of type I stars is much more affected by gravitational encounters with cloud complexes than by encounters with the field stars. The purpose of the present paper is to point out that in a similar manner the disruption of clusters by successive encounters with extended interstellar clouds is a full order of magnitude greater than the disruption resulting from passing field stars. The analysis in the subsequent sections indicates that this effect may be of great importance in the evolution of all but the densest galactic clusters.

This analysis was undertaken in an attempt to explain the apparent rarity of old clusters, a fact recently emphasized by Oort (1957). According to Oort's survey of the data, the distribution of ages of galactic clusters, for clusters younger than $5 \times 10^8$ years, is about what we would expect on the hypothesis of a uniform rate of formation. The age of a galactic cluster is indicated by the spectral type of the earliest stars, with the age ranging from $10^7$ years for stars of type O to $5 \times 10^8$ years for types B9 to A0. However, clusters older than $5 \times 10^8$ years, in which the earliest stars are of type A1 or later, are almost completely lacking, although, on the assumption of a uniform rate of formation, such clusters should outnumber all others. Oort has suggested that some process leads to the disintegration of most galactic clusters within $10^9$ years after their formation. While conclusive results are not now possible, the present analysis strongly suggests that disruption by tidal encounters with interstellar clouds is just such a process.

In Section II the rate of increase of cluster energy, produced by passing clouds, is computed on the impulsive approximation, i.e., with the assumption that the position of the star in the cluster does not change appreciably during the passage of the cloud. The probable range over which this assumption is valid is considered in the next section, where the full time-dependent problem is solved by successive approximations in an idealized case. In Section IV these results are applied to a discussion of cluster evolution. Computations of the velocity-distribution function in an isolated cluster, to yield the rates of evaporation for stars of different masses together with the associated density distributions, are given in a separate paper (Spitzer and Härm 1958).

## II. Rate of Energy Increase on the Impulsive Approximation

As a cloud passes by a cluster, each star in the cluster experiences a tidal force relative to the center of gravity of the cluster. If the cloud passes by sufficiently rapidly, the velocity change, $\Delta v$, of each star may be computed as though the star were not moving. If an average is taken over all stars, the mean value of $(\Delta v)^2$ is finite and corresponds to an increase in the total kinetic energy, $T$, of the cluster. From the virial theorem it follows that this increase in the total energy, $U$, leads to a new equilibrium, in which the total gravitational energy is less negative than before, $T$ is reduced below its former value, and the cluster has expanded. We proceed now to compute the rate at which $U$ increases.

We define a co-ordinate system centered at the cluster center and with the $x$-axis pointing toward the cloud when it is at its distance, $p$, of closest approach to the cluster center. The $y$-axis is taken parallel to the velocity, $V$, which the cloud has, relative to the cluster. The $z$-axis, of course, is perpendicular to the $x$- and $y$-axes.

The tidal acceleration, $F$, is most conveniently expressed in a rotating reference frame, $x'$, $y'$, $z'$, in which $x'$ points to the instantaneous position of the cloud, $y'$ is in the $xy$-plane, and $z'$ is parallel to $z$. If we let $R$ be the instantaneous distance of the cloud from the cluster center, then we have the familiar results,

$$F'_x = \frac{2Gm_n x'}{R^3}, \qquad F'_y = -\frac{Gm_n y'}{R^3}, \qquad F_z = -\frac{Gm_n z'}{R^3}, \quad (1)$$

where $G$ is the gravitational constant, $m_n$ is the mass of the cloud or nebula (we use a subscript $n$ for the clouds throughout, reserving the subscript $c$ for properties of the cluster). Evidently, by a proper choice of time origin, we have

$$R^2 = p^2 + V^2 t^2. \tag{2}$$

The rotating co-ordinate frame is related to the fixed frame by the equations

$$x' = x \cos \theta + y \sin \theta,$$

$$y' = -x \sin \theta + y \cos \theta, \tag{3}$$

$$z' = z, \tag{4}$$

where $\tan \theta = Vt/p$. After some algebra, we obtain

$$\frac{dv_x}{dt} = \frac{Gm_n}{R^3}[x(2 - 3\sin^2 \theta) + 3y \sin \theta \cos \theta], \tag{5}$$

$$\frac{dv_y}{dt} = \frac{Gm_n}{R^3}[y(2 - 3\cos^2 \theta) + 3x \sin \theta \cos \theta], \tag{6}$$

$$\frac{dv_z}{dt} = -\frac{Gm_n z}{R^3}. \tag{7}$$

Integration of equations (5), (6), and (7), with the use of equation (2), yields[†]

$$\Delta v_x = \frac{2Gm_n x}{p^2 V}, \qquad \Delta v_y = 0, \qquad \Delta v_z = -\frac{2Gm_n z}{p^2 V}. \qquad (8)$$

It follows that $\Delta U$, the increase in total energy of the cluster, which is equal, of course, to $\Delta T$, is given by

$$\Delta U = \tfrac{1}{2} m_c \left( \frac{2Gm_n}{p^2 V} \right)^2 \tfrac{2}{3} r_c^2, \qquad (9)$$

where $m_c$ is the cluster mass and $r_c^2$ is the mean-square cluster radius.

The rate of change of $U$ is given by equation (9) multiplied by the number of collisions per unit time, per interval of $p$, and per interval of $m_n$, and integrated over $p$ and $m_n$. In this way we obtain

$$\frac{dU}{dt} = \frac{8\pi G^2 m_c r_c^2}{3V} \int_0^\infty n(m_n) m_n^2 \, dm_n \int_{R_n}^\infty \frac{dp}{p^3}, \qquad (10)$$

where $R_n$ is the radius of a cloud whose mass is $m_n$, and $n(m_n) \, dm_n$ is the number of clouds, or nebulae, per unit volume with a mass between $m_n$ and $m_n + dm_n$. Encounters for which $p$ is less than $R_n$ will increase $U$ somewhat, but we neglect them.

To evaluate equation (10) approximately, we see that

$$\frac{m_n}{R_n^2} = \tfrac{4}{3} \pi R_n \rho_{in}, \qquad (11)$$

where $\rho_{in}$ is defined as the internal density of the cloud, assumed to be a uniform sphere of radius $R_n$. We assume that this product $(R_n \rho_{in})$ is independent of the size of the cloud, corresponding to an optical extinction per cloud independent of cloud size. As pointed out by Bok (1946), observed clouds seem to have a tendency in this direction. The integral over $m_n$ then gives simply $\rho_{an}$, the total density of clouds, averaged over that volume of the Galaxy in which the clusters move.

---

[†] In equations (8), a typographical error has been corrected, with a minus sign inserted in the formula for $\Delta v_z$.

The final formula for $dU/dt$, then, is

$$\frac{dU}{dt} = \left(\frac{4\pi G}{3}\right)^2 \frac{m_c r_c^2 \rho_{an}(R_n \rho_{in})}{V}. \tag{12}$$

As $U$ increases, the cluster expands, increasing the rate of expansion. To evaluate this effect, we must know the dependence of $U$ on $r$. From the virial theorem we know that $U$ is $\Phi/2$, where $\Phi$, the total gravitational energy of the cluster, may be written

$$\Phi = -\frac{\gamma G m_c^2}{r_c}, \tag{13}$$

where, as before, $r_c^2$ is the mean-square value of $r$ for all the cluster stars and $\gamma$ is a numerical constant. For a homogeneous sphere, $\gamma$ is $(\frac{3}{5})^{3/2}$, or 0.465; and, for a polytropic sphere of index $n$, $\gamma$ varies from 0.469 to 0.564 as $n$ increases from 1 to 4; these values are based on radii of gyration computed for polytropes by Motz (1952). We shall let $\gamma$ be equal to 0.5 in the numerical application of these results.

If we now eliminate $r_c$ from equation (12) by means of equation (13) and integrate, we find that $U$ reaches zero at a "disruption time," $t_d$, given by

$$t_d = \frac{\gamma}{8\pi G} \frac{V}{\rho_{an}(R_n \rho_n)} \rho_c, \tag{14}$$

where $\rho_c$ is defined as the cluster mass divided by $4\pi r_c^3/3$. This equation will be used in the last section in a discussion of cluster evolution.

### III. Approximate Solution for Slow Encounters

The results in the preceding section, while they have the great advantage of simplicity, have the disadvantage that they are based on the assumption that the cluster stars move a negligible amount while the cloud passes by. Since this assumption is not realistic, it is necessary to investigate the more general case, in which the stellar motions are taken into account. Such an investigation will be carried through here with two other simplifying assumptions, as follows: (a) the tidal force of the passing cloud is small compared to the gravitational attraction of the cluster; and (b) the gravitational potential in the cluster varies as the square of the distance from the center.

Assumption $a$ is certainly valid, except for the closest encounters with the densest clouds. This assumption makes it possible to solve the equations by a perturbation method. Assumption $b$ provides an idealization of the problem, but it should not falsify the results for actual clusters. According to this assumption, each star moves in simple harmonic oscillation in each co-ordinate, simplifying the analysis very greatly. Actually, we shall consider here only those stars whose initial velocity lies in the $xz$- or the $yz$-plane. The analysis could be extended in a straightforward manner to stars with initial velocities in all three directions, but with considerably greater algebra.

We consider, first, the motion of a star in the $z$-direction; we let the acceleration in this direction equation equal $-\omega^2 z$. Let the perturbing tidal force be $\lambda z f(t)$, where $f(0)$ equals unity and $\lambda$ is assumed small compared to $\omega^2$. The equation of motion, then, is

$$\frac{d^2 z}{dt^2} + \omega^2 z = \lambda z f(t). \tag{15}$$

This equation may be solved as a power series in $\lambda$,

$$z = z_0 + \lambda z_1 + \lambda^2 z_2 \cdots. \tag{16}$$

As we shall see, the change in kinetic energy is of order $\lambda^2$, and hence we must compute $z_2$ as well as $z_1$.

The zero-order solution is obviously

$$z_0 = A_0 \cos \omega t + B_0 \sin \omega t. \tag{17}$$

The first-order function, $z_1$, is obtained by inserting $z_0$ on the right-hand side of equation (15), which is then a known function of $t$, and solving for $z_1$ by use of the variation of parameters. In this way we obtain

$$z_1 = A_1(t)\cos \omega t + B_1(t)\sin \omega t, \tag{18}$$

where

$$A_1(t) = -\frac{1}{\omega} \int_{-\infty}^{t} f(\tau)\sin \omega\tau (A_0 \cos \omega\tau + B_0 \sin \omega\tau)\, d\tau, \tag{19}$$

$$B_1(t) = \frac{1}{\omega} \int_{-\infty}^{t} f(\tau)\cos \omega\tau (A_0 \cos \omega\tau + B_0 \sin \omega\tau)\, d\tau. \tag{20}$$

The lower limit of integration has been determined by the condition that $A_1$ and $B_1$ both vanish when $t$ is $-\infty$.

Similarly, if $z_1$ from these equations is inserted by the right-hand side of equation (15), this equation may then be solved for $z_2$. The result is an equation similar to (18), with $A_2$ and $B_2$ replacing $A_1$ and $B_1$, where

$$A_2(t) = \frac{1}{\omega^2} \int_{-\infty}^{t} f(\tau)\sin \omega\tau \cos \omega\tau \, d\tau \int_{-\infty}^{\tau} z_0(u)f(u)\sin \omega u \, du$$

$$- \frac{1}{\omega^2} \int_{-\infty}^{t} f(\tau)\sin^2 \omega\tau \, d\tau \int_{-\infty}^{\tau} z_0(u)f(u)\cos \omega u \, du, \quad (21)$$

$$B_2(t) = -\frac{1}{\omega^2} \int_{-\infty}^{t} f(\tau)\cos^2 \omega\tau \, d\tau \int_{-\infty}^{\tau} z_0(u)f(u)\sin \omega u \, du$$

$$+ \frac{1}{\omega^2} \int_{-\infty}^{t} f(\tau)\cos \omega\tau \sin \omega\tau \, d\tau \int_{-\infty}^{\tau} z_0(u)f(u)\cos \omega u \, du.$$

$$(22)$$

In the previous section, $\Delta U$ was simply the change in kinetic energy produced by the encounter. When the encounter is gradual, this simplification cannot be made, and the total energy, including both potential and kinetic energies, must be considered. The energy in the $z$ coordinate may now be taken as the kinetic energy of motion in the $z$-direction when $z$ vanishes, and we have

$$\Delta\left[\tfrac{1}{2}m\left(\frac{dz}{dt}\right)^2_{z=0}\right] = \frac{m\omega^2}{2}\left[(A_0 + \lambda A_1 + \lambda^2 A_2 \cdots)^2 - A_0^2\right.$$

$$+ (B_0 + \lambda B_1 + \lambda^2 B_2 \cdots)^2 - B_0^2\Big], \quad (23)$$

evaluated at $t$ equal to $+\infty$. To obtain $\Delta U_z$, the total change in the energy of the cluster in the $z$-direction, we must multiply equation (23) by the number of stars in the cluster and average over all values of $A_0$ and $B_0$. Since the phases of the stars are at random, we may write, denoting by broken brackets an average over all stars,

$$\langle A_0 B_0 \rangle = 0, \quad (24)$$

$$\langle A_0^2 \rangle = \langle B_0^2 \rangle = z_c^2, \quad (25)$$

where $z_c^2$ is evidently the mean-square value of $z$ for the cluster stars. Because of equations (24) and (25), the terms in $\lambda$ in equation (23) cancel out. After some straightforward algebra, all the terms in $\lambda^2$ may be combined.

The double integrals may be evaluated on integration by parts. For a typical such integral we have, representing trigonometric functions by $F$ and $G$,

$$\int_{-\infty}^{+\infty} f(\tau)F(\omega\tau)\,d\tau \int_{-\infty}^{\tau} f(u)G(\omega u)\,du$$

$$= \left| \int_{-\infty}^{t} f(\tau)F(\omega\tau)\,d\tau \times \int_{-\infty}^{t} f(u)G(\omega u)\,du \right|_{-\infty}^{+\infty}$$

$$- \int_{-\infty}^{+\infty} f(t)G(\omega t)\,dt \int_{-\infty}^{t} f(\tau)F(\omega\tau)\,d\tau, \qquad (26)$$

and hence

$$\int_{-\infty}^{+\infty} f(\tau)F(\omega\tau)\,d\tau \int_{-\infty}^{\tau} f(u)G(\omega u)\,du$$

$$+ \int_{-\infty}^{+\infty} f(\tau)G(\omega\tau)\,d\tau \int_{\infty}^{\tau} f(u)G(\omega u)\,du$$

$$= \left[ \int_{-\infty}^{+\infty} f(t)F(\omega t)\,dt \right]\left[ \int_{-\infty}^{+\infty} f(t)G(\omega t)\,dt \right]. \qquad (27)$$

We obtain, finally for $\Delta U_z$, the change in $U_z$ in a single encounter,

$$\Delta U_z = \frac{m_c}{2}\,\lambda^2 z_c^2(I_c^2 + I_s^2), \qquad (28)$$

where

$$I_c = \int_{-\infty}^{+\infty} f(\tau)\cos 2\,\omega\tau\,d\tau, \qquad (29)$$

$$I_s = \int_{-\infty}^{+\infty} f(\tau)\sin 2\,\omega\tau\,d\tau. \qquad (30)$$

We now apply equation (28) to the case where $f(t)$ corresponds to the tidal force produced by a passing cluster. The co-ordinate system used is the same as in the first section, and we have, from equation (7),

$$\lambda = -\frac{Gm_n}{p^3}, \tag{31}$$

$$f(t) = \left(\frac{p}{R}\right)^3 = \left(1 + \frac{V^2t^2}{p^2}\right)^{-3/2}. \tag{32}$$

With this expression substituted in equations (29) and (30), we find that $I_s$ vanishes, since $f(\tau)$ is an even function of $r$, and equation (29) gives, for $I_c$,

$$I_c = \frac{2p}{V} \beta K_1(\beta), \tag{33}$$

where $\beta$, the effective duration of the encounter, is defined by

$$\beta \equiv \frac{2\omega p}{V}, \tag{34}$$

and $K_1(\beta)$ is the usual Bessel function of imaginary argument. Evidently, $\beta$ is essentially the time required for a star to pass by the cluster, divided by the oscillation period of a cluster star. These results, substituted in equation (28), give $\Delta U_z$. Motions of the star in the $x$- and $y$-directions do not affect this result, as the tidal force in the $z$-direction is independent of $x$ and $y$.

To compute $\Delta U_x$ and $\Delta U_y$, we must take into account that the $x$- and $y$-motions are coupled through equations (5) and (6). To simplify the mathematics, we compute $\Delta U_z$ on the assumption that $y_0$, the initial function $y(t)$ before the encounter, vanishes; however, $y_1$ contributes directly to $\Delta U_x$ and, by affecting $x_2$, contributes indirectly as well. A similar contribution to $\Delta U_y$ is computed on the assumption that $x_0$ vanishes. The final result, including acceleration in the $z$-direction also, may be written thus:

$$\Delta U_i = \frac{m_c}{2} \left(\frac{2Gm_n}{Vp^2}\right)^2 \frac{r_c^2}{3} L_i(\beta), \tag{35}$$

where $i$ stands for $x$, $y$, or $z$, and[†]

$$L_x(\beta) = \left[\beta K_1(\beta) + \beta^2 K_0(\beta)\right]^2 + \left[\beta^2 K_1(\beta)\right]^2, \tag{36}$$

[†]To correct a typographic error in equation (36), the factors $\beta^2$ and $\beta$ have been interchanged within the first square bracket. With this change, equation (36) yields the values in Table 1, confirmed by E. Knobloch, Ap. J. **209**, 411, 1976.

TABLE 1
Values of $L_i(\beta)$

| $\beta$ | $L_x(\beta)$ | $L_y(\beta)$ | $L_z(\beta)$ | $L_x + L_y + L_z$ |
|---|---|---|---|---|
| 0 | 1.000 | 0.000 | 1.000 | 2.000 |
| 0.1 | 1.029 | 0.010 | 0.971 | 2.010 |
| 0.2 | 1.088 | 0.041 | 0.912 | 2.042 |
| 0.3 | 1.158 | 0.091 | 0.841 | 2.089 |
| 0.4 | 1.229 | 0.154 | 0.763 | 2.146 |
| 0.5 | 1.294 | 0.225 | 0.686 | 2.205 |
| 0.6 | 1.347 | 0.298 | 0.611 | 2.256 |
| 0.7 | 1.386 | 0.370 | 0.541 | 2.296 |
| 0.8 | 1.409 | 0.435 | 0.475 | 2.319 |
| 0.9 | 1.417 | 0.492 | 0.416 | 2.325 |
| 1.0 | 1.409 | 0.540 | 0.362 | 2.311 |
| 1.2 | 1.352 | 0.602 | 0.272 | 2.226 |
| 1.4 | 1.254 | 0.624 | 0.202 | 2.079 |
| 1.6 | 1.130 | 0.611 | 0.148 | 1.889 |
| 1.8 | 0.993 | 0.574 | 0.108 | 1.674 |
| 2.0 | 0.854 | 0.521 | 0.078 | 1.453 |
| 2.5 | 0.543 | 0.365 | 0.034 | 0.942 |
| 3.0 | 0.318 | 0.228 | 0.015 | 0.561 |
| 3.6 | 0.175 | 0.132 | 0.006 | 0.313 |
| 4.0 | 0.092 | 0.072 | 0.002 | 0.166 |
| 4.5 | 0.047 | 0.037 | 0.001 | 0.085 |
| 5.0 | 0.023 | 0.019 | 0.000 | 0.042 |

$$L_y(\beta) = \left[ \beta^2 K_0(\beta) \right]^2 + \left[ \beta^2 K_1(\beta) \right]^2, \tag{37}$$

$$L_z(\beta) = \left[ \beta K_1(\beta) \right]^2. \tag{38}$$

The values of these three functions are given in Table 1, together with their sum in the last row.

Examination of Table 1 indicates the extent to which the impulsive approximation is valid in this idealized situation. When the effective duration time, $\beta$, vanishes, the stars may be taken as essentially motionless during an encounter, and we recover equation (9) of Section II. With increasing $\beta$, $L_x + L_y + L_z$ actually increases at first, finally decreasing again. This initial increase in $\Delta U$ with increasing $\beta$ is due to the possibility of resonance. For a fixed star the tidal force in the xy-plane changes its direction as the cloud passes. A moving star can be passing through the origin when the tidal force has one sign and be far from the cluster center when the force has the opposite sign. Evidently, the impulsive approximation somewhat underestimates $\Delta U$ for $\beta$ about equal to 1.0. Even for $\beta$ equal to 1.5, $L_x + L_y + L_z$ has almost the same value as for zero $\beta$. For a further increase in $\beta$, however, $\Delta U$ decreases sharply. When the encounter duration time exceeds four periods of oscillation and $\beta$ exceeds 4, the value of $\Delta U$ is reduced an order of magnitude or more below the impulsive approximation. In an actual cluster, where the period is not a constant for all stars but decreases with increasing distance from the cluster center, this decline in the $L$-functions with decreasing $\beta$ would not be so abrupt.

In conclusion, we may safely use the results of the impulsive approximation up to about $\beta$ equal 2 with an accuracy probably better than 20 percent, on the average. For greater values of $\beta$, the value of $\Delta U$ decreases rapidly and can probably be neglected.

## IV. Evolution of Galactic Clusters

The rate of disruption of a cluster depends inversely on the cluster density. The observational evidence on the density, $\rho_c$, of galactic clusters will first be summarized briefly. Reliable cluster masses can be obtained only from measured velocity dispersions and use of the virial theorem. Such measures are available only for the Pleiades and Praesepe clusters. Analyses of these data by Mineur (1939) and van Wijk (1949) give total masses of about 500 to $700 M_\odot$, respectively. The densities depend on the distribution of mass in each cluster.

According to an analysis by Oort (informal communication), the mean value of $r^2$ for those stars in the Pleiades whose apparent magnitudes lie between 6.0 and 9.9 is $2.9 (\deg)^2$. With the value of $0.011''$ for the parallax, given by Mineur (1939), the corresponding value of $r_c$ is 2.7 pc, giving $6.0 M_\odot/\text{pc}^3$ for $\rho_c$ in the Pleiades. The eleven stars brighter than the sixth magnitude are more concentrated toward the center but will not appreciably affect $r_c$ for the cluster as a whole. However, the true value of $r_c$ may be increased somewhat above 2.7 pc by the stars fainter than the tenth magnitude, for which the available data are not sufficiently complete to indicate the spatial distribution. Thus $6.0 M_\odot/\text{pc}^3$ may be regarded as an upper limit for $\rho_c$. If for the parallax of the Pleiades we take the recent spectroscopic value of $0.0079''$ found by Mitchell and Johnson (1957) instead of the value $0.011''$ adopted by Mineur, this upper limit on the density is reduced to $3.1 M_\odot/\text{pc}^3$, the cluster mass increasing to $700 M_\odot$. The star counts for Praesepe have not been analyzed for the spatial distribution.

For other clusters we must use star counts as a measure of the total mass. For the Pleiades and Praesepe the total mass estimated from star counts is about half the dynamical value, and we shall assume this same ratio in other clusters. Much denser clusters than the Pleiades certainly exist. For example, the value of $\rho_c$ in M67, if we assume that $r_c$ equals the radius containing half the mass, is about $70M_\odot/\mathrm{pc}^3$, according to van den Bergh (1957). Allowance for the difference between star counts and the true value for the total mass increases this density to about $140M_\odot/\mathrm{pc}^3$. On the other hand, less dense clusters are also known. The structure of the Hyades cluster has been extensively studied by van Bueren (1952). The root-mean-square radius, determined from his data, is 8 pc, giving a value of $0.15M_\odot/\mathrm{pc}^3$ for $\rho_c$. If this value is doubled, to allow for undetected masses, the resultant density is still only about $0.3M_\odot/\mathrm{pc}^3$, which is comparable with the density at which a cluster becomes unstable because of the shearing effect of galactic rotation. While Bok (1934) found instability for a density of $0.09M_\odot/\mathrm{pc}$, Mineur (1939), in a more elaborate analyses, taking internal motions into account, found that a homogeneous cluster would be unstable at a density of $0.6M_\odot/\mathrm{pc}^3$. Extension of the analysis to nonuniform clusters has been carried through by van Wijk (1949). Evidently, the Hyades cluster is near the minimum density possible for a cluster that can hold itself together for any appreciable length of time. We shall therefore consider clusters in which the mean density may lie anywhere from 0.3 to $100M_\odot/\mathrm{pc}^3$.

We pass on to the numerical application, for actual clusters, of the results in the preceding two sections. First, we consider the applicability of the impulsive approximation. We have seen in Section III that this approximation could certainly be used for $\beta$ equal to or less than 2. Equation (34) then yields the condition that $1/\omega$ be at least $p/V$. The greatest relevant value of $p$ is roughly the radius $R_n$ of the largest interstellar cloud, which we may here take to be about 50 pc. The value of the relative velocity, $V$, may be set equal to 10 km/sec, a conservative value for most clusters. The impulsive approximation is therefore valid if

$$\frac{1}{\omega} \geq 5 \times 10^6 \text{ years.} \tag{39}$$

In a cluster of uniform density $\rho_c$, $\omega$ is given by

$$\omega^2 = \frac{4\pi G\rho_c}{3}. \tag{40}$$

Substitution of equation (40) in (39) yields

$$\rho_c \leq 2.2 \, \frac{M_\odot}{\mathrm{pc}^3}, \tag{41}$$

corresponding to a value in c.g.s. units of $1.5 \times 10^{-22}$ gm/cm$^3$. Evidently, the impulsive approximation is entirely valid for the Hyades but is not always correct for M67.

Next we insert numerical values into equation (14) for the disruption time, $t_d$. The mean density of interstellar matter will be taken as $1.68 \times 10^{-24}$ gm/cm$^3$, corresponding to 1 H atom/cm$^3$. The value of $(R_n \, \rho_{in})$ may be taken from a recent survey of interstellar cloud types of Spitzer (1957); the data yield a value of $2.0 \times 10^{-3}$ gm/cm$^2$, corresponding to a cloud of radius 20 pc, with a mean density of 20 H atoms/cm$^3$. Inserting these values in equation (14) and letting $\gamma$ equal 0.5, with $V$ equal to $10^6$ cm/sec, we find

$$t_d = 1.9 \times 10^8 \rho_c \left( \frac{M_\odot}{\text{pc}^3} \right) \text{ years.} \tag{42}$$

We have already seen in equation (41) that the impulsive approximation, on which this result is based, is valid for $\rho_c$ up to $2.2 M_\odot/\text{pc}^3$. If $\rho_c$ is ten times this upper limit, the process is not of great interest, as the disruption time, $t_d$, is about $10^{10}$ years. We discuss briefly the value of $t_d$ to be expected for $\rho_c$ in the range from 2.2 to $22 M_\odot/\text{pc}^3$.

For clusters with $\rho_c$ in this range, the encounters with small clouds will be correctly given by the impulsive approximation, while encounters with large clouds, characterized by a value of $\beta$ substantially greater than 2, will produce only a small effect. In this case the integral over $m_n$ in equation (10) must extend only over those clouds which are sufficiently small that $\beta$ is less than 2. It is readily shown that this upper limit on $R_n$ is proportional to $1/\rho_c$. Moreover, if $\rho_{in} R_n$ is assumed constant, as before, the integral over $m_n$ is proportional to the total density of material in these smaller clouds. While it is not possible to state definitely how the integral will vary as the upper limit, or cutoff, on $R_n$ is decreased, it seems likely that the integral will not decrease vary rapidly. For example, if all the interstellar material were in small clouds 5 pc in diameter and these were gathered together in vast cloud complexes 50 pc in diameter, each containing some 100 small clouds, then cutting off the integral in equation (10) to exclude the cloud complexes, including only the small clouds, would reduce $dU/dt$ by a factor of only 2. It seems likely that, even for a density of $22 M_\odot/\text{pc}^3$, equation (42) is not in error by as much as an order of magnitude. For densities near the middle of the range, about $6 M_\odot/\text{pc}^3$, equation (42) is probably correct to within a factor of 2. We shall therefore use equation (42) for all densities, with the understanding that, when $t_d$ computed from this equation exceeds $10^9$ years, the true value is probably somewhat larger than the computed one.

It is evident from equation (42) that the disruption time is very short for the more extended clusters. The Hyades cluster, for example, will apparently be completely disrupted in about $6 \times 10^7$ years. For the Pleiades, on the other hand, the disruption time does not much exceed $10^9$ years and may be considerably less. While uncertainty concerning $\rho_c$ prevents very precise statements, it would appear that, for most galactic clusters, the disruption time is between $10^8$ and $10^9$ years. Only a relatively dense cluster, such as M67, can apparently survive disruption for a period as great as $5 \times 10^9$ years. It may be noted that for a cluster of 5 pc r.m.s. radius, the rate of disruption by these tidal encounters is about thirty times more rapid than the value computed by Bok (1934) for encounters with field stars.

These disruption times may be compared with the evaporation times computed by Spitzer and Härm (1958). From their equation (4) we find that the reference time, $T_R$, is given by

$$T_R = 8.3 \times 10^5 \; \frac{N^{1/2}[r_c(\text{pc})]^{3/2}}{(m_0/M_\odot)^{1/2}(\log_{10} N - 0.3)} \; \text{years}, \qquad (43)$$

where $N$ is the number of cluster stars, assumed to be virtually all of the same mass, $m_0$. The close numerical agreement between equation (43) and the equation for $\overline{T}_E$ given by Chandrasekhar (1942) is fortuitous, since the cutoff factor, $\ln \alpha$, used here is 1.5 times that used by Chandrasekhar. The evaporation time, $t_{ev}$, is defined as the time in which $N$ would decrease by $1/e$ through evaporation, if the relative rate of evaporation remained constant. Values of $t_{ev}/T_R$ are given in Table 1 of Spitzer and Härm.

It may be noted that the values of $t_{ev}$ are probably less reliable than those of $t_d$. While there are uncertainties in the size distribution of interstellar clouds which affect equation (42), the evaporation time is computed for a hypothetical uniform density, and unknown corrections are required for an actual cluster, where stars of relatively high energy per unit mass may spend most of their time far from the cluster center.

For a star of mass $m_0$ the evaporation time is $88T_R$. Hence for a typical cluster containing 300 stars of solar mass and with $r_c$ equal to 3 pc, $t_{ev}$ is $3.0 \times 10^9$ years, considerably larger than $t_d$. For a more concentrated cluster, such as M67, $r_c$ may be taken as 1.2 pc, giving $7.6 \times 10^8$ years for $t_{ev}$, considerably less than $t_d$. Whether a cluster expands by disruption or contracts by evaporation depends primarily on whether $r_c$ is greater or less than about 2 pc; the value at which $t_d$ equals $t_{ev}$ varies only as $N^{1/9}$.

We are now in a position to discuss the evolution of a cluster, formed at some epoch $t = 0$. From a dynamical standpoint, the first process to occur is the establishment of equipartition among stars of the average mass, $m_0$. This process takes a time $T_R$, about $10^7$ years. Next the stars of lighter mass approach equipartition with the heavier ones, their random velocities increasing correspondingly. The time required for this process has been investigated by Spitzer and Schwarzschild (1951), on the assumption that the stars of lighter mass interact not with one another but with larger masses, which possess a Maxwellian velocity distribution. For stars of mass much less than the average mass, $m_0$, numerical integration yields the approximate result for the mean-square velocity,

$$\frac{\langle v^2 \rangle}{v_0^2} = \left(1 + 3.2 \frac{t}{T_R}\right)^{2/5}, \tag{44}$$

where $v_0^2$ is the mean-square velocity of stars of mass $m_0$; at the time $t$ equal to zero, the lighter and heavier stars are all assumed to have a Maxwellian distribution with the same mean-square velocity. According to equation (44), the mean energy will double in only $1.5T_R$ and will triple in a total time equal to $5T_R$. Evidently, equipartition of energy for the less massive stars will be set up in substantially less than $10^8$ years in most galactic clusters.

The increase in velocity of the less massive stars will have two important effects. In the first place, these stars will evaporate more rapidly than those of average mass. As shown by Spitzer and Härm (1958), the evaporation rate increases by a factor of about 2 for each numerical decrease of $m/m_0$ by 0.2. In the second place, the mean-square distance of the less massive stars will presumably increase as the velocity increases, the stars of low mass essentially forming an extended aura around the cluster. While neither of these effects can be regarded as quantitatively established, the general character of the phenomena to be expected seems moderately clear.

If a substantial increase in $r^2$ for the less massive stars is accepted, an important result follows at once. The rate of energy input into the less massive stars, as a result of tidal forces produced by passing interstellar clouds, will be much greater for less massive stars than for those of average mass. If, for example, the mean value of $r^2$ for such stars is an order of magnitude greater than for the cluster as a whole, the energy of these stars may increase appreciably in $10^7$ years, the estimated time of relaxation. Under such conditions these lighter stars may gain energy more rapidly than they can lose it by encounters in the dense central region and will be lost by tidal disruption much more rapidly than the average stars.

As the distance of a star from the cluster center increases, the star can finally be pulled completely away by one of two effects. On the one hand, the tidal force of the galactic center can overcome the gravitational pull of the cluster at a distance $r$, if the mean density of the cluster in the sphere of radius $r$ is less than about $0.1 M_\odot/\mathrm{pc}^3$. In the Pleiades, for example, this occurs for a star more than about 11 pc away from the cluster center. On the other hand, the change in velocity resulting from a single passing cloud may exceed the velocity of escape from the cluster. From equation (8) it will be seen that escape in this manner is possible when the mean cloud density, within a sphere of radius $p$, exceeds the mean cluster density, within a sphere of radius $r$, by about the factor $pV^2/2Gm_n$, the square of the ratio of the relative velocity to the escape velocity from the cloud. This factor is probably between $10^2$ and $10^3$. Since the densities of interstellar clouds are typically within the range from 0.3 to $30 M_\odot/\mathrm{pc}^3$ (from 10 to $10^3$ H atoms/$\mathrm{cm}^3$), it is not clear which of these two effects is usually more important in finally detaching a star from a cluster.

Evidently, many quantitative details of the disruption process remain to be explored. In particular, whether the less massive stars are lost by evaporation or by rapid disruption of an extended aura cannot be indicated definitely at present. The general conclusion may be drawn, however, that all galactic clusters with mean densities between 0.5 and $5 M_\odot/\mathrm{pc}^3$ will be completely disrupted by successive tidal disturbances in $10^8$–$10^9$ years.

This theoretical result may account in part for the apparent rarity of old clusters (Oort 1957). A more detailed survey would be required, however, to indicate whether or not the present theory is consistent with the observations.

### References

Ambarzumian, V. A. 1938, *Ann. Leningrad State U.*, No. 22 ("Astronomical Series," Issue 4).

Bergh, S. van den. 1957. *A.J.*, **62**, 100.

Bueren, H. G. van. 1952, *B.A.N.*, **11**, 385 (No. 432).

Bok, B. J. 1934, *Harvard Circ.*, No. 384.

——. 1946, *Centennial Symposia* ("Harvard Obs. Monographs," No. 7), chaps. 1–5.

Chandrasekhar, S. 1942, *Principles of Stellar Dynamics* (Chicago: University of Chicago Press), chap. v.

——. 1943*a*, *Ap. J.*, **97**, 263.

——. 1943*b*, ibid., **98**, 54.

——. 1943*c*, *Rev. Mod. Phys.*, **15**, 1.

Mineur, H. 1939, *Ann. d'ap.*, **2**, 1.

Mitchell, R. I. and Johnson, H. L. 1957, *Ap. J.*, **125**, 414.

Motz, L. 1952, *Ap. J.*, **115**, 562.

Oort, J. H. 1957, *Stellar Populations* (Rome: Pontifical Academy of Sciences), discussion in Session on Star Clusters.

Spitzer, L. 1940, *M.N.*, **100**, 396.

——. 1957, *Stellar Populations* (Rome: Pontifical Academy of Sciences) (in press).

Spitzer, L., and Härm, R. 1958. *Ap. J.*, in press.

Spitzer, L., and Schwarzschild, M. 1951, *Ap. J.*, **114**, 385.

Wijk, U. van. 1949, *Ann. d'ap.*, **12**, 81.

# EQUIPARTITION AND THE FORMATION OF COMPACT NUCLEI IN SPHERICAL STELLAR SYSTEMS

(ASTROPH. J. LETT. 158, L139, 1969)

## *Commentary*

THIS research in the dynamical evolution of stellar systems was a natural outgrowth of my continuing interest in the topic, focused by two important conferences (see Commentary #18). This lack of an equilibrium solution for certain two-component model clusters, often referred to as a "mass stratification instability," is based on simple concepts and seems well established, though somewhat idealized. This commentary deals first with certain counterexample models, which seem to be valid equilibria even though they violate the theoretical upper limit on $M_2/M_1$, the ratio of total masses in heavy and light stars. Second, recent work is discussed indicating that the models in these counterexamples tend to be unstable, and hence cannot represent realistic clusters. The final paragraph considers the overall importance of this instability in the evolution of globular clusters.

The counterexamples[1,2] are all based on model clusters in which the velocity distribution function, $f$, for individual stars depends only on $E$ (defined here as the total energy of a star per unit mass), and is set equal to a lowered Maxwellian; i.e., proportional in the simplest case to $\exp(-BE) - \exp(-BE_e)$. Evidently $f$ vanishes at $E = E_e$, the energy of a star which just reaches the cluster boundary, and at all greater $E$. In these models equipartition appears in the assumption that $B$ is proportional to $m$, the mass of an individual star. This assumption usually gives equipartition of kinetic energy rather closely for stars in the main peak of the Maxwellian distribution, but not for stars slightly inside the cluster boundary. Heavier and lighter stars in this region will require the same velocities to just reach the boundary, and will have the same mean square velocities, violating equipartition.

This significant departure from equipartition for stars with a potential energy not far below the escape energy is shown by the counterexample models, which possess very extensive halos. As the constant $\beta$ in equation (6) below increases much above $\beta_{max} = 0.16$, the halos become more and more exaggerated, and the models become rather unrealistic, with less and less true equipartition.

More recently, these models with exaggerated halos, and with $M_2/M_1$ significantly exceeding the value corresponding to $\beta = \beta_{max}$, have been shown[2] to be unstable if equipartition is established at the center. More specifically, if the average kinetic energy $M_2 E_2$ of the heavier stars in the core exceeds by no more than 15 percent that of the adjacent lighter stars, the systems are generally stable only if $\beta$ lies below the range 0.15 to 0.20. The analysis does not indicate which type of instability may occur for larger $\beta$, but the gravothermal collapse seems likely (see Paper #20).

While mass stratification can certainly promote core collapse, if $\beta$ is sufficiently high, the importance of this effect for globular clusters is difficult to establish. Since the distribution of masses in most clusters is not accurately known, it is difficult to determine whether or not $\beta$ exceeds $\beta_{max}$ in a particular cluster. More important, according to the idealized theoretical models the gravothermal collapse acting alone is more rapid than that produced by mass stratification only. For a realistic cluster model it is difficult to untangle the contribution of these two processes to the overall collapse rate. Certainly the rapid concentration of the more massive stars toward the cluster core helps to increase the core density early in the evolution of the system, leading to the gravothermal collapse substantially earlier than would occur in a system with stars all of the same mass. However, it is doubtful whether an increase of $\beta$ above $\beta_{max}$ can have a dramatic effect on the overall collapse rate once the gravothermal collapse is fully underway.

REFERENCES

1. Merritt, D.—*Astron. J.* **86**, 318 (1981).
2. Katz, J. and Taff, L. G.—*Astroph. J.* **264**, 476 (1983).

# 17

## Paper

### ABSTRACT

In a spherical system composed of stars of two masses, $m_1$ and $m_2$, with $m_2$ greater than $m_1$, the heavier stars will lose kinetic energy with a time constant about twice the equipartition time, $t_{eq}$, and will gravitate toward the center. It is shown that if $M_2$, the total mass of the heavier stars, exceeds a certain critical value, the self-attraction of these stars requires such a large rms velocity dispersion that equipartition with the lighter stars becomes impossible. This critical value equals $\beta(m_1/m_2)^{3/2}M_1$, where $M_1$ is the corresponding total mass of the lighter stars and $\beta$ is a numerical constant less than unity that equals 0.16 when $m_2/m_1$ is large. When $M_2$ exceeds this value, the subsystem of heavy stars continues to contract with a time constant about equal to $t_{eq}$, forming a dense nucleus within the core of the system. Application of this criterion to the mass distribution for newly created stars suggests that equipartition is impossible initially for all young systems. The formation and continuing contraction of a dense nucleus of heavy stars is likely to be important in the dynamical evolution of galactic and globular clusters and of galactic nuclei.

## I. INTRODUCTION

The long-term dynamical evolution of a spherical stellar system, in which the stars may be regarded as mass points, is dominated by the tendency of encounters to establish kinetic equilibrium, in which stellar velocities have a Maxwellian distribution. Since the mean kinetic energy of the stars in such a system is one-fourth the mean energy required to escape from the cluster entirely, encounters tend to produce escape, or "evaporation," of stars, as pointed out by Ambartsumain (1938) and Spitzer (1940). A more detailed analysis by Hénon (1960) has shown that the dominant effect is to expel stars from the core of the system into an extended halo, with only a small fraction actually escaping completely if the system is isolated. In any case the evaporation of stars from the core leads to accelerated evaporation, and no steady state is possible.

If stars of different masses are present, encounters will tend to establish equipartition of kinetic energy. The lighter stars with accelerate and will evaporate from the core at a rate which exceeds that for the heavier stars and which approaches a limiting value for decreasing mass (Spitzer and Härm 1958). The heavier stars, with lower velocities, will, of course, tend to settle to the center of the system. It does not seem to have been generally realized that, if the fraction of massive stars exceeds a critical value, these heavier stars cannot reach a condition of approximate equipartition, and the subsystem of such stars will continue to contract rapidly at the center of the system. The present Letter analyzes in a simplified case the conditions required for this effect to occur and the rate at which such a subsystem will contract.

Numerical integrations by von Hoerner (1960, 1963), van Albada (1968), and Aarseth (1963, 1966) for spherical systems containing between 10 and 100 stars have shown that a dense nucleus tends to develop at the very center of the core, and the present Letter provides a theoretical understanding of some conditions under which this effect will occur. The rate of contraction of this compact subsystem can significantly exceed the rate of contraction of a stellar core when all stars have the same mass. A preliminary discussion of the possible importance of this phenomenon in the evolution of galactic nuclei and stellar clusters is given in §IV below.

## II. Conditions for Equipartition

We derive a criterion for the possibility of approximate equipartition between stars of different masses in a spherical stellar system. For simplicity, only two stellar types will be considered, with masses $m_1$ and $m_2$; we denote the larger mass by $m_2$. The total masses of the two types will be denoted by $M_1$ and $M_2$, equal to $m_1 N_1$ and $m_2 N_2$, respectively. In accordance with the usual virial theorem, we have for $\langle v_2^2 \rangle$, the mean-square velocity for stars of mass 2,

$$\langle v_2^2 \rangle = \frac{0.4 G M_2}{r_2} + \frac{G}{M_2} \int_0^\infty \frac{\rho_2 M_1(r)\, dV}{r}, \tag{1}$$

where $\rho_2$ is the density of stars of mass $m_2$ at the distance $r$ from the center, $M_1(r)$ is the corresponding total mass of the lighter stars interior to $r$, and $dV$ is the volume element. In the first term on the right-hand side, representing the self-gravitational energy, $r_2$ is the radius containing half the mass; the constant of probability in this term lies between 0.42 and 0.38 for polytropes of index between 3 and 5, respectively, and the value of 0.4 adopted is thus a good approximation. If the subscripts 1 and 2 are interchanged, a corresponding equation is obtained for

$\langle v_1^2 \rangle$. The condition of equipartition is simply

$$m_1 \langle v_1^2 \rangle = m_2 \langle v_2^2 \rangle. \tag{2}$$

In applying these equations, we assume now that $m_2$ appreciably exceeds $m_1$ and that the heavier stars are concentrated at the center of the system. Hence, on the right-hand side of equation (1), we may replace $M_1(r)$ by $4\pi\rho_{01}r^3/3$, where $\rho_{01}$ is the central density for stars of mass $m_1$. The resulting integral may then be expressed in terms of $r_{s2}^2$, the mean value of $r^2$ for stars of mass $m_2$. We ignore the corresponding term in the equation for $\langle v_1^2 \rangle$, which is permissible provided that the total mass $M_2$ is small compared with $M_1$ and the mutual interaction between the two types of stars has a consequently negligible effect on $\langle v_1^2 \rangle$. We denote by $\rho_{m1}$ the mean density for the stars of mass $m_1$ within the sphere of radius $r_1$ (containing half of the lighter stars), yielding

$$\rho_{m1} = 0.5 M_1 / \tfrac{4}{3}\pi r_1^3, \tag{3}$$

and similarly for $\rho_{m2}$. If now in equation (2) we substitute equation (1) for $\langle v_2^2 \rangle$ and the corresponding equation for $\langle v_1^2 \rangle$, and express $r_2/r_1$ in terms of $\rho_{m2}/\rho_{m1}$, we obtain

$$\frac{M_2 m_2^{3/2}}{M_1 m_1^{3/2}} = \frac{(\rho_{m1}/\rho_{m2})^{1/2}}{(1 + \alpha\rho_{m1}/\rho_{m2})^{3/2}}, \tag{4}$$

where the constant $\alpha$ is given by

$$\alpha = \frac{5\rho_{01}}{4\rho_{m1}} \left( \frac{r_{s2}}{r_2} \right)^2 = 5.6. \tag{5}$$

The ratio of $r_2$, the radius containing half the mass, to $r_{s2}$, the rms radius, has been set equal to 0.89, its value for a Maxwellian distribution in a parabolic potential well. The ratio of the central density, $\rho_{01}$, to $\rho_{m1}$ has been set equal to 3.5, as compared with values between 2.5 and 4.5 for polytropes with $n$ between 3 and 5. The value of $\alpha$ appears to be insensitive to the detailed density distribution.

The right-hand side of equation (4) has a maximum value $0.38\alpha^{-1/2}$. Hence the conditions required for equipartition can be satisfied only if

$$M_2/M_1 < \beta(m_1/m_2)^{3/2}, \tag{6}$$

where $\beta$ equals 0.16 when equation (5) is used for $\alpha$. The physical explanation of this requirement is that, for constant $M_2$, $\langle v_2^2 \rangle$ in equation (1) has a minimum value for $\rho_{m2}$ somewhat less than $\rho_{m1}$. For smaller values of $\rho_{m2}$, $r_2$ is increased and the second term on the right-hand side of equation (1) increases as $r_2^2$. For larger values of $\rho_{m2}$, $r_2$ decreases and the first term is greater, representing increased self-attraction of the heavy stars. Evidently, if this minimum value of $\langle v_2^2 \rangle$ exceeds the value required by equation (2), equipartition is excluded.

This analysis is based on the assumption that $m_2/m_1$ is large and that as a result $M_1/M_2$ must also be large. As $m_2/m_1$ decreases toward unity, a reconsideration of the analysis indicates that inequality (6) remains valid, with $\beta$ approaching unity.

## III. Rate of Contraction by Nucleus of Massive Stars

When $E_2$, the mean kinetic energy of the massive stars, exceeds the corresponding energy $E_1$ for stars of mass $m_1$, the massive stars lose energy at the rate given by

$$dU_2/dt = (E_1 - E_2)/t_{\text{eq}}, \tag{7}$$

where $U_2$ refers to the mean total energy, potential plus kinetic, and $t_{\text{eq}}$ is given by (Spitzer 1940, 1962)

$$t_{\text{eq}} = \frac{(\langle v_1^2 \rangle + \langle v_2^2 \rangle)^{3/2}}{8(6\pi)^{1/2} \rho_{01} G^2 m_2 \ln N_1}. \tag{8}$$

The stellar density has been set equal to its value at the center of the system, and the argument of the logarithmic term has been evaluated approximately. The ratio of $t_{\text{eq}}$ to a relaxation time $T_{R1}$ for stars of mass $m_1$ is given by

$$\frac{t_{\text{eq}}}{T_{R1}} = \frac{3\pi^{1/2}}{16} \frac{m_1}{m_2} \left( 1 + \frac{\langle v_2^2 \rangle}{\langle v_1^2 \rangle} \right)^{3/2}. \tag{9}$$

The energy-exchange time, $T_E$, introduced by Chandrasekhar (1942) and the self-collision time, $t_c$, defined by Spitzer (1962) equal 0.71 and 0.64 times $T_R$, respectively. From equation (9) it follows that, if the rms velocities of the two stellar types are equal, $t_{\text{eq}}/T_{R1}$ is about equal to $m_1/m_2$.

We compute the contraction in two simple cases. First we assume that $\rho_{m2}$ is less than $\rho_{m1}$, but that the heavier stars are concentrated in the central region of the system, where $\rho_1$ is nearly constant and equal to $\rho_{01}$. Then the potential is parabolic, the kinetic energy equals the potential energy (relative to the cluster center), and $U$ equals $2E$. If $E_2$ much exceeds $E_1$, $E_2$ decreases exponentially with a time constant $2t_{eq}$, and[†]

$$r_2 \propto \epsilon^{-t/t_{eq}}. \tag{10}$$

When $\rho_{m2}$ much exceeds $\rho_{m1}$, on the other hand, the self-potential dominates, and, expressed relative to its value some distance from the nucleus, the potential energy is twice the kinetic energy and opposite in sign. In this case $U_2$ equals $-E_2$, and one readily obtains the result that $E_2$ increases exponentially, with a time constant $t_{eq}$. Since the potential energy varies as $1/r_2$, relation (10) is still valid. This result cannot be used at the beginning of the contraction, if both types of stars have the same initial distribution; both $\rho_1$ and $\langle v_1^2 \rangle$ will then vary over the dynamical trajectory of each massive star. Nor is relation (10) valid when the contraction has progressed so far that $\langle v_2^2 \rangle$ exceeds $\langle v_1^2 \rangle$, which will occur when $\rho_{m2}/\rho_{m1}$ exceeds $(M_1/M_2)^2$; the density distribution of the less massive stars will then be strongly modified near the center of the system, and again a more complex analysis is required. Finally, as $\rho_2$ increases, $T_{R2}$ will finally decrease below $t_{eq}/100$, and evaporation of stars from the compact nucleus will begin to have an evolutionary effect comparable to the loss of energy to the lighter stars.

## IV. Discussion

These results may be compared with the numerical dynamical integrations that have been carried out. Most such integrations have assumed a distribution of stellar masses, and in all cases $Mm^{3/2}$ increases with increasing mass; thus in the computations by Aarseth (1963, 1966) the total mass, $M$, in each of the four groups considered was assumed to increase slightly as $m$ was progressively about doubled, while Wielen (1968) assumed that $M$ decreased approximately as $m^{-1/4}$. We may apply the theoretical results obtained above to the two groups with the largest and the next-to-largest stellar masses, since the mass density of

---

[†]Since $E_2$ varies as $v_2^2$ in this case, the time constant for $v_2$ should be $4t_{eq}$, a correction pointed out by R. H. Durisen. The uncorrected equation (10) is valid in the next case, provided $t_{eq}$ is approximately constant.

the lighter stars is negligible in the inner cluster regions occupied by the heavier stars. Even if $\beta$ is set equal to its upper limit of unity, inequality (6) is clearly not satisfied; thus equipartition cannot be expected, and in fact the computations show that a central nucleus of the heaviest stars forms within the core. Relation (10) is not applicable, since mass groups other than the heaviest expand rapidly, and $t_{eq}$ increases with time. However, the contraction of the nucleus is clearly more rapid than would be predicted from evaporation of stars of identical mass, since the time scale for evaporation exceeds the equipartition time until $\rho_2$ exceeds $\rho_{01}$ by about 100.

Model calculations for globular clusters (von Hoerner 1957; Oort and van Herk 1959) have assumed equipartition between stars of different mass. The mass distributions assumed, together with the central densities obtained from these models, appears to be roughly consistent with equation (4).

As suggested by the theoretical arguments of von Hoerner (1968) and Lynden-Bell and Wood (1968), it is possible that a central cusp will also form in a spherical system even if all the stars have the same mass. Exact computations by Wielen (1969) for 100 stars indicate that such an effect must be relatively slow. However, detailed computations by Hénon (1969), using a Monte Carlo method to follow the evolution of a spherical cluster under the effect of binary encounters, indicate that in time a dense central cusp does in fact appear when all stars have the same mass. Further computations are required to assess the rate of formation and contraction of a central cusp when there is no dispersion of stellar masses.

Further analysis is also needed before the dynamical effects of a wide distribution of stellar masses can be assessed for actual systems, in which a continuous distribution of stellar masses is present. However, the mass distribution for newly formed stars seems to be a much flatter function of mass than would appear consistent with equation (4). If $\theta(m)d \log m$ represents the mass of new stars formed per unit volume within an increment $d \log m$, then, according to Salpeter (1955), $\theta(m)$ is nearly independent of $m$ over the range from 1.6 to 12 $M_\odot$. In a subsequent more detailed study by Limber (1960), $\theta(m)$ was found to vary about as $m^{-1/2}$, decreasing more steeply than $m^{-3/2}$ only for $m$ exceeding some 70 $M_\odot$. Similar conclusions have been reached by Reddish (1965). Inequality (6) suggests that equipartition is not possible in a newly created stellar system in which $\theta(m)$ is so flat for $m$ much exceeding the average mass.

Thus it seems likely that any newly formed spherical system of stars, including galactic clusters, globular clusters, and the original galactic nucleus itself, will tend to develop at its center a contracting nucleus of the more massive stars. The development of such a compact nucleus would depend, of course, on the evolution of the stars themselves; the more massive stars have a limited life and presumably lose mass either by a gradual process of mass ejection or by a supernova explosion. In a galactic nucleus much of the escaping gas may remain confined in the system, fall toward the center, and form new stars, as discussed by Spitzer and Saslaw (1966); the more massive of these new stars might form a denser, contracting system within the nucleus. Thus this process may contribute significantly to the formation of a central nucleus with so high a density of stars that direct collisions produce the phenomena observed in the quasi-stellar sources.

In a globular cluster the mass ejected from stars will presumably escape from the cluster entirely, and the nucleus will be transformed into a collection of dead stars, either degenerate dwarfs or neutron stars; while the dynamical evolution of such a nucleus is unclear if the stellar mass does not much exceed the average stellar mass in the cluster, von Hoerner's (1957) analysis of observed cluster density distributions seems to suggest the presence of a nonluminous central mass at the very center.

In a galactic cluster the value of $t_{eq}$ is between $10^7$ and $10^8$ years, but, because of the relatively small number of stars involved, the contraction of the nucleus would be expected to terminate with the formation of a stable binary or multiple star system at the center, as is found in the numerical integrations; when the binary loses most of its mass, the process is presumably repeated. These various possibilities require further study before even tentative conclusions can be drawn, but it seems clear that the impossibility of equipartition for the massive stars in young systems is likely to have important consequences in the dynamical evolution of such systems.

It is a pleasure to acknowledge a number of suggestions by J. P. Ostriker, a helpful discussion with J. E. Gunn, and important communication of unpublished results by M. Hénon and R. Wielen.

REFERENCES

Aarseth, S., 1963, *M.N.R.A.S.*, **126**, 223.
——, 1966, ibid., **132**, 35.
Ambartsumian, V. A., 1938, *Ann. Leningrad State Univ.*, No. 22.
Chandrasekhar, S., 1942, *Principles of Stellar Dynamics* (Chicago: University of Chicago Press), chap. 5.
Hénon, M., 1960, *Ann. d'ap.*, **23**, 668.
——, 1969 (informal communication).

Limber, D. N., 1960, *Ap. J.*, **131**, 168.

Lynden-Bell, D., and Wood, R., 1968, *M.N.R.A.S.*, **138**, 495.

Oort, J. H., and van Herk, G., 1959, *B.A.N.*, **14**, 299, No. 491.

Reddish, V. C., 1965. *Vistas in Astronomy*, Vol. 7, ed. A. Beer (New York: Pergamon Press), p. 173.

Salpeter, E. E., 1955, *Ap. J.*, **121**, 161.

Spitzer, L., 1940. *M.N.R.A.S.*, **100**, 396.

——, 1962, *Physics of Fully Ionized Gases* (2d ed.; New York: Interscience Publishers, Inc.), §5.3.

Spitzer, L. and Härm, R., 1958, *Ap. J.*, **127**, 544.

Spitzer, L. and Saslaw, W., 1966, *Ap. J.*, **143**, 400.

van Albada, T. S., 1968, *B.A.N.*, **19**, 479.

van Hoerner, S., 1957, *Ap. J.*, **125**, 451.

——, 1960, *Zs. f. Ap.*, **50**, 184.

——, 1963, ibid., **57**, 47.

——, 1968, *B.A.* (3d ser.), **3**, 147.

Wielen, R., 1968, *B.A.* (3d ser.), **3**, 127.

——, 1969 (informal communication).

# DYNAMICAL EVOLUTION OF DENSE, SPHERICAL STAR SYSTEMS

(IN: NUCLEI OF GALAXIES, PONTIFICAL ACADEMY OF SCIENCES
SCRIPTA VARIA NO. 35, ED. D.J.K. O'CONNELL, AMSTERDAM:
NORTH HOLLAND PUBL. P. 443, 1971)

## *Commentary*

THIS paper provides a good example of the great importance which conferences can have in the development of science. The particular concepts presented here came to my attention at the Thirteenth Solvay Congress, which I attended in Brussels during August 1964. At this meeting there was considerable discussion of quasi-stellar radio sources and their possible supplies of energy, including collisions between stars. It occurred to me that escape of stars by evaporation might drive a galactic nucleus toward higher and higher densities and that the resulting onset of star-star collisions could initiate quite a chain of processes. If such an evolutionary fate were in store for most galactic nuclei, this process should be worth investigating even if it were not responsible for the observed quasi-stellar sources.

Shortly after this meeting, a Princeton undergraduate, Bill Saslaw, and I carried out a very preliminary analysis of such an evolutionary process, starting with an evaporation-driven contraction of the nucleus, leading to a collisional phase, with new stars forming from the partial disruption of colliding stars.[1] A simple-minded method for analyzing stellar collisions, including grazing ones, was perhaps one of the most important contributions of this early paper.

Presumably because of this work, I was invited by the Pontifical Academy of Sciences to join a small meeting (a "Study Week" in Rome attended by some two dozen astronomers) in April 1970, on the general topic of galactic nuclei. In preparation for this meeting, another of our Princeton students, Bob Sanders, extended and refined[2] our earlier analysis, improving the estimates of gas ejected in disruptive collisions, taking account of coalescence[3] as well as disruption[1] of colliding stars, and investigating the detailed evolutionary history with a Monte Carlo calculation. In my own studies of the contraction due to Coulomb encounters, I came across the mass stratification instability (see Paper

#17). Paper #18, which I presented at the Rome meeting, gives a revised description of a possible evolutionary sequence.

Evidently Paper #18 gives only a rough general survey of this complex but potentially important topic. In 1970 information was lacking on the properties of galactic nuclei at the present era; hence no definite results could be given on the time required for present nuclei to enter the collision phase, although Table 2 shows the general trends in terms of $N$, the total number of stars, and $R$, the effective radius of the nucleus. In addition, the computations for star-star collisions were very approximate, though they now seem to be significantly better then anticipated; extensive numerical simulations[4] now give average rates of mass loss per collision which for fast collisions (relative velocity some two of three times the escape velocity from the stellar surface) between two stars of solar type are within a factor two of our early results.[1]

Our evolutionary calculations neglected a number of important effects. The evolution of the gaseous disk forming from the debris of stellar collisions will probably be affected by viscous processes, as in accretion disks generally. If the gaseous disk envisaged in a late stage of evolution condenses into stars, instability is likely,[5] and the resultant increase of random stellar velocities may lead to additional energetic collisions, and to an extended evolutionary phase, perhaps of higher luminosity than the earlier ones.

Knowledge of these and other relevant processes is now beginning to be nearly adequate for a detailed analysis of the different stages in the evolutionary sequence proposed here. The formation and growth of a massive black hole may turn out to be an important consequence of some such sequence. In any case, to determine the ultimate fate of a galactic nucleus is a fundamental scientific problem.

### References

1. Spitzer, L. and Saslaw, W. C.—*Astroph. J.* **143**, 400 (1966).
2. Sanders, R. A.—*Astroph. J.* **162**, 791 (1970).
3. Colgate, S. A.—*Astroph. J.* **150**, 163 (1967).
4. Lai, D., Rasio, F. A., and Shapiro, S. L.—*Astroph. J.* **412**, 593 (1993).
5. Begelman, M. C. and Rees, M. J.—Monthly Notices, *Roy. Astr. Soc.* **185**, 847 (1978).

# 18

## *Paper*

### ABSTRACT

Three physical processes that must affect the evolutionary history of a galactic nucleus are considered briefly. First, the ejection of mass from newly formed massive stars throughout a galaxy is discussed, together with the infall of gas toward the center and the formation of the next generation of stars. Second, a brief analysis is given of the exchange of kinetic energy between stars as a result of their mutual gravitational encounters; if the relative number of massive stars is sufficiently great, these heavy stars will form a secularly unstable system at the center which will contract rapidly. Third, direct physical collisions between individual stars are considered; these can lead to ejection of mass (disruption) or to coalescence of the two colliding stars, leading in turn either to formation of new stars at the center or to the building up of massive stars, respectively. A possible evolutionary sequence for a galactic nucleus is described, embodying these three processes. In particular, results are presented from a Monto Carlo calculation by Sanders which indicates the relative importance of collisional mass ejection and coalescence in systems of different total mass.

The observational evidence presented in the previous papers indicates that many complex phenomena must be taking place in galactic nuclei. Clearly the unravelling of these processes, or even the dominant ones only, presents theoretical astrophysics with a challenging task, one that may require many years for its completion.

The present paper does not attempt to explain any of the types of activity observed in galactic nuclei. Instead, we adopt a more deductive approach, and investigate what properties can be expected for an evolving spherical systems of stars. Whatever else may be happening in a galactic nucleus, such a system is generally assumed to contain many stars within a relatively small volume. An understanding of the properties to be expected of a dense aggregation of stars sets the stage, so to speak, for an analysis of the complex activity that we observe in galactic nuclei.

In the evolution of a spherical system of stars three effects must clearly be present. The first of these, which is treated in the following Section I, is the loss of mass from individual stars, either gradually or by supernova explosions; if much of the ejected gas is retained in the system, as seems not unlikely for a galactic nucleus, this material may form new stars near the center. Section II considers a quite different effect, the changes of each star's kinetic energy and angular momentum because of the graininess of the total gravitational field. This process, which is usually analyzed in terms of two-body encounters between mass points, leads to virtual expulsion or evaporation of some stars from the core of the system, the contraction of the remaining system as a whole, and under some conditions to the development of a central core of relatively high star density. The third process that must be considered is the direct physical collision of two or more stars. Such catastrophic events, which are considered in Section III, can produce loss of mass from the colliding stars, a process called disruption, and also coalescence of the two stars, with the formation of a more massive object.

The evolution to be anticipated for a galactic nucleus, under the influence of these three effects—mass loss from single stars, distant gravitational encounters and direct impacts between stars—is treated in Section IV. This final section is somewhat hypothetical, partly because the effects of individual collisions cannot be accurately predicted, and especially because the fate of gas ejected from the stars is unclear; in particular one does not know to what extent this gas forms new stars of familiar types.

## I. Mass Loss from Stars

The death of individual stars naturally affects the evolution of any stellar system. For a star with a mass no greater than the Sun's, death produces a contraction of the star to the white dwarf phase, and a large decrease in luminosity. Since little if any mass loss is believed to be involved in the death of such low-mass objects, the dynamical behavior of a stellar system will be unaffected.

The death of a star much more massive than the Sun leads to quite different results. It is generally believed that such objects, when they run out of nuclear fuel, lose much of their mass, either by gradual ejection in the red giant phase, or by a catastrophic supernova explosion. For a system of stars all formed at about the same time, the total mass of gas lost by the stars since they were formed may be determined from the initial luminosity function, as measured from stars forming recently in the solar neighborhood. If we take the form of this function as determined by Limber[1] and assume that every star when it dies leaves a remnant (white dwarf or neutron star) of one $M_\odot$, ejecting the rest into space, we obtain the results in Table 1. Evidently between $10^7$ and $10^8$ years about a fifth of the total mass present initially in stars is ejected into space, with a rapidly falling ejection rate in subsequent years.

TABLE 1
Fraction of Mass Ejected as a Function of Age

| Age (years) | $3 \times 10^6$ | $10^7$ | $3.6 \times 10^7$ | $1.4 \times 10^8$ | $1.6 \times 10^9$ | $1.2 \times 10^{10}$ |
|---|---|---|---|---|---|---|
| Fraction ejected | .0012 | .068 | .134 | .20 | .27 | .29 |
| Greatest Remaining Mass ($M_\odot$) | 200 | 16 | 6.3 | 3.2 | 1.58 | 1.00 |

The assumption that the initial process of star formation produced the same distribution as we observe for stars formed recently in the solar neighborhood is, of course, entirely arbitrary. Truran and Cameron[2] have recently pointed out that the chemical composition of early-type stars is better explained if the more massive stars predominated in the early days of the Galaxy, a conclusion pointed out many years ago by Spitzer and Schwarzschild.[3] However, even on the rather conservative assumption underlying Table 1 the fraction of mass ejected is already large.

The fate of the gas ejected is uncertain. In a globular cluster, where the escape velocity is only a very few kilometers per second, it seems unlikely that the gas can be contained. If the ejected gas does not stream out of the system immediately, it is likely to evaporate from the cluster at the expected temperatures. In a galaxy, however, where the escape velocities exceed a hundred kilometers per second, containment of an appreciable fraction of this gas seems very probable. For example, in a galaxy with a mass of $10^{11} M_\odot$, there will be an appreciable number of supernovae each year during the first $10^8$ years, and collisions between the expanding envelopes will dissipate much of the directed momentum, heating the gas to very high temperatures. Even in the absence of heavy elements, the gas would cool by free-free collisions of electrons with protons in less than $10^7$ years if $T$ is less than $10^8$ degrees and the proton density exceeds $10$ cm$^{-3}$, corresponding to a mass of $10^8$ suns within a sphere of radius 300 pc. While a more detailed theory is required to indicate what fraction of the ejected gas escapes the galaxy entirely and what fraction is retained, we shall assume here that a substantial fraction of the gas is in fact retained.

Ultimately any gas retained must cool and fall toward the center of the system. During the phase of violent mass ejection, the gas may conceivably be kept sufficiently hot to extend throughout much of the galaxy; the r.m.s. velocities required are an appreciable fraction of the stellar velocities and the kinetic temperature, correspondingly high. When the period of activity ends, however, and possibly well before this, the gas will cool and start to contract.

If the galaxy in question has no angular momentum, the gas will tend to collect at the center. As pointed out some 30 years ago,[4] if the mass of the gas is a minute fraction of the stellar mass, the gas can be in stable equilibrium in the stellar gravitational field. However, if the mass of the gas exceeds a certain critical fraction of the total mass, equal to about $4 \times 10^{-4}$ if the ratio of root mean square velocities for atoms and gas is about 0.1, then the self-attraction of the gas is so large that no equilibrium is possible if isothermal conditions are assumed; the gas will presumably collapse gravitationally. We assume that stars will form from such a gas cloud. Before the critical amount of mass is passed, the ratio of gas cloud radius to the radius of the stellar system is somewhat less than the ratio of root mean square velocities, or several percent. The system of new stars that form at the center will presumably have about the same radius. While the details of the process are obscure, it would seem that gas ejection and star formation can produce a subsystem of relatively high density at the center of a galaxy. Thus if 10 percent of the mass collapses to the center and forms new stars with a radius of 5 percent of the original system radius, an increase of density by about $10^3$ occurs.

This process can presumably occur again; the new stars again evolve, eject 20 percent of their mass in $10^8$ years, and a subsystem of still larger density, but less total mass, forms at the center. Thus with three successive stages of contraction a spherical galaxy with $10^{11}M_\odot$ in a radius of 10,000 pc might form first a subsystem of $10^{10}M_\odot$ with a radius of 500 pc, next a tighter subsystem of $10^9M_\odot$ with a radius of 25 pc, followed by a core of $10^8M_\odot$ with a radius of about 1 pc. This process of gas ejection, infall, and star formation could be nearly continuous, though star formation may well occur in a great many separate spurts as gas collects at the center, exceeds the critical mass, and condenses into new stars. As the number of new stars formed at the center increases, the radius of a gas cloud of small mass in equilibrium decreases, and successive generations of new stars are formed closer and closer to the center.

The assumption that the infalling gas forms new stars is, of course, arbitrary, and the mass distribution is particularly uncertain. We shall assume throughout that the gas reaching the center condenses into new stars with the same distribution of masses observed in the solar neighborhood[1] and discussed above. It is equally conceivable that a single supermassive object could form directly, a possibility to which we return again below.

If the stellar system possesses angular momentum, these phenomena are altered. The gas in this case will presumably contract to a disc, and star formation may require more mass than in the case considered above. Discussion of the angular momentum problem and of the general manner in which these processes may have entered into the history of our Galaxy and similar systems is postponed to Section IV.


## II. Dynamics of Spherical System of Point Masses

We consider now the dynamical evolution of a spherical system of stars under the assumption that each star can be treated as a point of constant mass. It is well known that such a spherical system of self-gravitating constant masses cannot be in a state of long-term equilibrium. From the virial theorem it is easily shown that in a system that is changing only slowly, the r.m.s. velocity of the individual masses is about one-half the r.m.s. escape velocity of an individual mass from the system. The irregularities in the gravitational field, resulting from the particulate nature of the gravitating masses, lead to an exchange of kinetic energy between stars, tending to produce a Maxwellian velocity distribution. Evidently any particle in the Maxwellian tail, with a velocity more than twice the r.m.s. value, will escape from the system.

The rate at which particles escape will be determined, of course, by the time required for the exchange of energy between the stars to be comparable with the stellar kinetic energy. In an actual cluster this time will depend markedly on the orbit of the star, since deviations from a smooth potential function are much greater near the center of the system where the density is greater and encounters between stars are more frequent. To obtain a first approximation, it is frequently useful to consider an idealized cluster of constant density, situated in a potential well with a flat bottom and steep sides. Then conditions are essentially uniform through the cluster and the relaxation time may be set equal to a reference time $r_R$, defined by Spitzer and Härm[5] as[†]

$$t_R = \frac{(2/3)^{3/2} v^3}{2\pi G^2 m^2 n \, ln \mathrm{N}} \tag{1}$$

where $m$ and $n$ are the mass and particle density of the stars, $v$ is the r.m.s. velocity, and N is the total number of stars in the system. If the virial theorem is used to evaluate $v$, and the cluster radius, R, is

---

[†]A typographic error has been corrected in equation (1), where $(3/2)^{3/2}$ has been replaced by $(2/3)^{3/2}$.

expressed in parsecs, equation (1) gives

$$t_R \approx \frac{8 \times 10^5 \, N^{1/2}[R(pc)]^{3/2}}{[m/m_\odot]^{1/2} \log_{10} N} \text{ years.} \tag{2}$$

A more detailed discussion of relaxation times has been given earlier by Chandrasekhar.[6] Since in a Maxwellian velocity distribution about 1 percent of all the stars have a velocity exceeding the r.m.s. value, one may then infer that 1 percent of the stars escape during the time $t_R$, a conclusion first pointed out by Ambartsumian[7] and Spitzer.[8]

If the escaping stars are assumed to escape with no appreciable kinetic energy, the total energy of the remaining system must be constant, and $N^2/R$ is unchanged; R decreases with time and the density decreases sharply as $1/N^5$. According to equation (2), $t_R$ decreases as $N^{7/2}$, and evaporation accelerates. On these assumptions, the cluster contracts to a point in about $40t_R$, as shown by King.[9]

Evidently it is most unrealistic to assume that $n$, the particle density of stars, is uniform throughout the system. Henon[10, 11] has shown that a consideration of the radial density gradient significantly alters the conclusions based on the approximate constant-density model. Exchange of energy between stars takes place mostly in very small amounts, and many orbits of a star are required before its energy or angular momentum are significantly changed. Stars are likely to gain appreciable energy only if their orbit passes through the dense core of the system, where the rate of encounters and of energy interchange is greatest. As such stars gain more and more energy, they will move in more and more elongated orbits passing through the central core more and more rarely. Thus as a star approaches the zero total energy needed for escape, its effective relaxation time increases steadily. The actual rate of escape is in consequence much reduced below the result obtained on the simplified constant-density model.

However, during a time of $100t_R$ a substantial fraction of the stars in the central core will gain sufficient energy so that their orbits become markedly elongated, and the stars become essentially part of an extended "halo" surrounding the system. While these stars have not escaped, in the sense that their orbits continue to pass through the dense central nucleus, such central passages occupy a small fraction of their time. Thus the simple theory may be used to give the rate not of actual escape but of accumulation in the outer regions. This ejection of stars must give rise to a contraction of the central core. In place of the uniform radial contraction given by the constant density theory it is clear that the rate of loss of energy and the resultant rate of contraction will be greatest where the density is greatest, at the center of the cluster. This tendency may produce a relatively dense core at the center of the system, a possibility discussed by Lynden-Bell and Wood.[12] The time scale for this development is relatively long, much greater than $t_R$.

The presence of relatively more massive stars is probably a much more important factor in producing a central high-density core in a galactic nucleus. As shown recently elsewhere,[13] a sufficiently massive core of such stars should contract at a relatively rapid rate. In the simple case of a system with two types of stars present, lighter ones with mass $m_1$, and heavier ones of mass $m_2$, the condition for the development of a contracting core can be stated very simply, provided $m_2$ is much greater than $m_1$. It is evident physically that encounters will tend to establish equipartition of kinetic energy so that

$$m_1 v_1^2 = m_2 v_2^2 \tag{3}$$

where $v_1$ and $v_2$ are the r.m.s. velocities of the two types of stars. As the heavier stars lose energy, they will fall toward the center.

If the total mass of the heavier stars is only a small fraction of the total mass of the system, these stars can form a low-velocity subsystem at the very center. The situation is similar to the distribution of low-temperature gas at the center of a stellar system, which was discussed above. In either case, equilibrium is possible with predetermined low velocities if the density of the low-velocity stars or atoms is negligible, and the gravitational field is essentially that of the high-velocity stars. If the density of the more massive stars much exceeds the density of the lighter stars at the center of the system, then the self-gravitational energy of the heavier stars is great and the random velocities of these stars must become large and cannot satisfy equation (3). Thus $v_2^2$ will be above its equipartition value, and as a result of energy exchange in collisions the subsystem of heavy stars will continually lose kinetic energy to the lighter stars in the vicinity. This loss of energy by the self-gravitating system leads to a contraction of the system and a continual increase of $v_2^2$, which is essentially a form of secular instability. This collapse of the heavy star subsystem will occur[13] if $M_2/M_1$ exceeds $\beta(m_1/m_2)^{3/2}$ where $\beta$ is a constant of order unity; if $m_1$ is much less than $m_2$, an approximate theory indicate that $\beta$ equals 0.16.

The time scale for this resultant contraction is roughly the equipartition time, $t_{eq}$, about equal to $t_R \, m_1/m_2$, where $t_R$ is computed for stars of mass $m_1$. Thus the rate of contraction of the central subsystem is between one and two orders of magnitude more rapid than the rate of contraction computed from the evaporation of stars. As the central subsystem contracts, the relaxation time for the heavier stars will decrease in accordance with equation (2), and will ultimately become less than $t_{eq}$. Evidently a dense core at the center of the system will develop with a time scale about equal to $t_{eq}$; this conclusion has been confirmed by detailed dynamical calculations, which will be published elsewhere.

The finite lifetime of the more massive stars will clearly modify any results involving the concentration of massive stars. These stars will eject matter on a time scale discussed in the preceding section. If the system is sufficiently compact so that $t_R$ is comparable to the stellar lifetime, relaxation effects will be important for these new stars. As a result of the tendency toward equipartition, stars of average mass will tend to acquire the same velocity as the average stars in the cluster, and will consequently diffuse out from the center. Stars of high mass will tend to be concentrated near the center, and as these accumulate they will tend to form a contracting subsystem exactly as before. Thus the rebirth of massive stars may lead to a continuing process of contraction until a very dense core is reached. This process is discussed again in Section IV.

### III. COLLISIONS BETWEEN STARS

As the density increases in an evolving stellar system, direct collisions between stars become inevitable. The number of such collisions can be computed approximately if the stars are assumed to remain spherical before collision. On this assumption, the value of the impact parameter, $p$ (defined as the distance of closest approach of the centers of the two stars in the absence of gravitational forces) for a just grazing collision may be computed from the conservation of energy and momentum, and becomes

$$\pi p^2 = \pi (r_1 + r_2)^2 \left\{ 1 + \frac{2G(m_1 + m_2)}{(R_1 + R_2)V^2} \right\} \tag{4}$$

where V is the initial relative velocity, while $r_1$, $r_2$, $m_1$, and $m_2$ are the radii and masses of the two stars. For slow collisions, in which the initial kinetic energy is much less than the potential energy at closest approach, tidal deformation may be significant and equation (4) is not very precise.

TABLE 2
Relaxation and Collision Times[†]

| No. of Stars | Radius, R(pc) | 0.1 | 1 | 10 | 100 |
|---|---|---|---|---|---|
| $N = 10^6$ | Stellar velocity, $v$ | 147 | 47 | 14.7 | 4.7 |
| | Relaxation time, $t_R$ | $4.6 \times 10^6$ | $1.46 \times 10^8$ | $4.6 \times 10^9$ | $1.46 \times 10^{11}$ |
| | Collision time, $t_c$ | $3.2 \times 10^8$ | $1.09 \times 10^{11}$ | $3.5 \times 10^{13}$ | $1.09 \times 10^{16}$ |
| $N = 10^8$ | Stellar velocity, $v$ | 1470 | 470 | 147 | 47 |
| | Relaxation time, $t_R$ | $3.4 \times 10^7$ | $1.08 \times 10^9$ | $3.4 \times 10^{10}$ | $1.08 \times 10^{12}$ |
| | Collision time, $t_c$ | $2.8 \times 10^6$ | $5.2 \times 10^9$ | $3.1 \times 10^{12}$ | $1.09 \times 10^{15}$ |
| $N = 10^{10}$ | Stellar velocity, $v$ | 14700 | 4700 | 1470 | 470 |
| | Relaxation time, $t_R$ | $2.7 \times 10^8$ | $8.6 \times 10^9$ | $2.7 \times 10^{11}$ | $8.6 \times 10^{12}$ |
| | Collision time, $t_c$ | $3.1 \times 10^3$ | $9.7 \times 10^6$ | $2.8 \times 10^{10}$ | $8.5 \times 10^{13}$ |

[†] Values of $v$ and of times are in units km/sec and years.

For any one star the mean time, $t_c$, between collisions is roughly given by

$$t_c = 2^{1/2}/n\pi p^2 v \qquad (5)$$

where $n$ is the number of stars per unit volume and $v$ is the r.m.s. velocity, equal on the average to $2^{1/2}$ times the r.m.s. value of the relative velocity V. If we evaluate $v$ from the virial theorem and substitute from equation (4) for $\pi p^2$, equation (5) gives

$$t_c \approx \frac{10^{22} \, R^{7/2}}{N^{3/2}(m^{1/2}r^2/m_\odot^{1/2}r_\odot^2)(1 + 8.8 \times 10^7 \, Rr_\odot/Nr)} \text{ years} \qquad (6)$$

where $m$ and $r$ are again the mass and radius of the stars and R is the radius of the cluster in parsecs. Table 2, taken from Ref. [16], compares values of $t_c$ for different values of N and R.

The detailed processes that occur during a direct impact of two stars are obviously very complex and difficult to predict. To trace the effect of such collisions on the evolution of the stellar system there are two questions of particular importance that need to be answered. First, how much gas is ejected from the two stars? Second, will the two stars separate or will they coalesce to form a single star, as proposed by Colgate?[15] To answer these two questions, at least on an approximate basis, a very simplified theory has been developed.[16] In this theory the two stars are assumed to move in straight lines, as in Figure 1. The assumption is then made that the collision is strictly one-dimensional, and that the mass elements in one star interact only with the mass elements toward which they are moving; all lateral effects are entirely

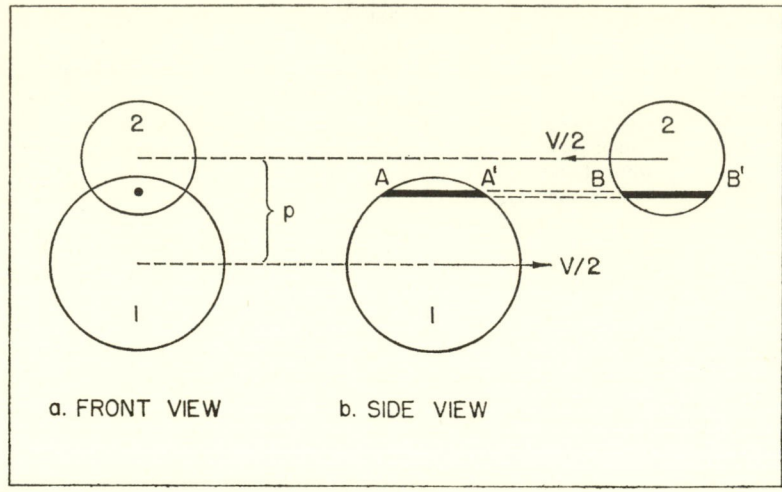

a. FRONT VIEW     b. SIDE VIEW

FIG. 1. Front and side view of two colliding stars, showing colliding mass elements in each star.

ignored. Thus the mass elements in the cylinder A–A', extending through star 1, in a direction parallel to the relative velocity, V, are assumed to collide only with the mass-elements in the coaxial cylinder B–B' extending through star 2. The collision of these mass elements is assumed to be completely inelastic.

To determine the mass of gas ejected, escape of the mass element A–A' is assumed if the thermal energy which it gains during the collision exceeds the total binding energy of these mass elements in star 1, or the corresponding binding energy of the mass elements B–B' in star 2, whichever is the greater. The answer to the second question, on the possible coalescence of the two stars, is determined from the total loss of kinetic energy in the collision, summed over all colliding mass elements. If this loss of kinetic energy exceeds the initial kinetic energy of relative motion, when the stars are far away from each other, then it is assumed that the two stars do not have sufficient energy to escape from each other's gravitational field; if they do not coalesce immediately, they will form a binary which, after repeated collisions at each periastron passage, will ultimately coalesce to form a single star.

Evidently these assumptions represent a very crude approach to a complex physical problem. They probably yield an upper limit on the number of collisions leading to coalescence; this application of the principle of momentum conservation would seem to give a reliable upper limit on the amount of kinetic energy transformed into heat. If this upper limit is less than the kinetic energy associated with the relative velocity, V, at infinity, it is difficult to see how coalescence is possible. On the other hand, the amount of gas escaping from the two stars, in this process which we may call "disruption," is not well determined; the thermal energy in a gas element is not likely to be converted entirely into the potential energy of the same element in the gravitational field of one of the two stars. Some of the thermal energy will certainly be transferred elsewhere by shocks, adiabatic expansion, etc., and may lead to expulsion of a small amount of gas at large energy, or to heating of a gas that is securely bound gravitationally. On the other hand, in the time-dependent gravitational field of the two colliding stars the amount of energy required for a mass element to escape from both stars may be much less than assumed.

This simple model should give correctly the dominant feature of grazing collisions, in which the dense central cores do not collide directly. These cores presumably move past each other without much change, while the parts of the envelopes that collide will be violently heated, with some gas escaping completely.

For direct central collisions, on the other hand, this model does not give anything very useful if the two stars have the same mass. In this situation all the kinetic energy of relative motion is converted into heat, and all mass elements receive the same kinetic energy per unit mass. On the approximations made here, none of the mass escapes unless most of it does. The exact computations for central collisions between identical stars, carried through by Seidl and Cameron,[17] show that even when V, the relative velocity at infinity, is zero, shock waves accelerate a small amount of gas to rather high kinetic energies, so that 4 percent of the mass of each star is ejected. This fraction rises to 17 percent at 1000 km/sec. At 2000 km/sec, when the total energy is positive and complete disruption of both stars would result according to the simple model, only 60 percent of the mass is found to escape. Such central collisions are relatively improbable and for the more important grazing collisions the model described here gives the best results presently available.

Detailed calculations of mass loss and coalescence probability have been made by Sanders,[18] on the basis of the simple model, for a wide variety of stellar mass ratios and velocities. The results have been used in an investigation of how a galactic nucleus is likely to evolve in time. We return to this topic again in the next section.

## IV. Evolutionary History

We attempt now to reconstruct the evolution of a galactic nucleus, taking into account the physical processes discussed in the preceding sections. Primary emphasis will be placed on systems without angular momentum, though we treat briefly the possible extensions of the theory required to explain spiral systems like our own.

### a) Formation of the Nucleus

The first stage may be called the formation of the nucleus; this stage may be regarded as part of the formation of the galaxy out of inter-galactic gas. One may visualize a relatively homogeneous but turbulent cloud which condenses into stars as it is collapsing toward the center. Numerical computations by Henon[19,20] and others show that even if appreciable turbulence is present, the resultant system shows strong central condensation; for example, the ratio of the radii containing 90 percent and 10 percent, respectively, of the mass is 8 in a typical case, as compared with about 4 for an $n = 3$ polytrope and 2.2 for a uniform sphere. However, the central concentration in a galactic nucleus is enormously greater, with an appreciable fraction of the mass in less than 1 percent of the radius. To explain so great a concentration it appears necessary to invoke the mass ejection from stars discussed in Section I.

This strong central condensation could also be achieved by the infall of gas that had not yet condensed into new stars. Some uncondensed gas is certainly likely to be present. The time taken for such gas to fall some 10,000 pc in the gravitational field of a galaxy is about $10^8$ years, about equal to the time for significant mass loss by the heavy stars—see Table 1. Thus the infall of uncondensed gas and of gas ejected from stars are dynamically rather similar in the early stages of contraction; the ratio of these two sources of gas in the inner regions will determine the ratio of heavy elements to hydrogen in these regions. If a sequence of mass ejection from stars, formation of new stars, and subsequent mass ejection again is the dominant process in the formation of a galactic nucleus, the relative abundance of heavy elements might be expected to increase systematically with decreasing distance from the galactic center.

There is little evidence on how the chemical composition of elliptical galaxies varies with distance from the center. For spiral systems, on the other hand, there is substantial evidence on chemical composition as a function of position. The paper by Spinrad, presented earlier during this Study Week, indicates that the stars in the nuclei of spirals have systematically higher metal abundances than the Sun. Similarly, Peimbert[21] has concluded that the tenfold increase of N II emission lines relative to H $\alpha$, observed within about a hundred parsecs of the center of spiral nuclei, must be explained by a higher ratio of nitrogen to hydrogen than is observed in the solar neighborhood. If gas in the galactic nucleus represents material ejected from evolved stars, progressively less and less diluted with primordial hydrogen at decreasing distances from the center, these observations find a ready explanation.

One naturally enquires how gas can concentrate toward the center of a spiral system, in which so much angular momentum is present. Let us consider what happens to gas ejected from a spheroidal distribution of stars, such as the halo of our Galaxy. Stars at some 10,000 pc from the axis of rotation possess an average angular momentum corresponding to equilibrium circular orbits at about 5000 pc from the center. Stars much closer to the axis will have a mean angular momentum varying about as the square of their distance from the axis if we make the plausible assumption that in such a spheroidal system, where the random velocities much exceed the circular velocity, the mean rotational velocity varies linearly with distance from the axis. If the circular velocity is assumed constant with distance, as is nearly the case in our Galaxy at present,[22] the radius of the equilibrium orbit into which any ejected gas must fall as it cools will vary as the square of the initial distance from the axis. In consequence, gas ejected from within 2000 pc of the axis of rotation, containing about 6 percent of the mass, can condense to within about 200 pc of the center. One can visualize gas ejected by the stars in the galactic halo condensing into a disc at radii between 1000 and 5000 pc, but forming in addition a relatively dense spheroidal system at the center. A repetition of this same process will produce an even smaller system of stars in the galactic center. As compared with a spherical system, where all the gas ejected presumably condenses into an inner spheroidal system, the amount of gas in a spiral system which reaches the inner regions of a nucleus may be reduced by several orders of magnitude. In any case these arguments would indicate that relatively dense nuclei, with some $10^5$ and $10^8$ stars within a radius of a few parsecs, may be expected to form early in the life of most galaxies.

### b) Relaxation and Stratification

We consider next the development of a nucleus when it has reached a sufficiently compact stage so that the relaxation effects discussed in Section II become important. From equation (2) and Table 2 we see that for a system of $10^8$ stars, R must be less than about 5 pc if $t_R$ is to be less than $10^{10}$ years. For dynamical relaxation to have a large effect,

$t_R$ must not exceed $10^9$ years, requiring a system either with a smaller radius or a lower mass. While such small radii cannot be measured directly, Stratoscope observations[23] have set an upper limit of 5 pc on the radius of the nucleus in the Seyfert galaxy NGC 4151, with preliminary analysis of new data[24] suggesting an even lower limit of 2 pc. As we have seen above, there is some theoretical reason to expect such compact structures to arise naturally, early in the history of a galaxy.

As relaxation affects become important, the more massive stars will tend to fall toward the central regions of the nucleus. If $t_R$ exceeds $10^9$ years, then after one time of relaxation there will be no stars heavier than $1.5M_\odot$. Whether the remaining distribution of mass is secularly unstable cannot be determined without more detailed information; it is possible that a relatively dense central core of the more massive stars may form, become secularly unstable, and contract further. In accordance with the discussion in Section II such a core will initially contract through loss of energy to the lighter stars, but as the central density increases, evaporation of some of the heavy stars from the central core might provide an additional effect in decreasing the energy per unit mass. In any case, the contraction will accelerate, and the collisions discussed in the next section will soon become important.

If the relative number of massive stars is insufficient to produce secular instability, the evolution of the nucleus during the relaxation phase will be significantly slower. In this case the stratification of stars of different mass will be secularly stable, and the nucleus as a whole will lose mass by evaporation into the halo. The total energy of the remaining nucleus will not be much altered, and hence $M^2/R$ will remain constant. As pointed out above, the density, $n$, will vary as $1/M^5$ and will rise relatively rapidly as evaporation decreases M. This process is relatively slow, since the time scale for evaporation is about $100t_R$, while for the concentration of massive stars the time scale is more nearly equal to $t_R$.

### c) Collisions, Disruption, and Coalescence

As collisions become important, the nucleus enters its active phase, in which the stars themselves are profoundly altered and vast energies are released. In the absence of angular momentum it appears inevitable that this activity must escalate without limit, with the core of the nucleus contracting at an accelerated rate until at least the core itself disappears into the Schwarzschild singularity, possibly followed by many of the remaining stars outside the core. We shall not attempt to follow the nucleus very far in this apocalyptic course to self-destruction, but shall consider only the beginning of this final collisional stage, treating such matters as time scale, changes in the stars involved, and the energy released.

The time scale given by equation (5) is best expressed as a function of $v$, the r.m.s. stellar velocity, since the nature of the collisions is strongly a function of $v$. If we express $n$ in terms of M and $v^2$, eliminating R through the dependence of $v^2$ on M/R, equations (4) and (5) yield

$$t_c \propto \frac{M^2}{v^7 \left(1 + \beta \dfrac{v_\infty^2}{v^2}\right)} \tag{7}$$

where $\beta$ is a coefficient of order unity and $v_\infty^2$ is the mean square velocity of escape from the stellar surface. Evidently the collision rate accelerates enormously as $v$ increases. The rate of energy released goes up even more rapidly. Comparing two systems of different mass, the time scale for evolution evidently varies as $M^2$. The rate of energy release is proportional to the number of stars and hence varies as $M/t_c$ or as $1/M$. The total energy released from the system, during a fixed number of collisions per star, must evidently vary as M.

We consider next the changes produced by collisions. As we have seen in Section III, either partial disruption or coalescence or some combination of both is to be expected. The calculations carried through by Sanders[18] indicate that for V much less than 1000 km/sec, coalescence occurs with relatively high probability, but for V much in excess of 1000 km/sec, disruption is the more likely. Thus if $v$ is less than 700 km/sec, we may say that the system is in the "coalescence regime," while greater V corresponds to the "disruptive regime." The fate of the coalesced stars depend on the competition between their own evolution, which accelerates with increasing mass, and collisions with other stars, which in the coalescence regime tend to increase the mass still further. The time scale for evolution of a massive star is less than $10^8$ years. For comparison, the collision time, $t_c$, for a system with $10^8$ stars of solar mass, is $5 \times 10^9$ years when $v$ is 500 km/sec, well within the coalescence regime—see Table 2 or equation (6). Even at 700 km/sec $t_c$ is $5 \times 10^8$ years, still less than the lifetime of the more massive stars. Evidently in such a system any relatively massive star formed by coalescence will tend to burn its fuel and become a supernova before it increases its mass much further by more coalescence. In a system with $10^7$ solar masses, however, the collision times for the same values of $v$ are less by two orders of magnitude, and become less than the evolution times. As such a system evolves, stars of rather large mass should be possible.

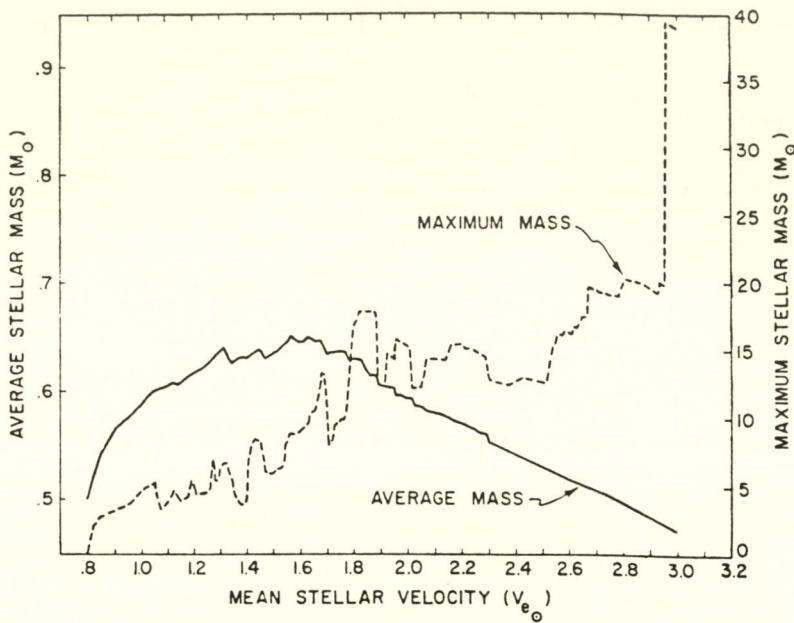

FIG. 2. Average stellar mass and maximum stellar mass in a system with a mass of $10^8 M_\odot$, shown as a function of ratio of r.m.s. stellar velocity to the velocity of escape from the solar surface.

These expectations are confirmed by detailed Monte Carlo computations by Sanders,[18] tracing the history of a galactic nucleus. He considered an initial population of several thousand stars, regarded as a sample from a much larger system, and followed the fate of each star with a computer, determining by random choices the types of collisions which it undergoes with the other stars of the sample. The cluster was assumed uniform and homogeneous, with no consideration of stratification effects; dynamical effects between collisions, such as evaporation and equipartition of kinetic energy, were introduced analytically rather than through the Monte Carlo approach. Figure 2 shows his results on the mass distribution in a system with $10^8 M_\odot$, with all stars assumed to have one-half a solar mass initially. The mean stellar velocity is expressed in terms of the escape velocity from the solar surface, equal to 620 km/sec; thus $v$ ranges from 500 km/sec initially to 1860 km/sec at the end, after an evolution time of $1.4 \times 10^{10}$ years. The initial and final radii of the nucleus were 0.88 and 0.048 pc, respectively. This figure shows that in the coalescence regime ($v_\infty$ less than 1.6 times the escape velocity for the Sun) the mean mass increases slightly and the maximum mass reached does not become very great. For larger velocities the

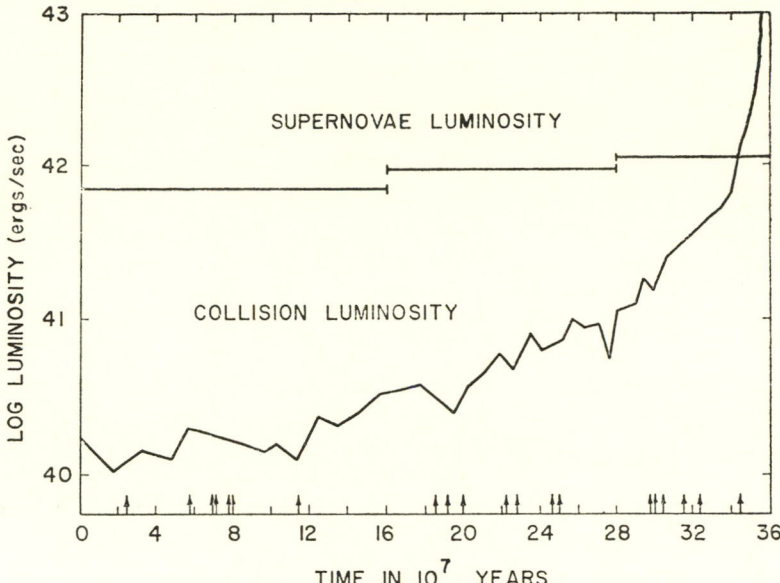

FIG. 3. The luminosity resulting from stellar collisions and from coalesced stars exploding as supernovae in the late evolution of a $10^8$ $M_\odot$ system. The small arrows on the time axis represent individual supernovae in the sample group of 4000 stars.

mean mass declines, as coalescence becomes less probable, but collisions become so rapid that a very few stars can build up to a large mass by successive coalescence before they run out of fuel and explode.

The energy radiated by this model is shown in Figure 3 during the final $3.5 \times 10^8$ years covered by the integration. The value of $v$ during this period increases up to 1860 km/sec, at which point the calculation ceased. The supernova luminosity does not increase significantly with time during this period; while the rate of collisions increases dramatically, the percentage of collisions producing coalescence declines correspondingly. The "collision luminosity" is the energy released by the gas ejected from the colliding stars; this gas is assumed to cool and fall to the center. In the computer simulations it was assumed that this gas promptly formed new stars all of mass $M_\odot/2$, which shared kinetic energy with the old stars. Since this energy release varies somewhat more rapidly than $v^9$, a continuation of this process to velocities as great as 3700 km/sec would increase the computed luminosity to $10^{46}$ ergs/sec. The assumptions underlying the theory certainly break down well before so high a luminosity can be reached; for example, the time

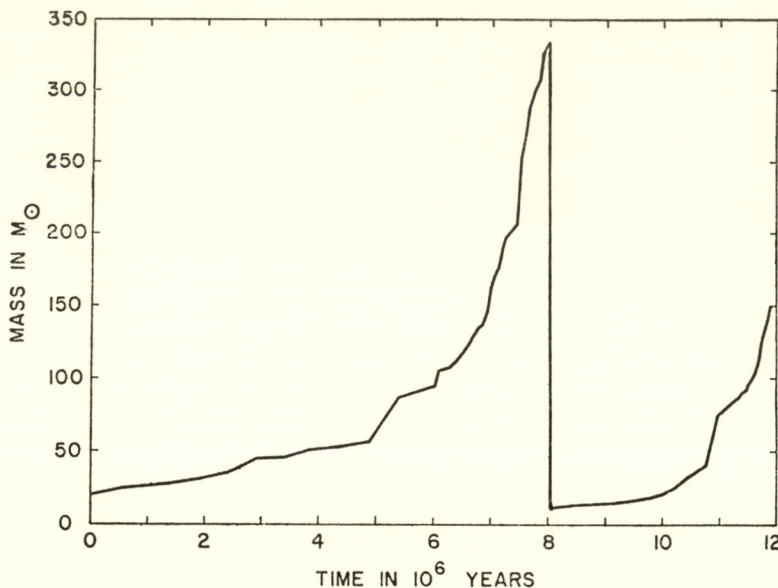

Fig. 4. Growth of the maximum stellar mass in a sample group of 2500 stars in the late evolution of a system whose mass is $10^7 M_\odot$. At $8 \times 10^6$ years the star of maximum mass is arbitrarily destroyed.

required for new stars to exchange kinetic energy with the old stars becomes so long[25] that the process visualized in these computations would clearly be altered.

Computations for a system with $10^7$ stars were also carried out by Sanders, extending over a period of $1.4 \times 10^8$ years, during which the r.m.s. velocity increased from 500 to 900 km/sec. At the end of this period, a star of several hundred solar masses was formed; when it was arbitrarily removed, a similar massive star began to form as shown in Figure 4. Evidently the growth of massive stars is a stable characteristic of this type of model. No supernova explosions occur, since the interval between successive collisions is small compared to the evolution time, and each coalescence is assumed to mix up the stars, prolonging their lives before a burnt out core leads to a supernova explosion. In the actual nucleus there would be many such stars of several hundred solar masses, and they would tend to coalesce, forming even more massive stars. Stratification of the more massive objects near the center would enhance this tendency. It is quite possible that such very massive stars would be unstable; Schwarzschild and Härm[26] have found that main sequence stare are pulsationally unstable if their mass exceeds $60 M_\odot$.

Conceivably, however, this process may provide a method for concentrating appreciable mass in a few massive stars, possibly in one single very massive object, of the sort discussed by Hoyle and Fowler[27] and others.

Obviously if the system possess no angular momentum, no steady state appears possible, and there seems no escape from the conclusion that at least the inner core and probably, in time, most of the entire system as well must inevitably contract so much that it disappears from the visible universe, approaching the Schwarzschild singularity. Problems associated with this singularity will be discussed by Wheeler in a later paper of this series. The latter stages of this catastrophe are difficult to follow analytically. The stars will collide at such high velocities that nuclear detonations may aid in the complete disruption of the stars; the presence of normal stars becomes improbable when $v$ is as great as 10,000 km/sec, though white dwarfs, with their small cross-sections, might survive to even later stages, and neutron stars might well persist until they approached the Schwarzschild singularity.

In the more likely situation, some angular momentum will be present, and as with the galaxy as a whole the material ejected from stars will tend to collect in a disc, with a small amount of material contracting further in a nearly spherical configuration. The amount of material that could disappear down the Schwarzschild singularity would presumably be much reduced. It is not clear what would happen to the material in the disc, or even whether such a disc could be stable. The total amount of energy that could be released from gravitational contraction of a system with a mass of $10^8 M_\odot$ is presumably limited by the angular momentum present, if the limiting disc is stable; if stability is not possible, presumably oscillatory phenomena could eject some mass outwards, carrying away the bulk of the angular momentum, leaving the rest of the system free to contract.

Evidently these last few paragraphs are entirely speculative. The tentative theory outlined here, based on work by many different astronomers and physicists, seems to account for the development at the galactic center of a compact core of many millions of stars, in which collisions are rapidly changing the nature of the system and where further developments are bound to be fascinating and important.

## References

1. Limber, N., *Ap. J.*, **131**, 168 (1960).
2. Truran, J. W. and Cameron, A.G.W., *Nature*, in press (1970).
3. Schwarzschild, M. and Spitzer, L., *Obs*, **73**, 77 (1953).
4. Spitzer, L., *Ap. J.*, **95**, 329 (1942).

5. Spitzer, L. and Härm, R., *Ap. J.*, **127**, 544 (1958).
6. Chandrasekhar, S., *Principles of Stellar Dynamics* (U. of Chicago Press) (1942).
7. Ambartsumian, V. A., *Ann. Leningrad State Univ.*, No. 22 (1938).
8. Spitzer, L., *M.N.R.A.S.*, **100**, 396 (1940).
9. King, I., *A.J.*, **63**, 114 (1958).
10. Hénon, M., *Ann. d'Astroph.*, **23**, 668 (1960).
11. Hénon, M., *Astron. and Astroph.*, **2**, 151 (1969).
12. Lynden-Bell, D. and Wood, R., *M.N.R.A.S.*, **138**, 495 (1968).
13. Spitzer, L., *Ap. J.*, **158**, L139 (1969).
14. Salpeter, E. E., *Ap. J.*, **121**, 161 (1955).
15. Colgate, S. A., *Ap. J.*, **150**, 163 (1967).
16. Spitzer, L. and Saslaw, W. C., *Ap. J.*, **143**, 400 (1966).
17. Seidl, F. and Cameron, A.G.W., informal communication.
18. Sanders, R. H., *Ap. J.*, **162**, 791 (1970).
19. Hénon, M., *Ann. d'Astroph.*, **27**, 83 (1964).
20. *Symposium on Computer Simulation of Plasma and Many-Body Problems*, NASA, p. 349 (1967).
21. Peimbert, M., *Ap. J.*, **154**, 33 (1968).
22. Schmidt, M., *Galactic Structure*. Vol. 7 of *Stars and Stellar Systems* (U. of Chicago Press), p. 527 (1965).
23. Danielson, R. E., Savage, B. D., and Schwarzschild, M., *Ap. J.*, **154**, L117 (1968).
24. Schwarzschild, M., informal communication (1970).
25. Spitzer, L. and Stone, M. E., *Ap. J.*, **147**, 519 (1967).
26. Schwarzschild, M. and Härm, R., *Ap. J.*, **129**, 637 (1958).
27. Hoyle, F. and Fowler, W. A., *M.N.R.A.S.*, **125**, 169 (1963).

# RANDOM GRAVITATIONAL ENCOUNTERS
# AND THE EVOLUTION OF SPHERICAL
# SYSTEMS. I. METHOD

### (WITH M. H. HART)

### (ASTROPH. J. 164, 399, 1971)

## *Commentary*

THIS PAPER was submitted for publication about six months after the Rome meeting referred to in the previous commentary. My plan was to prepare a series of papers to test by Monte Carlo calculations the effects of relaxation processes in stellar systems. Such processes, whose effects were discussed from a theoretical standpoint in Paper #18, included evaporative escape of stars and mass stratification.

Michel Hénon had faced a similar problem a few years earlier. He had computed analytically a model for an evolving, idealized cluster subject to the tidal attraction of the Galaxy (see Commentary, Paper #20). To extend his work to less idealized model clusters he developed a Monte Carlo method for following the dynamical evolution caused by distant gravitational encounters. A useful check on this technique for following diffusion in velocity space was provided by comparison with the exact Fokker-Planck equation in the absence of a large-scale gravitational field (see Hénon 1967 referred to in Paper #19).

Following the same general direction, but along a path quite different in detail, we developed the Princeton Monte Carlo scheme outlined in Paper #19. For application to spherical clusters we integrated numerically the equations of motion for each group of stars, moving in the smoothed potential field of the cluster, and at periodic intervals we applied in a random fashion the velocity perturbations resulting from two-body distant encounters. This approach seemed to us a transparent, physically understandable technique. We checked our method in the same way followed by Hénon; Section III, which has been omitted here, indicates the satisfactory agreement obtained.

It was not until Paper #19 was in press and our subsequent paper was essentially complete that we learned of Hénon's later modification[1] of his Monte Carlo technique for application to spherical clusters. This approach involved perturbations of the two orbital elements, energy E and angular momentum J, with no dynamical integration of the equations of motion. Our method of following many different orbital trajectories numerically requires considerably more computing time than Hénon's but during the early 1970s at Princeton there was no financial charge or fixed limit for our use of the University's central computer. Equations (16) and (17) below represent additional approximations in our method, but about a year later these were dropped, and the exact expressions for the diffusion coefficients were used. With this modification, the two methods shared a virtually identical set of physical principles, though the actual computations were entirely different. It is gratifying that the two techniques gave nearly identical results for an isolated cluster with a specified initial equilibrium (see Figure 4, Paper #10).

The one important difference between the two sets of results, as pointed out in Paper #20 and elsewhere[2] is in the rate of escape from an isolated cluster. Theory gives the probability of escape in a single period, i.e., in a single passage through the cluster core. The Hénon method does not consider the orbital periods and yields no escape of stars at all as a result of distant encounters but instead a steadily increasing population in the far halo of the cluster. For tidally limited clusters this difference between the two methods disappears.

During the late 1970s several other important methods were developed[3] for following the dynamical evolution of clusters, including numerical solution of the orbit-averaged Fokker-Planck equation, and modification of the Hénon method to include the effects of orbital period for a star approaching the escape energy. The results for cluster evolution obtained with the various available methods are briefly summarized in Paper #20.

During the last decade the methods which theorists have used to follow the dynamical evolution of clusters have not greatly changed, though increasing computer power has significantly increased the number of stars included in numerical simulations. However, the growth of observational knowledge concerning clusters has greatly expanded the list of physical processes that are relevant to this topic. As pointed out in the following commentary, direct physical collisions between stars in dense clusters must now be considered in attempts to explain the fate of these systems, just as for galactic nuclei (see Paper #18). Thus for dense clusters, dynamical and stellar evolution become closely linked, and methods must be developed for dealing with both of these topics together, perhaps with stochastic methods for introducing collisions into the computations of dynamical evolution. Recent numerical simulations[3,4] of collisions between stars are now providing some of the basic information needed for such studies.

## References

1. Hénon, M.—In: Gravitational N-body Problem, IAU Colloq. No. 10, ed. M. Lecar, Dordrecht: Reidel 406 (1972); *Astroph. Space Science* **14**, 151 (1971).
2. Spitzer, L.—Dynamical Evolution of Globular Clusters, Princeton: Univ. Press §4.1 (1987).
3. Benz, W. and Hills, J. G.,—*Astroph. J.* **389**, 546 (1992).
4. Lai, D., Rasio, F. A., and Shapiro, S. L.—*Astroph. J.* **412**, 593 (1993).

# *Paper*[†]

ABSTRACT

A modified Monte Carlo method is described for following the dynamical evolution of a spherical stellar system. In this method, designed for use with a digital computer, the trajectory of each star is computed numerically through time on the assumption that the gravitational potential is spherically symmetric. This procedure, used earlier by Campbell and Hénon, becomes exact if each star is taken to represent a spherically symmetric shell composed of many stars, all characterized by the same values of the mass, of the distance $r$ from the center, of the radial velocity $v_r$, and of the transverse velocity $v_t$; however, the orientation of the orbital planes is assumed to be random for the stars at each location in the shell. To take into account irregular variations in the gravitational field, produced by encounters with passing stars, the velocity of each star is perturbed at intervals of time $(\Delta t)_p$. The perturbations are chosen to give the appropriate diffusion coefficients for stars within each small range in $r$. To expedite the computations, the diffusion coefficients are simplified, with $\langle (\Delta v_\parallel) \rangle v$ taken to be constant, with $\langle (\Delta v_\perp)^2 \rangle$ set equal to $2\langle (\Delta v_\parallel)^2 \rangle$ and also assumed constant, and with these constant values set equal to the exact values of these coefficients for $v^2$ equal to the mean square velocity. In the case of an isotropic velocity distribution in which all velocities are initially equal, this method of computing the effect of gravitational encounters is shown to be in satisfactory agreement with a full numerical solution of the Fokker-Planck equation. For reliable results, $(\Delta t)_p$ must not exceed about one-fifth of the relaxation time.

## I. INTRODUCTION

To discuss the formation and development of open and globular clusters and of galactic nuclei one must analyze the effect of gravitational encounters on the evolution of a spherically symmetric stellar system. The distribution of velocities and the corresponding density distribution as a function of distance from the center will in general be markedly changed by such encounters. With modern electronic computers, rather full information can now be obtained on this complex problem.

Straightforward dynamical integrations of the equations of motion have been carried out for values of $N$, the total number of mass points

---

[†] Section III, Comparison with Exact Theory, has been omitted here.

in the system, as great as 100–250 (see Aarseth 1968; Wielen 1967, 1968). This approach is difficult to extend to higher values of $N$, since the amount of computing required for one relaxation time varies about as $N^3$. The evolutionary development of the system depends markedly on $N$, at least when $N$ is small, and thus the results obtained in these direct dynamical integrations cannot be applied directly to globular clusters and galactic nuclei.

Let us consider more specifically how the dynamical evolution depends on the value of $N$. We adopt the point of view that the gravitational field may be represented as a smooth, average field plus the perturbations resulting from two-body encounters. A single such encounter is characterized by an impact parameter $p$, defined as the distance of closest approach if there were no mutual gravitational attraction between the two passing stars. We denote by $p_0$ the value of $p$ such that the deflection of the two stars in their relative orbit equals $\frac{1}{2}\pi$; as is well known, $p_0$ is given by

$$ p_0 = \frac{G(m_A + m_B)}{V^2}, \tag{1} $$

where $m_A$ and $m_B$ are the masses of the two stars and $V$ is the initial relative velocity. From the virial theorem we may write for the mean square stellar velocity, which we denote by $v_m^2$,

$$ v_m^2 \equiv \langle v^2 \rangle = \frac{0.4GM}{R_h}, \tag{2} $$

where $M$ is the total mass of the system, $R_h$ is the radius containing half the mass, and the constant 0.4 provides a reasonably good approximation for most systems (Spitzer 1969). If we assume that all stars have the same mass, and set $V^2$ equal to its mean value, $2v_m^2$, we find

$$ \frac{p_0}{R_h} = \frac{1}{0.4N}, \tag{3} $$

where $N$ is the total number of stars. Equation (3) may be used to give an approximate result even where a dispersion of masses exists within the system. The dynamical time $t_{dh}$ we define as

$$ t_{dh} \equiv \frac{R_h}{v_m} = \frac{1.58R_h^{3/2}}{(GM)^{1/2}}. \tag{4} $$

If we take the relaxation time defined in equations (30) below and apply it to average conditions for the inner half of the system's mass, setting the particle density $n$ equal to its mean value interior to the radius $R_h$, we obtain, setting $\psi$ equal to unity, substituting for $\kappa$ from equation (23), omitting the subscript $f$,

$$t_{rh} = \frac{v_m^3}{15.4 G^2 m^2 n \ln(0.4N)} = \frac{0.060 \, M^{1/2} R_h^{3/2}}{m G^{1/2} \log(0.4N)}. \qquad (5)$$

Hence we have

$$t_{dh}/t_{rh} = 26 \log(0.4N)/N, \qquad (6)$$

where $\log N$ denotes a logarithm to the base 10. The subscript $h$ in $t_{dh}$ and $t_{rh}$ emphasizes that these two quantities refer to $R_h$, and that the dynamical times and the relaxation times vary greatly through the cluster.

Equation (6), with a different numerical constant, has been given much earlier by Chandrasekhar (1942; see eq. [5.227]). Evidently for $N$ as low as 100, $t_{dh} = 0.42 t_{rh}$, while for $N$ as great as $10^8$, $t_{dh}/t_{rh}$ falls to $2 \times 10^{-6}$. It should be emphasized that this ratio is only a convenient parameter. At different locations in the system both $t_d$ and $t_r$ may be very different from their mean values; in particular, because of the great increase of density toward the center, the rate of approach to a Maxwellian distribution in the core may be one or more orders of magnitude greater than its mean value for the inner half of the mass.

One of the ways in which the value of $N$, and of $t_{dh}/t_{rh}$, affects the evolution of a stellar system is through the rate of formation of binary stars. These form rapidly in a cluster containing only a few stars. In general, the only binaries that can have a major effect on the dynamical evolution of a spherical stellar system are those which are tightly bound, in the sense that their relative orbital velocity is comparable to or greater than the rms velocity $v_m$ of the stars in the system. If the orbital velocity is small compared with $v_m$, the binding energy of all such binaries will be unimportant compared with the binding energy of the entire system of stars. In addition, the tendency toward equipartition of kinetic energy will lead to an increase of energy of the loosely bound binaries, and hence to their disruption. Tightly bound binaries, on the other hand, may become more tightly bound and can provide a significant sink of energy for the system as a whole. With the use of equations (1)–(3) it is readily shown that a pair of stars will be tightly bound if $a$, the semimajor axis of the orbit, is small compared with $p_0$.

We now show that the rate of formation of tightly bound binaries will become negligibly small in a spherical system of stars as $N$ becomes very large. The following argument is due essentially to von Hoerner (1969), modified in accordance with comments by Hénon (1970). We assume that a tightly bound system can be formed only if three stars approach each other within a distance $\alpha_1 p_0$, where $\alpha_1$ is a constant of order unity. If we consider a single star, the number of close two-body encounters within the relaxation time $t_{rh}$ is independent of $N$ (we ignore here the slow dependence of $t_{rh}$ on $\ln N$), and thus in the system as a whole the number of encounters between two stars with $p$ less than $\alpha_1 p_0$ during a time interval $t_{rh}$ is proportional to $N$. The fraction of these which are triple encounters in that a third star is also located within the distance $\alpha_1 p_0$, is proportional to $N(\alpha_1 p_0)^3/R^3$, or to $1/N^2$, if we use equation (3). Hence the total number of binaries formed within the system per relaxation time $t_{rh}$ varies as $1/N$. This result is consistent with the fact that in the direct numerical integrations binaries form frequently if $N$ is between 10 and 30, but for $N$ between 100 and 250 they form only occasionally and at the center of the system, where the relaxation time is appreciably decreased. Evidently if $N$ is as great as $10^4$, the fraction of stars which become binaries will be entirely negligible during the 100 or so relaxation times at most that the system will experience before the end of its life as a bound collection of mass points. Similar conclusions have been reached by van Albada (1968) and others.

Other ways in which the value of $t_{dh}/t_{rh}$ influences the evolution and equilibrium of the system will be discussed in a subsequent paper. The numerical method described here lends itself readily to an investigation of this problem, since the ratio $t_{dh}/t_{rh}$ may easily be changed arbitrarily in the computations; however, because of practical limitations on computing time available, the present technique is not well suited to investigate very small values of this ratio.

## II. PRINCIPLES AND APPROXIMATIONS

The basic procedure followed in this study is to compute numerically the orbits of $\mathfrak{N}$ stars in a spherically symmetric system containing a much greater number, $N$, of stars.

This procedure rests on two basic approaches. First, in a system containing $N$ stars, the stellar orbits are calculated by numerical integration on the assumption that the potential field is spherically symmetric; i.e., with this assumption the velocity and the radius of each star are computed as functions of the time. Second, the velocities of all these $\mathfrak{N}$ stars are perturbed slightly at intervals to give statistically the same results as would be anticipated from random encounters with neighboring stars.

The numerical integrations of the stellar orbits follow the method applied by Hénon (1964), based on earlier work by Campbell. In this approach, the $N$ stars are divided into $\mathfrak{N}$ subgroups, each of which has a number of stars all characterized by the same mass $m$, the same radial velocity $v_r$, the same transverse velocity $v_t$ normal to the radius vector, and all situated at the same distance $r$ from the center. Thus each subgroup is essentially a shell, with stars uniformly distributed over the spherical area. Not only is each shell assumed to be spherically symmetric but in addition the distribution of the direction of the transverse velocities is isotropic; i.e., at each point in the shell, the directions of the transverse velocities $v_t$ for the different stars in the shell are uniformly distributed in the plane perpendicular to the radius vector.

Let us now arrange the $\mathfrak{N}$ shells in the order of increasing distance $r_j$ from the center, and number them from $j = 1$ up to $j = \mathfrak{N}$. Then the equations of motion become

$$dr_j/dt = v_{rj}, \tag{7}$$

$$\frac{dv_{rj}}{dt} = \frac{J_j^2}{r_j^3} - \frac{G(j - \frac{1}{2})\mathfrak{M}}{r_j^2}, \tag{8}$$

where $J_j$ is the angular momentum per unit of the stars in shell $j$, and $\mathfrak{M}$ is the shell mass, taken to be identical for all shells; thus the number of stars in each shell is inversely proportional to the stellar mass $m_j$. In the numerical work we choose units such that $G\mathfrak{M}$ equals unity. The quantity $\frac{1}{2}$ in the last term in equation (2) takes into account the self-attraction of each shell. In the absence of encounters, $J_j$ is constant for each shell, and equal to $r_j v_{tj}$.

We pass on now to a consideration of collisional perturbations in the velocities $v_{rj}$ and $v_{tj}$. As usual, we make the familiar general assumptions that only two-body encounters need to be considered, that these are random in space and time, and that they occur locally, with each encounter confined to a time interval short compared with the orbital period; this last assumption is not followed consistently since the maximum impact parameter is set equal to the radius of the system, but the error introduced in this way is believed to be small.

Let us consider a gravitational encounter between two stars, denoted by $A$ and $B$, with velocities $v_A$ and $v_B$ before the encounter. The velocity changes, $\Delta v_A$ and $\Delta v_B$, resulting from the encounter will be a function of three quantities: the relative velocity $v_A - v_B$, the angle $\Theta$ between the orbital plane and the plane containing $v_A$ and $v_B$, and the impact parameter $p$, defined as the distance of closest approach in the absence of collisions (Chandrasekhar 1942). In a full Monte Carlo approach to the problem, one might consider individual encounters with $v_A$, $v_B$, $\Theta$, and $p$ all chosen at random. This approach would be subject to the objection that an enormous number of encounters would have to be considered, since the cumulative effect of many very small changes in velocity is responsible for collisional relaxation.

A modified Monte Carlo method has been adopted by Hénon (1967), who selects $v_A$, $v_B$, and $\Theta$ at random, but chooses $p$ so that the mean values of $\Delta v$ and $(\Delta v)^2$ per unit time, integrated over all $p$, are correctly given. This approach does not give the correct distributions of $\Delta v$ and $(\Delta v)^2$; hence this method does not give the occasional large changes in velocity that result from occasional close encounters. However, the effect of small encounters is taken into account precisely by this method, which should become exact as $\ln N$ becomes large.

The approach followed in this paper is an even simpler Monte Carlo method, designed to give approximately correct results with a minimum of computations. The averaging procedure introduced by Hénon is carried further, and for a star of velocity $v_A$, the values of $\Delta v$ and of $(\Delta v)^2$ are averaged not only over the impact parameter $p$ but over $\Theta$ and $v_B$ as well. This procedure requires that specific velocity distributions be assumed; in most situations the assumption of a Maxwellian distribution is a good approximation. Finally, the mean values obtained for $\Delta v$ and $(\Delta v)^2$ are simplified somewhat to permit more rapid computations.

We now proceed to give the detailed equations used. Let us consider a particular star, of velocity $v$, whose change of velocity is to be considered in successive interactions with many other stars in the system; such a star is called a "test star" by Chandrasekhar (1942). We denote by $\Delta v_\parallel$ and $\Delta v_\perp$ the collisionally induced changes of $v$ in directions parallel and perpendicular, respectively, to the initial value of $v$. The mean values of $\Delta v_\parallel$, $(\Delta v_\parallel)^2$, and $(\Delta v_\perp)^2$, summed over all collisions in a unit of time, are called diffusion coefficients (Spitzer 1962) and are denoted by $\langle \Delta v_\parallel \rangle$, $\langle (\Delta v_\parallel)^2 \rangle$, and $\langle (\Delta v_\perp)^2 \rangle$, respectively. The stars which a test star encounters will be called "field stars," following Chandrasekhar (1942), and their properties will be denoted with a subscript $f$. If we assume that the distribution of $v_f$ is Maxwellian, then integration over $p$, $\Theta$, and $v_f$ gives (Spitzer 1962)

$$\langle \Delta v_\parallel \rangle / v = -A_D l_f^3 (1 + m/m_f) G(x)/x, \qquad (9)$$

$$\langle (\Delta v_\parallel)^2 \rangle = A_D l_f G(x)/x, \qquad (10)$$

$$\langle (\Delta v_\perp)^2 \rangle = A_D l_f (\phi(x) - G(x))/x, \qquad (11)$$

where

$$A_D \equiv 8\pi G^2 m_f^2 n_f \ln(0.4N), \qquad (12)$$

$$l_f^2 \equiv 3/(2 v_{mf}^2), \qquad (13)$$

and

$$x \equiv l_f v. \qquad (14)$$

The quantity $\phi(x)$ is the usual error function; the quantity $G(x)$, defined originally by Chandrasekhar (1942), is tabulated by Spitzer (1962). In equations (12) and (13) $n_f$ and $m_f$ are the number density and mass of the field stars, $v_{nf}^2$ is the mean square velocity of the field stars, and $N$ is the total number of stars in the system. The argument of the natural logarithm in equation (12) is simply $R/p_0$, evaluated with the aid of equation (3). It follows from the analysis by Cohen, Spitzer, and Routly (1950) that to evaluate the diffusion coefficients in the gravitational case the integrals over the impact parameter must be extended over the entire stellar system. The radius $R$ containing half the mass is used here as a convenient cutoff point in these integrations; for computing the diffusion coefficients in a dense central core the radius of the core might be a more appropriate cutoff, but we omit here this possible refinement.

The diffusion coefficients above are computed from a Maxwellian distribution of field-star velocities. Hence if the velocity distribution of the test stars is also Maxwellian, equilibrium should be reached and $\partial f/\partial t$ should vanish. From the Fokker-Planck equation (Cohen et al. 1950; Rosenbluth, MacDonald, and Judd 1957), which becomes relatively simple when the velocity distribution is assumed isotropic, one obtains the following condition on the diffusion coefficients when $\partial f/\partial t$

TABLE 1

Values of $G/x$ and $(\phi - G)/2x$

| | | | | | $x$ | | | | | |
|---|---|---|---|---|---|---|---|---|---|---|
| Function | 0 | 0.1 | 0.2 | 0.4 | 0.6 | 1.0 | 1.2 | 1.5 | 2.0 | 2.5 |
| $G/x$ | 0.376 | 0.374 | 0.367 | 0.342 | 0.305 | 0.214 | 0.171 | 0.117 | 0.060 | 0.032 |
| $(\phi - G)/2x$ | 0.376 | 0.375 | 0.373 | 0.365 | 0.351 | 0.315 | 0.294 | 0.264 | 0.219 | 0.184 |

vanishes:

$$\frac{\langle \Delta v_\| \rangle}{l_f} + \frac{1}{x}\left(\frac{\langle (\Delta v_\perp)^2 \rangle}{2} - \langle (\Delta v_\|)^2 \rangle\right) + \frac{mx}{m_f}\langle (\Delta v_\|)^2 \rangle$$

$$-\frac{1}{2}\frac{d}{dx}\langle (\Delta v_\|)^2 \rangle = 0. \tag{15}$$

With direct substitution we may verify that equations (9), (10), and (11) satisfy equation (15).

Table 1 gives values of the two functions appearing in the diffusion coefficients, when the field stars have an isotropic and Maxwellian distribution of velocities. We see that for small $x$, $G(x)/x$ is nearly equal to $(\phi - G)/2x$, and hence $\langle (\Delta v_\|)^2 \rangle$ is about equal to $\frac{1}{2}\langle (\Delta v_\perp)^2 \rangle$. It is evident physically that for small velocity the mean square change in velocity will be the same in all directions; the factor $\frac{1}{2}$ appears because $(\Delta v_\perp)^2$ includes velocity changes in two directions perpendicular to the test star velocity $v$.

Table 1 shows also that the values of the diffusion coefficients do not change dramatically with $v$. As part of the basic approximation scheme adopted here, we shall set $\langle \Delta v_\| \rangle$ and $\langle (\Delta v_\|)^2 \rangle$ equal to their values at $x = 1.225$, corresponding to $v = v_{mf}$; at this value of $x$, $G(x)/x = 0.167$, or about 0.44 times its value at $x = 0$. Since $\langle \Delta v_\| \rangle$ and $\langle (\Delta v_\|)^2 \rangle$ play a more important role in the transfer of energy between particles than does $\langle (\Delta v_\perp)^2 \rangle$, which produces primarily deflections, we shall take this same value of 0.167 for $(\phi - G)/2x$, and set

$$\langle (\Delta v_\|)^2 \rangle = \frac{1}{2}\langle (\Delta v_\perp)^2 \rangle = 0.167 A_D l_f. \tag{16}$$

In order to satisfy equation (15), we must then write

$$\langle V_\| \rangle / v = -l_f^2 \langle (\Delta v_\|)^2 \rangle m / m_f. \tag{17}$$

It may be noted that equation (17), which is required for self-consistency if the diffusion coefficients are constant, with $2\langle(\Delta v_\parallel)^2\rangle$ equal to $\langle(\Delta v_\perp)^2\rangle$, differs from the relationship satisfied by the exact diffusion coefficients.

The assumption that the mean square change in velocity is the same as all directions, even for large $v$, materially expedites the computation of the velocity perturbations, and the assumed lack of dependence of the diffusion coefficients on $v$ is also a substantial simplification.

Finally, we must sum equations (16) and (17) over each type $k$ of star present, each with a particle density $n_k$, a mass $m_k$, and a mean square velocity $v_{mk}^2$. To simplify the numerical computations we perform this sum analytically, computing the total diffusion coefficients for a one-component system with particle density $n_f$, mass $m_f$, and mean square velocity $v_{mf}^2$; these quantities are defined by

$$n_f = \sum_k n_k, \tag{18}$$

$$m_f = \sum_k n_k m_k / n_f, \tag{19}$$

$$v_{mf}^2 = \sum_k n_k m_k v_{mk}^2 / n_f m_f. \tag{20}$$

If equipartition of energy is established, and $m_k v_{mk}^2$ is the same for all $k$, it follows from equations (16) and (17) with straightforward algebra that

$$\langle(\Delta v_\parallel)^2\rangle = \tfrac{1}{2}\langle(\Delta v_\perp)^2\rangle = \kappa \psi n_f m_f^2 / v_{mf} \tag{21}$$

and

$$\frac{\langle\Delta v_\parallel\rangle}{v} = -\frac{3}{4}\frac{m}{m_f v_{mf}^2}\langle(\Delta v_\parallel)^2\rangle, \tag{22}$$

where

$$k = 0.167 \times 8\pi(1.5)^{1/2} G^2 \ln(0.4N) = 11.8 G^2 \log(0.4N) \tag{23}$$

and the factor $\psi$ is given by

$$\psi = \frac{\sum n_k m_k^{5/2}}{n_f m_f^{5/2}}. \tag{24}$$

If $v_{mk}^2$ is the same for all stellar types, equation (24) is no longer valid; in fact, for this case there is no value for $\psi$ that gives correct results in equations (21) and (22). However, setting $\psi$ equal to unity gives the correct value of $\langle \Delta v_\| \rangle$ in this case. In the numerical computations actually carried out, equipartition was never in fact approached, and $\psi$ has been set equal to unity. The maximum resulting error in $\langle (\Delta v_\|)^2 \rangle$ is about equal to the maximum value of $\psi$ given by equation (24) for the case of equipartition; for a system with stars of two stellar masses differing by a factor of 5, the maximum value of $\psi$ found from equation (24) is 2.88, which is probably a generous upper limit on the error in $\langle (\Delta v_\|)^2 \rangle$. Under most situations one component will make the dominant contribution to the diffusion coefficients, in which case setting $\psi$ equal to unity gives the correct value of $\langle (\Delta v_\|)^2 \rangle$ as well as $\langle \Delta v_\| \rangle$. While it would be entirely possible to compute separately the velocity perturbations produced by each stellar type, by using equations (16) and (17) instead of (21) and (22), this additional complication does not seem worthwhile in view of the approximations already made.

It may be noted that $\langle (\Delta v_\|)^2 \rangle$ is independent both of the mass and of the velocity of the test star. Lack of dependence on the mass is a general result—see equations (10) and (11)—but the independence of the velocity, of course, results from the particular assumptions made here.

If equations (21) and (22) are assumed, it can be shown that $(dE_k/dt)_e$, the rate of change of kinetic energy resulting from encounters, is given by

$$\left( \frac{dE_k}{dt} \right) = \frac{3m_k \langle (\Delta v_\|)^2 \rangle}{2} \left( 1 - \frac{E_k}{E_f} \right), \qquad (25)$$

where

$$E_f = \tfrac{1}{2} m_f v_{mf}^2. \qquad (26)$$

Equation (25) takes into account only those interactions between stars of mass $m_k$ and field stars all of mass $m_f$. When only two components are present, denoted by subscripts 1 and 2, equation (25) may be expressed in terms of the usual equipartition time $t_{eq}$ (Spitzer 1962, eq. [5-30]), and for component 2 we have

$$t_{eq} = \frac{E_1 - E_2}{(dE_2/dt)_e}. \qquad (27)$$

If we use equations (20) and (26) to express $E_f$ in terms of $E_1$ and $E_2$, then substitution of equation (25) into (27) yields

$$t_{\text{eq}} = \frac{n_f m_f v_{mf}^2}{3 n_1 m_2 \langle (\Delta v_\parallel)^2 \rangle}. \tag{28}$$

If $n_2/n_1$ is small, the value of $t_{\text{eq}}$ given by equation (28), with $\langle (\Delta v_\parallel)^2 \rangle$ found from equation (21), equals the exact value for two interacting groups of stars, each with a Maxwellian distribution (Spitzer 1962, eq. [5-31]), when $v_{m2}^2$ is equal to $0.73 v_{m1}^2$. While equation (25) does not conserve kinetic energy exactly for the system as a whole, this condition is enforced by the correction factor $R_c$, discussed below.

As a measure of the relaxation time for a particular component $k$, we shall define $t_{rk}$ by the relationship

$$t_{rk} = \frac{v_{mk}^2}{3 \langle (\Delta v_\parallel)^2 \rangle}. \tag{29}$$

During the time interval $t_r$, the mean value of $\Sigma (\Delta v_\parallel)^2$ equals $\frac{1}{3} v_{mk}^2$, which for an isotropic velocity distribution is the mean square velocity in any one direction. As an overall measure of the relaxation time in a particular region we may define $t_{rf}$ as the value of $t_{rk}$ when the mean square velocity $v_{mf}^2$ for all the stars in the region is substituted for $v_{mk}^2$ in equation (29). From equation (21) we then obtain

$$t_{rf} = \frac{v_{mf}^3}{3 \kappa \psi n_f m_f^2}. \tag{30}$$

This quantity, $t_{rf}$, if 0.75 times the reference time $T_R$ introduced by Spitzer and Härm (1958), if all stars have the same mass.

We are indebted to R. H. Durisen for important assistance in the programming and to J. P. Ostriker for several helpful suggestions. One of us (M.H.H.) was supported by a National Science Foundation pre-doctoral fellowship while this work was in progress.

## REFERENCES

Aarseth, S. J., 1968, *Bull. Astr.*, Ser. 3, **3**, 105.
Chandrasekhar, S., 1942, *Principles of Stellar Dynamics* (Chicago: University of Chicago Press), chap. 5.
Cohen, R. S., Spitzer, L., and Routly, P. McR., 1950, *Phys. Rev.*, **80**, 230.

Hénon, M., 1964. *Ann. d'ap.*, **27**, 83.

——, 1967, *Bull, Astr.*, Ser. 3, **2**, 91.

——, 1970, informal communication.

Hoerner, S. von., 1969, informal communication.

Rosenbluth, M. N., MacDonald, W. M., and Judd, D. L., 1957, *Phys. Rev.*, **107**, 1.

Spitzer, L., 1962, *Physics of Fully Ionized Gases* (2nd ed.; New York: Interscience Publishers), §§ 5.2 and 5.3.

——, 1969, *Ap. J.*, **158**, L139.

Spitzer, L. and Härm, R., 1958, *Ap. J.*, **127**, 544.

van Albada, T. S., 1968, *B.A.N.*, **19**, 479.

Wielen, R., 1967, *Veroff. Astr. Rechen Inst., Heidelberg*, Vol. 2, No. 19.

——, 1968, *Bull. Astr.*, Series 3, **3**, 127.

# DYNAMICS OF GLOBULAR CLUSTERS

(SCIENCE, 225, 465, 1984)

## Commentary

FOLLOWING the development of our Monte Carlo method, described in Paper #19, we used this technique to trace the evolution of model clusters with a range of properties. Seven more papers appeared during a decade in collaboration with graduate students, usually a different one for each paper. Each of these research assistants made modifications and additions to our computing program, which by the end of this period had become so involved that an overall comprehension of the details became difficult!

Paper #20 is included in this volume to provide a brief review of the results on cluster evolution which we and others obtained during this period. Since I was preparing a more detailed review[1] for a symposium on this topic, I was very glad to write this shorter version, designed for scientists not familiar with the details of astronomy. A number of important papers by various authors, not discussed in Paper #20, are included in more complete surveys.[1,2] Since 1984, studies of cluster evolution have expanded rapidly. Two of the general research topics which have received particular attention are discussed here.

### EVOLUTION OF CLUSTERS

After the initial collapse of a cluster core seemed largely understood, interest shifted to the subsequent expansion phase, including the oscillations which the expanding cores unexpectedly displayed in theoretical models.[2,3] A theoretical stability analysis[4] has clarified the conditions under which such oscillations might occur, but more research is needed for a full understanding.

Of the various cluster evolution models computed during the last decade, two are referred to here, each beginning with the initial equilibrium state of a tidally limited globular cluster and ending with final disappearance. Stars are all assumed to have the same mass, and the tidal field is taken to be constant, as for a cluster in a circular orbit around a spherical nucleus. Formation and evolution of binaries is considered, and thus passage through core collapse and expansion is included.

The prototype of such configurations is the classic self-similar model of Hénon,[5, 6] referred to in the previous commentary. This idealized model is based on an exact solution of the Fokker-Planck equation, subject to two major approximations: an isotropic velocity distribution and a spherical density distribution, with stars escaping at radii exceeding the tidal radius, $r_t$. Any energy released from binaries or other processes is considered to produce a flux from a central singularity. The final solution is beautifully simple. The mean density remains constant as stars escape, while the cluster mass decreases linearly with time. At any epoch in the evolution, the interval remaining before the cluster disappears equals 22 times the value of $t_{rh}$ at that epoch; the corresponding escape probability $\xi_e$ per unit $t_{rh}$ is about an order of magnitude greater than that found above for an isolated cluster.

The more recent, more physical models, based on the same two approximations, approach[7, 8] this Hénon model after the collapse ends. The initial phases differ greatly from the Hénon model; as a result, the initial lifetime exceeds 22 times the initial $t_{rh}$ by a factor of about 50. For a cluster containing $N$ stars, this total lifetime is[8] about $500(N/10^5)P$, where $P$ is the period of the cluster's orbit around the Galaxy.

As pointed out in Paper #20 below, most clusters in orbit around the Galaxy will experience transient variations in the ambient gravitational field; the resulting perturbations are often referred to as gravitational shocks. Monte Carlo methods have computed the rate of star escape as a result of such shocks. The increased random diffusion of velocities is found to increase the rate of escape of stars about as much as does the accompanying systematic gain in energy, a result recently analyzed quantitatively in detail.[9] Such perturbations, which vary strongly from one cluster to another, can significantly decrease the lifetime of globular or galactic clusters.

Interest in cluster evolution has also been directed toward the early life of the system, shortly after birth. The early evolution of massive stars and the attendant loss of mass from the system can fully disrupt many young clusters;[10] young systems of low density can also be disrupted early by the transient perturbations discussed in the preceding paragraph. An important purpose of such studies is to deduce from present clusters what the conditions were at their birth, and thus to learn more about the early history and evolution of our Galaxy.

## BINARY STARS

Binary systems have been of particular importance in cluster studies during the last ten years. Since binaries are thought to dominate the post-collapse evolution of a cluster, this topic somewhat overlaps the first. A major development in this area has been the mounting evidence[3] that binaries have been present in clusters from the beginning, and that, contrary to earlier beliefs, the abundance of these "primordial" binaries, relative to single stars, may be nearly the same in globular clusters as in the disk of the Galaxy. A dynamical simulation, treating evolution of binaries with a Monte Carlo technique, outlines[11] how such a subsystem of primordial binaries could evolve with time.

A major development concerning cluster binaries has been the discovery of bright X-ray sources and millisecond pulsars, which, quite unexpectedly, are relatively much more numerous in globular clusters than elsewhere in the Galaxy.[3] Most of these objects are presumably binaries, including a neutron star. The prevalent theory for the origin of these double-star systems is that the neutron stars have undergone exchange reactions with a less massive star in a primordial binary. In addition to the constraints which these binary systems provide on the past evolution of a cluster, the precise periods of a pulsar permit a direct determination of the local stellar density and gravitational potential gradient.[12]

The final sections of Paper #20 as originally published give a considerable discussion of binaries, summarizing various ways in which these systems could affect globular clusters. All this material is omitted here, except for the first four paragraphs, giving basic physical principles, which are still valid. The subsequent text, which attempted to apply these physical principles to the evolution of actual clusters, is now largely out of date, superseded by more modern research.[3]

Future research on cluster evolution may well place substantial emphasis on physical collisions between stars (including grazing and very close encounters) during collapse of the core to very high densities. The capture of one star by another in a close encounter, which dissipates kinetic energy by exciting strong tidal oscillations in both stars, is a good example of an important collisional process. The close binary pairs formed in this way may have a major effect in terminating core collapse and influencing the post-collapse phase.[8] Tidal-capture binaries may also include some of the observed X-ray binaries and millisecond pulsars.

Some other consequences of direct collisions between two stars have been discussed in Paper #18, with reference to galactic nuclei instead of globular clusters. Theorists will presumably wish to determine what types of stellar structure will result from various types of collisions and what evolutionary scenario might follow. The numerical simulations of star-star collisions referred to in the Commentary, Paper #19, make possible a preliminary approach to this problem. More efforts seem likely in the future on the details of such collisions as well as on the consequent cluster evolution.

## REFERENCES

1. Spitzer, L.—In: Dynamics of Star Clusters, IAU Symp. No. 113, eds. J. Goodman and P. Hut, Dordrecht: Reidel 109 (1985).
2. Spitzer, L.—Dynamical Evolution of Globular Clusters, Princeton: Univ. Press (1987).
3. Hut, P. et al.—*Publ. Astron. Soc. Pacific* **104**, 981 (1992).
4. Goodman, J.—*Astroph. J.* **313**, 576 (1987).
5. Hénon, M.—*Annales d'Astroph.* **24**, 369 (1961).
6. Spitzer, L.—Dynamical Evolution of Globular Clusters, Princeton: Univ. Press §3.2b (1987).
7. ——, ibid. §5.1b.
8. Lee, H. M. and Ostriker, J. P.—*Astroph. J.* **322**, 123 (1987).
9. Kundić, T. and Ostriker, J. P.—*Astroph. J.* **438**, 702 (1995).
10. Weinberg, M. D. and Chernoff, D. F.—In: Dynamics of Dense Stellar Systems, ed. D. Merritt, Cambridge: Univ. Press 221 (1989).
11. Hut, P., McMillan, S., and Romani, R. W.—*Astroph. J.* **389**, 527 (1992).
12. Phinney, E. S.—*Phil. Trans. Roy. Soc. London*, Series A **341**, 39 (1992); In: Structure and Dynamics of Globular Clusters, eds. S. G. Djorgovski and G. Meylan, San Francisco: Astron. Soc. Pacific 141 (1993).

## Paper[†]

ABSTRACT

In their attempt to reach kinetic equilibrium, through gravitational encounters between separate stars, globular clusters are driven to destruction, with their cores collapsing and their outer regions expanding. The effects of core collapse, which apparently produces x-ray sources, are not yet fully understood, but white dwarfs and neutron stars, probably in binary systems, are thought to be involved, and possibly black holes as well.

Globular clusters are nearly spherical stellar systems associated with many galaxies and generally containing from $10^5$ to $3 \times 10^6$ stars. Figure 1 is a photograph of Messier 19 (No. 19 in a catalog of some hundred diffuse objects compiled by Charles Messier late in the 18th century), a conspicuous such cluster in our own Galaxy, at a distance of some 3000 parsecs (1 parsec = 3.26 light years). Typically the relatively dense central core of a globular cluster, with a radius of about 1 parsec, contains some $10^4$ stars, while the outer regions of the cluster extend with much diminished density out to distances of roughly 25 to 100 parsecs. Studies of the stellar spectra indicate that these systems within our own Galaxy were formed early in the life of the Universe, about $10^{10}$ years ago, not very long after the initial Big Bang.

The dynamical evolution of these beautifully symmetrical, very ancient systems has provided astrophysicists with an intriguing and challenging problem, which so far is only partly solved. Even in the cores, the average distance between neighboring stars is generally more than $10^4$ times the radii of even the giant stars and direct collisions between stars are extremely rare. Thus the cluster stars move as mass points under their mutual gravitational attraction, with random velocities of some tens of kilometers per second. The large value of $N$, the total number of stars, can be expected to average out any large statistical fluctuations, and the way a spherical cluster evolves with time, as a result solely of Newton's laws of motion, appears deceptively simple.

Although the general principles underlying this evolution have been known for some time, it is only within the last decade that theoretical analyses, supported by high-speed computers, have provided a detailed understanding of the later evolutionary phases. As we shall see below, these involve actual collapse of the central core and lead to the

---

[†] The final sections on binary stars have mostly been omitted here.

FIG. 1. Photograph of globular cluster M19 (NGC 6273) with the 3.9-m Anglo-Australian telescope.

occurrence of new physical processes not important at the earlier
stages. During all this activity the outer regions of the cluster gradually
expand. The evolution of the cluster in the post-collapse phase is an
active research field. The x-ray sources observed in the cores of some of
the more centrally condensed clusters may well result from processes
occurring during and after the core collapse.

In this article, as in several general surveys,[1,2] first the physical
principles affecting the early evolution of the cluster and the detailed
evolutionary models based on these principles are outlined. Then additional physical processes that become important during the collapse
phase, such as formation of binary systems, both by tidal capture in a
close two-body encounter and by direct three-body encounters, are
discussed.

## PHYSICAL PRINCIPLES

In discussing stellar motions in a globular cluster, we first separate the
gravitational potential energy, $\phi(\mathbf{r}, t)$ into the sum of two terms. The
first is a smoothed, spherically symmetric potential obtained by averaging $\phi(\mathbf{r}, t)$ over a time interval including several orbital periods of the
stars. A star moving at 10 kilometers per second goes 1 parsec in $10^5$
years, and the time required to travel back and forth across a cluster is
generally less than $10^6$ years, which in turn, is a small fraction of the
evolution time. The average of $\phi(\mathbf{r}, t)$, over roughly $10^6$ years we denote
by $\phi_A(r)$, assumed to be spherically symmetric. This smoothed potential
will change slowly as the cluster evolves. In a zero-order approximation
each star moves in this spherical potential, with constant energy $E$ and
angular momentum $\mathbf{J}$, both measured per unit mass.

On this approximation no evolution occurs. A basic constraint on a
cluster in this approximation is that the average smoothed stellar
density, $\rho_A(r)$, must be consistent with Poisson's equation

$$\nabla^2 \phi_A(r) = 4\pi \rho_A(r) \qquad (1)$$

where $\rho_A(r)$ is averaged over the same time interval used in determining
$\phi_A(r)$. Many equilibrium solutions are possible.

The difference in potential $\phi(\mathbf{r}, t) - \phi_A(r)$ results from the granularity of the gravitational field. It is generally assumed that this granularity
can be represented in a first approximation by two-body encounters
between stars, and that the effects of such encounters in altering $E$ and
$\mathbf{J}$ of each star can be computed as though the two stars involved were
moving in a hyperbolic path relative to each other, unaffected by other
stars. The effects of such encounters have been computed in detail.[3,4]

The results can be used to follow the way in which the distribution of stars among different orbits is changed and thus how the cluster evolves. It should be emphasized that, for the cluster as a whole, the effects produced by stellar encounters occur very slowly in comparison with the time for a star to move across the cluster. Essentially the mean free path is many orders of magnitude greater than the dimensions of the cluster, and thousands of cluster crossings are required for appreciable evolution.

One important feature of two-body gravitational encounters is that the cumulative effect of many distant encounters, each of which produces only a small change in stellar velocities, tends to outweigh the less frequent close encounters, in which the stars are deflected by some $90°$ or more. Thus the velocity of a star is subject to a diffusion process, and similar diffusion occurs in $E$ and $\mathbf{J}$. Changes in the velocity distribution of stars are governed by the integrodifferential Fokker-Planck equation[5].

While the details of these dynamical interactions are somewhat complex, the general physical tendency is clear. Encounters between stars will tend to increase entropy, evolving the stellar system toward a state of higher probability. The distribution of stars can be described by the density $f(\mathbf{r}, \mathbf{v})$ in phase space; $f(\mathbf{r}, \mathbf{v})$, multiplied by the phase space volume element $dx\,dy\,dz\,dv_x\,dv_y\,dv_z$, is the number of stars within this volume element centered at $\mathbf{r}, \mathbf{v}$. In the local state of highest probability, toward which the cluster evolves, the phase space density $f(\mathbf{r}, \mathbf{v})$, which we designate simply by $f$, is given by

$$f = Ke^{-\beta E} \tag{2}$$

where $E$ is the energy of each star per unit mass, $\beta$ is inversely proportional to the average energy, and $K$ is a normalization constant. For stars within a small region of space the potential energy is constant, and only the kinetic energy, $mv^2/2$, need be considered; Eq. 2 then gives $f_M$, the usual Maxwellian distribution function

$$f_M = \frac{n}{v_m^3} \left( \frac{3}{2\pi} \right)^{3/2} e^{-3v^2/2v_m^2} \tag{3}$$

where $v_m$ is the root-mean-square (rms) velocity, and $n$ is the number of stars per unit volume of physical space. Multiplication of $f_m$ by $4\pi v^2 dv$ gives the number of stars per unit volume whose total velocity lies between $v$ and $v + dv$.

It is this tendency toward a more probable state, as in thermodynamic equilibrium, which leads the cluster straight to catastrophe. The volume of accessible phase space per energy increment is greatest for stars which are at the greatest distances from the cluster center, especially those which escape the cluster entirely and have an entire galaxy to roam around in. On the other hand, $f$ in Eq. 2 is maximized if some of the cluster stars are very close together, giving a large negative potential energy $E$. Thus velocity perturbations lead to an expansion of some regions of the cluster and contraction of others. Analysis of the various ways in which these processes occur in a spherical star cluster provides a challenging task, whose status is summarized in this article.

## CATASTROPHES WITH SIMPLE MODELS

The simultaneous processes of expansion and contraction to which star clusters are subject can be understood physically from very simple models. While precise numerical results can be obtained only from the realistic, detailed calculations discussed later in this article, these simple models are helpful in understanding and interpreting the more complex calculations. Three of these simple models, each of which leads the cluster to catastrophe in a different way, are presented below.

In the first model, discussed some 40 years ago,[6,7] the cluster is regarded as a uniform sphere, whose density $\rho$ and rms velocity $v_m$ are constant. The total mass is $M$, and all stars are taken to have the same mass, $m$. We make use of the virial theorem, which states that for an isolated system of self-gravitating mass points in equilibrium

$$2T = Mv_m^2 = -W \qquad (4)$$

where $T$ is the total kinetic energy and $W$ is the total gravitational energy. Thus the average kinetic energy per star is half the corresponding average gravitational binding energy. However, the average change of potential energy involved in removing one star initially from the cluster is twice the average potential binding energy of all the stars; this may be seen if one computes the energy required to disassemble the entire cluster—the energy required per star declines steadily as the remaining mass decreases, with the initial value twice the average value. It follows that the average energy for escape of the first few stars is not twice but four times the average kinetic energy; if we denote the escape velocity by $v_{esc}$, we obtain the general result for any isolated stellar system

$$\langle v_{esc}^2 \rangle = 4v_m^2 \equiv 4\langle v^2 \rangle \qquad (5)$$

where the brackets denote average values over all the stars.

The Maxwellian distribution in Eq. 3 can be used to compute the fraction, $\xi_e$, of stars for which $v^2 > 4v_m^2$, giving $\xi_e = 7.4 \times 10^{-3}$. Encounters between stars will tend to establish a Maxwellian velocity distribution during some time interval, which is called the time of relaxation and is denoted by $t_r$. If we assume that for velocities exceeding $4v_m^2$, $f$ approaches its Maxwellian value in time $t_r$ and that all particles escape if their kinetic energy exceeds four time the average, we obtain

$$\frac{1}{M}\frac{dM}{dt} = -\frac{\xi_e}{t_r}.$$ (6)

For the relaxation time we adopt the value (8)

$$t_r = \frac{v_m^3}{15.4G^2m^2n\ln(0.4N)}$$ (7)

where $v_m$ is again the rms stellar velocity, $m$ the stellar mass, $n$ the density of stars per unit volume, and $N = M/m$, the total number of stars in the system. The general form of Eq. 7 follows from the fact that the cross section for a 90° deflection in the relative orbit is of order $\pi(Gm/v^2)^2$; the numerical constant and the logarithmic term are obtained from the detailed theory of stellar encounters. For the uniform sphere considered here $t_r$ is independent of position in the cluster.

The assumption that the fraction of stars escaping during the time $t_r$ is given so directly by $f_M$ is, of course, a simplification. A solution of the Fokker-Planck equation for a system of stars in a hypothetical square-well spherical potential (constant inside the cluster and zero outside) gives[9] Eq. 6 with the constant $\xi_e$ now equal to $8.5 \times 10^{-3}$. The stars which diffuse to values exceeding $v_{esc}$ leave the cluster with very little excess energy. As a result, the total energy of the cluster, proportional to $M^2/R$, where $R$ is the cluster radius, remains constant as $M$ decreases. Hence $v_m^2$ varies as $1/M$, $n$ varies as $M^{-5}$, and Eqs. 6 and 7 may be integrated approximately to yield[10]

$$M(t) = M(0)\left(1 - \frac{7}{2}\frac{\xi_e t}{t_r(0)}\right)^{2/7}$$ (8)

where $M(0)$ and $t_r(0)$ are the initial values of $M$ and $t_r$. Evidently, evaporation of stars produces a collapse of the cluster, with cluster mass $M$ and radius $R$ approaching zero together after a time interval equal to $2t_r(0)/7\xi_e$.

Equation 8 is applicable not only to this idealized homogeneous cluster but also to any cluster of constant total energy, $E_T$, which undergoes homologous contraction; that is, a cluster in which the smoothed density is a function of $r/r_c(t)$, where $r_c(t)$ is some characteristic time-dependent cluster dimension, either the outer radius of a uniform cluster or the radius of a compact central core.

For homologous contraction the structure of the system, including the spatial variation of $\rho$ and $v_m$, remains constant except for time-dependent scale factors. If the evaporating stars carry away appreciable energy, diminishing $E_T$, and if $\zeta$ is the ratio of the fractional loss of energy to the fractional loss of mass, then

$$\zeta = \frac{dE_T}{E_T\, dt} \bigg/ \frac{dM}{M\, dt} \tag{9}$$

and Eq. 8 is replaced[10] by

$$1 - (3.5 - 1.5\zeta)\frac{\xi_e t}{t_r(0)} = \left[\frac{M(t)}{M(0)}\right]^{3.5-1.5\zeta}. \tag{10}$$

In addition, $r_c$ becomes proportional to $M^{2-\zeta}$, and $v_m$ to $M^{(\zeta-1)/2}$.

We turn now to a second model, in which the cluster is replaced by an isothermal sphere, whose equilibrium structure has been extensively studied. Since in such a sphere $\rho(r)$ varies asymptotically as $1/r^2$, the mass is infinite if the radius is infinite. To give a model with finite mass, the sphere is truncated at some radius $R$ with a hypothetical rigid, confining shell. For large $R$, the phenomena of interest occur well inside this confining surface, which does not much affect the results. Since the central regions of a cluster are in fact nearly isothermal, this model is much more realistic than the first.

This model is subject to the remarkable "gravothermal" instability,[11] associated with the negative specific heat of self-gravitating stellar systems. According to the virial theorem in Eq. 4, the total energy $T + W$ is, of course, negative and equal to $-T$. Thus if the total energy is increased (becomes less strongly negative), $T$ will decrease. For example, if a small satellite loses energy as it orbits around the Earth (from frictional retardation by the Earth's atmosphere), it spirals inward, accelerating its motion, so that the centrifugal force remains nearly in balance with the gravitational force.

Since Eq. 4 applies to isolated systems confined by their self-gravitational attraction, it is not strictly valid for a system confined by a rigid wall. Nevertheless, results based on this equation provide a good first approximation for the compact core of a bounded isothermal sphere, in view of the dominant self-attraction of this core.

Consequently, the core of an isothermal sphere can contract, heat up, and release energy, which flows to the outer regions. The outer regions, being less bound gravitationally, will tend to have a positive specific heat; but if the sphere is sufficiently condensed at the center, the core temperature will increase faster than the temperature of the outer regions, and the temperature gradient will increase, accelerating the core collapse. The rate of the collapse will be limited only by the rate at which heat can flow outward. Analysis shows that if the velocity distribution is nearly isotropic, the gravothermal instability can occur[11,12] if the density at the center exceeds the density at the assumed bounding shell by a factor of 709.

A detailed time-dependent solution for such a collapsing sphere has been found[13] on the assumption that the contraction is homologous, as defined above. The result does not apply exactly to actual clusters, since the mean free path is assumed to be short, and the velocity distribution is consequently nearly isotropic. In fact, stars on radial orbits, which pass frequently through the heated collapsing core, will have a higher kinetic energy than stars in outer circular orbits, which are less immediately affected by the process of collapse. In the short-mean-free-path approximation, Eq. 10 is valid, with $M$ replaced by $M_c$, the mass in the core, which remains essentially isothermal. The value of $\zeta$ is found to be 0.74, giving $v_m$ varying very slowly with the core density $\rho_c$ (as $\rho_c^{0.047}$). The density distribution outside the core differs slightly from that of an isothermal sphere in equilibrium, with $\rho$ varying asymptotically as $r^{-2.21}$ instead of $r^{-2}$. While the inner regions have an inward velocity, the outer regions move outward, with the velocity vanishing at the radius where $\rho(r)/\rho(0) = 0.0071$. As we shall see below, the properties of this theoretical model are in general agreement with those obtained from more detailed, more realistic models.

A third simplified model considers effects associated with stars of two different masses, which tend toward equipartition of energy as a result of mutual encounters. In this model, the system of heavier stars must inevitably collapse if their relative number exceeds a small limiting value.[14] We omit the detailed analysis but derive this result from simplified physical arguments. First we assume that $\rho_2(0)$, the smoothed density of heavy stars at the cluster center, is small compared to $\rho_1(0)$, the corresponding density for the lighter stars. We can then assume that the gravitational potential is entirely produced by the lighter stars. We take this potential, $\phi(r)$, to be zero at $r = 0$. Then in equilibrium, the radial distance attained by stars of each type is determined by the condition that the mean potential energy is proportional to the mean kinetic energy. If $\rho_1(r)$ is constant with $r$, the gravitational potential varies as $r^2$ according to Eq. 1, and we may write

$$\frac{v_{2m}^2}{v_{1m}^2} = k \frac{r_{2m}^2}{r_{1m}^2} \qquad (11)$$

where $v_{2m}^2$ and $v_{1m}^2$ are the mean square velocities for stars of the two types, and $r_{2m}^2$ and $r_{1m}^2$ are the mean square distances from the center. The numerical constant $k$ is needed because in determining $r_{1m}^2$ one cannot neglect the decrease of $\rho_1$ with increasing $r$, and the effect of this change on $\phi(r)$. This effect is negligible for the heavier stars, provided the mass $m_2$ of such a star appreciably exceeds $m_1$.

In equipartition, $v_{2m}^2/v_{1m}^2$ equals $m_1/m_2$. Evidently as $m_2/m_1$ becomes larger, the equilibrium condition (Eq. 11) requires that $r_{2m}^2/r_{1m}^2$ decrease; as the velocities of the heavier stars become smaller, because of equipartition, the radial distance out to which they can rise, against the gravitational attraction of the lighter stars, decreases in proportion.

However, equilibrium becomes impossible if the ratio $\rho_2(0)/\rho_1(0)$ becomes too great, since in this circumstance the self-attraction of the heavier stars becomes appreciable, and the value of $v_{2m}^2$ required for equilibrium consequently increases as $r_{2m}$ decreases. Thus if the total mass $M_2$ of the heavier stars is sufficiently large compared to $M_1$, the total mass of the lighter stars, there is no equilibrium distribution of heavy stars in which $v_{2m}$ is much less than $v_{1m}$. From Eq. 11, plus the assumed equipartition of kinetic energies, we find that

$$\frac{\rho_2(0)}{\rho_1(0)} = \kappa k^{3/2} \frac{M_2}{M_1} \left(\frac{m_2}{m_1}\right)^{3/2} \qquad (12)$$

where $\kappa$ is another numerical constant relating $\rho(0)$ to $M/r_m^3$. Determination of the critical value of $\rho_2(0)/\rho_1(0)$, below which equilibrium is possible, and of the constants $\kappa$ and $k$ shows[14] that $(M_2/M_1)(m_2/m_1)^{3/2}$ must be less than 0.16 for equilibrium. For higher values, the loss of kinetic energy to the lighter stars will lead to continuing contraction of a dense system of the heavier stars, which, as we have seem above, will heat up as they lose energy, another example of the negative specific heat of a self-gravitating system.

In the realistic models described below, the three effects shown here separately all occur together, each contributing to the cluster collapse.

## DETAILED MODELS OF GLOBULAR CLUSTERS

To follow the dynamical evolution of a spherical cluster a number of detailed numerical calculations have been made. While the procedures have varied, all have considered the motion of point-mass stars in the smoothed potential, $\phi_A$, given by Eq. 1, with perturbations of these motions by two-body encounters. We discuss first the analyses of systems that are (i) isolated from other gravitating masses and (ii) composed of stars all of the same mass. While these two assumptions are unrealistic, they simplify the problem and provide a clear indication of the physical processes involved.

Two different approaches have been followed. In the first, the orbits of stars in the smoothed potential field are considered, and the changes in energy, $E$, and angular momentum, $J$, resulting from stellar encounters are considered. This approach has been adopted in Monte Carlo computations,[15, 16] with a number of representative stars followed through time, with frequent small changes in $E$ and $J$ computed in accordance with the appropriate probability distributions. The Fokker-Planck equation, transformed to give the diffusion of stars in $E, J$ space, has also been solved numerically.[17] In the second approach,[8, 18] the motions of 1000 representative stars in the potential field $\phi_A(r)$ are followed by numerical integration; frequent small changes in velocity, produced by two-body stellar encounters, are obtained with the usual Monte Carlo techniques. In both approaches, changes of the smoothed potential with changing density are, of course, taken into account.

The results obtained by these different methods are in close agreement. The various models show that whatever its initial origin, the spherically symmetrical system develops a core-halo structure, with a nearly isothermal central region surrounded by a halo in which the orbits are mostly radial. The resultant structure is shown in Fig. 2, where the computed values of the smoothed density, $\rho$, are plotted against radius, $r$ (both in dimensionless units). This particular system began as a homogeneous sphere, shown by the dotted line, with all stars in circular orbits about the cluster center but with random orientation. Evidently for $r$ less than about 50, the density profile in the evolved system is close to that of an isothermal sphere. At larger $r$, the orbits are more nearly radial, and the density approaches the theoretically anticipated[19] relation $\rho \propto r^{-3.5}$ for an isolated cluster, shown by the dashed line.

Before discussing further the results obtained with these numerical models, we introduce the reference relaxation time $t_{rh}$, which is a convenient measure of a cluster's evolutionary age; the quantity $t_{rh}$ is defined as the value of Eq. 7 when $\rho \equiv mn$ is set equal to the mean

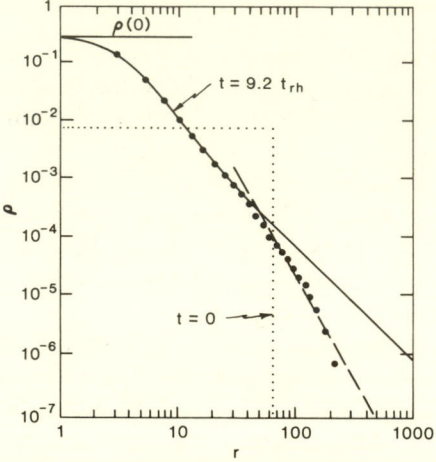

FIG. 2. Structure of an evolving globular cluster. The dotted line shows the initial density, $\rho$, as a function of radius, $r$, at $t = 0$, when the sphere is in equilibrium, with all orbits circular. The plotted points[18] show $\rho(r)$ at $t = 9.2t_{rh}$, where $t_{rh}$ is the reference relaxation time (see text). The solid curve represents the theoretical relationship for an isothermal sphere, and the dashed straight line represents $\rho$ varying as $r^{-3.5}$, the behavior predicted theoretically in the halo, where the orbits are predominantly radial.

density inside the radius $r_h$, containing half the cluster mass, and $v_m$ is set equal to the rms velocity for the entire cluster. During the evolution of the cluster, $t_{rh}$ usually remains relatively constant. If the virial theorem is used to equate $v_m^2$ and $0.4GM/r_h$ (a reasonably accurate approximation for $-W/M$), we obtain

$$t_{rh} = \frac{0.060M^{1/2}r_h^{3/2}}{mG^{1/2}\log(0.4N)}.$$ (13)

If we express $r_h$ in parsecs and $m$ in terms of the solar mass, $M_\odot$, we obtain for $t_{rh}$ in years

$$t_{rh} = \frac{0.90(r_h)^{3/2}N^{1/2}}{(m/M_\odot)^{1/2}\log(0.4N)} \times 10^6.$$ (14)

For most clusters $t_{rh}$ is from $10^8$ to $10^{10}$ years. For comparison, the ratio of the period of a circular orbit at the half radius, $r_h$, to $t_{rh}$ equals $148 \log(0.4N)/N$ and is less than $1/147$ for $N > 10^5$. This small ratio is an example of the general result, referred to earlier, that the mean free path for a star, before encounters strongly modify its velocity, much exceeds the dimensions of a typical cluster.

The system shown in Fig. 2 is rather advanced in its evolution, which has proceeded for a time interval $9.2t_{rh}$ since the origin of the uniform system at $t = 0$. As we see below, the collapse of the core to a central singularity occurs at $12.1t_{rh}$ for this particular model, only $2.9t_{rh}$ after the state shown in Fig. 2.

Another important characteristic of a cluster, in addition to its density profile, is the variation of the phase space density function $f(r, v)$ and, in particular, how this differs from Eq. 2, valid for an isothermal sphere. In the central regions, the velocity distribution is isotropic and $f$ is a function of $E$ only. Values of $f(E)$ near the center of a model cluster,[17] relatively late in its evolution, are plotted in Fig. 3; both $f$ and $E$ are in dimensionless units. Since the relaxation time much exceeds the orbital period of a cluster star, the phase space density is relatively constant along each orbit in the cluster. Hence these values of $f(E)$ refer to all radial orbits, at any distance from the cluster center. This figure shows that, as expected, for appreciable $-E$, encounters between stars establish the exponential form of $f(E)$ in Eq. 2. However, $f(E)$ must clearly vanish for positive energy, since unbound stars escape rapidly. As shown by the upper curve in Fig. 3, the computed values of $f$ can be fitted reasonably well with the "lowered Maxwellian" distribution, given by

$$f(E) = K(e^{-\beta E} - e^{-\beta E_0}) \quad \text{for} \quad E < E_0 \quad (15)$$

and vanishes otherwise. For an isolated cluster the energy $E_0$ at which $f(E)$ is taken to vanish is zero.

Last, we discuss the evolutionary changes shown by these numerically derived globular clusters. Not surprisingly, the numerical models demonstrate the three effects found earlier with simpler models—escape of stars, gravothermal instability, and mass segregation; the relative importance of each in the final collapse is demonstrated by these more realistic models. The observed rate of escape from the cluster may be used in Eq. 6, with the reference relaxation time $t_{rh}$ replacing $t_r$, to determine an effective value of $\xi_e$. The escape of stars from an isolated cluster results from the energy change of a halo star, with an energy only slightly negative, in its passage through the high-density central region where encounters are important. Thus the period of these halo

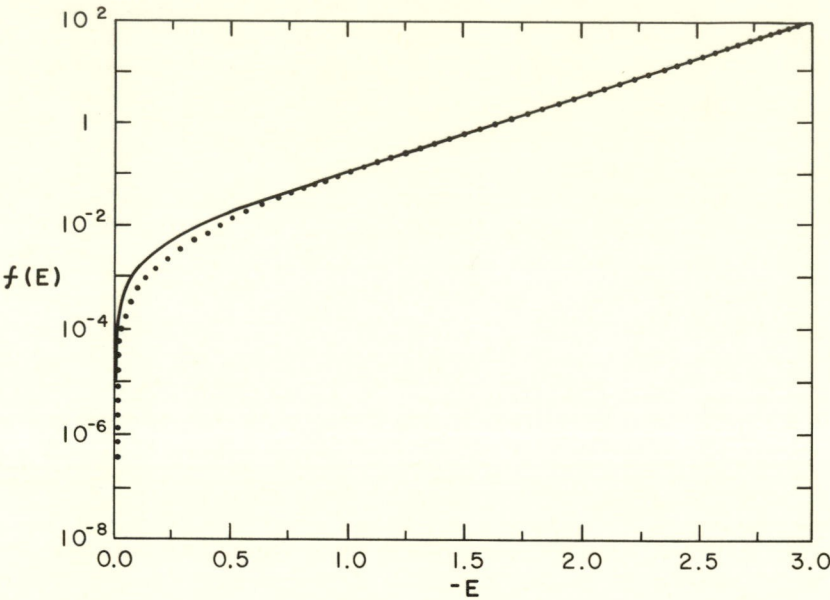

FIG. 3. Phase space density for radial orbits in an evolved cluster. The points represent $f$, the density in phase space for stars at zero distance $R$ from the center of a highly evolved cluster.[17] These values also apply to all radial orbits passing through the central core. The upper curve represents a lowered Maxwellian, given by Eq. 15 with $E_0 = 0$.

stars plays an important part in the escape rate. As a result, models for computing the diffusion of orbits in $E, J$ space which ignore the values of the orbital periods give no escape at all from an isolated cluster. The models integrating the detailed equations of motion yield "observed" values of $\xi_e$ of about $3 \times 10^{-3}$. Although this evaporation process, together with the gradual accumulation of stars in the far halo, is responsible for the initial contraction of the cluster core, as in the simple model, it can apparently not explain the later evolutionary stages of isolated clusters.

The nature of this final evolution is shown in Fig. 4, where the radii containing indicated percentages of the total initial mass are shown plotted against the time.[1] The initial state of the system was the same uniform sphere in equilibrium whose structure is plotted in Fig. 2. The radii are expressed in units or $r_h$, the radius initially containing half the mass, while one unit of the dimensionless time indicated equals about $9.1 t_{rh}$. For comparison, extrapolation of the results indicates that the core approaches a singularity at a "collapse time," $t_{coll}$ equal to $12.1 t_{rh}$.

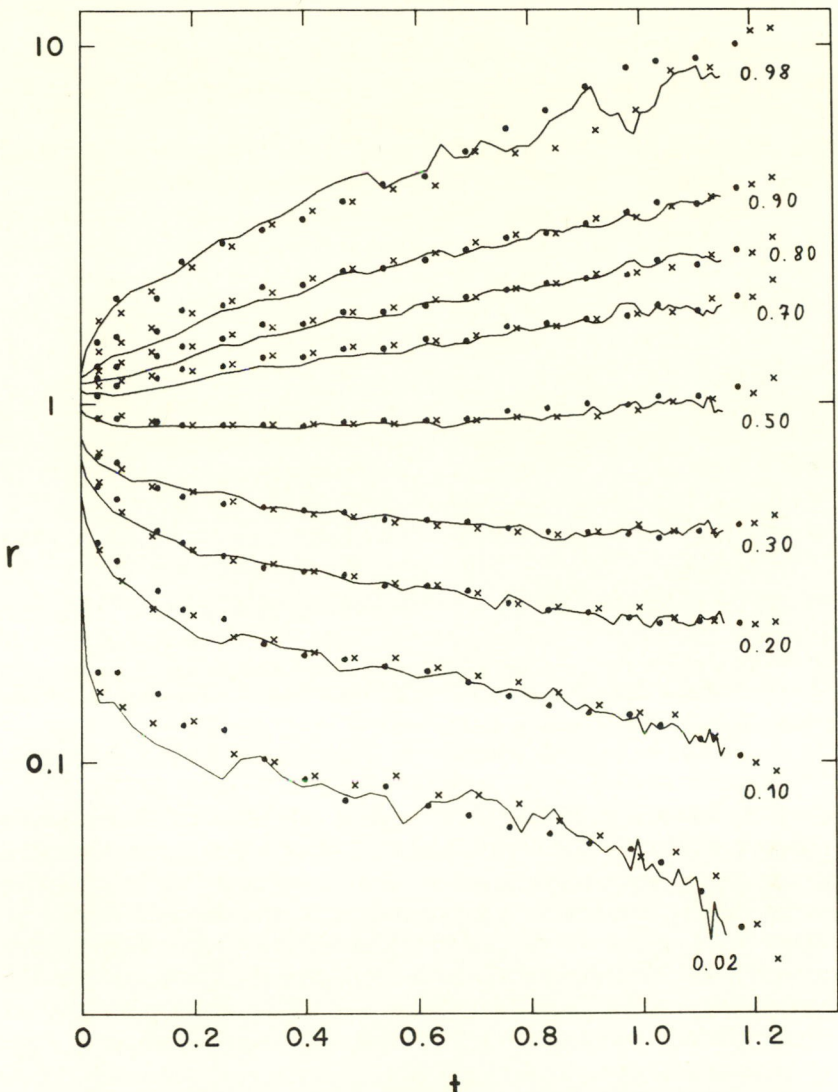

FIG. 4. Time development of an evolving globular cluster. The ordinate shows the values of the radius $r$, containing the fractions of the total mass on the right. The model is the same as the one portrayed in Fig. 2. The solid line represents the results of Monte Carlo computations based on the diffusion of orbits in $E, J$ space, while the dots and crosses represent results obtained from direct integration of stellar orbits, with diffusion in velocity.[1]

The agreement between the solid line (obtained from the diffusion of orbits in $E$ and $J$) and the plotted points (obtained from integration of dynamical trajectories) is excellent. The figure shows that the outer half of the cluster mass expands almost from the very beginning, while the radii containing less than half the mass first contract, then expand; the late expansion for the radii containing 2 and 10 percent of the mass is shown in calculations[16] that extend closer to the final collapse.

The general evolutionary behavior of these models is remarkably close to the behavior predicted for the gravothermal instability in a gaseous sphere. In the contraction of the central core, $r_c^2$ is observed to decrease nearly linearly with $t_{coll} - t$, permitting an accurate determination of time $t_{coll}$ at which the density becomes infinite; in a theoretical model $r_c^{1.9}$ varies linearly with $t_{coll} - t$. Similarly, $v_m$ increases as $\rho_c^{0.05}$, in agreement with a theoretical variation as $\rho_c^{0.047}$. Finally the density at the radius where contraction stops and expansion begins is about two orders of magnitude less than the central density, as compared with the theoretical ratio 0.0071. Exact agreement is not to be expected, in view of the approximations in the instability theory, especially the assumption of a short mean free path and a consequently isotropic velocity distribution. However, there seems little question that the collapse found in the model clusters must be due to the gravothermal instability of an isothermal sphere.

The numerical models also show the expected mass segregation when stars of differing masses are assumed to be present. This segregation occurs relatively rapidly, within a time of 1 to $2t_{rh}$. For example, a two-component model was computed with a stellar mass ratio of 5 to 1, and with 10 percent of the cluster mass in the heavier stars, uniformly distributed initially. After a time interval of only $0.81t_{rh}$, at the cluster center the more massive stars provide 62 percent of the smoothed stellar density, a dramatic increase in the relative densities in the two stellar components. While the initial collapse rate for these models results from the tendency towards equipartition, at a later time this process slows down, since the relative number of lighter stars near the center becomes progressively smaller. Hence it seems likely, though not yet proven, that the final collapse of multicomponent models is due to the gravothermal instability.

Several other effects must be taken into account before a detailed comparison can be made between any of these models and the observations. The most important of these is the gravitational force of the Galaxy. The tidal force produced by the mass in the inner regions of our Galaxy can draw some stars out of the globular clusters if their distance from the cluster center exceeds the "tidal cutoff," $r_t$. While the dynamics of the cluster stars in the presence of such a tidal force form a complex problem, a simple first approximation is that the cluster remains spherical, with vanishing density for $r > r_t$. A theoretical model for such a system can be computed[20] if a lowered Maxwellian distribution function, given in Eq. 15, is assumed, with $E_0$ (per unit stellar mass) set equal to $-GM/r_t$, where $M$ is again the cluster mass. Such "King models" have been widely used for comparison with observed cluster data; with two parameters, $r_c$ and $r_t$, they usually provide a good fit to the observed surface density profiles. The evaporation probability $\xi_e$ per time interval $t_{rh}$ for these tidally truncated models can exceed by more than an order or magnitude the corresponding value for an isolated cluster, with a major effect on the evolution.

Other important physical effects are the gravitational perturbations produced when a cluster crosses the galactic plane; such perturbations heat the cluster, leading to the escape of halo stars and usually (if somewhat paradoxically) a more rapid collapse of the central core.[1] It has even been suggested that most of the high-velocity subdwarf stars in the Galaxy may have been formed in globular clusters that were subsequently dissipated by the increased exporation rate resulting from this process. This scenario requires[21] that most of these early clusters had mean densities one or two orders of magnitude smaller than those observed in present clusters.

Mass loss from individual cluster stars also affects the dynamics of the system, although at the present epoch, when giant stars lose mass only shortly before their death, this effect may be somewhat minor. The presence of a massive black hole at the center would certainly affect the cluster dynamics,[2, 16] although the x-ray evidence discussed below does not support this possibility. Finally, binary star effects may be important.

## FORMATION AND EVOLUTION OF BINARIES

Binaries are potentially very important in cluster evolution because they can give up energy to passing stars and become more and more tightly bound. The energy available is more than enough to slow down or even reverse the core collapse discussed above. If we regard the two stars in a binary system as mass points, with masses $m_A$ and $m_B$, the total energy, denoted by $-x$, can be written as

$$x = Gm_A m_B/2a \qquad (16)$$

where $a$ is the semimajor axis of the binary orbit. The factor $1/2$ in Eq. 16 results from Eq. 4, according to which the average kinetic energy is half the average negative gravitational energy.

When a single star encounters a binary, the net binding energy, $x$, may be changed. The result depends critically on whether the mean stellar kinetic energy of translation is large or small compared to $x$. If the former, the binary is called soft; the velocities of the stars in the binary orbit are less than those of passing stars, and one would expect from equipartition arguments that encounters will impart energy to the binary, decreasing $x$ on the average.[22] If $x$ exceeds the mean stellar kinetic energy, the binary is called hard; the equipartition argument is now less directly applicable, since a passing star is itself accelerated as it approaches the binary, but $x$ will, in fact, increase on the average.[23,24] Thus there is a "watershed" value of $x$, comparable with the mean kinetic energy of single stars; binaries with greater $x$ will become more tightly bound, on the average, giving up energy to the system.

While the energy absorbed by soft binaries is negligibly small, that given up by hard binaries may strongly influence cluster evolution. In any one encounter, a hard binary may change its binding energy by any amount, from zero to infinity, but on the average the rate of change of $x$ is given by

$$\left\langle \frac{dx}{dt} \right\rangle = \frac{4 \times 3^{1/2}A}{35} \times \frac{G^2 m_s^3 n_s}{v_{sm}} \tag{17}$$

where $m_s$ is the stellar mass, here assumed the same for all stars, while $n_s$ and $v_{sm}$ are the particle density and the random rms velocity of the single stars. The constant $A$, found by averaging $10^6$ numerical orbits, is between 30 and 35 for $x$ some one to two orders of magnitude greater than $mv_{sm}^2/2$.[25] While $\langle dx/dt \rangle$ is nearly independent of $x$, the increase of $x$ per close encounter averages $0.4x$. As a result, when a binary gradually hardens, it loses energy less and less frequently but in progressively large increments. When $x$ becomes substantially greater than the energy required to escape from a globular cluster, interaction with a passing star can result in the ejection of both the star and the binary.

Hard binaries, with their important dynamical effects, may appear in globular clusters through three different routes. ($i$) They may be primordial—that is, present in initial stellar population from which the cluster formed. ($ii$) They may be formed by three-body encounters between cluster stars. ($iii$) They may be produced by dissipative two-body collisions between these stars. Detailed computations of formation rates have been carried out for the second[24] and third[26] of these processes. For globular clusters these formation rates are negligible under normal conditions but can become important during core collapse.

Grateful acknowledgements are due to various colleagues, especially J. Goodman, D. C. Heggie, P. Hut, J. P. Ostriker, and S. Tremaine, for helpful comments on this paper, and to D. Malin for the photograph of M19, obtained with his unsharp masking technique to show both the bright core and the faint outer region.

## REFERENCES

1. Spitzer, L. in *Dynamics of Stellar Systems* (IAU Symposium Number 69), A. Hayli, Ed. (Reidel, Dordrecht, Netherlands, 1975), p. 3.
2. Lightman, A. P. and Shapiro, S. L. *Rev. Mod. Phys.* **50**, 437 (1978).
3. Chandrasekhar, S. *Principles of Stellar Dynamics* (Univ. of Chicago Press, 1942).
4. ——, *Astrophys. J.* **97**, 255 (1943).
5. Rosenbluth, M. N., MacDonald, W. M., and Judd, D. L. *Phys. Rev.* **107**, 1 (1957).
6. Ambartsumian, V. A. *Ann. Leningrad State Univ. (Astron. Ser.), No. 22* (1938): *Tr. Astron. Obs.*, Issue 4, p. 19.
7. Spitzer, L. *Mon. Not. R. Astron. Soc.* **100**, 396 (1940).
8. Spitzer, L. and Hart, M. H. *Astrophys. J.* **164**, 399 (1971).
9. Spitzer, L. and Härm, R. Ibid. **127**, 544 (1958).
10. King, I. R. *Astron. J.* **63**, 114 (1958).
11. Antonov, V. A. *Vestn. Leningr. Univ. Mat. Mekh. Astron.* **7**, 135 (1962).
12. Lynden-Bell, D. and Wood, R. *Mon. Not. R. Astron. Soc.* **138**, 495 (1968).
13. Lynden-Bell, D. and Eggleton, P. P. Ibid. **191**, 483 (1980).
14. Spitzer, L. *Astrophys. J. Lett.* **158**, L139 (1969).
15. Hénon, M. *Astrophys. Space Sci.* **13**, 284 (1971): Ibid. **14**, 151 (1971).
16. Marchant, A. B. and Shapiro, S. L. *Astrophys. J.* **239**, 685 (1980).
17. Cohn, H. Ibid. **234**, 1036 (1979).
18. Spitzer, L. and Thuan, T. X. Ibid. **175**, 31 (1972).
19. Spitzer, L. and Shapiro, S. L. Ibid. **173**, 529 (1972).
20. King, I. R. *Astron. J.* **71**, 64 (1966).
21. Ostriker, J. P., Spitzer, L., and Chevalier, R. A. *Astrophys. Lett.* **176**, L51 (1972).
22. Gurevich, L. E. and Levin, B. Yu. *Astron. Zh.* **27**, 273 (1950); *NASA Technical Translation TT F-11, 541* (1968).
23. Heggie, D. C. in *Dynamics of Stellar Systems* (IAU Symposium Number 69), A. Hayli, Ed. (Reidel, Dordrecht, Netherlands, 1975), p. 73: see also the discussions on pp. 93 and 94.
24. ——, *Mon. Not. R. Astron. Soc.* **173**, 729 (1975).
25. Hut, P. *Astrophys. Lett.* **272**, L29 (1983).
26. Press, W. H. and Teukolsky, S. A. *Astrophys. J.* **213**, 183 (1977).

# PART THREE

# Space Science

# ASTRONOMICAL ADVANTAGES OF AN EXTRA-TERRESTRIAL OBSERVATORY

(PROJECT RAND REPORT, DOUGLAS AIRCRAFT CO., 1946;
REPRINTED IN ASTRON. QUARTERLY, 7, 131, 1990)

## Commentary

DURING the final years of World War II, I was in Washington, D.C., applying to undersea warfare the knowledge gained on the physical properties of underwater sound (see Commentary, Paper #29). One day a friend asked me whether I would be interested in consulting for the RAND Corporation, an Air Force "think tank" for which he was working. He told me that RAND was carrying out a design study for an earth-circling Air Force satellite; they wanted someone to analyze whether such a satellite might have important uses in scientific research. Since most of the RAND activities were secret, he had already checked my Navy clearance. Incidentally, this clearance had come through only after a long delay, because my cousin, Anna Louise Strong, a great devotee of communism, had emigrated to Russia!

With my early concentrated reading in science fiction and my later experience in astronomical research, I was much attracted by this opportunity. I went out to the RAND headquarters at Santa Monica, and we discussed their detailed satellite plans. In the following year, while I was finishing up at the Navy and then getting settled back at Yale, I found time to write the report reprinted below.

It is perhaps as well that this enthusiastic discussion was not widely distributed in 1946, since the lack of explicit description of possible research programs, or of likely research results, might have made the report quite unconvincing to most astronomers. In any case, astronomers in general were very skeptical at that time as regards research from space vehicles. The practical difficulties seemed enormous, the possibility of obtaining useful science, vanishingly small. Indeed, the frequent delays and failures of research efforts based on small sounding rockets rather discouraged hopes of success with larger, more complex instruments. Even several years later, when I reported to an astronomer friend our plans for interstellar matter research with a 32-inch ultraviolet telescope-spectrometer on a satellite (later christened *Copernicus*), he replied, "Lyman, you're young, you'll live to see it fail!"

Writing this 1946 paper convinced me that a large space telescope could revolutionize astronomy. Working for such a project became for me a major long-term objective, as evident in Paper #31. Over the years enthusiasm for the concept discussed in this early paper gradually spread among a small group of astronomers. In 1966, when the Space Science Board, National Academy of Sciences, decided to set up a committee for detailed study of this project, eight astronomers enthusiastically agreed to serve in the group. In contrast, major efforts were required to persuade the wider astronomical community that such a large, costly instrument should have very high priority. A chronicle of such efforts, involving an increasing number of astronomers, and leading finally to a 1977 Congressional go-ahead for the program, is presented in Paper #23 and in the commentary which accompanies it.

Now that the Hubble Space Telescope (HST) is orbiting and operating as planned, I have been pleased and a little surprised by the correspondence between some of the leading research programs underway and our earlier expectations, as expressed in Paper #21 and in the somewhat expanded discussion given in our National Academy of Sciences 1969 report.[1] Scientific programs which were emphasized in these reports and which are now being actively pursued with HST observations include: determination of distances to other galaxies by measurements of stars with known absolute magnitudes, such as Cepheid variables; study of galactic structure and evolution by high-resolution images of galaxies at different red shifts; determination of globular-cluster structure through resolution of many more stars than had previously been possible; and study of weather on other planets by means of synoptic high-resolution images. I cannot claim great credit for the success of these long-range predictions. Any scientist familiar with physical principles and with a good background in astronomy could have identified these major research areas in which high optical resolution might be of crucial importance.

The imaging performance predicted[1] for the large space telescope included photometry with hour-long exposures on stars as faint as $V \approx 27$ mag. (Use of image tubes and an overall photon detection efficiency of 10% were assumed.) One-hour exposures on Cepheids[2] with the WF/PC 2 of HST show random errors of about 0.03 mag at $V = 27$, in agreement with this expectation. Detection of stars as faint as $V \approx 29$ mag, as also predicted,[1] appears feasible with longer exposures.

Since the 1977 go-ahead on HST, I have participated in various capacities for this program, serving on technical committees at NASA Headquarters, at several NASA centers, and at the Space Telescope Science Institute. These activities seem to me much less central in comparison with what I regard as my major contribution to HST—that decade after decade, before the official start, I continued actively to push this long-range program. In the summer of 1991, the HST Science Working Group dedicated to me a Special Issue of the Astrophysical Journal Letters;[3] their citation of my HST activities referred primarily to this early work, in the formative years of the project. I deeply appreciated this recognition of my protracted efforts toward the seductive but remote goal of a large space telescope.

## REFERENCES

1. Spitzer, L.—Science **161**, 225 (1968); Scientific Uses of the Large Space Telescope, Ad Hoc Committee on the LST, Space Science Board, Washington: National Academy of Sciences (1969).
2. Freedman, W. L. et al.—*Nature* **371**, 757 (1994).
3. HST Science Working Group—*Astroph. J. Lett.* **377**, i–iv (1991).

*21*

## *Paper*

This study points out, in a very preliminary way, the results that might be expected from astronomical measurements made with a satellite vehicle. The discussion is divided into three parts, corresponding to three different assumptions concerning the amount of instrumentation provided. In the first section it is assumed that no telescope is provided; in the second a 10-inch reflector is assumed; in the third section some of the results obtainable with a large reflecting telescope, many feet in diameter, and revolving about the earth above the terrestrial atmosphere, are briefly sketched.

It should be emphasized that this is only a preliminary survey of the scientific advantages that astronomy might gain from such a development. The many practical problems, which of course require a detailed solution before such a satellite might become possible, are not considered, although some partial mention is made of certain problems of purely astronomical instrumentation. The discussion of the astronomical results is not intended to be complete, and covers only certain salient features. While a more exhaustive analysis would alter some of the details of the present study, it would probably not change the chief conclusion—that such a scientific tool, if practically feasible, could revolutionize astronomical techniques and open up completely new vistas of astronomical research.

### I. Solar Spectroscopy with a Small Ultra-Violet Spectroscope

The simplest astronomical instrumentation for a satellite would be a small spectroscope, analyzing the ultra-violet radiation which it receives from any portion of the sky; in practice, this would be the solar spectrum whenever the sun was visible. Such a spectroscope could analyze either the light incident on a diffuse reflector or the light passing through a small LiF sphere, or bead. Such a system has the advantage that it would not need to be accurately oriented in any particular direction. The intensity in the spectrum could presumably be radioed down to earth. An instrument of this sort would have the following uses:

### Continuous Recording of the Solar Ultra-Violet Spectrum

The scientific and military importance of information on the sun's ultra-violet spectrum has already been pointed out.[1]

Occasional spectra of the sun in the far ultra-violet can presumably be obtained with high altitude rockets which subsequently fall to earth. However, for an adequate picture of the sun's probably large variability in ultra-violet radiation, more frequent measurements may be necessary. For a complete examination of the effect which solar disturbances produce on terrestrial phenomena, especially on conditions in the ionosphere, a relatively continuous portrayal of the sun's output of ultra-violet energy may be required. For example, if a radio fade-out occurs at some particular time, only a record of the solar spectrum during the time immediately preceding can show what the relationship between sun and earth was for that particular fade-out. More important still, for detailed predictions of ionosphere conditions, and thus for practical advance information on radio transmission conditions, daily measurements of the sun's ultra-violet spectrum are believed to be essential. These can probably be obtained most simply by a satellite observatory.

### Detailed Analysis of the Earth's Upper Atmosphere

As seen from the satellite, the sun will rise and set at frequent intervals. On each such occasion, the sun's ultra-violet light will change markedly as the sun's rays shine through atmospheric layers of changing height. By observing changes of the spectrum with time it would be possible to obtain a detailed picture of how the densities of different types of atoms in the earth's upper atmosphere change with changing height. While essentially similar information could be obtained from a rocket which rose out of the earth's atmosphere and then fell back to earth, the observations from a satellite could be obtained much more frequently. In view of the probable variability of the ionosphere, resulting from the variability of the sun's ultra-violet radiation, rather frequent spectrographic observations of the structure of the ionosphere, as well as of the sun's ultra-violet spectrum, are probably required to indicate exactly what is happening. It may well be the case that this information can be

---

[1] "The Importance of High Altitude Spectroscopy," by L. Goldberg and L. Spitzer, Jr., July 15, 1946.

obtained at less cost with such a satellite than with a series of rockets of lower velocity.

## II. Spectroscopy of the Sun and Stars with a 10-Inch Reflecting Telescope

To obtain information about the ultra-violet spectrum of the stars, or to analyze in detail the sun's surface as seen in ultra-violet light, a telescope is required, together with means for orienting the instrument in any desired direction. Orientation might be accomplished in principle by reducing the angular momentum of the satellite to zero by means of external jets; thereafter the satellite could be rotated by internal means to any particular direction, and would point in that direction indefinitely unless hit by a meteorite. Since the telescope would be designed for spectroscopic purposes only, the shape of the mirror would not need to be highly accurate.

A 10-inch reflecting satellite telescope, equipped with one or more spectroscopes, would be a powerful astronomical tool. While it would intercept less light than the large reflecting telescopes on earth, it would have the advantage that the background light from the night sky would be much reduced, provided that the satellite was above the atmospheric layers responsible for this night illumination; 500 miles should be adequate for this purpose. Thus the faintest star which could be reached with such a telescope might be as faint as that which can just be photographed with the 100-inch telescope, provided that photocell techniques can reach the point where they are as effective as the photographic plate. A photon counting technique, with the use of long "exposures" or, more appropriately, "counting intervals," would probably serve this purpose. Such a telescope-spectroscope combination could measure the spectra of stars, planets, etc., down to at least 1000 Å and also out to the infra-red, without the absorption of the earth's atmosphere, which blots out all the ultra-violet and obscures many regions in the infra-red. Listed below are some of the astronomical uses of such an instrument.

It may be noted that practical uses of this instrument would not be immediate; this would be an instrument which might be expected to increase very basically our understanding of what goes on in the stars and in the spaces between them. Since in this way we obtain information on the behavior of matter under conditions not attainable in the laboratory, knowledge of fundamental physics would thereby be enhanced.

## Detailed Information on Solar Meteorology

With a reflecting telescope and accessory equipment, sunspots, promi-nences, and other types of storms on the sun could be examined in ultra-violet light of different wavelengths. In particular, the behavior of the resonance line of hydrogen (Lyman) at 1216 Å would give basic information on the nature of these puzzling and complicated distur-bances, which are related to the variability in the output of ultra-violet radiation from the sun.

## Composition of Planetary Atmospheres

The small amount of $O_2$ and $H_2O$ present in the atmosphere of Mars and Venus cannot be detected spectroscopically because of the absorp-tion produced by these same molecules in our own atmosphere. A spectroscopic satellite telescope could observe the spectra of planetary atmospheres without any such interferences, and could supplement observations in the infra-red with equally useful ultra-violet data.

## Structure of Stellar Atmospheres

Among the most abundant elements in typical stars are hydrogen, helium, carbon, nitrogen, and oxygen. The absorption lines produced by these atoms in their lowest states (called "resonance lines") all lie in the ultra-violet; the absorption lines of these atoms in the visible spectrum all arise from states whose excitation potential is at least 7 volts; since few atoms are so highly excited, the visible absorption lines produced by these atoms are all very weak, except for hydrogen, whose great abun-dance makes up for its high excitation potential. Thus practically no direct evidence is available on the behavior of helium, carbon, nitrogen, or oxygen in most stars. While the resonance lines of helium lie in the far ultra-violet at about 500 Å, those of carbon, nitrogen, and oxygen all lie between 1000 and 2000 Å; the resonance lines of these three elements are unquestionably very strong in the spectra of most stars, and should be readily observable with a satellite spectroscopic tele-scope. Such observations should indicate any differences in composition between different stars—these differences are important in stellar evolution and stellar energy generation. In addition, the nature of unusual stellar atmospheres—expanding, rapidly rotating, etc.—would be more clearly indicated by information on the behavior of such abundant elements as carbon, nitrogen and oxygen as well as by the behavior of the resonance lines of hydrogen.

## Color Temperatures of Hot Stars

For stars hotter than about 15,000°C, the color of the star, as measured in visible radiation, is independent of temperature. Measurements in the ultra-violet would help to determine the surface temperatures of hot stars, a basic item in astrophysical research.

## Bolometric Magnitudes

The determination of the total energy radiated by a star depends on the measurement of the total heat energy reaching the earth from the star; i.e., on the "bolometric magnitude." For stars whose surface temperature is similar to that of the sun, corrections for infra-red and ultra-violet absorption in the earth's atmosphere are not too serious, but for very cool or very hot stars the result depends heavily on the assumed corrections. Bolometric measurements made on a satellite observatory would give bolometric magnitudes directly for stars nearby, unobscured by interstellar dust.

## Analysis of Eclipsing Binaries

Much of our present information about the masses, radii, and structure of stars has been derived from eclipsing binaries. Measurements in the ultra-violet would be a powerful new tool in such research. For example, to determine stellar masses it is necessary to observe the Doppler shifts in the lines produced by each of the two stars, and in this way to measure the velocity of each. When the stars are of unequal luminosity this is difficult. However, the less luminous star is frequently smaller and hotter. In ultra-violet radiation the smaller star will frequently be more luminous, and from a satellite observatory its ultra-violet spectrum could be observed, and its velocity thus determined. Changes in the shape of the light curve during eclipse with changing frequency would also give important information on the structure of the atmosphere and on the nature of the opacity of matter in the stars.

## Absolute Magnitudes and Stellar Distances

If the surface temperature of a star is approximately known from its spectrum, its absolute magnitude can be found if its radius can be estimated. Since the surface gravity and resulting pressure decrease together with increasing radius, a measurement of pressure suffices to give the absolute magnitude, which in turn gives the distance of the star. Observations of visible stellar spectra have given extremely important results along this line by determining the relative numbers of neutral and ionized atoms, which depend on the pressure. Measurements in the ultra-violet would yield data on the presence of highly ionized atoms, not detectable in visible radiation, and would greatly increase the sensitivity of this method for determining stellar brightnesses and distances.

## Composition of Interstellar Gas

Interstellar atoms and molecules are known to be present between the stars, and to have a total aggregate mass about equal to that of the stars. Such particles are all in their ground state; hence observations of stellar spectra in the visible give no information on the presence of many of the atoms and molecules that may be expected. Measurements in the ultra-violet would give information on the density of interstellar hydrogen in space near the sun, and would indicate how much if any of this material was in the form of molecules. Such measurements would also indicate how much carbon, nitrogen, and oxygen was present. Detailed information on the nature of interstellar gas may be important in understanding the origin of stars and of cosmic rays, which may both be produced from interstellar matter.

## Properties of Interstellar Absorbing Grains

In addition to atoms and molecules, small grains of matter, about $10^{-5}$ cm in diameter, absorb starlight in space. This absorption, generally important only for distant stars, is greater for shorter wavelengths. The distribution of these grains is known to be very uneven. Measurement of stellar spectra in the ultra-violet should therefore provide a very sensitive indication of the presence of these obscuring particles; comparison of this absorption with that in the visible region of the spectrum should yield information about the composition of these particles, which is an important item in the evolution of interstellar matter and in related cosmogonic problems.

## Nature of Supernovae

These exploding stars must be the result of some gigantic cataclysm, possibly a chain reaction involving the entire star. The spectrum of the brighter supernovae is a complete puzzle. Measurements in the ultra-violet would be difficult to obtain with a 10-inch reflector, owing to the great distance and resultant faintness of these objects, but if obtainable might yield an important clue to the nature of the processes involved.

### III. Astronomical Research with a Large Reflecting Telescope

The ultimate objective in the instrumentation of an astronomical satellite would be the provision of a large reflecting telescope, equipped with the various measuring devices necessary for different phases of astronomical research. Telescopes on earth have already reached the limit imposed by the irregular fluctuations in atmospheric refraction, giving rise to "bad seeing." It is doubtful whether a telescope larger than 200 inches would offer any appreciable advantage over the 200 inch instrument. Moreover, problems of flexure become very serious in mounting so large an instrument. Both of these limitations disappear in a satellite observatory, and the only limitations on size seem to be the practical ones associated with sending the equipment aloft.

While a large reflecting satellite telescope (possibly 200 to 600 inches in diameter) is some years in the future, it is of interest to explore the possibilities of such an instrument. It would in the first place always have the same resolving power, undisturbed by the terrestrial atmosphere. If the figuring of the mirror could be sufficiently accurate, its resolving power would be enormous, and would make it possible to separate two objects only .01″ of arc apart (for a mirror 450 inches in diameter); an object on Mars a mile in radius could be clearly recorded at closest opposition while on the moon an object 50 feet across could be detected with visible radiation. This is at least ten times better than the typical performance of the best terrestrial telescopes. Moreover, in ultra-violet light the theoretical resolving power would of course be considerably greater; ideally an object 10 feet across could be distinguished on the moon with light of 100 Å wavelength. In addition, with such a large light-gathering surface and such low background light, the positions and spectra of stars and galaxies could be analyzed out to much greater distances than is now possible. If the shape of mirror could not be figured so accurately without excessive effort, a large spectroscopic satellite telescope would still have many important uses.

The practical problems of operating such a large installation would of course be enormous. Telemetering back to earth the two-dimensional picture obtainable with such an instrument would involve many problems. With such high angular resolutions, some guiding of the telescope might be necessary to correct for changes in the aberration of light during the satellite's orbit. Absorption and radiation of the light received from both sun and earth would require careful consideration to ensure a constant temperature in the mirror and its mounting (to reduce distortion of the mirror's shape by thermal expansion and contraction) and to give a very low temperature in the photo-electric measuring equipment (to reduce the background of thermal emission from the photo-sensitive surface). To provide for a leisurely orbit and thus for relatively constant conditions, such an observatory should preferably be some distance away from the earth, probably as far as telemetering techniques and celestial mechanics might allow.

Most astronomical problems could be investigated more rapidly and effectively with such a hypothetical instrument than with present equipment. However, there are many problems which could be investigated only with such a large telescope of very high resolving power. A few of these problems are given in the following partial and tentative list. It should be emphasized, however, that the chief contribution of such a radically new and more powerful instrument would be, not to supplement our present ideas of the universe we live in, but rather to uncover new phenomena not yet imagined, and perhaps to modify profoundly our basic concepts of space and time.

### Extent of the Universe

The 200-inch telescope is designed to push back the frontiers of explored space. It is not likely that this instrument will reach to the greatest distance possible. Further measurements with the more powerful instrument envisaged here would help answer the questions whether space is curved, whether the Universe is finite or infinite. This instrument would help in particular to resolve individual stars in a distant galaxy and to analyze their spectra, thus identifying particular stars of known absolute magnitude and in this way determining accurately the distance to the galaxy. At present the distances of most galaxies are known only very approximately.

### Structure of Galaxies

With such great resolving power, such an instrument could explore the details of the structure of galaxies, individual stars could be resolved, and the nature of the as yet enigmatic spiral arms could be investigated. Measurement of radial velocities by spectral analysis would yield velocities of rotation in a number of galaxies and thus provide direct information about their masses—information now available for only a few galaxies.

### Structure of Globular Clusters

These objects contain so many stars that resolution of individual stars has been possible only for the brighter members. With such great resolving power a much greater percentage of the individual stars could be resolved, some spectra and radial velocities obtained, and a serious attempt made to explore the structure of these stellar aggregations.

## Nature of Other Planets

The controversy as to the presence of intelligent life on Mars could perhaps be settled by measurements with such a giant telescope. Similarly the type of surface detail present on the other planets could be accurately explored with such high resolving power and invariably "perfect seeing."

# INTERPLANETARY TRAVEL BETWEEN
# SATELLITE ORBITS*

### (JOUR. AMERICAN ROCKET SOC., 22, 92, 1952)

## *Commentary*

FOR THE NEXT ten years following Paper #21 nothing much was happening on the large space telescope front (see Paper #23). However, active research with sounding rockets was alive and flourishing. While I was following this latter program with interest and even taking a small part in it, I found myself imagining what exciting possibilities nuclear fission energy could provide for long-range space travel within the solar system; some rather brief studies led to Paper #22.

As pointed out in note 3 below, I discovered only after this paper was complete that some of the concepts presented had already been discussed by other space enthusiasts. In particular, the acceleration of positive ions electrically to achieve high rocket exhaust velocities for long-range space travel had been analyzed by H. Oberth in his 1923 book[1] (where he considered space telescopes also). Others have also discussed[2] the possibility of ion rockets. Apparently the one really novel item in my paper was the 100-km distance between the nuclear reactor and the manned control car, with wires holding the two together! NASA has since experimented with such "tethered" payloads, fastened by wires (much shorter than 100 km) to the Space Shuttle. In addition, my emphasis on using nuclear power for travel between circular orbits around different planets was perhaps both new and important.

I have included this paper because the topic still fascinates me, and because I was pleased to find that all the physical constraints could apparently be satisfied in a configuration which seemed practicable. More recent studies[3] have considered that hydromagnetic acceleration of a plasma might be preferable to electrostatic acceleration. For the more ambitious goal of travelling outside the Solar system, a fusion reactor has been proposed[4] in a large space ship, with a mass of order 5,000 tons.

---

*A paper read before the Second International Congress on Astronautics in London, September 3–8, 1951.

In view of the wide experience and impressive successes gained with chemical rockets, these are likely to remain for some years as the primary thrusters in most space programs. In the long run, however, the basic advantage of high-speed plasma beams may give them some role, at least as alternative thrusters for certain space applications where high rocket exhaust velocities would be important.

REFERENCES

1. Oberth, H.—English translation: Man into Space, New York: Harper and Row (1957).
2. Heller, G. B.—In: Space Age Astronomy, eds. A. J. Deutsch and W. B. Klemperer, New York: Academic Press 288 (1962).
3. Jahn, R. G.—Physics of Electric Propulsion, New York: McGraw-Hill (1968).
4. Teller, E. et al.—Fusion Technology 22, 82 (1992).

# Paper

## ABSTRACT

An analysis is given of the performance to be expected of a rocket powered by nuclear energy, and utilizing an electrically accelerated ion beam to achieve a gas ejection velocity of 100 km/sec without the use of very high temperatures in the propellant gases. While such a rocket would have much too low a thrust to take off from the surface of a planet, it would appear to be capable of traveling from a circular orbit about the Earth to a circular orbit about any other planet in the solar system. Gases obtained from planetary atmospheres could be used for the propellant, and the only refueling required from the Earth would be supplies for the crew and small amounts of fissionable material. Preliminary indications are that such a rocket could feasibly be constructed and operated at the present time.

## INTRODUCTION

During the past decade much thought has been given to the problem of sending a rocket up to a circular orbit some three hundred miles above the earth, and thus creating an artificial satellite. It appears that this problem can be solved by present techniques. With the use of chemical fuels a multistage rocket designed to launch a man-made satellite now seems perfectly possible in principle, although much engineering development would be needed to realize this possibility in practice. The ratio of take-off weight to the weight of the satellite would be large, probably several hundred or so, but not prohibitive.

The next step beyond an artificial satellite has seemed much more difficult. To travel from a close circular orbit around the Earth to the Moon and back or to another planet and back requires almost prohibitive amounts of fuel with conventional rockets. In theory, many Earth-launched rockets could be used, each one carrying fuel to one of several refueling stations for an eventual interplanetary trip. However, the many analyses made of this technique indicate that a discouragingly large number of very large rockets would be required to make possible even one trip to the Moon and back. For a planetary trip the difficulties are even greater.

While the possible use of nuclear energy may revolutionize our concepts of space travel, the use of nuclear power simply to heat up the propellant gases from a rocket would not make interplanetary travel particularly feasible. It is well known that one of the chief limitations on a conventional rocket is the temperature which the rocket tubes can tolerate without melting or evaporating. A nuclear-powered rocket using heated propellant gases could not use a higher gas temperature than a chemically powered rocket, and its only advantage would lie in the use of lighter propellant gases, with higher velocities at a particular temperature. A nuclear-powered rocket making use of molecular hydrogen as propellant would need somewhat less propellant than would a conventional rocket using the best chemical fuels, but the gain would fall far short of the orders of magnitudes of improvement needed to make interplanetary travel feasible. Moreover, the rapid transfer of enormous quantities of heat from a uranium pile to a propellant gas offers difficulties that might well prove insuperable. We may conclude, as others have already done, that the use of nuclear power in an otherwise conventional rocket does not seem to provide an ideal solution to the problem of interplanetary flight, though it might be of use in the trip from the Earth's surface to a close satellite.

This entire picture changes if we shift our point of view and consider space ships which need not land on planets but instead travel from one circular orbit to another. To take off from the Earth and reach a satellite orbit requires not only an enormous amount of energy, but also requires that the energy be delivered in a very short time. A rocket designed to take off from the Earth's surface must obviously have a thrust greater than the rocket's weight, and this requires that the take-off period be short; i.e., the power, or energy per unit time, must be very great. A conventional chemical rocket is the only practical means yet found to produce such enormous powers without excessive weight.

For an interplanetary space ship, however, a large thrust is not required. Such a ship, traveling between circular orbits about different planets, can accelerate relatively slowly, and while the total amount of energy required is still large, the power needed may be reduced by about 1/100, and instead of the 10,000 hp used in a large chemical rocket, a few thousand hp is sufficient. The present paper describes a low-thrust, low-power ship of this type, designed to travel from one planet to another, without ever landing. The several thousand hp needed is generated in a uranium or plutonium pile, converted into electrical energy, and used to accelerate a stream of ions by purely electrical means to a speed of about 100 km per sec. Different components of this ship are discussed in the following sections, with the necessary mathematical analysis given in the appendixes.

The ideas outlined here are probably not new; possibly the extensive literature on space flight contains some reference to a system similar to that proposed in the present paper. No search of the literature has been made.[1] The chief purpose of this paper is not to claim priority for any ideas but to focus attention on what promises to be the most practical means for interplanetary flight in the near future.

## POWER SUPPLY

The power-weight ratio needed to give a ship an acceleration depends only on the acceleration itself multiplied by the velocity of the ejected propellant gas. As a reasonable acceleration we may take $0.3$ cm/sec$^2$, which would make it possible to escape from a close circular orbit about the Earth in a few weeks and to reach Mars a few months later. The velocity of the ejected gas may be taken as 100 km/sec. A higher value would require more power, while a lower value would yield a less favorable mass ratio. With this choice of acceleration and gas velocity, the power needed is about 1/10 hp per lb, as shown in Appendix 1. We shall consider a ship whose gross weight is about 10 tons, since this is probably the smallest-size ship that could carry a uranium or plutonium pile. The total useful power must therefore be roughly 2000 hp. If an overall efficiency of one-third is assumed in the conversion of heat into useful power, a heat source of 6000 hp must be provided.

How can this power be produced? A normal pile weighs many hundreds of tons and could presumably generate much larger powers. For an interplanetary spaceship one may envisage a pile constructed with U 235, the lighter isotope of uranium, or with plutonium, Pu 239. These materials are the ingredients of an atomic bomb, and a pile made with these fissionable materials could have a much smaller size and weight than one using ordinary uranium. The use of atomic fission for power production is still in the development stage, but one may assume that within the next ten years the design and operation of a small pile, weighting about a ton and producing some 6000 hp of heat, will become entirely practical. Such a pile would consume about 2 kg of U 235 or Pu 239 in a year of steady operation, and nuclear fuel for 50 years of continuous operation would weigh only 100 kg, or a tenth of a ton!

---

[1] The author is much indebted to Dr. L. R. Shepherd for calling attention to some earlier work along the present lines. Oberth in his book, "Wege zur Raumschiffahrt," proposes a rocket propelled by an ion beam, powered by solar energy. The possible performance of an ionic rocket powered by uranium or plutonium has been considered by Shepherd and Cleaver (*J. Brit Interplanetary Soc.*, vol. 8, p. 59, 1949); their pessimistic conclusions regarding such a device result from their assumed acceleration of 0.01 gravity, some 30 times the acceleration assumed in this paper.

What about the dangerous effects of so much radio-activity on the crew aboard this ship? The usual pile is surrounded by concrete and lead, to shield personnel from the neutrons and gamma rays. Such shielding is very heavy and would be practical only in an interplanetary ship whose size and weight were already very large. Thus, one could design and probably build a space ship weighting 10,000 tons, powered with a uranium pile weighing perhaps 1000 tons, and generating 2,000,000 hp of useful energy. Such a ship could carry thousands of people and vast supplies anywhere in the solar system, and could even navigate to other stars, though many generations would be born, grow up, and die on shipboard before such a journey were complete. However, launching such a ship from the Earth's surface to a close circular orbit would be a tremendous undertaking. With the use of chemical fuels, such a launching would require a rocket of some million tons gross weight, an achievement that would seem far, far in the future.

Returning, then, to our smaller interplanetary ship, we still have the problem of how the personnel are to escape the hazards of nuclear radiation. The easiest way to achieve safety from neutrons and gamma rays is simply by distance. One may envisage the main interplanetary ship as a pilotless engine room, with a light control car, containing the crew, attached by wires some 100 km in length. Power and communications could be sent over the connecting wires, with probably also radio communication and an auxiliary power supply in the control car. Appendix 2 shows that at 100 km from a pile generating 6000 hp of nuclear energy, the flux of neutrons and gamma rays would be down to a safe value. If a shorter towing distance were desired, the pile might be designed as a long thin cylinder, with the axis pointed at the control car, and some shielding provided at the end of the cylinder nearest the control car.

If the car containing the crew and control equipment weighed 2 tons, the wires pulling the car with an acceleration of 0.3 cm per sec would be subject to a force of 1.2 lb. With a safely factor of 100, the two connecting wires need each have a diameter of only 0.02 in. and would each weigh about 300 lb, not an excessive contribution to the mass of the ship. Careful precautions would be needed to maintain tension of the wires at all times to avert sudden acceleration and breaking of the wires. There seems no reason why this technique could not be used, although so great a distance between passengers and power plant may seem somewhat unconventional. If repairs to the main ship became necessary during flight, the power could be turned off and the control car could be drifted up to the space ship.

## GENERATION OF ELECTRICAL POWER

A uranium or plutonium pile generates power in the form of heat, and this must be converted into electrical energy. One may visualize a small pile, with heavy water used as a moderator, in which the heavy water is heated up, turned to steam, and used to drive a steam turbine. This turbine then drives a d-c generator, whose output is used to accelerate the propellant gases, as discussed in the next section.

Since no problems of shielding of personnel are involved in the space ship, the design of such a system is somewhat simplified, but many problems of detail remain. The neutrons will produce nuclear transformations all through the ship, with possibly unfavorable results on the operation of the electrical and other equipment. The gamma rays will produce ionization in all materials, again creating problems of equipment performance and maintenance. Considerable research on such problems is presumably underway in connection with present piles, and one may hope that solutions to these problems will be available shortly. There appears to be no reason in principle why such problems cannot be solved.

If the radiation hazard to the ship itself can be overcome, two other problems remain. First, there is the matter of cooling. To operate a heat engine, the heat produced at a high temperature must be given up in a condenser at a low temperature, and the only way for the condenser in a spaceship to lose this heat is by radiation. The rate at which a solid body radiates heat varies as the fourth power of the temperature. Thus we find ourselves in something of a dilemma. If the condenser temperature is made low, about room temperature, the rate of radiation is so low that impractically large radiating surfaces are required. If the condenser temperature is made very high, then the temperature at which the heat is generated must be even higher, and materials tend to melt or lose their strength, especially when submitted to nuclear bombardment. This subject is analyzed in Appendix 3, where the temperatures of heat input and heat output are taken to be 900°K and 450°K, respectively, or 1160°F and 350°F. With these temperatures a heat engine has an ideal efficiency of one-half, and at a pressure of about 10 atm the steam will liquefy in the condenser. Even at this relatively high temperature, however, the radiating surface required is about a thousand square yards. One may visualize a thin fin some 30 yd square extending on one side of the ship, radiating the heat produced by the pile.

Secondly, there is the problem of the weight of all the material required to produce some 2000 hp of useful electric power. An overall weight-power ratio 10-lb per hp is in the same general neighborhood as the weight-power ratio for Diesel-electric railroad locomotives and for bombing planes. It is not unreasonable to assume that about this same value can be achieved for an interplanetary space ship.

## PROPULSION OF THE SHIP

It now remains to convert the electrical power, whose generation was discussed in the preceding section, into useful work. In particular, the only way a spaceship can be propelled is by ejection of a stream of gases, and the electrical power must be used for this purpose. By use of electrostatic forces to accelerate a beam of ions, very high gas velocities can be achieved without the use of very high temperatures. The production of intense ion currents has been extensively studied in the past decade, and the acceleration of a spaceship by an ion beam seems to offer no particular difficulties.

The electrical voltage to be applied depends only on the mass of the ion to be used and the velocity desired. The equations, given in Appendix 4, show that to accelerate nitrogen ions to a speed of 100 km/sec requires a potential of 730 volts. Nitrogen is indicated as a propellant, since this gas is readily obtained from the Earth's atmosphere, and the ship can therefore obtain propellant gases in its circular orbit; this avoids the necessity of bringing tons of propellant up from the Earth's surface for every trip. Since nitrogen is probably also abundant in the atmospheres of Mars and Venus and possibly other planets, propellant gases could also be obtained at various points in the solar system.

At the relatively low voltage of 740 volts required to accelerate nitrogen ions, space-charge effects limit the total ion current that can be accelerated. Appendix 4 shows that an accelerating area of about 7 sq yd would be needed to produce an ion beam of the necessary 2000 amp and 1500 kw. This result assumes that the accelerating voltage is applied over a distance of only a millimeter. Two fine-mesh wire screens, made with wires of very small diameter, could be placed this far apart and given the requisite potential difference. Thermionic emission from the wires in the outer screen could add electrons to the beam so that the ejected gases and the ship would remain electrically neutral.

It may be remarked that if the accelerating voltage were increased to 100,000 volts, the ion velocities would be about 1000 km per sec, and a spaceship could in theory attain a speed of this order after about a hundred years of acceleration. At such a speed about a thousand years would be required to reach the nearest star.

## LAUNCHING AND USE OF THE SHIP

The preliminary analysis presented here indicates that there is every reason to believe that an interplanetary spaceship could be built with essentially present techniques. Such a ship could not by itself land or take off on any major planet or satellite, although it could readily land on a small asteroid or on one of the tiny moons of Mars, where the small weight of the ship could easily be balanced by the rocket's thrust. However, the ship could proceed from a circular orbit about the Earth to a similar orbit about any other body in the solar system.

How could such a ship be used? Since we are all situated on the Earth's surface, a ship of this type could be employed only if some means for traveling up to a circular orbit were available. We have already indicated that with the use of chemical rockets this problem is, in principle, solved. One may envisage short-range chemical rockets, with a starting weight of at least several hundred tons, sending a final weight of about a ton up to a circular orbit, where contact would be established with an interplanetary ship.

Probably the most difficult problem of this sort would be the initial launching of the interplanetary ship itself. While a voyage of many hundred million miles in space could readily be achieved by this ship, ascent of the first few hundred miles to a circular orbit would definitely require a booster of some sort. In this case the booster would weigh initially some hundred times as much as the ship, or a few thousand tons. This is definitely a case where the first few hundred miles are the hardest, and in fact the design and construction of such a large "launching rocket" might well be more difficult than the design and construction of the long-range spaceship.

Another problem, almost as difficult as that of the take-off, is that of landing. Air resistance seems the obvious way to slow down a ship returning to the Earth's surface. However, a body in a circular orbit has a vast amount of energy, and disposing of all this energy by air resistance without vaporizing the material is no trivial problem. There are a number of ways in which glided flight down from a circular satellite might seem to be feasible. Since this subject is dealt with in another paper in this symposium it will not be discussed here.

One possibility in this connection should be noted, however. Gliders constructed for descending from a satellite will probably require wings and bodies of appreciable weight. It might be cheaper in the long run to fabricate these in space, from the nickle-iron presumably available on asteroids, rather than haul them up to the circular orbit by chemical rockets. An interplanetary spaceship could readily land on a small asteroid and might conceivably carry the tools necessary to fabricate out of the iron in the asteroid at least the wings needed; these might then be attached to the rockets shot up from the surface.

Landing on another planet would be more of a problem, since a chemical rocket with large quantities of fuel would presumably be needed for the return trip. Possibly such a rocket could be hauled up from the Earth's surface to the circular orbit without its fuel, and the gaseous fuel collected by long-range spaceships from the atmosphere of various planets—possibly oxygen from the Earth and hydrogen or methane from Titan, the satellite of Saturn. All this material could then be glided down to the surface of Mars. After exploration of the planet, the chemical rocket could be assembled and ascent to a circular orbit made.

Evidently the nuclear-powered electrical rocket could open up many possibilities for interplanetary travel. Only the future can reveal to what extent these possibilities may be realized.

## APPENDIXES

### Power–Weight Ratio

We consider the power consumed in an ideal rocket in which the atoms in the rocket jet have no random motion. Let $N$ atoms per sec, each of mass $m$, be propelled away from the rocket at a speed $v_a$. Then the total backward momentum imparted to the atoms per sec will be $Nmv_a$; if $M$ is the mass of the ship, and $a$ its acceleration, then

$$Ma = Nmv_a. \tag{1}$$

The energy imparted to each atoms is $mv_a^2/2$. Hence the power $P$, in watts, is given by

$$P = \tfrac{1}{2}Nmv_a^2 \cdot 10^{-7}. \tag{2}$$

Combining Equations (1) and (2) we have

$$P = \tfrac{1}{2}Mav_a \cdot 10^{-7}. \tag{3}$$

Equation (3) gives the power–weight ratio for a given acceleration and atomic velocity $v_a$. For an acceleration of 0.3 cm/sec$^2$, sufficient to attain a velocity of 15 km/sec in two months, and for a value of $v_a$ equal to 100 km/sec, sufficient for a total velocity change of 30 km/sec with a mass ratio of about 1.3, we find

$$\frac{P}{M} = 0.15 \text{ watt/gram} \tag{4}$$

corresponding to a weight–power ratio of 11 lb per hp. Thus, for a ship with a mass of 10 tons, the power capacity must be roughly 2000 hp, or about 1500 kw.

### Radiation Shielding

We compute the distance at which personnel will be safe from the neutrons produced in a uranium or plutonium pile, generating 1,500,000 watts. With an ideal efficiency of one-half and an actual efficiency of perhaps one-third, the total heat generated will be three times the electrical power produced. Let $N_n$ be the number of neutrons liberated per sec. Since one neutron per fission must go to sustain the reaction,

$$N_n = \frac{3P(y - 1)}{10^{-7}E_f} \tag{5}$$

where $y$ is the number of neutrons per fission, and $E_f$ is the energy released per fission, in ergs. While precise values of $y$ are not apparently available, this quantity is somewhat greater than unity. The energy $E_f$ is about $3 \cdot 10^{-4}$. If none of the excess neutrons liberated were absorbed, the flux of neutrons $F_n$ per sq cm at a distance $r$ centimeters would be given by

$$F_n = \frac{N_n}{4\pi r^2}. \tag{6}$$

If we combine these equations we find that at a distance of 100 km from a reactor with an electrical power of $1.5 \times 10^6$ watts,

$$F_n = 1.3 \cdot 10^2(y - 1) \text{ neutrons/cm}^2\text{sec}. \tag{7}$$

If $y - 1$ is about one, this value of $F_n$ is about the safe dosage for continuous exposure to fast neutrons.[2] For comparison, a lethal dose within a short period would be $5 \cdot 10^{11}$ neutrons per sq cm. At the rate obtained from Equation (7) about a hundred years would be required

---

[2] The author is indebted for this figure to L. R. Shepherd, who kindly pointed out an inaccuracy in an earlier version of this paper.

for a dose of this magnitude. It is well known that over a long period of time the human body can stand without any injury many small successive doses of radiation, which would be fatal if received all at once. Consideration of additional neutron absorption within the pile would presumably reduce the neutron flux even further.

The computation of the gamma-ray flux is somewhat more complicated. The lethal dosage of gamma rays corresponds to about the same flux of photons per sq cm as the lethal dose of neutrons, in neutrons per sq cm. Since one would expect the number of gamma rays and neutrons produced in fission to be roughly equal, one may conclude that at 100 km the gamma-ray flux will also be below the safe limit. A more detailed analysis of gamma rays would be necessary for conclusive results, however, especially since the absorption of gamma rays within the pile will probably be less than the absorption of neutrons.

### Problems of Cooling

The generation of electrical power by means of a heat engine requires that the heat produced at a temperature $T_1$ be conveyed to a "sink" at a temperature $T_2$. The ideal efficiency is $(T_1 - T_2)/T_1$. In interplanetary space the only way heat can be dissipated is by radiation, and the radiating surface $A_r$ must be sufficient to radiate at the temperature $T_2$ the heat flowing to it. If $Q$ represents the heat radiated per sec, measured here in watts,

$$\frac{P}{Q} = z\left(\frac{T_1 - T_2}{T_1}\right) \tag{8}$$

where $z$ is the ratio of the actual efficiency of power generation to the ideal efficiency. The heat radiated per sec from the surface, if we assume this is a perfect absorber, is given by

$$Q = \sigma A_r T_2^4, \tag{9}$$

where

$$\sigma = 5.7 \times 10^{-12} \text{ watts/cm}^2 \text{ deg}^4 \text{ sec.} \tag{10}$$

Combining Equations (8) and (9) we obtain for $A_r$

$$A_r = \frac{PT_1}{\sigma z(T_1 - T_2)T_2^4}. \tag{11}$$

We shall assume here that $T_1$ is 900 K (627 C) and $T_2$ is 450 K (177 C). If we let $z = 2/3$, the overall efficiency of the heat engine $= 1/3$ and, if $P$ if 1,500,000 watts, we find, for the assumed 10-ton ship,

$$A_r = 1.9 \times 10^7 \text{ cm}^2. \tag{12}$$

This would yield a cooling area roughly 30 meters square, radiating on both sides.

### Electrical Acceleration of the Propellant

We compute the electrical current required to accelerate a beam of ions for rocket propulsion. The current is related to the power $P$, in watts, and the potential $V$ in volts by the relationship

$$P = iV. \tag{13}$$

The potential $V$ across which the ions are accelerated is determined from the energy equation

$$\frac{eV}{300} = \tfrac{1}{2}mv_a^2 \tag{14}$$

where $e$ is the charge on each ion, equal to $4.8 \times 10^{-10}$ electrostatic units if each ion is singly charged, and $m$ is the mass per ion. For nitrogen atoms, $m$ is $2.32 \times 10^{-23}$ gm, and if we again let $v_a$ equal $10^7$ cm/sec, we find

$$V = 730 \text{ volts.} \tag{15}$$

Thus, for a power of 1,500,000 watts, we have

$$i = 2000 \text{ amp.} \tag{16}$$

With so large a current, space-charge limitations must be considered. If the potential drop $V$ occurs across a gap $x$, the maximum current density, $j$ amp per cm$^2$, is given by

$$j = \frac{2^{1/2}}{9\pi x^2 \times 3.10^9} \left(\frac{e}{m}\right)^{1/2} \left(\frac{V}{300}\right)^{3/2}. \tag{17}$$

The area $A_i$ of the ion source is evidently given by

$$A_i = \frac{i}{j}. \tag{18}$$

If we substitute $i$ from Equation (13), for $j$ from Equation (17) and for $V$ from Equation (14), we find

$$A_i = \frac{36\pi x^2 e^2 P 10^7}{m^2 v_a^5}.$$  (19)

For a power of $1.5 \times 10^6$ watts, a velocity $v_a$ of $10^7$ cm/sec, an accelerating gap of 0.1 cm, and for nitrogen ions as the propellant, we find

$$A_i = 7.2 \times 10^4 \text{ cm}^2.$$  (20)

This area of about 7 square meters required for the ion source is much smaller than the area required for radiative cooling.

# HISTORY OF THE LARGE
# SPACE TELESCOPE

(In: Large Space Telescope—a New Tool for Science,
12th Aerospace Sciences Meeting, New York: AIAA, 3, 1974)

## *Commentary*

The American Institute of Aeronautics and Astronautics (AIAA) invited me to attend an Aerospace Sciences meeting devoted to the "Large Space Telescope" (LST). It occurred to me that this would be a good opportunity to summarize the history of this project to date—how the concept arose and developed, and how support for the program gradually grew among astronomers. So I accepted the invitation with enthusiasm. I wrote this paper in detail, which was fortunate, since a slight illness prevented me from attending the meeting. Instead, my colleague, Bob Danielson, read the paper. Bob's contribution to plans for imagery with HST were of central importance, and his death two years later was a great loss to the program.

As I read this Paper now, one major criticism comes to me. In my review of the various stepping stones en route to the LST, I may have placed undue emphasis on activities by the Princeton group. It is not surprising that I would naturally think in terms of work with which I was already familiar. If I were now rewriting this paper, I would, for example, put more emphasis on space observations of the Sun, including both ultraviolet light and X rays. By 1962 this field of solar research had been developing for some years, and had much increased our knowledge of the Sun.[1] The experience gained in operating complex instrumentation in space and the important scientific results obtained must have contributed to growing confidence among astronomers generally as to the practical possibility of LST.

This Commentary is chiefly devoted to brief summaries of two historical developments in the history of LST after 1974, when Paper #23 was written. A full account[2,3] of the LST project has covered most of this later period in considerable detail. My comments here concentrate on those events in which I was personally involved, and they represent a somewhat different point of view.

The first topic is the three-year campaign[4] to win full Congressional approval of the LST. This was a logical consequence, in a way, of the strong endorsement achieved for the LST among astronomers by 1974. Nevertheless, official approval took much continuous pressure and several major campaigns to convince Congress of the strong, widespread support by most of the astronomical community. Of course, by the standards of 1994, when government support of expensive scientific programs is more difficult to obtain, three years from first request to final approval seems a short time!

In 1974, NASA requested that its 1975 budget include $6,200,000 for detailed planning of LST. The House deleted all these funds, but the Senate was more sympathetic. Astronomers seemed willing, even anxious, to express their support for this enterprise. John Bahcall and I appointed ourselves the codirectors of a campaign to "let Congress know your views on the LST." Friends were urged to persuade their friends, and their friends' friends too, to write or telephone their Representatives or Senators, especially those on the two relevant Congressional committees. George Field and others joined us in this effort. To explain what the telescope was for and what sort of results it might achieve, both John and I went back and forth to Washington to talk with members of Congress and their staffs. They told us that the outpouring of support they had received from so large a fraction of the nation's astronomers had never before occurred in any scientific field. In the end, the House and Senate compromised by approving about half the planning funds requested.

The next big effort came two years later. In the meantime, NASA, in response to Congressional pressure for economy, had reduced the overall cost of the telescope in two ways—by negotiating for some support by the European Space Organization, and by reducing the mirror diameter from 3 to 2.4 meters, a change reluctantly endorsed by our committee of astronomers advisory to NASA.[2] This time, NASA requested full approval for the entire project. Another tussle between House and Senate led to an additional campaign by astronomers to educate Congress, which finally voted full approval in the following summer of 1977. The project was officially underway.

During the next decade, I had little responsibility for the LST, by then renamed the Hubble Space Telescope (HST). I had hoped for some significant connection with the design and fabrication of the telescope, but my specific plans to this end did not work out. However, as Chairman for nine years of AURA's Council for oversight of the Space Telescope Science Institute (STScI), I enjoyed working part time with the President of AURA (first John Teem and then Goetz Oertel) and with the Director of the STScI (Riccardo Giacconi), as well as with other Council members helping to keep the wheels of top management well oiled.

Watching the launch of HST, on the fine clear morning of April 24, 1990, was tremendously exciting for all of us present. My next close contact with technical problems of HST followed the discovery of the spherical aberration, a few months later. The phone call to me, passing along the news of this flaw in the primary mirror's figure, was an unexpected and certainly a most unwelcome present on my 76th birthday! Right away I began trying to think of ways to modify HST in orbit, to restore the optical capability originally planned. My Princeton colleague, Jim Gunn, was most helpful in using his powerful optical computing program to design the optimum lenses for placement in various locations in the telescope. None of these seemed adequate, and they each offered considerable difficulty in placement by astronauts. The next thought was, could a set of small lenses be used, one for each optical channel, to correct the aberration for that channel? This line of though seemed to offer promise, though it was not clear how astronauts would position these small lenses precisely at the entrance apertures of various spectroscopes and cameras. I forwarded these conclusions to NASA, urging that they be considered.

Other scientists had already proposed[3] that broad consideration be given to all possible methods for restoring the full optimal capability of HST. As a result, in mid-August, 1990, the Space Telescope Science Institute (STScI) set up a Strategy Panel[5] for this purpose; Holland Ford and Bob Brown, who had pushed for such action, were cochairmen. At our first Panel meeting I presented an analysis of individual correcting lenses, one for each optical channel in each instrument. In response, Murk Bottema, the brilliant optical designer at Ball Aerospace (who unfortunately died shortly before COSTAR was successfully installed in HST), pointed out that use of two-mirror correctors instead of single lenses would have the great advantage of achromatism as well as wider fields; he discussed possible designs for such correctors. The first of the two mirrors would image the HST primary mirror on the second correction mirror, which would be figured to cancel out the spherical aberration of the primary and to yield a sharp image at the focal surface of the instrument.

At a subsequent Panel meeting, Jim Crocker, a STScI staff member, advanced the key suggestion which made such modifications practical; he proposed that all the correctors be placed in a standard HST experiment box for an axial instrument, with each corrector fastened to an arm mounted on a movable optical bench. In orbit, control from the ground would be activated; the optical bench would move forward, and the arms would all unfold, placing the correcting mirrors in their proper positions; if necessary, the mirrors could be tilted slightly for optical alignment. This was the device later christened COSTAR.

Before COSTAR could be approved, the Strategy Panel gave full consideration to some 30 other imaginative proposals.[5] More detailed analyses of COSTAR optics and mechanisms were also carried out and numerous modifications suggested. Until our final session there was some question as to whether the Panel would approve COSTAR. Several Panel members felt, not unreasonably, that this complex device, with its somewhat intricate motions, would be very unlikely to operate properly in a space environment. However, no more promising alternative seemed available, and in the end there was unanimous approval. I believe that NASA in general and the Goddard Space Flight Center (GSFC) in particular must have gone through similar agonizing debates before approving COSTAR.

COSTAR was designed and fabricated by Ball Aerospace, under the direction of GSFC, advised by STScI. It is a tribute to these three organizations that the work was carried out on schedule and within budget.

WF/PC 2 was planned as a back-up for WF/PC 1 even before the launch of HST, and was still being fabricated when the spherical aberration was recognized. This device had internal mirrors, which were then reconfigured to correct imagery, just as in COSTAR. Hence, the array of COSTAR mirrors does not include any for this camera. Substitution of WF/PC 2 for WF/PC 1 in HST was endorsed by the Strategy Panel at the same time that COSTAR was approved.

As is well known, COSTAR and WF/PC 2 were both installed in HST during the spectacularly successful Maintenance and Repair Mission of December 1993. The behavior of these devices in orbit conforms very closely to their design objectives. Thanks to these and other modifications skillfully installed by astronauts, the scientific performance of HST now conforms to the idealized vision advanced in 1946. As the telescope moves on from one fascinating discovery to the next, the nature of a possible next great mission in optical space astronomy is already under vigorous debate.

REFERENCES

1. Friedman, H.—*Annu. Rev. Astron. Astroph.* **1**, 59 (1963).
2. Smith, R. W.—The Space Telescope, Cambridge: Univ. Press (1989).
3. ——, paperback updated version, see pp. 421–23 (1993).
4. Spitzer, L.—*Quarterly J. Roy. Astron. Soc.* **20**, 29 (1979).
5. Brown, R. A. and Ford, H.—A Strategy for Recovery, Baltimore: Space Telescope Science Inst. (1991).

# 23

## Paper

### Abstract

The concept of a powerful telescope in earth orbit has developed gradually during the last few decades. The development of plans for such an instrument is traced here from the tentative discussions of the early postwar years to the more detailed recent studies. The growing realization by astronomers generally of the unique role which such a telescope could play in many branches of astronomical research has provided an essential driving force for the Large Space Telescope Program.

### I. Introduction

More than fifty years ago it was pointed out[1] that an astronomical telescope in orbit around the Earth, far above the atmosphere, would have three tremendous advantages over ground-based instruments. First, the images produced by such an instrument in space would be free of the fluctuations and distortions produced by variable refraction in the Earth's atmosphere, an effect known to astronomers as variable "seeing"; more simply the stars would not twinkle, stellar images would be steady, and ideally sharp pictures could be obtained, subject only to the basic limitation imposed by the diffraction of light waves. Second, electromagnetic radiation of most wavelengths could reach a satellite from a distant star without the nearly complete absorption that prevents γ-rays, X-rays, ultraviolet light, and some infra-red wavelengths from reaching the Earth's surface. Third, a telescope in a satellite would be virtually weightless, permitting a high stability of optical figure and mechanical alignment, free from the perturbations produced by gravitational flexure.

These advantages are, of course, very obvious to anyone familiar with astronomical observations and their limitations, provided, of course, that one takes seriously the engineering possibilities of launching equipment into space and operating it there. However, early suggestions along these lines encountered a general skepticism. Largely for this reason, the few professional astronomers interested in such possibilities did not generally publish any detailed analyses of this topic, but argued for their point of view in popular talks,[2] science fiction journals,[3] and project reports.[4] With the successful launching of satellites by the USSR starting in 1957 and by this country starting in 1958, and with the gradual growth and success of space astronomy, early skepticism gradually changed to general acceptance, and the magnificent advantages of telescopes in space, as well as their disadvantages in great cost and relative lack of flexibility, are now well known.

The present paper traces the growth of one particular concept in this field, the so-called Large Space Telescope (LST), defined as an optical space telescope with an aperture roughly three meters in diameter, observing wavelengths from about 1000 Å to a few millimeters, and capable of operation for many years. The gradual realization by the astronomical community of the central importance which this instrument could have for future astronomical research forms an interesting chapter in the history of science.

## II. STUDY GROUPS AND THE LST

The first decade following World War II, from 1946 to 1956, was marked by great activity in the sounding rocket field, and by active planning for the small satellites launched in the ensuing decade. There was little discussion of anything so ambitious as the LST. A symposium organized in 1956 by the Upper Atmosphere Rocket Research Panel, on the occasion of its tenth anniversary meeting, was devoted[5] to research use of satellites, but since precise guidance was not considered, none of the principal programs now envisaged for the LST was discussed.

During the second postwar decade, 1956–1966, this situation changed dramatically following the successful launching of satellites from 1957 on. The National Aeronautics and Space Administration (NASA) was organized in 1958, and soon began planning for three Orbiting Astronomical Observatories (OAO's), each designed to carry out astronomical research at ultraviolet wavelengths, using telescopes up to a meter in diameter. In 1959 the American Astronomical Society, with support from the National Science Foundation, sponsored a symposium[6] on space telescopes and the research which they could make possible. Several surveys of space research[7, 8] at about this same time also dealt with these topics. During this period of lively activity and confident expectations, two summer study groups were organized by the Space Science Board, National Academy of Sciences (NAS), at NASA request, to make recommendations on future space research. The first[9] of these was held at the State University of Iowa, the second[10] at Woods Hole, Massachusetts. It was at these two study sessions that serious consideration of the LST began.

When the first of these sessions was held, in 1962, the first Orbiting Solar Observatory had been launched and the OAO program was well into the hardware phase. The conferees naturally considered the programs that should be planned for the post-OAO period, and pointed out[9] that "a much larger instrument, say 100 inches or more, would represent a truly enormous investment for astronomy. For this reason, it is vital that its scientific justification receive the most careful and comprehensive consideration by the astronomical and related scientific communities." The establishment of a small study group was officially recommended to explore the scientific uses and technical problems associated with such a large instrument.

However, this recommendation was not unanimous. A minority report stated, "At a time when not a single image of a celestial body has been obtained in a satellite, it is premature to convene a group to study a space telescope larger than the 38-inch telescope of the OAO." At the time, this dissenting opinion struck me as a bit pusillanimous, but in view of the subsequent major difficulties encountered both with Stratoscope II (a 36-inch balloon-launched optical telescope) and with the OAO's before successful operation yielded scientific results, this more conservative opinion now rather impresses me with its wisdom! In any case no action was taken to establish the proposed study group.

The next general space science study session, in 1965, returned again to this subject of a large space telescope. One problem particularly discussed was whether a large telescope of conventional optical design but located out in space would continue to be an indispensable tool for high-resolution studies for many decades, or whether new techniques might give diffraction limited images much more economically from the ground. It was clear that theoretical possibilities exist for circumventing atmospheric seeing and gravitational flexure when imaging bright objects, such as stars and planets. However, there seemed little prospect, in principle, that such techniques could be applied generally to the very faint stars and diffuse objects such as remote galaxies. Partly for this reason, agreement was reached at this study session in favor of one larger general-purpose telescope instead of many smaller ones.

While there had been no very spectacular achievements in space astronomy during the three years since the 1962 Iowa summer study, confidence in the feasibility of a large instrument had grown by the summer of 1965, thanks to a number of developments. Detailed engineering studies[11] had come to the conclusion that a large, high-resolution, general-purpose telescope in orbit seemed entirely feasible. Astronomical research on the Sun had achieved results with equipment of increasing sophistication, and in the field of stellar astronomy spectra of bright stars with moderate resolution (about 1 Å) had been achieved with gyro-stabilized sounding rockets.[12] The progress in fabricating and testing OAO equipment also suggested that rapid progress could be expected. After some discussion of various technical problems of what was then referred to as the Large Orbital Telescope, including aperture (preferably as large as feasible), role of man (required for maintenance and updating), and location (a choice between low Earth orbit at several hundred kilometers altitude, a 24-hour Earth orbit, or a site on the lunar surface) the group at Woods Hole adopted the following recommendations:[10]

> "We conclude that a space telescope of very large diameter, with a resolution corresponding to an aperture of at least 120 inches, detecting radiation between 800 Å and 1 mm, and requiring the capability of man in space, is becoming technically feasible, and will be uniquely important to the solution of the central astronomical problems of our era."

A preliminary schedule was also outlined, with 11 years elapsing between the start of detailed design in 1968 and a 1979 launch.

### III. AD HOC COMMITTEE FOR THE LST

An Ad Hoc Committee for further consideration of such an instrument was appointed by the Space Science Board in due course, and functioned for several years. This Committee devoted considerable time to technical aspects of the proposed large telescope. Partly because of some possibility that a location on the Moon might be indicated, if a substantial lunar base were to be developed, the instrument was designed as a space telescope rather than an orbital telescope, and generally denoted by the initials LST. The Committee stressed the importance of certain developmental items particularly critical to the program, especially the technology for obtaining large mirror blanks and for figuring them to diffraction-limited performance.

It is of interest that the discussions by this Committee and by subsequent groups produced very few changes in the general performance characteristics of the LST as first envisaged in the Iowa and Woods Hole summer studies. Following the usual practice with large telescopes on the ground, the instrument has generally been regarded as a multipurpose device, to be used for a variety of astronomical observations, including particularly both imagery and spectrophotometry in a variety of wavelength bands. The size of the LST has generally been taken to be significantly greater than the 1-meter diameter class of the OAO's, and limited by cost and technical problems to a diameter of perhaps 3 meters. Scientific requirements have indicated an optical quality giving images as close to the diffraction limit as cost will permit and as wide a wavelength range as feasible. Continuing studies[13] during this period confirmed earlier indications that the problems of optical imagery and precise guidance associated with an orbiting telescope seemed to offer no insuperable difficulties.

From the beginning it has usually been assumed that maintenance by man would be required to keep such a major and costly instrument as the LST in operation for an indefinite period, exceeding 10 years as a minimum. Almost as important would be the possibility of "updating" the instrument; i.e., replacing equipment modules[14] in orbit to permit changes in the scientific program and the use of improved instrumentation. More controversial has been the question of whether operation of the LST by astronauts within or close to the instrument capsule might be desirable, at least under some conditions. In recent years the advocates of unmanned operation, with control of the LST from the ground, have had a clear majority, and in any case absence of immediate plans for a permanently manned space station has brought this particular dispute to a close.

Probably the most important work of the Ad Hoc Committee on the LST was its detailed discussions and its brief written report[15] on the specific research programs which could be carried out with this instrument. Not only was this final report helpful in later studies and discussions, but the analyses of specific research problems provided useful education for the substantial number of astronomers involved with this Committee, either as members or as consultants. Through discussions organized by the Committee, astronomers engaged in observational research with ground-based telescopes became familiar with the great potential of the LST in may central areas of astronomy. At the same time, enthusiasts for space astronomy became convinced of the close interrelations between ground-based and space-based astronomy, with strong development of each required for a balanced and effective overall research program.

## IV. The LST Today

The last five years have seen increasing support for the LST from scientists in general and from the astronomical community in particular, with strong endorsement by a variety of committees, including: the NASA Astronomy Mission Board[16] in 1969; the Woods Hole Study on Priorities in the Space Science and Earth Observations,[17] organized in 1970 by the Space Science Board, NAS; and the NAS Astronomy Survey Committee,[18] chaired by J. L. Greenstein, in 1972. The report from this last group, which considered broadly the needs of astronomy as a whole, stated:

> "The Committee feels that the LST has extraordinary potential for a wide variety of astronomical uses and believes it should be a major goal in any well-planned program of ground and space-based observations."

It is no coincidence that these strong recommendations came at a time when large optical space instruments were achieving important scientific results. The Stratoscope II 36-inch balloon telescope had very successful flights in 1970 and 1971, achieving photographs[19] of planets, satellites, and galactic nuclei with an instrumental profile about 0.2 arcsecond in diameter at half maximum intensity. The Wisconsin and Smithsonian telescopes on OAO-2, launched in December 1968, were operating successfully throughout this period, obtaining important ultraviolet observations[20] on a wide variety of objects for many different astronomers. More recently, the successful operation[21] of the Princeton telescope on the OAO-3 *Copernicus* satellite has given added confidence in the approach to the LST.

A number of significant technological advances directly related to the LST have also occurred during this latest five-year period. Important planning efforts have been undertaken by NASA. Of the many relevant engineering developments both at NASA centers and at university and industrial laboratories, only two can be mentioned here. First, the development of the Space Shuttle promises to provide a relatively inexpensive launch of the LST and, more important yet, offers the prospect of regular visits by trained astronauts, who could maintain and update the instrument. Second, the development of image tubes with electrical readout, and their use for astronomical research,[22] makes it now seem nearly certain that operable tubes of this sort, which are vital for effective instrumentation in satellite telescopes, will be available initially for the LST, though superior tubes, with a more nearly uniform response, higher spatial resolution, and a much larger dynamic range will likely be available in future years. Replacement of image tubes with later, improved versions would be an important mission for astronauts on one of the periodic Space Shuttle visits to the LST.

Both as regards its scientific importance and its technical feasibility the LST seems to be an idea whose time has come, although its launching date may be uncertain. The growth of this idea has been very exciting to follow during the last twenty-five years. The most exciting chapter of this history, culminating in the revolutionary scientific observations we shall obtain with this great instrument, lies ahead of us.

### References

1. Oberth, H., *Die Rakete zu den Planetenräumen* (R. Oldenbourg-Verlag, Munich), 1923.
2. Stewart, J. Q., Lecture to Brooklyn Academy of Science, April 11, 1929.
3. Richardson, R. S., "Lunar Observatory No. 1." *Astounding Science Fiction* (Street and Smith), Feb. 1940, p. 113.

4. Spitzer, L. Jr., "Astronomical Advantages of an Extra-Terrestrial Observatory" Project RAND report (Douglas Aircraft Co.), Sept. 1, 1946.
5. *Scientific Uses of Earth Satellites*, edited by J. A. Van Allen (U. of Michigan Press), 1956.
6. "Conference on Astronomical Observations from above the Earth's Atmosphere," *A.J.*, **65**, 239ff, 1960.
7. *Science in Space*, edited by L. V. Berkner and H. Odishaw (McGraw-Hill, New York), 1961; see esp. Part 6, "The Stars," by L. Goldberg and E. R. Dyer, Jr.
8. *The Challenges of Space*, edited by H. Odishaw (U. of Chicago Press), 1962.
9. *A Review of Space Research*, Publ. No. 1079 of the National Academy of Sciences-National Research Council, 1962; Chapter 2, Astronomy.
10. *Space Research: Directions for the Future*, Publ. No. 1403 of the National Academy of Sciences-National Research Council, 1966; Part 2, Astronomy and Physics.
11. *A System Study of a Manned Orbital Telescope*, Boeing Co., Aerospace Group, Report D2-84042-1, 1965.
12. "Line Spectra of Delta and Pi Scorpii in the Far Ultraviolet," D. C. Morton and L. Spitzer, *Ap. J.*, **144**, 1, 166.
13. *Advanced Princeton Satellite Study*, Perkin-Elmer Corp., Electro-Optical Div., Reports 8346, 1966, and 8688, 1967.
14. *Study of Telescope Maintenance and Updating in Orbit*, Itek Corp., Optical Systems Div., Report 68-8599-1, 1968.
15. *Scientific Uses of the Large Space Telescope*, Space Science Board, National Academy of Sciences-National Research Council, 1969; see also the summarizing article by L. Spitzer, Jr. in *Science*, **161**, 225, 1968 (July 19).
16. *A Long-Range Program in Space Astronomy*, Position Paper of the Astronomy Mission Board, edited by R. O. Doyle, NASA SP-213, 1969; see Section III.
17. *Priorities for Space Research*, 1971–1980, Space Science Board, National Academy of Sciences, 1971.
18. *Astronomy and Astrophysics for the 1970's*, Astronomy Survey Committee, National Academy of Sciences, 1972.
19. "High-Resolution Imagery of Uranus Obtained by Stratoscope II," R. E. Danielson, M. G. Tomasko, and B. D. Savage, *Ap. J.*, **178**, 887, 1972; "The Nucleus of M31," E. S. Light, R. E. Danielson, and M. Schwarzschild, *Ap. J.*, **194**, 257, 1974.
20. *The Scientific Results from the Orbiting Astronomical Observatory (OAO-2)*, Proceedings of a symposium sponsored by NASA and the American Astronomical Society, Amherst, Mass., Aug. 23–24, 1971, NASA SP-310, 1972.
21. "Spectrophotometric Results from the *Copernicus* Satellite. I. Instrumentation and Performance," J. B. Rogerson, L. Spitzer, J. F. Drake, K. Dressler, E. B. Jenkins, D. C. Morton, and D. G. York, *Ap. J. (Lett.)*, **181**, L97, 1973.
22. "The Spectrum of the Quasi-Stellar Object PHL 957," J. L. Lowrance, D. C. Morton, P. Zucchino, J. B. Oke, and M. Schmidt, *Ap. J.*, **171**, 233, 1972.

# PART FOUR

# Plasma Physics: Controlled Fusion

# THE ELECTRICAL CONDUCTIVITY OF AN IONIZED GAS*

(WITH R. S. COHEN AND P. McR. ROUTLY)

(PHYS. REV. 80, 230, 1950)

## Commentary

AFTER World War II, on my return to Yale, I was looking forward to resuming the interstellar studies on which I had been engaged before the War. As part of a continuing effort to establish the physical properties of interstellar matter, I gave some thought to the theory of electrical conductivity within a fully ionized gas. The existing theories of conductivity had been developed for partially ionized gases; while these theories gave approximate results for a fully ionized gas, their treatment of encounters between particles was not based on the relevant physical process; in a gas composed entirely of charged particles, it is the cumulative effect of many distant encounters that is important, not the close encounters which produce the large deflections (see Paper #13). Accordingly, I decided to attempt a new formulation of conductivity theory for an electron-ion gas, based on distant encounters between the charged particles. Bob Cohen, a graduate student in the Yale physics department, agreed to tackle the analytical portion of this program as his Ph.D. thesis.

His work was approaching completion when I left Yale to accept Princeton's offer (see Paper #31), and I remember returning occasionally to New Haven in 1947–48 to consult with Bob on his analysis. During the following year, Paul Routly, one of our early graduate students at Princeton, agreed to join the effort as my Research Assistant, and to seek a numerical solution of Bob's equations. By present standards the electro-mechanical calculator he used was primitive indeed, but in time it gave results.

After our rather lengthy paper had been submitted, its publication was somewhat delayed by the referee, who stipulated that our discussion of physical principles should be clarified and expanded substantially, but that the overall length of the paper should be reduced by one-half! This seemed to us a rather drastic modification, but in

* This work has been supported in part by the ONR.

retrospect I agree that omitting so much mathematical detail made the paper much more readable, though it robbed us of a convenient source for intermediate results, such as series expansions, that were used in subsequent researches.

At a late stage in the evolution of this paper, possibly after acceptance for publication, I realized that a term omitted in our analysis was in fact an important one; I had mistakenly concluded that this term vanished "because of symmetry." My physical intuition sometimes gives me important clues as to what happens in various situations, but usually I do not trust such clues fully until their results are confirmed by a quantitative analysis. Naturally I was stunned to find that our detailed numerical results were invalid. We did not withdraw the paper, since we felt that the physical arguments and the technique of solving the equations were both of interest. Revised results, with this additional term taken into account, are given in Paper #25.

In some ways the most important contribution of Paper #24 to theory is the evaluation of $b_{max}$, the maximum value of the impact parameter, $b$, out to which distant encounters need to be considered. Many researchers had set $b_{max}$ equal to $b_i$, the mean distance between ions, since it was not obvious that the theory of separate two-body encounters could be applied when $b$ exceeded $b_i$ and a number of such encounters were occurring simultaneously. Our statistical analysis provided an argument for considering such very remote simultaneous encounters in the same way as the somewhat closer separate ones. In the paper below, $b$ is bounded by $h$, the Debeye shielding distance. For gravitating forces between particles, $h$ is irrelevant, and $b$ can be bounded only by some physical dimension of the gravitating system, such as $R_h$ (see Paper #19). This result was used in a number of our later papers on stellar dynamics.

Recently, very detailed $N$-body dynamical simulations have confirmed[1] that $b_{max}$ is indeed proportional to the system radius, $R$, rather than to $b_i$, which equals about $R/N^{1/3}$. These simulations make use of equation (6), Paper #19, which shows that in a cluster ln $\Lambda$, defined as $\ln(b_{max}/b_{min})$, (which equals $\ln(h/b_0)$ in Paper #24), is proportional to $Nt_d/t_r$; $t_d$ and $t_r$ are the dynamical (or crossing) time and the relaxation time, respectively. Each simulation follows the dynamical evolution of an isolated cluster, with stars of two different masses present; only a small fraction of the cluster mass is in the lighter stars. The velocity dispersion $v_m$ is initially the same for the two mass groups, with the same initial spatial distribution. The relaxation time $t_r$ is then measured by the rate at which the total energy of the lighter stars increases, a result of interactions with the heavier stars. The dependence of ln $\Lambda$ on $b_{max}$, as $N$ is varied from run to run, is much increased by assuming a softened gravitational potential, which increases $b_{min}$

from the usual value for a 90° deflection up to the much larger assumed value for the softening radius, $c$. The different runs, with $N$ varying from a hundred to a few thousand, show clearly that $b_{max}$ varies as $R_h$, not as $R_h/N^{1/3} \approx b_i$.

While the Fokker-Planck equation is now well established for solving a variety of problems involving either charged particles or gravitating masses, one must not forget that this technique entirely neglects the close encounters, which produce deflections through large angles and result in relatively large changes in particle energies. For many situations, especially those dealing with situations close to kinetic equilibrium, such close encounters are unimportant. This is generally the case for various situations discussed in this book (i.e., in Papers #13–20 dealing with stellar dynamics and #24–25 concerned with electrical conductivities). However, when some group of particles is moving at high speed, $V$, through a group of "field particles" with a relatively small velocity dispersion, $v_m$, close encounters can be important.

Such an appreciable enhancement of the role which can sometime be played by close encounters has recently been pointed out and quantified.[2] A detailed computation of the velocity distribution function, $f(v)$, was carried out for a group of initially monoenergetic "test particles," slowed down by interactions with the field particles. Since the analysis indicates a basic inaccuracy possible in some Fokker-Planck calculations, we summarize here the result of this calculation.

When $V/v_m$ equals 25, for example, close encounters gradually decelerate more and more test particles to low velocities, producing a low-velocity tail of $f(v)$, greatly enhanced over what the Fokker-Planck interactions, acting by themselves, would yield. These slower particles then slow down even further at a relatively rapid rate because of their own interactions with the relatively slow field particles. As a result of this somewhat complex sequence, the time required for all the original test particles to reduce their kinetic energy by a factor five, on the average, is 63 percent of that required if distant encounters alone are assumed present[2]—an appreciable reduction though not a large one. Moreover, in a plasma, the unduly weak low-velocity tail of $f(v)$ given by the Fokker-Planck equation could lead to appreciable errors in the computed excitation and ionization of atoms as the energetic ions slow down. Early analyses[3] of interactions between interstellar atoms and energetic charged particles (such as cosmic rays and their secondaries) have generally not considered this effect.

REFERENCES

1. Farouki, R. T. and Salpeter, E. E.,—*Astroph. J.* **427**, 676 (1994).
2. Shoub, E. C.—*Astroph. J.* **389**, 558 (1992).
3. Spitzer, L. and Scott, E. H.—*Astroph. J.* **158**, 161 (1969).

## *Paper*

### ABSTRACT

The interaction term in the Boltzmann equation for an ionized gas is expressed as the sum of two terms: a term of the usual form for close encounters and a diffusion term for distant encounters. Since distant encounters, producing small deflections, are more important than close encounters, consideration of only the diffusion term gives a reasonably good approximation in most cases and approaches exactness as the temperature increases or the density decreases. It is shown that in evaluating the coefficients in this diffusion term, the integral must be cut off at the Debye shielding distance, not at the mean interionic distance.

The integro-differential equation obtained with the use of this diffusion term permits a more precise solution of the Boltzmann equation than is feasible with the Chapman-Cowling theory. While one pair of coefficients in this equation has been neglected, the remaining coefficients have all been evaluated, and the resultant equation solved numerically for the velocity distribution function in a gas of electrons and singly ionized atoms subject to a weak electrical field. Special techniques were required for this numerical integration, since solutions of the differential equation proved to be unstable in both directions. For high temperatures and low densities the computed electrical conductivity is about 60 percent of the value given by Cowling's second approximation.

### INTRODUCTION

Quantitative analyses of non-uniform gases have naturally been developed along lines relevant to laboratory experiments. The theory of Enskog and of Chapman, systematically expounded by Chapman and Cowling,[1] is primarily concerned with the properties of gases composed predominantly of neutral atoms. While this theory has been applied[1-3] to the conductivity of a completely ionized gas (a gas containing no neutral atoms), the theory is in fact not well suited to handle inverse-square forces between the particles in a gas, and the accuracy of the results obtained is uncertain. In view of the great astrophysical importance of completely ionized gases, as, for example, in stellar interiors,

---

[1] Chapman, S. and Cowling, T. G.—*The Mathematical Theory of Non-Uniform Gases* (Cambridge University Press, London, 1939).
[2] Cowling, T. G.—*Proc. Roy. Soc.* A, **183**, 453 (1945).
[3] Landshoff, R.—*Phys. Rev.* **76**, 904 (1949).

stellar envelopes, and interstellar matter, a reconsideration of this problem has been undertaken.

A new approach to this subject is provided through the work of Chandrasekhar[4] on stellar dynamics. This work is based on the fact that when particles interact according to inverse-square forces, the velocity distribution function is affected primarily by the many small deflections produced by relatively distant encounters. There will be many such encounters during the time a particle travels over its mean free path, and the change in the particle velocity can be computed in the same way as in the change of the position of a particle in Brownian motion. On the assumption that the large deflections produced by the relatively close encounters may be neglected, Chandrasekhar therefore employs a diffusion equation for the velocity distribution function, similar to the equations describing the spatial distribution function in Brownian motion. A similar but incomplete approach was made somewhat earlier by Landau.[5] As we shall see below, the appropriate generalized diffusion equation may be solved numerically when a completely ionized gas is subject to a small electric field or a small temperature gradient.

In Part I, prepared by L. Spitzer, the basic principles of the present paper are developed. Part II, prepared by R. S. Cohen, applies this analysis to a singly ionized gas in a weak electric field, and evaluates certain coefficients in the appropriate integro-differential equation. In Part III, prepared by P. McR. Routly, the numerical solution of the resultant equation is briefly summarized. The final formulas for the electrical conductivity are given in Part IV.

## I. General Principles

The velocity distribution function $f_r$ for particles of type $r$, interacting with particles of different types $s$, is determined by Boltzmann's equation (reference 1, Eq. (8.1$_1$))

$$\frac{\partial f_r}{\partial t} + \sum_i v_{ri} \frac{\partial f_r}{\partial x_i} + \sum_i F_{ri} \frac{\partial f_r}{\partial v_{ri}} = \sum_s \left( \frac{\partial_e f_r}{\partial t} \right)_s, \qquad (1)$$

where the notation is similar to that used by Chapman and Cowling, except that $v_{ri}$ is used for the component of velocity of an $r$th particle in direction $i$; it should be noted that $F_r$ is the force per unit mass on a

[4] Chandrasekhar, S.—*Astrophy. J.* **97**, 255, 263 (1943).

[5] Landau, L.—*Physik Zeits. Sowjetunion* **10**, 154 (1936). In this reference, the important terms representing dynamical friction, which should appear in the diffusion equation, are set equal to zero as a result of certain approximations.

particle of type $r$. The quantity $(\partial_e f_r / \partial t)_s$ gives the change in $f_r$ produced by encounters of $r$ particles with particles of type $s$.

## 1. Evaluation of $\partial_e f_r / \partial t$ for Inverse-Square Forces

In the classical theory of nonuniform gases, the assumption is made that only the relatively close encounters are important, and that the forces between particles at greater distances have no effect. On this basis, $(\partial_e f_r / \partial t) \, dv_x \, dv_y \, dv_z$ is the sum of two terms, one representing the number of encounters which place particles of type $r$ in the volume element of velocity space $dv_x \, dv_y \, dv_z$, the other, the number taking particles out of this volume element. Thus we have (reference 1, Sections 3.5 and 3.52)

$$(\partial_e f_r / \partial t)_s = \iiint (f_r' f_s' - f_r f_s) g b \, db \, d\epsilon \, dv_s, \qquad (2)$$

where $g$ is the relative velocity $|v_r - v_s|$ of the two types of particle before the encounter, $b$ is the so-called impact parameter,—the distance of closest approach if no interaction forces are present—and $\epsilon$ is the angle between the orbital plane and the plane containing the velocities of the two particles before encounter. The corresponding quantities in Chandrasekhar's[6] analysis of stellar dynamics are $V$, $D$, and $\Theta$. The quantities $f_r'$ and $f_s'$ are the values of $f_r$ and $f_s$ for velocities such that a particle of type $r$ will be left after the encounter within the volume element $dv_x \, dv_y \, dv_z$.

When the force between two particles varies as the inverse square of their mutual separation, Eq. (2) is no longer appropriate. As has been shown by Jeans,[7] the cumulative effect of the weak deflections resulting from the relatively distant encounters is more important that the effect of occasional large deflections (relatively close encounters). To illustrate this effect, one may compute the cumulative squared value of the deflection angle $\chi$ produced during the time $\Delta t$ by all those encounters for which the impact parameter $b$ is less than some upper limit $b_1$. For collisions of electrons with heavy ions, whose space density is $n_i$ per

[6] Chandrasekhar, S.—*Principles of Stellar Dynamics* (University of Chicago Press, Chicago, Illinois, 1942), Chapter II.

[7] Jeans, J. H.—*Astronomy and Cosmogony* (Cambridge University Press, London, 1929), p. 318.

TABLE 1
Cumulative Mean-Square Deflection Produced by Encounters with $b < b_1$

| Impact Parameter $b_1 / b_0$ | 0 | 1 | 2 | 4 | 10 | $10^2$ | $10^4$ | $10^6$ |
|---|---|---|---|---|---|---|---|---|
| Mean-square deflection in arbitrary units | 0.00 | 0.19 | 0.81 | 1.89 | 3.63 | 8.21 | 17.4 | 35.8 |

cubic centimeter, it may be shown that

$$\left\langle \sum_{b<b_1} \sin^2 \chi \right\rangle_{Av} = 4\pi g n_i b_0^2 \, \Delta t \left\{ \ln\left(1 + \frac{b_1^2}{b_0^2}\right) - \frac{1}{1 + b_0^2/b_1^2} \right\}, \quad (3)$$

where $b_0$ is the value of $b$ for which $\chi$ equals $\pi/2$, and is given by

$$b_0 = Z_i e^2 / g^2 m, \quad (4)$$

where $m$ is the electron mass. Values computed from Eq. (3) are given in Table 1. It is evident from Table 1 that the relatively distant encounters outweigh the closer ones. For encounters between charged particles of comparable masses, the formula for $\sin^2 \chi$ considered by Chandrasekhar,[6] becomes much more complicated, but the general behavior shown in Table 1 is not altered.

While Eq. (2) could possibly be salvaged in this case, this equation is not appropriate for inverse-square forces, and obscures the true physical situation. When $\partial_e f_r / \partial t$ is produced by many small deflections, the total deflection produced in an interval of time is similar to the total distance traveled by a particle in Brownian motion, and the change of $f_r$ by such small collisions is described by a diffusion equation of the Fokker-Planck type.[8] In fact, the value of $\partial_e f_r / \partial t$ resulting from the relatively distant encounters depends almost entirely on the first and second derivatives of $f_r$, not on values of $f_r$ over the entire range of velocities.

If occasional large deflections were entirely negligible, $\partial_e f_r / \partial t$ would be entirely given by a diffusion equation. Actually, Table 1 shows this is not the case. If we define as a close encounter one for which $b$ is less than some critical value $b_c$, then the error introduced by the neglect of these encounters will be appreciable for low values of $b_m/b_0$ but will gradually decrease as $b_m/b_0$ increases.

The effects produced by the close encounters are best described by an equation of the form (2), with $b$ integrated only up to $b_c$. The relatively distant encounters are best described by the Fokker-Planck equation.[8] Thus we have finally

$$(\partial_e/f_r/\partial t)_s = -J(f_rf_s) - K(f_rf_s),\tag{5}$$

with

$$J(f_rf_s) \equiv \int_{\mathbf{v}_s=0}^{\infty} \int_{\epsilon=0}^{2\pi} \int_{b=0}^{b_c} (f_rf_s - f_r'f_s')gb\,db\,d\epsilon\,d\mathbf{v}_s,\tag{6}$$

and

$$K(f_rf_s) \equiv \sum_i \frac{\partial}{\partial v_i}(f_r\langle\Delta v_{i,s}\rangle) - \frac{1}{2}\sum_{i,j}\frac{\partial^2}{\partial v_i\,\partial v_j}(f_r\langle\Delta v_{i,s}\Delta_{j,s}\rangle),\tag{7}$$

where, in general, for any quantity $x$, $\langle x_s\rangle$ is defined by

$$\langle x_s\rangle = \int_0^{\infty} gf_s\,dv_s \int_0^{2\pi} d\epsilon \int_{b_c}^{b_m} xb\,db;\tag{8}$$

evidently $\langle x_s\rangle dt$ represents the mean value of $x$ resulting from all distant encounters with particles of type $s$ during the time interval $dt$. The definition of $J(f_rf_s)$ differs from that given by Chapman and Cowling (reference 1, Eq. $7.11_2$) in that only the close encounters are considered, with distant encounters entering into $K(f_rf_s)$.

The terms of third and higher order in $\Delta v$ have been neglected in Eq. (7). It is readily shown that these terms are relatively small if $b_c$ much exceeds $b_0$. Thus Eqs. (5), (6), and (7) should give moderately high accuracy in the evaluation of $\partial_e f_r/\partial t$.

To obtain the velocity distribution function in a first approximation $J(f_rf_s)$ may be neglected. This will involve an appreciable error, as shown in Table 1, but should at least provide a considerably more accurate determination of $f_r$ than has hitherto been available. It is probably best in this case not to neglect the close encounters altogether, and we shall therefore let $b_c$ equal zero in the computation of $K(f_rf_s)$. To obtain a higher approximation, a finite $b_c$ could be retained, and $J(f_rf_s)$ could be introduced as a small perturbation to the solution found below.

---

[8] A thorough survey of such processes has been given by S. Chandrasekhar, *Rev. Mod. Phys.* **15**, 1 (1943).

Accordingly, we derive an equation for $f_r$ on the assumption that $(\partial_e f_r / \partial t)_s$ in Eq. (1) may be set equal to $-K(f_r f_s)$. We shall assume that the gas is in a steady state with no systematic motion, but with an electrical field $\mathbf{E}$ and a temperature gradient $\nabla T$. Following Chapman and Cowling (reference 1, Section 7.1), we shall write

$$f_r = f_r^{(0)} + f_r^{(1)}, \tag{9}$$

where $f_r^{(0)}$ is the Maxwellian velocity distribution function, and obtain, finally (by use of expressions $(8.3_{8,\,10})$ of reference 1),

$$f_r^{(0)} \left\{ \frac{m_r v_r^2}{2kT} - \frac{5}{2} \right\} \sum_i v_{ri} \frac{\partial T}{T \partial x_i} - f_r^{(0)} \frac{eZ_r}{kT} \sum_i E_i v_{ri}$$

$$+ \sum_s K(f_r^{(1)} f_s^{(0)}) + \sum_s K(f_r^{(0)} f_s^{(1)}) = 0. \tag{10}$$

The electrical field $E_i$ and the electrostatic charge $e$ are in e.s.u.; thus for electrons $Z_r$ is $-1$. Quantities involving the square of $f^{(1)}$ have been neglected in Eq. (10). When Eq. (10) is applied to an electron gas, we shall omit all subscripts from quantities referring to electrons, such as $f^{(0)}$, $f^{(1)}$, $m$, and $v$.

## 2. The Cut-Off Parameter $b_m$

The quantities $\langle \Delta v_{i,s} \rangle$ and $\langle \Delta v_{i,s} \Delta v_{j,s} \rangle$ in Eq. (7) are expressible in terms of integrals over encounters with different values of $b$. As is well known, these integrals diverge logarithmically, and the integration must be terminated at some maximum $b_m$ to give a finite result. According to Cowling,[2] Chandrasekhar,[4] Spitzer,[9] and others, $b_m$ should be set equal to the interionic distance but according to Persico,[10] Landau,[5] and others, $b_m$ should equal the Debye distance, $h$, at which the electron-ion plasma shields any particular charge; this cut-off has also been discussed and used by Bohm and Aller.[11] For a gas composed of electrons with a particle density $n_e$ and of ions with an average charge $Z_i e$

$$h^2 = kT / \left[ 4\pi n_e e^2 (1 + Z_i) \right]. \tag{11}$$

The factor $1 + Z_i$ in the denominator takes into account shielding by heavy ions as well as by electrons.

[9] Spitzer, L., Jr.,—*M.N.R.A.S.* **100**, 396 (1940).
[10] Persico, E.—*M.N.R.A.S.* **86**, 294 (1926).
[11] Bohm D. and Aller, L. H.—*Astrophys. J.* **105**, 131 (1947).

We shall show that $h$ is almost certainly the proper cut-off distance. Let us consider the mean-square value of the velocity change $\Delta v$ for a single electron during the time $\Delta t$. We shall consider that the electrons and ions all move in straight lines, a legitimate assumption for the distant encounters. Under these assumptions we shall then show that for a particle initially at rest $\langle (\Delta v)^2 \rangle$ is given by an equation of the form (3), even for values of $b$ arbitrarily large compared to the interionic distance. It follows that some further assumption is needed to give a finite result, and the introduction of shielding gives $h$ as the natural value of $b_m$ to be used.

To obtain this result we consider the general statistics of the electrostatic field, a subject similar to that already treated by Chandrasekhar and von Neumann.[12] If $\Delta v$ is the change in velocity experienced by an electron of charge $-e$ and mass $m$ during the time interval $\Delta t$, then

$$(\Delta \mathbf{v})^2 = (e^2/m^2) \int^{\Delta t} \int_0 \mathbf{E}(t) \cdot \mathbf{E}(t') \, dt \, dt' \qquad (12)$$

where $\mathbf{E}(t)$ is the electrical field acting on the particle. If now $\Delta t$ becomes large and both sides are averaged over all complexions of the gas, the integrand of Eq. (12) is seen to involve the autocorrelation coefficient of the electrical field $\mathbf{E}(t)$. If each electron is assumed to move in a straight line, this autocorrelation coefficient can be evaluated exactly; averaging also over a Maxwellian velocity distribution for the electrons, we have, in an obvious notation,

$$(\mathbf{E}(t) \cdot \mathbf{E}(t + r))_{\mathrm{Av}} = 8\pi^{1/2} n_e e^2 j / |\tau|, \qquad (13)$$

where $n_e$ is the particle density of electrons, and where

$$j^2 \equiv m/2kT. \qquad (14)$$

Finally we obtain for Eq. (12)

$$\langle (\Delta v)^2 \rangle = (16\pi^{1/2} n_e e^1 j / m^2) \ln(\tau_2/\tau_1). \qquad (15)$$

Since the integral of the autocorrelation coefficient over $d\tau$ diverges at both limits of integration, we have replaced the limits 0 and $\infty$ by $\tau_1$ and $\tau_2$, respectively. Apart from the argument of the logarithm, Eq. (15) agrees exactly with the $\langle (\Delta v)^2 \rangle$ found by Chandrasekhar[13] for a particle

[12] Chandrasekhar, S. and von Neumann, J.—*Astrophys. J.* **95**, 489 (1941); **97**, 1 (1943).

[13] Reference 6. The sum of the two Eqs. (5.724) gives $\langle (\Delta v)^2 \rangle \, dt$; $(G + H)/v_2$ equals $2j/\pi^{1/2}$ when $v_2$ is small.

at rest. It is evident from the derivation that Eq. (15) is valid for particles whose distance of closest exceeds the interionic distance, and the correct upper limit for $\tau$ must be about the period of a plasma oscillation. The divergence for low $\tau_1$ is the natural result of extending the assumption of straight-line motion to the very close encounters; evidently $\tau_1$ should be about $b_0/v$.

A simple physical argument shows the reasonableness of this result. The effects produced by electrons beyond the interionic distance may be attributed to statistical fluctuations in the electron density; i.e., the electron density is greater on one side than on the other. For electrons outside a sphere of radius $r$, the effective number of electrons which contribute to the force at the center of the sphere will be proportional to $n_e r^3$, and the fluctuations in this number will vary as $n_e^{1/2} r^{3/2}$, yielding a net electrical field at the center which is proportional to $e n_e^{1/2} r^{-1/2}$, but which is random in direction. This field will not change appreciably in direction for a time somewhat less than $r/v$, but after a time somewhat greater than $r/v$ will be in a new random direction. Greater fields last a shorter time, while fields produced by more distant electrons are smaller. Thus this field from electrons outside a sphere of radius $r$ will be primarily responsible for the value of the autocorrelation coefficient of $\mathbf{E}$ when $t$ equals $r/v$. It follows that this autocorrelation coefficient is proportional to $(e n_e^{1/2} r^{-1/2})^2$ or to $e^2 n_e / vt$, in agreement with Eq. (13).

While the proof has been carried through only for electrons at rest, it seems most improbable that the results will be qualitatively different for electrons in motion. We shall therefore set $b_m$ equal to $h$ in all integrals for the diffusion coefficients, and these coefficients will then contain as a factor $\ln(h/b_0)$. While the quantity $h/b_0$ will in fact vary with velocity, the effect of such variations is no greater than that of other neglected terms. For the interaction of electrons with particles of average charge $Z_i e$ we may write

$$\ln(h/b_0) = \ln(qC^2) \qquad (16)$$

where $C^2$ is the mean square electron velocity and

$$q \equiv (m/2e^3 Z_i)[kT/\pi n_e (1 + Z_i)]^{1/2}. \qquad (17)$$

At high temperatures one must consider the wave character of the electrons, an effect pointed out in this connection by Marshak.[14] An electron passing through a circular aperture of radius $a$ will be spread

---

[14] Marshak, R. E.—*Ann. N.Y. Acad. Sci.* **41**, 49 (1941).

out by diffraction through angles of about $\lambda/2\pi a$, where $\lambda$ is the electron wavelength $h/mv$. If this angle exceeds the classical deflection angle for an electron passing by at a distance $a$ from an ion of charge $Z_i e$, the deflections produced by the most distant encounters will be materially increased. The ratio of the quantum mechanical to the classical deflection is $v/(2Z_i \alpha c)$, where $\alpha$ is the fine-structure constant $1/137$. If $Z_i$ is unity, this ratio equals one for a velocity of $4.4 \times 10^8$ cm/sec, corresponding to an electron temperature of about $4 \times 10^5$ degrees. For lower temperatures one may conclude that the classical formulas are valid. The corrections introduced by quantum mechanics will not be large except at temperatures substantially above $10^6$ degrees.

### 3. Diffusion Equation in Spherical Coordinates

Before Eq. (10) can be solved, $K(ff)$ must be expressed in spherical coordinates. The derivation of the Fokker-Planck equation by Chandrasekhar is readily carried over to the spherical case, provided we substitute for the particle density $f$ the quantity $h$, defined by

$$h(\mathbf{v}) \equiv f(\mathbf{v}, t)v^2 \sin \theta. \tag{18}$$

By standard methods it follows that

$$K(ff) = -[\partial_e f(\mathbf{v},t)]/\partial t = \frac{1}{v^2 \sin \theta} \left\{ \sum_i \frac{\partial}{\partial x_i}(h(\mathbf{v})\langle x_i \rangle) \right.$$

$$\left. - \frac{1}{2} \sum_{i,j} \frac{\partial^2}{\partial x_i \partial x_j}\left(h(\mathbf{v})\langle x_i, x_j \rangle\right) \right\}, \tag{19}$$

where $x_i$ stands for the changes in the three coordinates; i.e., for $\Delta v$, $\Delta \theta$, and $\Delta \phi$ as $i$ goes from 1 to 3.

To apply Eq. (19), the coefficients $\langle x_i \rangle$ must be expressed in terms of the velocity shifts in rectangular coordinates, since it is these which can be evaluated by the theory of binary encounters. We consider rectangular axes $\xi$, $\eta$, and $\zeta$, where $\xi$ is in the direction of the velocity $\mathbf{v}$ before the encounter, while $\eta$ and $\zeta$ are in the directions of increasing $\theta$ and $\phi$, respectively. Thus the $\eta$, $\zeta$ axes are tangent to the circles of constant $\theta$ and constant $\phi$ at the point $\mathbf{v}$. We let $\Delta_\xi$, $\Delta_\eta$, and $\Delta_\zeta$ represent the velocity displacements in these local rectangular coordinates.

The complicated general relations between $\Delta v$, $\Delta \theta$ and $\Delta \phi$ on the one hand and $\Delta_\xi$, $\Delta_\eta$, and $\Delta_\zeta$ on the other simplify when these quantities are averaged over all collisions. When a small electrical field is present, for example, then (reference 1, Section 8.31) $f^{(1)}(v)$ varies as $\cos \theta$, where $\theta$ is the angle between $\mathbf{v}$ and $\mathbf{E}$. In such a case it is evident from the symmetry of the problem that the coefficients $\langle \Delta_\zeta \rangle$, $\langle \Delta_\zeta \Delta_\xi \rangle$ and $\langle \Delta_\zeta \Delta_\eta \rangle$ all vanish.

We shall also show that under these conditions $\langle \Delta_\zeta^2 \rangle$ equals $\langle \Delta_\eta^2 \rangle$. Consider encounters between particles of velocity $\mathbf{v}$ and those of velocity $\mathbf{v}_1$. We shall keep the angle between these velocities fixed, but shall vary the angle $\Theta$ between the fundamental plane, containing $\mathbf{v}$ and $\mathbf{v}_1$, and the $\xi, \eta$ plane fixed by the direction of $\mathbf{v}$ and $\mathbf{E}$. Let $\Delta_F$ be the change of $\mathbf{v}$ perpendicular to the original value of $\mathbf{v}$ and lying in the fundamental plane, and let $\Delta_G$ be the change of $\mathbf{v}$ perpendicular to the fundamental plane. Evidently,

$$\Delta_\eta = \Delta_F \cos \Theta + \Delta_G \sin \Theta \qquad (20)$$

$$\Delta_\zeta = \Delta_F \sin \Theta - \Delta_G \cos \Theta. \qquad (21)$$

As $\Theta$ varies, $\Delta_F$ and $\Delta_G$ per encounter remain unchanged since the relative velocity $g$, the impact parameter $b$, etc., are all unaffected. It follows that $\langle \Delta_\eta^2 \rangle$ will equal $\langle \Delta_\zeta^2 \rangle$ provided that $\langle \sin 2\Theta \rangle$ and $\langle \cos 2\Theta \rangle$ are both zero. Since $f^{(1)}$ will have components varying only as $\cos \Theta$, these averages are in fact zero, and the result follows. Similar results hold when a small temperature gradient is present.

The relations we require then reduce to the simple form

$$\langle \Delta v \rangle = \langle \Delta_\xi \rangle + \left( \langle \Delta_\eta^2 \rangle / v \right),$$

$$\langle \Delta \theta \rangle = \frac{\langle \Delta_\eta \rangle}{v} - \frac{\langle \Delta_\xi \Delta_\eta \rangle}{v^2} + \frac{\cot \theta}{2v^2} \langle \Delta_\eta^2 \rangle,$$

$$\langle (\Delta v)^2 \rangle = \langle \Delta_\xi^2 \rangle, \qquad (22)$$

$$\langle (\Delta v)(\Delta \theta) \rangle = \langle \Delta_\xi \Delta_\eta \rangle / v,$$

$$\langle (\Delta \theta)^2 \rangle = \langle \Delta_\eta^2 \rangle / v^2.$$

If we substitute relations (22) into Eq. (19), we have, after some rearrangement,

$$K(ff) = \frac{1}{v^2} \frac{\partial}{\partial v}\left[v^2 f\left\{\langle\Delta_\xi\rangle + \frac{1}{v}\langle\Delta_\eta^2\rangle - \frac{1}{2v^2}\frac{\partial}{\partial v}\left(\partial^2\langle\Delta_\xi^2\rangle\right)\right\}\right]$$

$$+ \frac{1}{v\sin\theta}\frac{\partial}{\partial\theta}\left[f\sin\theta\left\{\langle\Delta_\eta\rangle - \frac{1}{2v}\frac{\partial}{\partial\theta}\langle\Delta_\eta^2\rangle\right\}\right]$$

$$- \frac{1}{2v^2}\frac{\partial}{\partial v}\left[v^2\langle\Delta_\xi^2\rangle\frac{\partial f}{\partial v}\right] - \frac{1}{2v^2\sin\theta}\frac{\partial}{\partial\theta}\left[\sin\theta\langle\Delta_\eta^2\rangle\frac{\partial f}{\partial\theta}\right]$$

$$- \frac{1}{v^2\sin\theta}\frac{\partial}{\partial\theta}\left[f\sin\theta\langle\Delta_\xi\Delta_\eta\rangle + \frac{\partial}{\partial v}\left(vf\sin\theta\langle\Delta_\xi\Delta_\eta\rangle\right)\right].$$

$$(23)$$

Equation (23) will give $K(f_r^{(n)}f_s^{(m)})$, provided that on the right-hand side $f_r^{(n)}$ replaces $f$ wherever this occurs explicitly, and the averages of $\Delta_\xi$, $\Delta_\zeta$ etc., are evaluated over $f_s^{(m)}$ rather than over $f$.

When the analysis was first carried out, it was thought that the cross-product term $\langle\Delta_\xi\Delta_\eta\rangle$ would have no effect on the velocity distribution function, and this term has been ignored throughout the remainder of this paper. It now appears that this term may be appreciable; to evaluate this term, however, an extension of Chandrasekhar's analysis of two-body encounters does not suffice, and a new approach to the statistics of such encounters is required.

## II. DERIVATION OF EQUATION

When the average energy imparted to the electrons between encounters is small compared with their kinetic energy, we may write (reference 1, Sections 7.31 and 8.31)

$$f^{(1)}(\mathbf{v}) = f^{(0)}(v)D(jv)\cos\theta. \qquad (24)$$

The function $f^{(0)}(v)$ is the Maxwellian distribution function, given by the equation

$$f^{(0)}(v) = (n_e j^3/\pi^{3/2})\exp(-j^2 v^2), \qquad (25)$$

where $n_e$ is the number of electrons per cm$^3$ and $j$ is defined by Eq. (14). For subsequent convenience, we define $D$ to be a function of the dimensionless variable $jv$. The quantity $\theta$ is again the polar angle measured from an axis parallel to the electric field $\mathbf{E}$.

For an electron-proton gas, $Z$ is $-1$ for electrons, and if no temperature gradient is assumed, Eq. (10) becomes, in the present notation,

$$(2j^2ef^{(0)}/m)Ev\cos\theta + K(ff_p) + K(ff) = 0. \tag{26}$$

## 1. Electron-Proton Interaction

The proton interaction term $K(ff_p)$ is found from Eq. (23). We assume the protons are at rest; all the terms in Eq. (23) for $K(ff_p)$ then cancel out or vanish except one, and we have

$$K(ff_p) = -\frac{1}{2v^2\sin\theta}\frac{\partial}{\partial\theta}\left(\sin\theta\langle\Delta^2_{\eta,p}\rangle\frac{\partial f}{\partial\theta}\right). \tag{27}$$

The diffusion coefficient $\langle\Delta^2_{\eta,p}\rangle$ may be taken from Chandrasekhar.[15] If now we substitute Eq. (9) and (24) into (27), and carry out the differentiation with respect to $\theta$, we have

$$K(ff_p) = [3Lf^{(0)}D(jv)\cos\theta]/2v^3, \tag{28}$$

where we have written

$$L \equiv (8\pi e^4 n_e/3m^2)\ln(qC^2). \tag{29}$$

## 2. Electron-Electron Interaction

Derivation of $K(ff)$ from Eq. (23) is more lengthy. Adopting the terminology of Chandrasekhar,[16] we denote as "test" particles, with a velocity $v$, those electrons whose change of velocity is being considered, and as "field" particles, with velocity $v_1$, those of all velocities whose perturbing effect on the test particle is being investigated. The velocities of the field particles fall in the two ranges $0 < v_1 < v$ and $v < v_1 < \infty$.

There are nine terms in Eq. (23), each involving one diffusion coefficient. The first, second, third, and sixth terms, upon use of (24), will each yield one part due to interaction of $f^{(0)}(v_1)$, the spherically symmetric component of the field particles' velocity distribution, with

---

[15] See reference 6, Eq. (5.724); the quantity $\Sigma \Delta\tau_{\perp}^2/dt$ equals $\langle\Delta_\zeta\rangle + \langle\Delta^2_\eta\rangle$ and is thus twice $\langle\Delta^2_\eta\rangle$. For electron-proton interaction $x_0$ is large and $H(x_0)$ equals unity.

[16] See reference 6, paragraph 2.3.

the asymmetric component, $D(jv)$, of the test distribution, and a second part due to the corresponding interaction of $D(jv_1)$ with the symmetric component $f^{(0)}(v)$ of the test distribution. The second of these parts each, in turn, consists of integrals over $dv_1$ of $D(jv_1)$, multiplied by functions of $v_1$. The fifth and seventh terms each yield one part only, and the eighth and ninth terms are neglected.

The fourth term can apparently not be evaluated directly by any simple extension of Chandrasekhar's analysis. To evaluate this term by an indirect method, we multiply Eq. (23) by $\cos \theta \sin \theta \, d\theta$ and integrate over all $\theta$. If we first subtract a term $f\langle \Delta_\xi \rangle / v$, adding this same quantity to the first term, then the fourth term becomes $-P/v$, where

$$P \equiv \int_0^\pi f \sin \theta \, d\theta \{\cos \theta \langle \Delta_\xi \rangle - \sin \theta \langle \Delta_\eta \rangle\}, \qquad (30)$$

which will be evaluated separately.

### 3. Evaluation of the Diffusion Coefficients

We follow Chandrasekhar's general method,[17] but we replace his spherically symmetric distribution $f^{(0)}$ by the modified function $f^{(0)}(1 + D \cos \theta)$, in accordance with Eqs. (9) and (24). The integrals obtained are all straightforward. If we let

$$\langle \Delta_\xi^2 \rangle = p_0 + p_1 \cos \theta,$$

$$\langle \Delta_\eta^2 \rangle = q_0 + q_1 \cos \theta, \qquad (31)$$

$$\langle \Delta_\xi \rangle = r_0 + r_1 \cos \theta,$$

we have, first, Chandrasekhar's results

$$p_0 = (3LjG(x)/x);$$

$$q_0 = (3LjH(x)/2x); \qquad (32)$$

$$r_0 = -6L_j^2 G(x);$$

where $G(x)$ and $H(x)$ are functions defined by Chandrasekhar.[18] The

---

[17] See reference 6, Chapter II, Section 2.3. We use, in place of his gravitational factor, $G^2 m_1^2$, the electrical analog, $e^4/m^2$. In place of his $q$, Eq. (17) is used.

[18] See reference 6, pp. 63 and 73.

terms $p_1$, $q_1$, and $r_1$ may be expressed in the form

$$p_1 = \frac{12}{5\pi^{1/2}} Lj\left[\frac{I_5(x)}{x^4} + x\{I_0(\infty) - I_0(x)\}\right]; \tag{33}$$

$$q_1 = \frac{2}{\pi^{1/2}} Lj\left[\frac{I_3(x)}{x^2} - \frac{3I_5(x)}{5x^4} + \frac{2x}{5}\{I_0(\infty) - I_0(x)\}\right]; \tag{34}$$

$$r_1 = \frac{4}{\pi^{1/2}} Lj^2\left[\frac{2I_3(x)}{x^3} - I_0(\infty) + I_0(x)\right]; \tag{35}$$

the quantities $x$ and $I_n(x)$ are defined by

$$x \equiv jv, \tag{36}$$

$$I_n(x) \equiv \int_0^x y^n D(y) \exp(-y^2)\, dy. \tag{37}$$

The integral $P$ which arises from the fourth term of Eq. (23) can be interpreted as $2/m$ times the rate of transfer of momentum per second from the test particles of velocity $v$ in a unit volume of velocity space to field particles of all velocities; the momentum in the direction of the electrical field is considered, and the rate of momentum transfer is averaged over the polar angle $\theta$ between **v** and **E**. This rate of momentum transfer can be expressed in terms of $D(jv)$, $D(jv_1)$ and the value of $\langle \Delta_\xi \rangle$ found for a spherical distribution of field particles. To obtain this result we note that no momentum transfer arises from the interactions of $f^{(0)}(v)$ and $f^{(0)}(v_1)$, and also that the interactions of $f^{(1)}(v)$ and $f^{(1)}(v_1)$ are assumed small and are neglected. The interaction of $f^{(1)}(v)$ with $f^{(0)}(v_1)$ can be computed by finding the rate of momentum change for a single test particle on interaction with a spherically symmetrical distribution of field particles, and then integrating this rate over $\theta$. Similarly, the interaction of $f^{(0)}(v)$ with $f^{(1)}(v_1)$ can be obtained by considering the rate of momentum change for a single field particle interacting with a spherically symmetrical distribution of test particles of velocity $v$, and then integrating over both $\theta_1$ and $v_1$. Since these rates involve $\langle \Delta_\xi \rangle$ only, we can thus evaluate $P$ without evaluating $\langle \Delta_\zeta \rangle$ explicitly. These calculations are much simplified by the fact, first noted by Chandrasekhar,[19] that a test particle loses no momentum on interaction with a spherically symmetrical distribution of

[19] See reference 4, p. 260.

field particles if the velocities of the field particles exceed that of the test particle. We obtain

$$P = 4Lj^2 f^{(0)}\left\{-D(x)G(x) + \frac{2}{\pi^{1/2}}[I_0(\infty) - I_0(x)]\right\}. \quad (38)$$

### Final Equation for D(x)

If now Eqs. (28), (30)–(35), and (38) are substituted into Eq. (26), suitably integrated over $\cos\theta \sin\theta\, d\theta$, we obtain a final equation for $D(x)$. Before writing this equation we first express the infinite integral $I_0(\infty)$ in closed form. We multiply Eq. (26) by $2\pi v^3 \cos\theta \sin\theta\, dv\, d\theta$ and integrate over all $\theta$ and $v$. The three terms in (26) then give the total change of momentum arising from the electric field, electron-proton interactions, and electron-electron interactions, respectively. The last term, involving $K(ff)$, must give zero on integration, since the mutual electronic interactions cannot change the total momentum of the electrons; the actual cancellation of all the component parts of this term, on integration, provided a check on the detailed form for $K(ff)$. The second term yields $I_0(\infty)$, and the integration of the first term, representing the effect of the electrical field, is simple; we find, after some straightforward substitution,

$$I_0(\infty) = \int_0^\infty D(x)\exp(-x^2)\,dx = (3\pi^{1/2}A)/8 \quad (39)$$

where

$$A \equiv -mE/[2\pi j^2 e^3 n_e \ln(qC^2)]. \quad (40)$$

If Eqs. (39) and (40) are used, the final equation for $D(x)$ becomes

$$D''(x) + P(x)D'(x) + Q(x)D(x) = R(x) + S(x), \quad (41)$$

where

$$p(x) = -2x - \frac{1}{x} + \frac{2x^2\Phi'(x)}{\Lambda}, \quad (42)$$

$$Q(x) = \frac{1}{x^2} - 2\,\frac{1 + \Phi(x) - 2x^3\Phi'(x)}{\Lambda}, \quad (43)$$

$$R(x) = \frac{-2.4x^2 + 1.6x^4 - 2.4x^6}{\Lambda} A, \tag{44}$$

$$S(x) = \frac{16}{\pi^{1/2}\Lambda} \left\{ xI_3(x) - 2\left(\frac{1 + x^2 + x^4}{5x^3}\right)I_5(x) \right.$$

$$\left. +x^2\left(\frac{2 - 3x^2 + 2x^4}{5}\right)I_0(x)\right\}. \tag{45}$$

This quantity $\Phi(x)$ is the usual error function, while $\Lambda(x)$ is defined by

$$\Lambda \equiv \Phi(x) - x\Phi'(x). \tag{46}$$

### 5. Behavior of D(x) for Small and Large x

When $x$ is small, $\Phi$ and $\Phi'$ may be expressed as power series, and the system of Eqs. (41) to (45) admits a special series solution in ascending powers of $x$, which we shall denote by $D_{a1}(x)$. For large $x$, on the other hand, $\Phi'$ may be set equal to zero, while $\Phi$ equals unity, and we have a special series of solution $D_{a2}(x)$ in descending powers of $x$. Both these series are asymptotic, and will diverge after a certain number of terms.

To obtain general solutions for $D(x)$ in each of these regions, solutions of the homogeneous equation, with $R(x) + S(x)$ set equal to zero, must be added. As will be evident from the analysis in the next section, such solutions are of two sorts. When $x$ is small, for example, one of these solutions goes to infinity as $\exp(\alpha/x^{1/2})$, yielding an infinite conductivity, and obviously cannot represent a physical solution. The other goes to zero more rapidly than $D_{a1}(x)$, and therefore becomes negligible as $x$ decreases. Similarly, for large $x$ one of the solutions cannot represent reality, while the other goes to zero more rapidly than the leading term of $D_{a2}(x)$. Hence the boundary conditions on Eq. (41) to (45) are that $D(x)$ approach $D_{a1}(x)$ and $D_{a2}(x)$, respectively, as $x$ approaches zero or infinity.

For a Lorentz gas,[20] in which electron-electron interactions are entirely neglected and the protons are again assumed at rest, $K(ff)$ may be ignored in Eq. (26). In this case we obtain the usual result

$$D(x) = Ax^4. \tag{47}$$

---

[20] Lorentz, H. A.—*Proc. Amst. Acad.* **7**, 438 (1905).

### III. Solution of Equation

We wish to find the solution to Eq. (41) which is of physical interest and which therefore agrees with the asymptotic series $D_{a1}(x)$ and $D_{a2}(x)$ at zero and infinity, respectively. This solution will be denoted by the superscript $c$. The complexity of Eq. (41) is such that a closed analytical solution cannot be expected. It might appear that Eq. (41) could be solved by direct numerical integration, since starting values for small $x$ are known from the asymptotic series and the integrals in $S(x)$ could be evaluated as the integration proceeded. Actually such a direct integration is not possible. The integration of Eq. (41) is in fact unstable for both increasing and decreasing $x$; i.e., a small deviation from the correct solution increases so very rapidly in the course of integration that any trace of the correct solution soon disappears. This behavior is associated with the singularity of $Q(x)$, which varies as $1/x^3$ for small $x$. A similar instability is introduced by the dominant term in $Q(x)$ for large $x$. To obtain $D^c(x)$ the approach described below was developed.

### 1. Decomposition of Basic Equation

To overcome the difficulties associated with the instability of the basic equation, we let

$$P(x) = P_0(x) + \Delta P(x); \quad Q(x) + Q_0(x) + \Delta Q(x), \quad (48)$$

where $P_0(x)$ and $Q_0(x)$ are chosen to represent the leading parts of $P(x)$ and $Q(x)$ and also to permit analytical solution of the following simplified reduced form of Eq. (41),

$$D''(x) + P_0(x)D'(x) + Q_0(x)D(x) = 0. \quad (49)$$

In general, Eq. (49) has two solutions, which we shall denote by $U(x)$ and $V(x)$. We now write

$$D(x) = g(x)U(x) + h(x)V(x) \quad (50)$$

where $g(x)$ and $h(x)$ are functions to be determined. The simultaneous differential equations for $g(x)$ and $h(x)$, which are easily obtained by the standard methods of variation of parameters, may then be solved numerically without any basic difficultly.

Because the leading terms of $P(x)$ and $Q(x)$ are different in the cases of small and large $x$, it was necessary to consider separately the two ranges of $x$, $0.10 \leq x \leq 0.80$ and $0.80 \leq x \leq 3.20$. The formulas used in each specific range are given below.

(i) *Range* $0.10 \leq x \leq 0.80$

All quantities peculiar to this particular range will carry the subscript 1. The functions $P_{01}(x)$ and $Q_{01}(x)$ are defined as follows:

$$P_{01}(x) = 2/x, \quad Q_{01}(x) = (-3\pi^{1/2}/2x^3) - 2/x^2; \tag{51}$$

Equation (49) than has the solutions

$$U_1(x) = x^{-(1/2)}I_3(\alpha/x^{1/2}), \tag{52}$$

$$V_1(x) = x^{-(1/2)}K_3(\alpha/x^{1/2}), \tag{53}$$

where

$$\alpha^2 \equiv 6\pi^{1/2}. \tag{54}$$

The numerical values of these functions and their derivatives were computed by interpolation with the aid of B.A.A.S. Tables.[21]

(ii) *Range* $0.80 \leq x \leq 3.20$

The corresponding quantities in this range carry the subscript 2 and have the following expressions.

$$P_{02}(x) = -2x, \quad Q_{02}(x) = -4, \tag{55}$$

$$U_2(x) = xe^{x^2}, \tag{56}$$

$$V_2(x) = 1 - 2xe^{x^2} \int_x^\infty e^{-y^2}\, dy. \tag{57}$$

## 2. Description of Integration

As a first step, a starting value of $g_1(x)$ at $x = 0.10$ was obtained by integration of the appropriate equation for $g_1(x)$ from $x = 0$, where $g_1(x)$ vanishes, to $x = 0.10$. In this integration the asymptotic series $D_{a1}$ was used to compute $g_1'$. The "correct" value of $g_1(x)$ at $0.10$ is denoted by $g_2^c(0.10)$. In the same way, the accurate determination of $g_2^c(3.20)$ was carried out, with $D_{a2}(x)$ used to compute $g_2'$.

---

[21] Our sincere thanks are due to Dr. W. G. Bickley of the Imperial College of Science and Technology, London, England, who very kindly sent us the proof sheets of these very complete Bessel function tables, prepared by the Committee for the Calculation of Tables of the British Association for the Advancement of Science.

The starting value $h_1^c(0.10)$ cannot be determined so simply. In general there will be one value of $h_1(0.10)$, which, together with $g_1^c(0.10)$, will yield on numerical integration the calculated value of $g_2(x)$ at $x = 3.20$; i.e., $g_2^c(3.20)$. This will be the correct solution. In order to obtain this correct solution, advantage was taken of the linear properties of the above equations. Two arbitrary starting values of $h_1(x)$ at $x = 0.10$ yield two linearly independent solutions, each with the correct starting value of $g_1(x)$ at $x = 0.10$. The linear combination of these two solutions, which at $x = 3.20$ gives $g_2^c(3.20)$, is then the correct solution.

The starting interval used in the integration was 0.01 and was doubled after every ten steps. Central difference formulas were employed throughout. From the difference tables it was possible to guess ahead the values of $S(x)$ and of the appropriate combination of $D(x)$ and $D'(x)$; then first approximations to $g(x)$ and $h(x)$ at the next integration point could be obtained. These values were then used to obtain more accurate values of $S(x)$, etc., and ultimately second approximations to $g(x)$ and $h(x)$. This cyclic process was carried out at each integration point, in some cases as many as four times, until the values of $g(x)$ and $h(x)$ arising from the last two cycles agreed to five significant figures.

The resulting values of $D^c(x)$ over the entire range from 0.10 to 3.20 are given in table 2. The maximum error in any of these values should not exceed unity in the last digit, except perhaps for the last ten values, where errors as great as two in the last digit are possible.

The integrals $I_n(x)$ were found in the course of numerical integration up to $x$ equal to 3.20. As $x$ increases further these integrals also increase slightly by amounts which obviously depend on the values of $D(x)$ for $x$ greater than 3.20. The special solution $D_{a2}(x)$ was not sufficiently accurate, and a term in $V_2(x)$ was added to yield the approximate general solution in this range. The following values were obtained.

$$I_0(\infty) = 0.66464A. \tag{58}$$

$$I_3(\infty) = 1.470A. \tag{59}$$

$$I_5(\infty) = 4.562A. \tag{60}$$

This computed value of $I_0(\infty)$ may be compared with the exact value, which according to Eq. (39) equals $3\pi^{1/2}A/8$ or $0.66467A$.

TABLE 2
Values of Velocity Distribution Function $D(x)$

| $x$ | $D(x)/A$ | $x$ | $D(x)/A$ | $x$ | $D(x)/A$ |
|------|----------|------|----------|------|----------|
| 0.10 | 0.01487 | 0.34 | 0.2079 | 1.12 | 1.790 |
| 0.11 | 0.01840 | 0.36 | 0.2324 | 1.20 | 2.031 |
| 0.12 | 0.02237 | 0.38 | 0.2579 | 1.28 | 2.291 |
| 0.13 | 0.02676 | 0.40 | 0.2844 | 1.36 | 2.573 |
| 0.14 | 0.03159 | 0.44 | 0.3401 | 1.44 | 2.878 |
| 0.15 | 0.03685 | 0.48 | 0.3994 | 1.52 | 3.208 |
| 0.16 | 0.04254 | 0.52 | 0.4620 | 1.60 | 3.567 |
| 0.17 | 0.04865 | 0.56 | 0.5278 | 1.76 | 4.380 |
| 0.18 | 0.05517 | 0.60 | 0.5967 | 1.92 | 5.343 |
| 0.19 | 0.06208 | 0.64 | 0.6687 | 2.08 | 6.484 |
| 0.20 | 0.06940 | 0.68 | 0.7438 | 2.24 | 7.842 |
| 0.22 | 0.08517 | 0.72 | 0.8219 | 2.40 | 9.437 |
| 0.24 | 0.1024 | 0.76 | 0.9033 | 2.56 | 11.34 |
| 0.26 | 0.1210 | 0.80 | 0.9876 | 2.72 | 13.54 |
| 0.28 | 0.1409 | 0.88 | 1.166 | 2.88 | 16.13 |
| 0.30 | 0.1621 | 0.96 | 1.359 | 3.04 | 19.05 |
| 0.32 | 0.1844 | 1.04 | 1.566 | 3.20 | 22.40 |

## IV. VALUES OF THE ELECTRICAL CONDUCTIVITY

The electrical conductivity $\sigma$ is simply the total current flowing per cm$^2$ divided by the electrical field strength $E$. Electrons with velocities between $v$ and $v + dv$ and with directions between $\theta$ and $\theta + d\theta$ and between $\phi$ and $\phi + d\phi$ will contribute $d\sigma$ to the conductivity per unit volume of physical space, where

$$d\sigma = [-f(v)ev \cos \theta/E]v^2 \sin \theta \, dv \, d\theta \, d\phi. \qquad (61)$$

The minus sign results from the negative electronic charge. According to Eqs. (9) and (24), $f(v)$ is the sum of a spherically symmetric term $f^{(0)}(v)$, which clearly makes no net contribution to the conductivity, and the term $f^{(0)}(v)D(jv)\cos \theta$. If we substitute this term for $f(v)$, integrate over $v$, $\theta$, and $\phi$, and substitute from Eq. (40) for $A$ we obtain

$$\sigma = [2m/3\pi^{3/2}e^2j^3 \ln(qC^2)][I_3(\infty)]/A. \qquad (62)$$

It is convenient to express $\sigma$ in terms of the conductivity in a Lorentz gas, multiplied by some constant $\gamma$. Since $I_3(\infty)/A$ for a Lorentz gas

TABLE 3
Constant in Electrical Conductivity Formula

| | |
|---|---|
| Lorentz gas | 1.000 |
| Reference 1, first approximation | 0.295 |
| Cowling, second approximation | 0.578 |
| Present work | 0.490 |

equals three, as may be seen by combining Eqs. (37) and (47), we have

$$\sigma = 2\gamma(2/3\pi)^{3/2} mC^3/[e^2 \ln(qC^2)], \qquad (63)$$

where

$$\gamma = [I_3(\infty)]/3A. \qquad (64)$$

It may be remarked that the mutual electronic interactions do not change the conductivity directly, since the total change of momentum in such interactions is zero. Nevertheless, they alter $D(x)$ and in this way modify the effect which electron-proton collisions have in impeding the current.

The values of $\gamma$ obtained from various theories are given in Table 3 above. The value given here, readily obtained on combining Eqs. (59) and (64), is some 15 percent smaller than that found by Cowling[2] in his second approximation.

In addition, the constant $q$ used here is greater than that used by Cowling, since, as was shown in Part I, the cut-off distance should be equated to the Debye shielding radius $h$, rather than the interionic distance. We shall denote Cowling's value, based on the interionic distance, by $q'$. If the usual formula for $C^2$ is taken, and the ionic charge $Z_i$ is set equal to unity, Eq. (17) yields

$$qC^2 = (3/\pi^{1/2} n_e^{1/2} e^3)(kT/2)^{3/2} \qquad (65)$$

while for $q'$ we have[22]

$$q'C^2 = 4kT/n_4^{1/3} e^2. \qquad (66)$$

Since $\ln(qC^2)$ appears in Eq. (63) for the conductivity, values of $\ln(qC^2)$ are given in Table 4, together with values of $\ln(q'C^2)$ for comparison.

[22] See reference 1, Section 10.33; $\ln(q'C^2)$ equals one-half the function $A_1(2)$ introduced by Chapman and Cowling.

TABLE 4
Values[a] of $\ln(qC^2)$ in $\ln(q'C^2)$

| Kinetic Temperature (°K) | Electron Density $n_e$ $(cm^{-3})$ | | | | |
|---|---|---|---|---|---|
| | $1$ | $10^6$ | $10^{12}$ | $10^{18}$ | $10^{24}$ |
| $10^2$ | 16.3 | 9.36 | — | — | — |
| | 12.4 | 7.78 | 3.2 | | |
| $10^4$ | 23.2 | 16.3 | 9.36 | — | — |
| | 17.0 | 12.4 | 7.78 | 3.2 | |
| $10^6$ | 30.1 | 23.2 | 16.3 | 9.36 | — |
| | 21.6 | 17.0 | 12.4 | 7.78 | 3.2 |
| $10^8$ | 37.0 | 30.1 | 23.2 | 16.3 | 9.36 |
| | 26.2 | 21.6 | 17.0 | 12.4 | 7.78 |

[a] For each pair of values of $x_e$ and $T$, the upper figure give $\ln(qC^2)$, the lower, $\ln(q'C^2)$; the logarithms are to the base $e$.

For high densities and low kinetic temperatures, $q$ falls below $q'$; the analysis leading to Eq. (65) breaks down, and $q'$ may be used. For still higher values of $n_e^{1/3}/T$, no values are given in the table; the present theory breaks down completely under such conditions and, moreover, the electron gas tends to become degenerate. It is evident from Table 4 that for low densities and high temperatures the change in the cut-off distance has a greater effect on the conductivity than does the change in the value of $\gamma$. For these conditions the resultant electrical conductivity is about 60 percent of the value obtained by Cowling.

It is hoped to extend these results in the near future to ionized gases with different average ionic charges, and also to compute thermal conductivities. Further analysis is needed, however, to evaluate the cross-product terms in Eq. (23) which have been neglected in the present work.

# TRANSPORT PHENOMENA IN A COMPLETELY IONIZED GAS[*]

(WITH R. HÄRM)

(PHYS. REV. 89, 977, 1953)

## Commentary

THIS PAPER gives final values for the electrical conductivity and other transport coefficients, corrected by inclusion of the one term neglected in Paper #24. These results are applicable to fully ionized gases, with a wide range of values for $Z$, the electrical charge on the positive ions.

The preparation of this material was a relatively easy task. Evaluation of the additional term was straightforward, as were the corresponding modifications required in the equations to be solved. The structure of these equations was similar to that in Paper #24, and the same methods of solution could be used. On the other hand, the machine computations required for determination of the four transport coefficients considered, with various values of $Z$, were extensive and time consuming.

Reprinted here are Sections I, II, and V of Paper #25. The numerical results may be useful as reference material. Sections III and IV in the Phys. Rev. published version discuss certain details of the equations and their solution. Apart from new equations for diffusion coefficients not previously evaluated (those involving the transverse velocity change, $\Delta_\eta$), these sections do not differ greatly from similar material presented in Paper #24.

The numerical accuracy of the final transport coefficients is attested by comparison with a later solution of the same equations, with use of different mathematical methods.[1] These later values tend to be slightly greater than those of Paper #25, but in most cases only by about 0.1 percent. In a very few cases (2 out of 16) the difference is as great as 1 percent. Similar confirmation of Paper #25 is provided in a subsequent determination[2] of the electrical conductivity in a relativistic electron-ion

[*] This work has been supported in part by the U.S. Atomic Energy Commission.

gas; the analysis, like ours, is based on the Fokker-Planck equation, linearized for small $E$. In the non-relativistic limit, the results are in close agreement with Table 3 below.

Two basic limitations may be noted on the validity of these results. First, the solutions assume that the electric current (or heat flow) varies linearly with the electric field (or temperature gradient). This assumption requires that the perturbations of $f(v)$ produced by the electric field (or thermal gradient) be relatively small compared to the Maxwellian function $f^{(0)}(v)$; i.e., in equation (24) of Paper #24, $D(jv)$ must be small compared to 1. This assumption in turn imposes an upper limit on $E/[j^2 n_e]$ through equation (40) in this same paper. Formally, the limitation on $D(jv)$ can never be satisfied fully, since, for any value of $E/[j^2 n_e]$, no matter how small, $D(jv)$ will become large for sufficiently high values of $v$. This difficulty can be avoided in practice if $D(jv)$ remains small out to sufficiently large values of $jv$ so that the number of electrons at such a velocity, way out in the tail of the Maxwellian distribution, is effectively zero.

If for some reason the electron velocity distribution has a non-Maxwellian tail, containing an appreciable number of electrons with $jv$ greater than 10, for example, these energetic electrons will experience only a slight retardation from encounters with low-velocity electrons and ions. Hence, even a slight electric field, too small to produce an appreciable $D(x)$, can accelerate such electrons at a constant rate. These electrons will soon be accelerated to higher energies, and are referred to as "runaways," well known in both laboratory and astrophysical plasmas. Theoretical analyses[3] have computed the rate at which electrons, treated as test particles, in a Maxwellian plasma of field particles, will "run away" when an $E$ field of intermediate strength is applied.

While an $E$ field can produce runaway electrons in the nonlinear regime, giving an increased electric current, a thermal gradient in the nonlinear regime produces a heat flux less than given by the linear Equation (26) above. This flux approaches saturation when the systematic drift velocity of electrons through the positive ions approaches $v_m$, the rms electron velocity dispersion. This regime is an important one in some astrophysical plasmas, such as the boundary layer between a warm cloud and a surrounding hot ambient gas. Methods have been developed[4] for estimating the heat flux in such situations.

A second limitation on the applicability of Paper #25 is the possible appearance of plasma instabilities of various types resulting from the presence of an electric current. Such instabilities usually reduce the electric current, but can increase it in certain complex situations. The behavior of plasma instabilities and the effects which they can produce are extensive topics, beyond the scope of this brief Commentary.

## REFERENCES

1. Kulsrud, R. and Sun, Y. C.—informal communication (1973).
2. Braams, B. J. and Karney, C.F.F.—*Phys. Fluids B*, **1**, 1355 (1989).
3. Kulsrud, R. M., Sun, Y. C., Winsor, N., and Fallon, H.—*Phys. Rev. Lett.* **31**, 690 (1973).
4. Cowie, L. L. and McKee, C. F.—*Astroph. J.* **211**, 135 (1977).

# *Paper*[†]

### ABSTRACT

The coefficients of electrical and thermal conductivity have been computed for completely ionized gases with a wide variety of mean ionic charges. The effect of mutual electron encounters is considered as a problem of diffusion in velocity space, taking into account a term which previously had been neglected. The appropriate integro-differential equations are then solved numerically. The resultant conductivities are very close to the less extensive results obtained with the higher approximations on the Chapman-Cowling method, provided the Debye shielding distance is used as the cutoff in summing the effects of two-body encounters.

## I. INTRODUCTION

A previous paper by Cohen, Spitzer, and Routly,[1] referred to hereafter as CSR, presented a new approach to the problem of transport phenomena in a completely ionized gas. In effect, the influence of mutual electron encounters on the velocity distribution function for electrons was considered as a problem of diffusion in velocity space. In particular, the electrical conductivity of an electron-proton gas was computed in this way. However, the results were not exact, since one term in the diffusion equation was neglected. In the present paper, a solution of the complete diffusion equation is obtained, and the results are extended to completely ionized gases with different mean nuclear charges. Computations are carried out for the thermal as well as the electrical conductivity.

In the first section below the general principles are explained and justified. Subsequent sections outline the derivation of the equations, the method of solution, and the results obtained.

## II. GENERAL PRINCIPLES

The velocity distribution function $f_r(v)$ for particles of type $r$ is determined by the familiar Boltzmann equation, basic in all studies of this

---

[†] Sections III and IV, presenting some details of the analysis, are omitted here.
[1] Cohen, Spitzer, and Routly, *Phys. Rev.* **80**, 230 (1950).

sort,

$$\frac{\partial f_r}{\partial t} + \sum_i v_{ri} \frac{\partial f_r}{\partial x_i} + \sum_i F_{ri} \frac{\partial f_r}{\partial v_{ri}} = \sum_s \left( \frac{\partial_e f_r}{\partial t} \right)_s, \qquad (1)$$

where the notation in CSR has been followed. The complexity of the problem arises entirely from the term $(\partial_e f_r/\partial t)_s$, which gives the change in $f_r$ produced by encounters of $r$ particles with particles of type $s$.

To visualize the physical situation more accurately, let us follow a single electron as it moves through the gas. The random electrical fields encountered by the electron will produce deflections and changes in velocity. To some extent these fields can be described in terms of separate two-body encounters; let $b$ be the impact parameter for such an encounter—the distance of closest approach between the two particles in the absence of any mutual force. The situation is characterized by the values of the following four distances: $d$, the mean distance from an electron to its nearest neighbor; $b_0$, the value of the collision parameter for which an electron is deflected 90° in an encounter with a stationary positive ion; $h$, the Debye shielding distance; and $\lambda$, the mean free path for a net deflection of 90°. It is readily verified that for virtually all situations of interest,

$$b_0 \ll d \ll h \ll \lambda. \qquad (2)$$

It is clear that encounters for which $b \ll d$ can be described adequately in terms of successive two-body encounters, since usually an encounter with one particle will be effectively over before another particle approaches to a distance less than $d$. These successive encounters may be divided into two classes. Those with $b \leq b_0$ produce large deflections, and will be termed "close" encounters. Those with $b_0 < b < d$ produce relatively small deflections, and will be called "distant" encounters. As shown in CSR and elsewhere, the cumulative effect of many distant encounters outweighs the effect of the less frequent close encounters, in the special case of inverse-square forces between the particles.

Encounters for which $d \leq b < h$ cannot be regarded as independent, since several such encounters will be taking place at the same time. More correctly, the deflection of a particular electron caused by such "encounters" must be attributed to statistical fluctuations of the electron density in a sphere of radius $b$. As shown in CSR, however, the mean square change of electron velocity produced by such fluctuations is correctly given if the formulas derived for successive two-body encounters are applied for $b > d$.

Particles passing at a distance large compared to $h$ produce a negligible effect. From the standpoint of the Debye shielding theory, the effective field of a charge in a plasma varies as $e^{-hr}/r$, where $h$ is given by

$$h^2 = \frac{kT}{4\pi n_e e^2(1 + Z)}. \tag{3}$$

If one considers rather the statistical fluctuations in electron density, Pines and Bohm[2] have shown that collective phenomena in a plasma reduce markedly the statistical fluctuations in electron density with wavelengths large compared to $h$, thus justifying the neglect of encounters such that $b > h$. There is some interaction between a single electron and the organized oscillations of the plasma—see Eq. (59a) of Pines and Bohm. However, comparison of their Eqs. (59a) and (59b) shows that for thermal electrons, with mean kinetic energies of the order $kT$, the rate of energy loss due to this process is less by a factor $1/\ln(h/b_0)$ than the energy loss due to random encounters such that $b < h$. The generation of plasma oscillations by a single thermal electron may therefore be neglected, together with a number of other terms of the same order.[3] Hence, we may neglect all interactions between electrons for which the distance of closest approach exceeds $h$.

Since $\lambda$ is much greater than $h$, it is evident that many small deflections will be experienced by a particle traversing its mean free path. It is also clear that these deflections are essentially independent of each other. Inasmuch as collective phenomena (oscillations) have been neglected, the random electrical fields encountered by an electron in one region will be completely independent from the fields in a similar region separated by a distance appreciably greater than $h$. Hence, the successive changes in velocity represent a Markoff process, and the change of the velocity distribution function may be found from the Fokker-Planck equation.[4] This equation neglects the close encounters; the relative error introduced is again of the order $1/\ln(h/b_0)$.

[2] Pines D. and Bohm, D., *Phys. Rev.* **85**, 338 (1952).

[3] The various terms of relative magnitude $1/\ln(d/b_0)$ have been called "nondominant" by Chandrasekhar, and are usually neglected in the computation of diffusion coefficients —see S. Chandrasekhar, *Principles of Stellar Dynamics* (University of Chicago Press, Chicago, 1939).

[4] See S. Chandrasekhar, *Revs. Modern Phys.* **15**, 1 (1943).

## V. Results

In the presence of a weak electrical field $\mathbf{E}$ and a small temperature gradient $\nabla T$, the current density $\mathbf{j}$ and the rate of flow of heat $\mathbf{Q}$ per unit area are given by

$$\mathbf{j} = \sigma \mathbf{E} + \alpha \nabla T, \tag{25}$$

$$\mathbf{Q} = -\beta \mathbf{E} - K \nabla T. \tag{26}$$

In terms of the velocity distribution function $D(x)$, $j$, summed over all electrons—see Eq. (61) of CSR—is given by

$$j = 2\pi^{-\frac{1}{2}} e n_e C (2/3)^{3/2} I_3(\infty), \tag{27}$$

while for the heat flow $Q$ we have

$$Q = \pi^{-\frac{1}{2}} m n_e C^3 (2/3)^{5/2} I_5(\infty), \tag{28}$$

where $e$ and $m$ are the electron charge and mass, and $C$ is the root mean square electron velocity. From the numerical values found for the integrals $I_3(\infty)$ and $I_5(\infty)$, values of the coefficients $\sigma$, $\alpha$, $\beta$, and $K$ may be determined.

It is convenient to express these transport coefficients in terms of their values in a Lorentz gas. In the case of an electrical field, we define

$$\gamma_E = Z I_3(\infty)/3A, \tag{29}$$

$$\delta_E = Z I_5(\infty)/12A. \tag{30}$$

When $Z = 1$, $\gamma_E$ is identical with the quantity $\gamma$ introduced in CSR. In the corresponding case where a temperature gradient is present, we write

$$\gamma_T = -4 Z I_3(\infty)/9B, \tag{31}$$

$$\delta_T = -Z I_5(\infty)/15B. \tag{32}$$

From Eqs. (22) and (23) above and the definition of $I_n(\infty)$—see Eq. (37)

TABLE 3

Values of Transport Coefficients

| | $Z = 1$ | $Z = 2$ | $Z = 4$ | $Z = 16$ | $Z = \infty$ |
|---|---|---|---|---|---|
| $\gamma_E$ | 0.5816 | 0.6833 | 0.7849 | 0.9225 | 1.000 |
| $\gamma_T$ | 0.2727 | 0.4137 | 0.5714 | 0.8279 | 1.000 |
| $\delta_E$ | 0.4652 | 0.5787 | 0.7043 | 0.8870 | 1.000 |
| $\delta_T$ | 0.2252 | 0.3563 | 0.5133 | 0.7907 | 1.000 |
| $\epsilon$ | 0.4189 | 0.4100 | 0.4007 | 0.3959 | 0.4000 |

of CSR—it is readily verified that for a Lorentz gas these four coefficients are unity. On elimination of the constants A and B from these equations, we obtain, after some substitutions,

$$\sigma = \frac{2mC^3}{e^2 Z \ln(qC^2)} \left( \frac{2}{3\pi} \right)^{3/2} \gamma_E, \tag{33}$$

$$\alpha = \frac{3mkC^3}{e^3 Z \ln(qC^2)} \left( \frac{2}{3\pi} \right)^{3/2} \gamma_T, \tag{34}$$

$$\beta = \frac{8m^2 C^5}{3e^3 Z \ln(qC^2)} \left( \frac{2}{3\pi} \right)^{3/2} \delta_E, \tag{35}$$

$$K = \frac{20m^2 kC^5}{3e^4 Z \ln(qC^2)} \left( \frac{2}{3\pi} \right)^{3/2} \delta_T. \tag{36}$$

Values of the four transport coefficients are given in Table 3 for various values of $Z$.

The quantity $qC^2$, which is essentially $h/b_0$, is given in Eq. (65) of CSR, while values of $\ln(qC^2)$ are tabulated in Table 4 of the same paper. The electron charge $e$ has here been taken in electrostatic units throughout. To obtain the conductivity in practical units (mho) the value found from Eq. (33) must be divided by $9 \times 10^{11}$.

It should be noted that if a temperature gradient is present in a steady state, but no steady current is flowing, the electrostatic field $E$ will build up to such a value that $j$ vanishes. This field then reduces the flow of heat, and $K'$, the effective coefficient of heat conductivity is readily shown to be

$$K' = \epsilon K, \tag{37}$$

TABLE 4

Comparison with Results Obtained by Landshoff for $Z = 1$

| $n$ | 1 | 2 | 3 | 4 | $\infty$ |
|---|---|---|---|---|---|
| $\gamma_E$ | 0.2945 | 0.5693 | 0.5743 | 0.5777 | 0.5816 |

where

$$\epsilon = 1 - 3\delta_{E\gamma T}/(5\delta_{T\gamma E}). \tag{38}$$

Values of $\epsilon$ are also given in Table 3.

It remains to compare these results with those of previous workers. When this work was undertaken, the best available results for the electrical conductivity of a completely ionized gas were those of Chapman and Cowling[5] and of Cowling,[6] who had obtained first and second approximations, respectively, for the conductivity. In terms of the present notation, the Chapman-Cowling method utilizes the expansion

$$D(x) = \frac{3\pi^{\frac{1}{2}}x}{4} \sum_{j=0}^{n-1} (-1)^i \frac{\Delta_{0j}^{(n)}}{\Delta^{(n)}} L_j^{(3/2)}(x^2), \tag{39}$$

where $L_j^{(3/2)}$ is a Laguerre polynomial, and the ratios $\Delta_{0j}^{(n)}/\Delta^{(n)}$ are determined from encounter theory and the Boltzmann equation; $n$ is the order of the approximation used. Since the value of $\sigma$ found by Cowling[6] with $n = 2$ was about twice the value obtained by Chapman and Cowling[5] with $n = 1$, it appeared that the present treatment, equivalent to letting $n = \infty$ in Eq. (39), might give a markedly different value.

More recently, this same problem has been considered by Landshoff,[7] using the Chapman-Cowling method, but with values of $n$ up to 4. From the values of $\Delta_{0j}^{(n)}/\Delta^{(n)}$ which he gives for $Z = 1$, the constant $\gamma_E$ has been computed, and is given in Table 4, together with the value found in the present work ($n = \infty$).

In view of the large difference between the first and second approximation, it is rather remarkable how close to the truth is the second approximation for $\gamma_E$. For thermal conductivity the convergence is somewhat less rapid, with the fourth approximation in close agreement with the value for $n = \infty$. Evidently the present results agree with

[5] Chapman S. and Cowling, T. G., The Mathematical Theory of Non-Uniform Gases (Cambridge University Press, Cambridge, 1939).

[6] Cowling, T. G., Proc. Roy. Soc. (London) A183, 453 (1945).

[7] Landshoff, T., Phys. Rev. 82, 442 (1951).

Cowling's[6] second approximation for the electrical conductivity, provided that for the cutoff in the integration over the impact parameter $b$, we take the Debye shielding distance $h$ rather than the electronic separation $d$ taken by Cowling. The value 0.490 obtained for $\gamma_E$ in CSR, in disagreement with Cowling's value 0.578, was the consequence of the neglect of the $\langle \Delta_\xi \Delta_\eta \rangle$ term in $K(ff)$; inclusion of this term removes virtually all the disagreement in $\gamma_E$.

It should be emphasized that the present theory considers only those terms in $\partial f_s / \partial t$ which are of order $\ln(h/b_0)$, and a variety of terms of order unity have been neglected, including, for example, the interaction between a high speed electron and its wake of plasma oscillation, an effect explored by Pines and Bohm.[2] Thus, the relative accuracy of the present theory does not exceed $1/\ln(h/b_0)$, or some 5 to 10 percent for most conditions of astrophysical interest. In view of the lack of observational data in this field, development of a more refined theory does not seem worth the very considerable effort required.

# THE STELLARATOR CONCEPT*

(PHYS. FLUIDS, 1, 253, 1958)

## *Commentary*

IT WAS an early morning in March 1951. My wife and I were looking forward with much pleasure to our departure the next day for a week of skiing in Aspen. My father phoned to wish us a pleasant vacation. He added, "I was surprised by today's paper—it seems the Argentines have scooped us in atomic energy." Startled, I brought in our N.Y. Times. There, sure enough, was Peron's announcement that Dr. Richter and his group had successfully achieved controlled release of thermonuclear energy in an Argentine laboratory. I was somewhat incredulous; according to the Times, a number of distinguished scientists had stated that such a thing was flatly impossible.

At Aspen the following week, I wondered how power for peaceful purposes might conceivably be obtained from fusion energy. The main Aspen ski lift then had single chairs, and the long ascents gave me time to think about this problem. Three years earlier we had profited from a visit by Hannes Alfvén, the father of magnetohydrodynamics (MHD—the effects of a magnetic field on ionized gases). During the 1951 spring term Tom Cowling was in Princeton, lecturing on this same topic. Using the principles learned from these visitors, it was relatively easy to work out in my head the MHD equations for confinement of a hot thermonuclear plasma column by a magnetic field. For simplicity, though, I did assume that the column and the lines of magnetic force were all straight, parallel, and infinitely long. Back in Princeton during the following week, I worked out a method for making the column finite, using the figure-8 geometry of what we then christened the "stellarator."

The Atomic Energy Commission, prodded by the same announcement that had started my thoughts, was interested in pushing research in controlled fusion, which they had decided must be kept closely secret. By July 1 we had their financial support for a year of theoretical studies. Fortunately, the University had just purchased a large research com-

*Supported by the U.S. Atomic Energy Commission under Contract AT(30-1)-1238 with Princeton University.

plex (which they renamed the Forrestal Research Center) from the Rockefeller Institute for Medical Research, which had developed the site for research on animals. This complex provided the space and facilities we needed, and conformed with the University's policy to keep secret research off the main campus. So Project Matterhorn, as we named our program, started right on schedule. We occupied a modest metal structure, built to house small animals. (For several years a separate section of Project Matterhorn, under John Wheeler, carried out theoretical research on hydrogen bombs.)

Our first addition to the scientific staff was Martin Kruskal. While waiting several months for his clearance to come through, he worked in the basement of the old Observatory. He did research on certain problems of rotational transforms (see Paper #26), with no idea as to what this was all for, but with brilliant results, important in the Matterhorn program.

The basic concepts outlined in Paper #26 were developed in the theoretical studies of controlled fusion carried out at Matterhorn during some half-dozen years. This paper and many others on controlled fusion, from many groups in different countries, were prepared for the 1958 Geneva Conference on Peaceful Uses of Atomic Energy. On this occasion all research in this field was declassified by the governments involved, secrecy requirements were at an end, and the extensive results already obtained were freely published.[1] To conform with this now completely open environment, Matterhorn was given the more academic title, Princeton Plasma Physics Laboratory.

Over subsequent years, additional theoretical research has enlarged and modified the concepts presented in Paper #26. Calculations of magnetic fields in specific devices have shown that the successive intersections of a line of magnetic force with some reference plane in a toroidal plasma can, indeed, produce regular closed curves in bounded areas; however, sometimes in neighboring areas chaos appears, with apparently irregular displacements of these successive intersection points. Many theoretical processes have been identified which impair plasma confinement, but apparently not yet all the important ones. The analysis of heating methods has increasingly concentrated on the injection of energetic neutral atoms (D or H), since this technique seems to produce the most rapid heating. Divertors, which are probably required in any power-producing fusion reactor to keep the impurity level under control, have been designed for a wide variety of toroidal plasma devices. In addition, a number of groups at controlled fusion laboratories have made major contributions to the basic physics of astrophysical and fusion plasmas.

Some remarks on initial Princeton research with early stellarators and on global plans for future large-scale devices appear in the subsequent Commentary, Paper #27.

REFERENCES

1. Allis, W. P.—Nuclear Fusion, Princeton: van Nostrand (1960).

# 26

## Paper

### Abstract

The basic concepts of the controlled thermonuclear program at Project Matterhorn, Princeton University are discussed. In particular, the theory of confinement of a fully ionized gas in the magnetic configuration of the stellarator is given, the theories of heating are outlined, and the bearing of observational results on these theories is described.

Magnetic confinement in the stellarator is based on a strong magnetic field produced by solenoidal coils encircling a toroidal tube. The configuration is characterized by a "rotational transform," such that a single line of magnetic force, followed around the system, intersects a cross-sectional plane in points which successively rotate about the magnetic axis. A theorem by Kruskal is used to prove that each line of force in such a system generates a toroidal surface; ideally the wall is such a surface. A rotational transform may be generated either by a solenoidal field in a twisted, or figure-eight shaped, tube, or by the use of an additional transverse multipolar helical field, with helical symmetry.

Plasma confinement in a stellarator is analyzed from both the macroscopic and the microscopic points of view. The macroscopic equations, derived with certain simplifying assumptions, are used to show the existence of an equilibrium situation, and to discuss the limitations on material pressure in these solutions. The single-particle, or microscopic, picture shows that particles moving along the lines of force remain inside the stellarator tube to the same approximation as do the lines of force. Other particles are presumably confined by the action of the radial electric field that may be anticipated.

Theory predicts and observation confirms that initial breakdown, complete ionization, and heating of a hydrogen or helium gas to about $10^6$ degrees K are possible by means of a current parallel to the magnetic field (ohmic heating). Flow of impurities from the tube walls into the heated gas, during the discharge, may be sharply reduced by use of an ultra-high vacuum system; some improvement is also obtained with a divertor, which diverts the outer shell of magnetic flux away from the discharge. Experiments with ohmic heating verify the presence of a hydromagnetic instability predicted by Kruskal for plasma currents greater than a certain critical value and also indicate the presence of other cooperative phenomena. Heating to very much higher temperatures can be achieved by use of a pulsating magnetic field. Heating at the positive-ion cyclotron resonance frequency has been proposed theoretically and confirmed observationally by Stix. In addition, an appreciable energy input to the positive ions should be possible, in principle, if the pulsation period is near the time between ion-ion collisions or the time required for a positive ion to pass through the heating section (magnetic pumping).

## 1. INTRODUCTION

The controlled release of thermonuclear power requires[1] the confinement of a hydrogen plasma at a temperature exceeding $10^8$ degrees for an appreciable fraction of a second. A strong magnetic field would appear to offer the only practical hope of achieving this objective. While an enormous variety of magnetic configurations are possible, two relatively simple geometries may be set apart at the outset, both involving infinite cylinders, with axial symmetry. In the first of these, the magnetic field is produced by an axial current flowing through the gas. This configuration is the so-called "pinch-effect," which has been extensively discussed in the recent literature. In the second simple geometry, the magnetic field is parallel to the axis, and is produced by external currents, flowing in solenoidal windings encircling the plasma. Such a straight cylinder, with an externally produced field, forms the basis of the magnetic mirror or "pyrotron" device, proposed by Post.[2]

The stellarator, like the pyrotron, utilizes an external magnetic field, produced by coils encircling a tube containing the heated gas. However, instead of a finite cylindrical tube, the stellarator employs a tube bent into a configuration topologically like a torus, without ends. Such a tube will be referred to as "toroidal." Relative to the pyrotron, this configuration has the advantage, in principal, of permitting more complete confinement, since end losses are eliminated. Relative to the pinch discharge the stellarator offers the advantage, again in principle, of permitting equilibrium in a steady state. Both these advantages might be of importance in a controlled thermonuclear reactor. The present paper outlines the basic concepts involved in the confinement and heating of a gas in a stellarator.

One basic complexity in the stellarator results from the fact that the simple torus, in which the magnetic lines of force are circles centered at the axis of symmetry, does not permit equilibrium confinement of a

[1] Post, R. F., *Revs. Modern Phys.* **28**, 338 (1956).
[2] Post, R. F. (to be published).

plasma in a straightforward manner. Microscopically this result follows at once from the particle drifts associated with the inhomogeneity of the magnetic field. These drifts, pointed out some 50 years ago by Thomson,[3] subsequently discussed by Gunn,[4] and recently analyzed in detail by Alfvén,[5] produce motions perpendicular both to $B$ and to $\nabla B$; these motions are in opposite directions for electrons and positive ions. The resultant separation of charges produces electric fields which sweep the ionized gas towards the wall. Macroscopically, this same result is obtained from the fluid equations in Sec. 3.1.

Basically, the confinement scheme in the stellarator consists of modifying the magnetic field so that a single line of force, followed indefinitely, generates not a single circle but rather an entire toroidal surface, called a "magnetic surface." The tube enclosing the gas is, ideally, one of these surfaces, enclosing an entire family of such magnetic surfaces. It is shown in Sec. 2 that such surfaces can be produced, to a high approximation. Section 3 discusses the confinement of a plasma in a system characterized by magnetic surfaces. Heating of the gas is treated in Sec. 4; the techniques considered involve heating to intermediate temperatures (about $10^6$ degrees K) by currents flowing parallel to the confining magnetic field, and subsequent heating to very high temperatures by means of a pulsating magnetic field, with a wide variety of possible pulsation frequencies.

These analyses do not take into account the complex time-dependent cooperative processes that normally occur in laboratory plasmas. Any attempt to predict theoretically how well a gas at thermonuclear temperatures might be confined within a stellarator is hindered at the present time by lack of information or understanding of such cooperative phenomena.

The present paper constitutes an introduction to the series of papers from Project Matterhorn printed in this Journal. Subsequent theoretical papers discuss the macroscopic equilibrium of plasma in the stellarator, treat in great detail the problem of hydromagnetic stability, and obtain quantitative estimates of the amount of heating to be anticipated with various methods. The experimental papers, to appear in the next issue of this Journal, present observational material on ionization and heating of a gas in the stellarator, with particular emphasis on the evidence for instabilities and other cooperative phenomena.

[3] Thomson, J. J., *Conduction of Electricity through Gases* (Cambridge University Press, New York, 1906), second edition, p. 109.

[4] Gunn, R., *Phys. Rev.* **33**, 832 (1929).

[5] Alfvén, H., *Cosmical Electrodynamics* (Clarendon Press, Oxford, England, 1950).

## 2. Rotational Transform and Magnetic Surfaces

In this section we consider certain properties of a stellarator magnetic field which do not depend directly on the presence of a plasma in the system. In particular, we are interested in how far a line of force in a stellarator tube may be followed before it intersects the tube wall. If we could show in all rigor that one and only one magnetic surface passed through each point within the tube, then no line of force would ever intersect the tube wall. While this result is not exactly true, it is apparently true to a high order of approximation. The demonstration of this result depends on a certain abstract property of the stellarator magnetic field, called a "rotational transform." We first discuss the properties of such transforms, and postpone until later a discussion of how they are produced.

### 2.1 Properties of Rotational Transforms

Let us pass a plane through the stellarator tube at some point. For convenience in representation, we shall take the plane to be perpendicular to the magnetic field, $\mathbf{B}$, in the central region of the tube, although this restriction is not required. We assume that the magnetic field is nonzero at all points in the cross-sectional plane. Then through any point, such as $P_1$, for example, a line of force will pass. This line of force may now be followed, in the direction of the magnetic field, along the stellarator tube, until it has completed one circuit of the toroidal tube and intersects the cross-sectional plane at a point $P_2$, as shown in Fig. 1. In the ideal torus $P_2$ coincides with $P_1$ and every line of force is closed after one circuit. If the degeneracy of the torus is removed, $P_2$ no longer coincides with $P_1$.

One might raise some question as to whether any of the points $P_2$ lie inside the tube wall. It is not difficult to construct a toroidal tube with a solenoidal winding such that most of the lines of force stay within the tube for at least one circuit. The interesting question is rather which lines intersect the tube wall after many circuits, and we may safely assume that for most points $P_1$, $P_2$ will exist and will lie inside the tube wall. If we exclude, for the moment, those areas for which $P_2$ does not lie within the tube wall, then for every point $P_1$, we have a point $P_2$; the transformation of the set of points $P_1$ into the set of points $P_2$ is called a "magnetic transform" of the cross-sectional plane.

A magnetic transform of this type is characterized by an important property. Let us take the density of points $P_1$ to be proportional to the component of $B$ normal to the cross-sectional plane. Since the plane is transformed into itself by the magnetic transform, the density of these

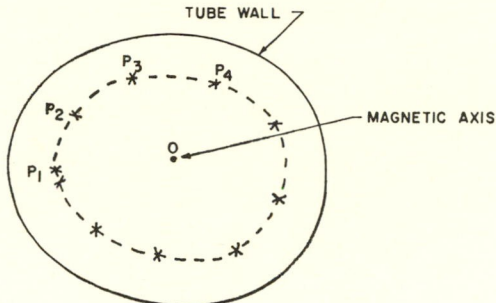

FIG. 1. Cross-sectional plane, showing intersection points produced by a single line of force.

points must be the same function of position before the transform as after. A transform with this property has been called "measure preserving" by Kruskal.

Let us now assume that the magnetic transform is also "primarily rotational." This condition states that at least the outer portions of the plane all rotate in the same direction in a single transformation. As we shall see subsequently, there are several ways of achieving this result in a toroidal system. According to Kruskal,[6] it follows from this assumption and from the Brouwer fixed-point theroem that there must be at least one point in the plane which is transformed into itself. In some types of stellarator the magnetic transform involves only small deformations of the plane, in addition to a general rotation. In such systems there will be only one point that transforms into itself, and only one line of force that is closed after a single circuit around the stellarator. This line is called the "magnetic axis."

A second basic result on measure-preserving rotational transforms, also established by Kruskal,[7] is that any point, other than one of the fixed points, when followed through successive transformations, will not move far from a single closed curve. This result is illustrated in Fig. 1, where the points $P_2$, $P_3$, $P_4$ generated by successive magnetic transforms of the point $P_1$, all lie close to a single closed curve. Thus a single line of force, after many circuits around the tube, generates a magnetic surface.

More precisely, let us introduce coordinates $r$ and $\theta$ in the cross-sectional plane depicted in Fig. 1; $r$ may be measured from the mag-

---

[6] Kruskal, M. D. (informal communication).

[7] Kruskal, M. D., U.S. Atomic Energy Commission Report No. NYO-998 (PM-S-5), 1952.

magnetic axis, denoted by the point 0. The value of $\Delta\theta$ between $P_1$ and $P_2$ is denoted by $\iota$, and is called the "rotational transform angle." Let $\theta$ equal 0 at $P_1$, and let us assume that $\theta$ equals $2\pi$ for the point $P_n$. The distance $\Delta r$ from $P_1$ to $P_n$ is called the "deviation from closure" of the point $n$. Evidently $\Delta r$ measures how far the line of force has strayed from a closed curve. Kruskal has shown that $\Delta r$ decreases more rapidly than any power of $1/n$. Hence, one may surmise that $\Delta r$ varies about as $\exp(-Kn)$, where $K$ is some dimensionless constant.

The physical reason for this result can best be understood in the special case that the normal component of the magnetic field is constant over the cross-section plane. The analysis of more general systems may be reduced mathematically to a consideration of this special situation. In this case the density of points in the plane must remain constant in successive transformations. Let us now draw in the cross-section plane a closed curve connecting point $P_1$ and its successive transformed points as smoothly as possible. Since the magnetic transform now preserves areas, the total area enclosed within this curve must remain constant in successive transformations. Hence all points on the curve cannot move inwards with successive transformations. If some move in, others must move out.

In the special case that the $\theta$ coordinate of every point returns to its original value after $n$ transforms, it is possible for some points on the curve, together with all their transformed points, to move steadily in, while the points between move steadily out. Thus the closed curve develops wrinkles in successive transformations; this rate of wrinkling decreases very rapidly with increasing $n$. In the more general case that the $\theta$ coordinate of a point never returns exactly to its initial value (to within a multiple of $2\pi$), one would expect a further averaging out of these radial motions to occur. Even if the transform angle, $\iota$, is not small, the deviation of a line of force from a magnetic surface should be small; if necessary, a group of $p$ transforms, which give an $\iota$ very nearly a multiple of $2\pi$, can be taken as the basic transform, and Kruskal's theorem used to demonstrate that a line of force never departs very far from a single closed surface. We shall therefore assume in the following discussions that each line of force does in fact generate a single magnetic surface, and that a single line, if followed sufficiently far, comes arbitrarily close to any point on this surface.

## 2.2 Methods for Producing a Rotational Transform

To produce a rotational transform in a vacuum field it suffices to twist a torus out of a single plane. Virtually any such distortion will remove the degeneracy of the ideal torus and produce a rotational transform.

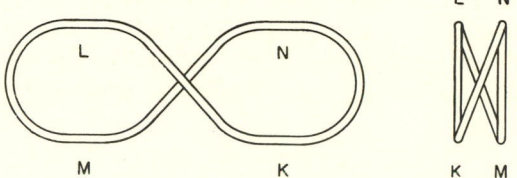

FIG. 2. Top and end views of a figure-eight stellarator.

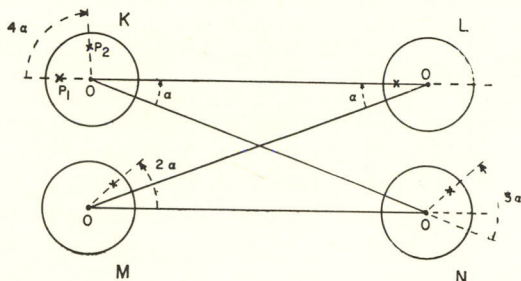

FIG. 3. Rotational transform in a figure-eight stellarator.

The simplest such system is the figure-eight, historically the first geometry proposed[8,9] for a stellarator. The topography is indicated in Fig. 2. The 180° curving end sections, $LM$ and $KN$, are in planes each tilted at an angle, $\alpha$, to the parallel planes in which the reverse-curvature sections $LK$ and $MN$ are placed. Figure 3 indicates that a rotational transform is present. This figure represents cross-sectional planes at $K$, $L$, $M$, and $N$, all as seen from one end of the device. The point 0 represents the magnetic axis, while $P_1$ and $P_2$ denotes the successive intersections of a single line of force with a cross-sectional plane at $K$. Intersections of this line of force with the other three planes are denoted by crosses. The solid lines represent the path followed by the magnetic axis. From $K$ to $L$ and from $M$ to $N$ there occur simple translations of the cross-sectional plane, while from $L$ to $M$ and from $N$ to $K$, a rotational about an axis, inclined at an angle $\alpha$ to the vertical, is involved.

Evidently the line of force which passes through $P_1$ in plane $K$, and is

[8] Spitzer, L. Jr., U.S. Atomic Energy Commission Report No. NYO-993 (PM-S-1), 1951.
[9] Spitzer, L. Jr., U.S. Atomic Energy Commission Report No. NYO-995 (PM-S-3), 1951.

Evidently the line of force which passes through $P_1$ in plane $K$, and is then followed through one circuit, through planes $L$, $M$, and $N$, intersects plane $K$ again in a point $P_2$, rotated by a rotational transform angle $\iota$. Examination of the figure shows that for this geometry,

$$\iota = 4\alpha. \tag{1}$$

Moreover, $\iota$ is independent both of distance, $r$, from the magnetic axis, and of angle, $\theta$. In an actual system, mutual interference between the stray fields of the curving sections $LM$ and $MN$ will modify these results slightly, but the general features remain unchanged. In more general twisted systems of this type, with a purely solenoidal field, $\iota$ is simply the integral of the torsion around the magnetic axis, a result first pointed out by Kruskal.[6]

A rotational transform angle may be produced in a variety of other ways. When a plasma current is flowing around the simple torus, a rotational transform appears, despite its absence in the vacuum field. If steady-state confinement is envisaged, however, a rotational transform must be present in the vacuum field, since a plasma current along the magnetic field cannot be maintained in a steady state. The most important alternative method for producing such a rotational transform is the use of a transverse magnetic field, whose direction rotates with distance along the magnetic axis. We shall follow a line of force and show that a transform angle appears. Let us consider an infinite cylinder, with coordinates $r$, $\theta$, and $z$. We consider the $r$ and $\theta$ coordinates of a single line of force as $z$ increases. The coordinates of points along such a line of force are related by the differential expression

$$\frac{dz}{B_z} = \frac{dr}{B_r} = \frac{r\,d\theta}{B_\theta}. \tag{2}$$

Suppose now that $B_z$ and $B_\theta$ are produced by $2l$ wires, wound helically on the outside of the cylinder, such that currents flow oppositely in adjacent wires, and with a pitch $2\pi l/h$. If we denote the tube radius by $r_1$, and if $hr$ is small compared to unity, then for small $r/r_1$ we have, from the appropriate solutions of Laplace's equation

$$B_r = Ar^{l-1}\sin(l\theta - hz), \tag{3}$$

$$B_\theta = Ar^{l-1}\cos(l\theta - hz), \tag{4}$$

where $A$ is a constant characterizing the strength of the transverse field.

There is also a component of $B_z$ associated with the current in the helical wires, but its magnitude is less than $B_r$ by a factor $hr$. We assume that $B_{z0}$, the component of $B_z$ produced by a separate solenoidal winding, is the dominant axial field.

Let us follow a line of force whose coordinates are $r_0$ and $\theta_0$ in the absence of the transverse field. Equation (2) may now be integrated by means of a power-series expansion in $A$, the coefficient in Eqs. (3) and (4). To first order in $A$, $r - r_0$ and $\theta - \theta_0$ vary as the cosine and sine, respectively, of $l\theta - hz$; to this order, the line of force is a helix, and its intersection with a plane moving along the $z$ direction is a circle. Solving to second order in $A$, we must take into account that for $l$ equal to 2 or more, $B_\theta$ is larger on the outside of the circle ($r$ greater than $r_0$), where $B_\theta$ is positive, than on the inside ($r$ less than $r_0$), where $B_\theta$ is negative. As a result, the positive values of $d\theta/dz$ in Eq. (2) more than offset the negative ones, and $\theta$ increases systematically with increasing $z$. A detailed integration by Johnson and Oberman[10] shows that $\iota_h$, the increase of $\theta$ in one period $2\pi/h$ of the helical field, is given by

$$\iota_h = \frac{\pi A^2 r^{2l-4}}{h^2 B_{z0}^2} \{2(l-1) + h^2 r^2 + O(h^4 r^4)\}. \tag{5}$$

The term in $(hr)^2$ is included to give results for $l$ equal to unity; in this case a rotational transform arises from the variation of $B_z$ with $r$. The configuration for which $l$ is unity, with a helical magnetic axis, and its use for confining a plasma were proposed by Koenig,[11] who first studied the use of helical fields in connection with the stellarator. For small $hr$, an appreciable $\iota$ is more readily obtained with transverse fields of higher multiplicity for which the transverse field vanishes at the magnetic axis. The properties of such fields, and of the magnetic surfaces associated with them, have been extensively studied[10] by the Matterhorn theoretical group, under E. A. Frieman.

In the experimental program at Matterhorn, described in the subsequent papers,[12-16] rotational transforms have been produced both with the figure-eight geometry and with transverse fields, with $l$ equal to 3. In either case, the existence of a rotational transform is readily confirmed

[10] Johnson, Oberman, Kulsrud, and Frieman, *Phys. Fluids* 1, 281 (1958).

[11] Koenig, H. R., U.S. Atomic Energy Commission Report No. NYO-7310 (PM-S-20), 1956.

[12] Coor, Cunningham, Ellis, Heald, and Kranz, *Phys. Fluids* (to be published).

[13] Kruskal, Johnson, Gottlieb, and Goldman, *Phys. Fluids* (to be published).

[14] Bernstein, Chen, Heald, and Kranz, *Phys. Fluids* (to be published).

[15] Burnett, Grove, Palladino, Stix, and Wakefield, *Phys. Fluids* (to be published).

[16] Stix, T. H. and Palladino, R. W., *Phys. Fluids* (to be published).

experimentally by observation of a narrow electron beam, which follows the line of force. As pointed out subsequently, the chief advantage of the multipolar transverse field over the figure-eight is the greater hydromagnetic stability which, in theory, it should yield.

## 3. CONFINEMENT

The objective of confinement theory is to demonstrate that the number of particles striking the tube wall is negligibly small. An exact proof would presumably require a detailed solution of the Boltzmann equation, together with the field equations. For the complex magnetic configuration of the stellarator this would be a difficult task indeed. An approximate treatment will be followed here.

First we shall use the macroscopic equations for the plasma, based on a number of simplifying assumptions. With these equations, together with the field equations, we can show that an equilibrium situation is possible in which the plasma is macroscopically confined. Since particles moving with some particular velocity might conceivably escape, even though the plasma as a whole were confined, we shall next consider the trajectories of single particles in the electric and magnetic fields determined from the macroscopic equations. These two approximate methods, taken together, indicate nearly perfect confinement, if collision and cooperative phenomena are ignored. The hydromagnetic stability of these equilibria has been extensively analyzed by Frieman, Kruskal, and their collaborators. The result of this analysis, summarized briefly at the end of Sec. 3.2, indicate stable equilibrium under certain conditions. All these results, taken together, encourage the belief that magnetic confinement in a stellarator may be adequate for a controlled thermonuclear reactor.

### 3.1 Macroscopic Equations

The two-fluid equations for the velocity and electric current of a fully ionized gas are well known.[17, 18] For the derivation of these equations and their application here the following nontrivial assumptions are made:

(a) The electrical resistivity $\eta$ is negligibly small, and the mean free path very long.

---

[17] Schlüter, A., *Z. Naturforsch.* **5a**, 72 (1950).

[18] Spitzer, L. Jr., *Physics of Fully Ionized Gases* (Interscience Publishers, Inc., New York, 1956).

(b) Over a distance of one radius of gyration the relative change of all macroscopic quantities is small.

(c) The transverse and longitudinal pressures, $p_\perp$ and $p_\parallel$ are equal.

(d) All macroscopic quantities are independent of time at each position.

(e) The mean macroscopic velocity, $\mathbf{v}$, vanishes.

Assumption (a) is clearly a good approximation for a rarefied gas at very high temperatures. Collisions between particles will produce some diffusion across the lines of magnetic force,[18] but this rate is so slow compared both with the diffusion rate observed[12] and with the diffusion rate that could be tolerated that we may neglect collisions entirely in most discussions of confinement. Once this first assumption has been made, assumption (b) is required for use of the macroscopic equations. This second assumption has a number of important consequences. Since the sheath thickness is generally much less than the radius of the gyration, the plasma must be characterized by approximate electrical neutrality, with the electron density $n_e$ closely equal to $Z$ times the ion density $n_i$, where $Z$ is the mean ionic charge. In addition, this assumption leads[19,20] to the result that the material stress tensor is diagonal, provided that the principal axis is parallel to the magnetic field; the three components are then $p_\parallel$ parallel to the field and $p_\perp$ in the two directions perpendicular to the field. In any device much larger than the radius of gyration, assumption (b) should be approximately valid in a steady state, except in a thin layer near the wall where this assumption must break down. Assumption (c) should also be valid in a system where any changes are even slower than the time between collisions, since collisions will clearly tend to equalize $p_\parallel$ and $p_\perp$.

Assumption (d) is of critical importance in the analysis. The assumption of a steady state immediately implies that the plasma is quiescent, and excludes turbulence, oscillations, instabilities, and other cooperative phenomena of the sort normally present in gaseous discharges. Any equilibria obtained on the basis of this assumption may not necessarily be stable against various types of disturbances. The last assumption replaces the more usual one (which partly results from assumption (b)) that quadratic terms in $\mathbf{v}$ and $\mathbf{j}$ are negligible. This more stringent condition is not so arbitrary as might first appear. In fact, the near vanishing of $\mathbf{v}$ is a simple consequence[21] of the equation of motion. The argument is that in most heating methods there are no appreciable forces tending to produce any momentum, and hence the macroscopic

[19] Watson, K. M., *Phys. Rev.* **102**, 12 (1956).

[20] Chew, Goldberger, and Low, *Proc. Roy. Soc.* (*London*) **A236**, 112 (1956).

[21] See reference 18.

velocity must be vanishingly small. A fuller analysis of effects associated with macroscopic velocities would be desirable.

On the basis of these assumptions, the equations of equilibrium become, in emu,

$$\mathbf{j} \times \mathbf{B} = \nabla p, \tag{6}$$

$$\nabla \times \mathbf{B} = 4\pi \mathbf{j}, \tag{7}$$

$$\nabla \cdot \mathbf{B} = 0. \tag{8}$$

In addition, the generalized Ohm's law determines the electric field, in terms of the pressure gradient for the positive ions, while Poisson's law then gives the charge density. Since neither of these quantities is of particular significance in the present analysis these equations may be omitted.

On the basis of these equations it was shown[22] several years ago that no simple equilibrium is possible in a torus if the lines of force are assumed circles centered at the axis of symmetry. We introduce cylindrical coordinates $R$, $\varphi$, and $z$, with $z$ taken along the axis of symmetry of the torus; we assume that only $B_\varphi$ differs from zero, and that all quantities are independent of $\varphi$. If we now take the curl of Eq. (6), eliminating $\mathbf{j}$ by means of Eq. (7), we obtain

$$\partial B_\varphi / R \partial z = 0. \tag{9}$$

Thus for equilibrium either $R$ must be infinite or the magnetic field (and pressure) must be independent of $z$. Equation (9) for toroidal fields ($B_R = B_s = 0$) corresponds to a theorem by Ferraro[23] for poloidal fields ($B_\varphi = 0$) in connection with the theory of magnetic stars. It may be remarked that the left-hand side of Eq. (9) is simply $-4\pi \nabla \cdot \mathbf{j}$; i.e., the currents required by Eq. (6) for the simple torus possess a nonvanishing divergence.

### 3.2 Plasma Equilibrium in the Stellarator

We next apply these equations to a stellarator, characterized by the existence of toroidal magnetic surfaces. From Eq. (6) it is evident that $\nabla p$ is perpendicular to $\mathbf{B}$ and hence $p$ is constant along a line of force and must therefore be constant over an entire magnetic surface. Simi-

[22] Spitzer, L. Jr., U.S. Atomic Energy Commission Report No. NYO-997 (PM-S-4), 1952.

[23] Ferraro, V.C.A., *Astrophys. J.* **119**, 407 (1954).

larly $\mathbf{j}$ is perpendicular to $\nabla p$ and, therefore, $\mathbf{j}$ must be parallel to the isobaric surfaces, and hence to the magnetic surfaces. We denote by $\mathbf{j}_\perp$ the component of $\mathbf{j}$ perpendicular to $\mathbf{B}$ (and to $\nabla p$) and by $\mathbf{j}_\parallel$ the component of $\mathbf{j}$ parallel to $\mathbf{B}$.

An important result, which permits plasma equilibrium in the stellarator, will now be established. Equation (6) determines $\mathbf{j}_\perp$ in terms of $\mathbf{B}$ and $\mathbf{p}$. In general, $\nabla \cdot \mathbf{j}_\perp$ will not be zero. In the torus the divergence of this current gives rise to charge separation which destroys equilibrium, since conditions are uniform along each line of force and an accumulation of electric charge cannot readily leak out across the lines of force. In the stellarator, currents along the lines of force are possible, and charges may flow from a region where $\nabla \cdot \mathbf{j}_\perp$ is positive to another where $\nabla \cdot \mathbf{j}_\perp$ is negative. The total divergence of $\mathbf{j}_\perp$, integrated over the volume element between two adjacent magnetic surfaces, must vanish, as may be seen by use of Gauss's theorem, together with the fact that $j_\perp$ is parallel to the magnetic surface. Hence currents flowing along the lines of force can cancel out the divergences in $\mathbf{j}_\perp$, and will lead to a current system in which $\nabla \cdot \mathbf{j}$ vanishes.

The existence of solutions to Eqs. (6)–(8), in a system characterized by magnetic surfaces, has been analyzed in an elegant manner by Kruskal and Kulsrud,[24] who also take diffusion into account. Here we follow an earlier and simpler treatment,[22] and demonstrate the existence of solutions to Eqs. (6)–(8) by the simple artifice of showing how to construct such solutions. We assume that $p$ is a small quantity, and obtain a solution by successive iteration. The zero-order solution is taken as the vacuum solution, with $j_0$ the current in the external coils, and $B_0$ the vacuum field, with its magnetic surfaces. The solution of order $n$ is then defined by the equations

$$\mathbf{j}_n \times \mathbf{B}_{n-1} = \nabla p_n, \tag{10}$$

$$\nabla \times \mathbf{B}_n = 4\pi \mathbf{j}_n, \tag{11}$$

$$\nabla \cdot \mathbf{B}_n = 0. \tag{12}$$

Evidently $p_n$ must be assumed constant on the magnetic surface obtained in the previous iteration, but is otherwise arbitrary. It may generally be assumed that $p_n$ is in each case a monotonically decreasing function of distance from the magnetic axis. If the solution converges, it must evidently yield a solution of Eqs. (6)–(8).

[24] Kruskal, M. D. and Kulsrud, R. M., *Phys. Fluids* 1, 265 (1958).

In principle this iteration scheme is straightforward, but in practice the algebra is cumbersome. It turns out that the chief obstacle to convergence is the distortion of the magnetic surfaces by the magnetic fields associated with the plasma current $j_{\parallel}$, along the magnetic field. Even though this current density is low, the currents must travel an appreciable distance, and even the weak magnetic field associated with these currents may distort the magnetic surfaces out of all recognition.

The approximate condition for convergence in a simple figure-eight device may be derived in the simplest manner. The transverse current $j_{\perp}$ is evidently about equal to $p/B_0 r$, where $B_0$ is the axial vacuum field and $r$ is the distance of the tube wall, or plasma boundary, from the magnetic axis. The divergence of $j_{\perp}$ results from the inverse proportionality between $B_0$ and $R$, and is about equal in magnitude to $p/B_0 rR$, where we take $R$ to be the radius of curvature of the magnetic axis. The divergence of $\mathbf{j}_{\parallel}$ is also equal to this quantity, and over a tube length equal to $R$, $j_{\parallel}$ will build up to about $p/B_0 r$. Hence $j_{\parallel}$ is of the same order of magnitude as $j_{\perp}$. Over a tube cross section, $j_{\parallel}$ will have opposite directions on opposite sides. The magnetic field on the axis due to this plasma current will be of the order $2\pi r j_{\parallel}$, or $2\pi p/B_0$. The new magnetic axis therefore will be inclined at an angle $2\pi p/B_0^2$ relative to its direction *in vacuo*. The condition for negligible distortion of the magnetic surfaces is that the deviation of the magnetic axis be small compared to $r$. If the currents $j$ flow along half the axial length, $L$, of the machine, on the average, before cancellation, this condition yields

$$\beta \ll 8r/L; \tag{13}$$

$\beta$, the ratio of material to magnetic pressure, is defined by

$$\beta = 8\pi p/B_0^2, \tag{14}$$

where $p$ is evaluated on the magnetic axis. A more precise discussion,[22] taking into account the detailed variation of $j_{\parallel}$ over the cross section, yields a coefficient $16/\pi$ instead of 8 in Eq. (13); this computation assumes that $L$ much exceeds $2\pi R$ and that the transform angle $4\alpha$ (see Fig. 3) is small. If inequality (13) is not satisfied, the method of iteration fails, and it is not known what type of solution, if any, may exist.

In an infinite cylinder, values of $\beta$ as great as unity might be envisaged. In the figure-eight stellarator, of the type shown in Fig. 2, $r/L$ can scarcely exceed 0.02, and $\beta$ must therefore be small compared to 0.1. If the rotational transform is produced by transverse fields, however, the transform angle, $\iota$, for the device may much exceed $2\pi$, in principle. It is readily shown that the upper limit on $\beta$ is proportional to $\iota^2 r/R$, in this situation, and hence equilibrium values of $\beta$ substantially greater than 0.1 should be possible, although at the cost of somewhat greater over-all axial length.

The question of the hydromagnetic stability of such configurations has been extensively studied by the Matterhorn theoretical group, under E. Frieman. Basic concepts and methods of analysis have been published by Bernstein, Frieman, Kruskal, and Kulsrud,[25] with application to the stellarator in the paper by Johnson, Oberman, Kulsrud, and Frieman.[10] Because of the importance of this work, the results will be summarized briefly here.

Instabilities tend to be most marked if the lines of force can interchange positions with the least possible bending. In the case of an axially symmetric field, if $B_\theta$ vanishes everywhere, the lines of force can interchange positions without bending, and if bulges are present in the field the plasma is unstable. However, if a $B_\theta$ component is present, so that lines of force are helices about the cylinder axis, and if $B_\theta/r$ increases with $r$ so that the pitch of the helices decreases with increasing $r$, then the outer and inner lines of force cannot be interchanged without appreciable bending, and the plasma tends to be stable. In the same way, if the transform angle, $\iota$, varies with $r$ in a stellarator, the outer lines of force are topologically different from the inner ones, and the plasma is stable against all hydromagnetic disturbances, provided that $\beta$ is less than some critical value $\beta_c$. Computations of $\beta_c$ for a cylinder, with helical transverse fields, with $l = 3$, added to an axial confining field, indicate[10] that values of $\beta_c$ as great as 0.1 could be obtained if the approximate theory could be trusted somewhat beyond its range of validity. There is some reason to believe that corrections for finite radius of gyration may increase $\beta_c$ by a factor of about two, although the theory is still very incomplete in this respect. An experimental test of this theory has not yet been obtained, although the corresponding theory applied to kink instability in the stellarator (see Sec. 4) is apparently in close agreement with the observations. The maximum value of $\beta$ for which a stellarator plasma is stable can probably best be determined by experiment.

### 3.3 Single Particles

In the absence of collisions, confinement of single particles will be shown to follow quite generally from the existence of a rotational transform and from the asymptotic behavior of a gyrating particle in a

---

[25] Bernstein, Frieman, Kruskal, and Kulsrud, *Proc. Roy. Soc.* (*London*) **A244**, 17 (1958).

strong magnetic field. It has been shown by Kruskal[26] that the magnetic moment, $\mu$ (about equal by $mw_{\perp}^2/2$), of a gyrating particle is constant to all orders of $ak$, where $a$ is the gyration radius and $k$ is about $|\nabla(\ln B)|$. Constancy of $\mu$ to first order in $ak$ had previously been demonstrated by Alfvén,[5] and to second order by Hellwig.[27] Similarly, Kruskal has also shown[26] that the motion of the guiding center is independent of the phase of gyration to all orders of $ak$. Kruskal's theory does not yield either a simple definition of $\mu$, to all orders in $ak$, nor yet a simple definition of the guiding center, but shows that such definitions must exist and how, in principle, to construct them. In consequence of these results we may assume that for each particle the total energy, $W$, is not the only integral of the motion, but there exists also a second integral, the magnetic moment, $\mu$.

Thanks to these results it may now be shown that successive intersections of particles with a particular cross-section plane, similar to that discussed in Sec. 2.1, produce a transformation similar in its properties to the magnetic transform generated by successive intersections of a line of force with this same plane. We restrict consideration to particles within ranges $dW$ and $d\mu$ centered at some energy, $W$, and some magnetic moment, $\mu$, and let the density in phase space, within these narrow ranges of $W$ and $\mu$, be constant everywhere. Within an interval of time $\Delta t$, the guiding centers of these particles will intersect the cross-sectional plane at a number of points, $P_1$. The density of such points will be a known function of position in the plane, depending on the magnetic field $B$, and the electric potential $\Phi$, through the two integrals of motion.

Each particle whose guiding center has intersected the plane at a point $P_1$ will ultimately cross the plane again with the guiding center intersecting at a point $P_2$. Normally, if a particle passes through a point, the three components of its velocity, $\mathbf{w}$ are arbitrary, and its subsequent trajectory is not determined. In the present case, the two integrals of motion determine $w_{\perp}$ and $w_{\parallel}$, and the third velocity component, the phase of gyration, has no effect on the motion. Hence to each point $P_1$, there corresponds one and only point $P_2$. Moreover, the particles which have produced all the points $P_1$ in the time $\Delta t$ will all produce points $P_2$ within the same time interval, and hence the density of intersection points in the transformed plane will be the same function of position as in the original plane. Thus this "particle transform" is measure preserving in the same sense as is the magnetic transform discussed earlier.

[26] Kruskal, M. D., *Proceedings of the Third International Conference on Ionization Phenomena in Gases*, Venice, 1957 (to be published).

[27] Hellwig, G., *Z. Naturforsch.* **10a**, 508 (1955).

From the same arguments as before it follows that free particles are confined in a stellarator to a very high approximation, provided that the particle transform is primarily rotational. Such a transform will be assured for particles whose velocity is mostly parallel to $\mathbf{B}$ ($w_\parallel > w_\perp$), so that no reflection can occur from regions of relatively high field. The rotational magnetic transform guarantees a rotational particle transform for such particles. Qualitative experimental confirmation of this prediction is obtained from observations of runaway electrons reported[12] by the Matterhorn experimental group under M. B. Gottlieb. Electrons traveling at speeds near the velocity of light are observed to make about $5 \times 10^5$ circuits around a stellarator, during the ten milliseconds or so after the applied voltage is reduced to zero, but the confining field is still moderately high.

For particles which are trapped between two regions of relatively high field, or which are moving at a relatively slow rate along the magnetic field, further arguments must be invoked to guarantee a primarily rotational particle transform. Two separate mechanisms are important. Firstly, the diamagnetic effect of the plasma produces a radial gradient of the axial field, and this inhomogeneity produces a drift of guiding centers about the magnetic axis. Secondly, a radial electric field will produce a similar rotation. Such an electric field is required by the assumption that the macroscopic velocity vanish, since only a radial electric field can cancel out, in a steady state, the velocity associated with a radial pressure gradient. It has been shown by Spitzer[22,28] that such a radial field arises naturally when the gas is ionized and heated. A more detailed analysis[29] of these phenomena indicates that these two effects produce adequate rotation about the magnetic axis to guarantee confinement for most particles with relatively low $w_\parallel$.

One important exception should be noted. For some particles the different effects producing rotation may cancel out, leaving only the unidirectional drift produced by the curvature of the field. The seriousness of this effect is reduced by two factors. As distance from the magnetic axis changes, the different mechanisms producing rotation will change in different ways, and the cancellation will disappear. Moreover, the cancellation will be exact only for a particular particle energy, $W$, and a particular magnetic moment, $\mu$; collisions will change these quantities, and restrict the time during which a unidirectional drift will occur. These effects have been discussed elsewhere[9,29]; the analysis, while admittedly approximate and incomplete, indicates that this pro-

[28] Spitzer, L. Jr., *Astrophys. J.* **116**, 299 (1952); see pp. 308–9.
[29] Spitzer, L. Jr., U.S. Atomic Energy Commission Report No. NYO-7316 (PM-S-26), 1957.

cess is not of great importance, although it may increase the diffusion rate somewhat above the value given by electron-ion collisions.

## 4. HEATING

A gas can be ionized and heated, in general, by energetic particles, by photons, or by electric fields. Consideration of heating in a stellarator has been limited to electric fields. Two general types of heating by electric fields may be distinguished, depending on whether the electric field, $E$, is parallel to or perpendicular to the magnetic field, $B$. In the configuration of the stellarator, an electric field parallel to $B$ can be produced only by induction, by changing the magnetic flux threading the toroidal tube. Since the current induced by this field heats the gas by ohmic, or Joule, losses, this process is known as "ohmic heating." An electric field perpendicular to $B$ could be produced electrostatically. However, a plasma shields itself so effectively against electrostatic fields that primary consideration has been given to electric fields induced by pulsating the magnetic field. Heating by this method is called "magnetic pumping" or, in the special case that the pulsation frequency is chosen close to the ion cyclotron frequency, "ion cyclotron resonance heating." The general principles involved in these heating methods are discussed in the following.

### 4.1 Ohmic Heating

In ohmic heating the only function of the magnetic field, in principle, is to prevent lateral diffusion, and the heating can be analyzed, to a first approximation, without regard to the magnetic field. The electrical breakdown of a gas is well known, and has been extensively analyzed elsewhere. Theoretical consideration at Princeton[30-33] has been devoted to the final ionization and heating of helium or hydrogen gas by means of a unidirectional pulse of constant voltage; the initial ionization level was assumed to be about 10 percent.

The problem of changes within a gas, when an electric field is applied, is very complicated in the general case, since both the radiation field and the electron velocity distribution may alter in complex ways. Under

[30] Berger, J. M. and Frieman, E. A., U.S. Atomic Energy Commission Report No. NYO-6046 (PM-S-16), 1954.

[31] Berger, J. M. and Goldman, L. M., U.S. Atomic Energy Commission Report No. NYO-7311 (PM-S-21), 1956.

[32] Berger, J. M., U.S. Atomic Energy Commission Report No. NYO-7312 (PM-S-22), 1956.

[33] Berger, Bernstein, Frieman, and Kulsrud, *Phys. Fluids* **1**, 297 (1958).

conditions of interest in the laboratory the radiation field is generally weak enough to be negligible. To make the problem tractable the electron velocity distribution may be assumed Maxwellian. The chief processes that need be considered other than straight heating, are excitation, ionization, charge exchange, etc., and in principle, at least, the rates of these processes are simple functions of the electron temperature.

The assumptions made in the theoretical work, and the results obtained, are given in the accompanying paper by Berger, Bernstein, Frieman, and Kulsrud.[33] The results indicate clearly that ionization and heating of hydrogen or helium by this technique should be entirely feasible, and temperatures of $10^6$ degrees K should be obtainable. At higher temperatures the plasma resistivity becomes so low that heating with practical currents becomes difficult. The detailed predictions of the theory are somewhat uncertain because the basic assumption of a Maxwellian distribution may not be entirely realistic. The mean free path of an electron increases so rapidly with increasing energy that electrons which are in the tail of the Maxwellian distribution and which are sufficiently energetic to excite and ionize atoms may gain very appreciable energies in one free path.

Before any experiments had been carried out on ohmic heating, it was pointed out by Kruskal[34] that discharges in the stellarator should be subject to kink instability. This hydromagnetic instability was predicted for heating currents greater than the critical current that will reduce the transform angle to 0 (or increase it to $2\pi$). This critical current is now generally known as the "Kruskal limit."

Extensive observations of ohmic heating have been carried out by the Matterhorn experimental group, under M. Gottlieb, and are reported in several subsequent experimental papers.[12-15] The data indicate clearly that nearly complete ionization is attained, with electron and ion temperatures in the neighborhood of $5 \times 10^5$ degrees K. The occurrence of the predicted kink instability, at currents above the Kruskal limit, is fully verified experimentally. However, the detailed predictions of the ohmic heating theory are not substantiated, presumably because of the non-Maxwellian distribution of electron velocities. In support of this hypothesis, intense x-rays from runaway electrons are observed, with energies up to $10^6$ ev.

These data emphasize the very great importance of impurities from the walls streaming into the discharge. In the early observations the carbon and oxygen ions presumably outnumbered the helium ions

[34] Kruskal, M. D., U.S. Atomic Energy Commission Report No. NYO-6045 (PM-S-12), 1954.

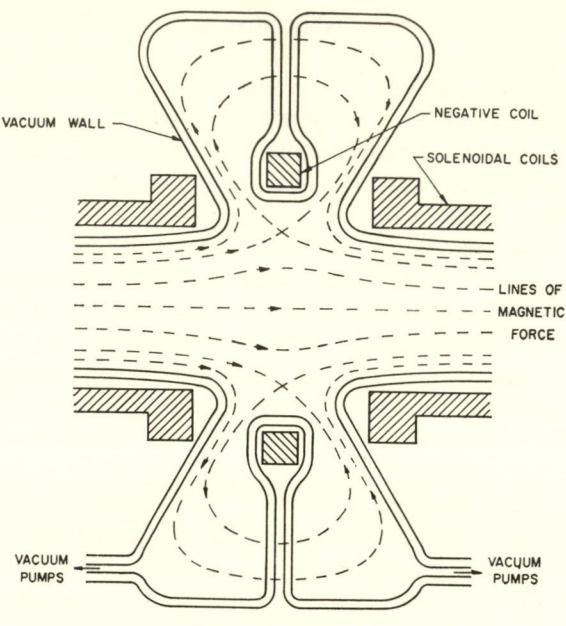

VACUUM WALL

NEGATIVE COIL

SOLENOIDAL COILS

LINES OF
MAGNETIC
FORCE

VACUUM
PUMPS

VACUUM
PUMPS

FIG. 4. Divertor.

during the later stages of the discharge, and sharply reduced the electron temperature. With the use of ultra-high vacuum techniques, resulting in base pressures below $10^{-9}$ mm of Hg and relatively clean surfaces, the efflux of wall impurities has been reduced by more than an order of magnitude.

Another method of reducing the impurity level has been use of a divertor. This device was proposed relatively early[8] to take away from the discharge the particles nearest the wall and to avert bombardment of the discharge tube by charged particles. In the divertor, an outer shell of flux is diverted or bent away from the main discharge into a large auxiliary chamber. Any impurities produced by wall bombardment in the divertor chamber return relatively slowly into the main discharge tube. A schematic diagram of a divertor is shown in Fig. 4; the device has cylindrical symmetry about the magnetic axis. The theory of this device, together with observations on its effectiveness in reducing the impurity level, without an ultra-high vacuum, is reported in the subsequent paper by Burnett, Grove, Palladino, Stix, and Wakefield.[15] Apparently the divertor reduces the ratio of impurity ions to helium ions by a factor of about one-fifth.

The most important new observational result that has emerged from these ohmic heating studies is the evidence on cooperative phenomena.

During ohmic heating the plasma is anything but quiescent. Runaway particles start abruptly to hit the tube wall, producing x-rays, sometimes in short bursts, at times dependent on the magnitude of the confining field. When voltage is applied around the stellarator, the plasma is not well confined by the magnetic field, and reaches the wall in about $10^{-4}$ sec, presumably because of cooperative phenomena of some sort. After ohmic heating, a current of runaway electrons, amounting to some 10 amp/cm$^2$ may persist for several milliseconds after the voltage is turned off, and then abruptly disappear, producing a burst of x-rays and additional ionization and excitation in the plasma. The detailed study of these phenomena should increase our understanding of plasma physics and enable one to predict how an ionized gas might behave in a full-scale thermonuclear reactor.

### 4.2 Magnetic Pumping

Pulsation of an axial confining field produces an oscillating electrical field, encircling the tube axis. This electric field can increase the energy of gyrating charged particles. If the pulsation frequency is much less than the cyclotron frequency, however, this increase in energy is computed most simply from the magnetic moment, $\mu$, which, according to the results by Kruskal[26] should be very accurately constant. Under these conditions, the increase in kinetic energy of motion, transverse to the field, is given by the usual formula for adiabatic compression of a gas, provided $\gamma$ is set equal to 2, corresponding to the presence of two degrees of freedom. Instead of thinking about the electric fields induced by the pulsating magnetic field, we may think of the external lines of force as constituting a piston, and the pulsation of the external field as providing a pumping action.

Evidently if the plasma were entirely adiabatic, or isothermal, the work done on the gas during the compression would just equal the work done on the piston during expansion. No heating would result. To obtain net heating in a pumping cycle there must be a phase lag between temperature and density. Such a phase lag may be produced in a variety of different ways, and hence there are many frequencies at which magnetic pumping can be effective.

One mechanism for producing such a phase lag is the effect of collisions in exchanging energy between motions parallel and perpendicular to the lines of force. This effect was analyzed early[35] at Project Matterhorn, as a possible substitute for ohmic heating, but was dropped

---

[35] Spitzer, L. Jr. and Witten, L., U.S. Atomic Energy Commission Report No. NYO-999 (PM-S-6), 1953.

because it was not an effective method for completing the ionization of a gas. More recently, this mechanism has been analyzed by Berger and Newcomb,[36] and, independently, by Schlüter,[37] with identical results. The analysis at Princeton is given in the subsequent paper by the Matterhorn theoretical group.[38] It is clear that magnetic pumping at the positive-ion collisional frequency can, in principle, heat a fully ionized gas to very high temperatures; however, the rate of heating falls off as $T^{-\frac{1}{2}}$ with increasing temperature, an inconvenient drop.

Another method of producing the desired phase lag is to pump in a short section of tube, with a pulsation period about equal to the time required for a positive ion to travel through the pumping section. In this situation the temperature lags because of loss of heat out the ends. If the mean free path is short compared to the length of the pumping section, magnetic pumping at this frequency produces acoustic waves, which travel along the magnetic field. For long mean free paths, the particles may be treated as free, and the energy is effectively thermalized by fine-scale mixing. The analysis[36] of this "transit-time heating," reported in a subsequent Matterhorn paper[38] shows that this method should be an effective means for heating a plasma to very high temperatures, particularly since the rate of energy input increases as $T^{\frac{3}{2}}$ with increasing temperature, if the frequency is optimized at each temperature.

Experimental verification of heating by magnetic pumping has not yet been possible at Project Matterhorn, since the radio-frequency power available has not been sufficient to balance the losses from the plasma, due to inadequate confinement and too high an impurity level.

### 4.3 Ion-Cyclotron Resonance Heating

Pulsation of the confining field at the cyclotron frequency of the positive ions should give very rapid heating at very low ion densities. The effect is most conveniently understood not as a macroscopic pumping but as a microscopic resonance between the oscillating electric field and the gyration of the positive ions. However, this type of heating produces separation of charges, and at appreciable plasma densities the resultant electrostatic fields prevent any appreciable heating in this way. It was pointed out by Stix that use of two adjacent heating sections, with

---

[36] Berger, J. M. and Newcomb, W. A., U.S. Atomic Energy Commission Report No. NYO-6046 (PM-S-13), 1954.

[37] Schlüter, A., Z. Naturforsch. **12a**, 822 (1957).

[38] Berger, Newcomb, Dawson, Frieman, Kulsrud, and Lenard, Phys. Fluids **1**, 301 (1958).

identical pulsation frequencies, but differing in phase by 180°, would make it possible for electrons to cancel out this separation of charges by flowing back and forth along the lines of force, and thus permit heating at the ion cyclotron frequency even at substantial plasma densities.

The axisymmetric free oscillations of a cylindrical plasma column, in a strong axial field, were analyzed in detail by Stix,[39] who found that indeed for sufficiently short wavelengths a plasma resonance existed close to the ion cyclotron frequency. Later analyses both by Stix[40] and by Kulsrud and Lenard[38] have considered the input of energy into the gas both at the exact ion cyclotron frequency and at the adjacent plasma resonance frequency. It appears that a substantial amount of energy can be fed into the gas at the plasma resonance frequency, and that thermalization of the energy can readily be achieved in a small system. At low $\beta$ (low ratio of material pressure to magnetic pressure in the vacuum field), this technique offers the great advantage, in principle, over magnetic pumping that the coupling between the external circuits and the plasma is very much better. For a large system, at moderate $\beta$, it is not obvious from theory alone which system would be most effective.

Detailed observational results on plasma heating at frequencies adjacent to ion cyclotron resonance are reported in the subsequent experimental paper by Stix and Palladino.[16] Measures on the external circuits at low power indicate that, in fact, resonant loading is observed at frequencies at and below the cyclotron frequencies both for hydrogen and helium. The energy fed into the gas exceeds the energy dissipated in the external circuit, a prerequisite for efficient heating. Measures at high power appear to be generally consistent with expectations, although direct evidence of plasma heating has not, as yet, been obtained.

[39] Stix, T. H., *Phys. Rev.* **106**, 1146 (1957).
[40] Stix, T. H., *Phys. Fluids* **1**, 308 (1958).

# PERSPECTIVES ON THE PAST, PRESENT, AND FUTURE OF FUSION RESEARCH

(J. FUSION ENERGY, 7, 221, 1988)

## *Commentary*

THIS paper, like the preceding Paper #26, was written for a scientific conference. The similarity extends no further. The conferences were three decades apart; the earlier one was in Geneva, the later one, in Princeton. The status of controlled fusion research, the subject of both conferences, had greatly changed, of course, in thirty years, and my relationship with the program had also changed from a central association in 1958 to an almost complete dissociation a decade or so later.

My brief remarks at the panel discussion on Past, Present and Future of Fusion are included here as a counterweight to the youthful enthusiasm of Paper #26. Also, I welcome the opportunity to supplement these remarks with discussions of two questions especially related to my own activities—what were the most important results of our early experimental work at Matterhorn (Princeton Plasma Physics Laboratory), and what are the prospects for a prototype fusion reactor to test the behavior of a thermonuclear plasma confined in such a device? The first of these two questions is discussed in detail elsewhere,[1,2] but my own abbreviated summary here may be of interest. I ignore here the interesting alternative research program on power from laser-induced microbombs, known officially as Inertial Fusion.

We began to think of an experimental program at Matterhorn after about a year of theoretical research. The Atomic Energy Commission (AEC) supported the idea,[1] after we had persuaded Jim Van Allen at Iowa to head this work for a few years. Van wisely suggested that we start with a simple, modest device. The resultant "table-top stellarator," our Model A, was indeed primitive. Martin Schwarzschild and I spent several weekends sitting on the floor of our rabbit hutch, winding flat copper wire around 2-inch diameter glass tubes. This experimental device[1] could be assembled either as a stellarator, with a rotational transform, or as a toroidal device with no rotational transform. The principal result obtained with Model A was that initial ionization of the gas, with an electric field E induced around the stellarator, was more easily achieved (i.e., with E less by an order of magnitude) with a rotational transform than without it.

There followed the more ambitious Model B, designed for a peak confining magnetic field of 50,000 Gauss; the vacuum tube diameter was still 2 inches (giving a 1-inch "minor plasma radius"), as in Model A. There were a number of successive modifications and new embodiments of this device,[1] which finally yielded several significant results. One was that the flood of impurities streaming off the wall during a discharge overwhelmed the hydrogen gas initially present; this problem was much reduced in subsequent experiments by the use of ultra-high vacuum techniques. Another was that the plasma behavior showed general agreement with the predictions of MHD theory; in particular, when the heating current in the gas exceeded the predicted values for onset of the lowest unstable modes, the plasma confinement was grossly impaired. A third result was that the time during which the plasma was confined was, at best, unexpectedly short, of order $10^{-4}$ s.

The next stage after our Model B stellarators was Model C. According to the sequence envisaged at the outset of our experimental work, Model C was supposed to follow after favorable results with the B models. This more powerful device was to be somewhat similar to the earlier B models but four times larger in linear dimensions. According to plausible scaling laws, the plasma confinement time, for the same magnetic field and temperature, should be 16 times longer in this larger device, giving plasmas which should be closer to conditions needed in a power-producing reactor. Model D, supposed to follow after favorable results with Model C, was to be a full-scale reactor, or at least a full-scale prototype, without the hardware needed for converting into electrical form any fusion power released. While our analysis of the various engineering challenges in Model D was a stimulating and even productive experience, this overall sequence now appears in retrospect as wildly optimistic. Model C was authorized and fabrication begun when early Model B results appeared indeed favorable. When improved experiments later showed that the confinement time was in fact unexpectedly short, the Atomic Energy Commission considered termination of the Model C program. However, work on this device and on the necessary buildings, motor-generator sets, capacitor banks, etc., was so far advanced that cancellation was judged impractical.

When Model C began experimental operation, in May 1961, the first observations immediately made clear that the confinement time was a full order of magnitude greater than in the B models—an important confirmation of our expectations. Precise measures showed that this time varied about as $a_p^2 B$, where $a_p$ is the minor radius of the plasma, and $B$ the confining magnetic field. Unfortunately, with the heating methods then available, this time was still not long enough to increase the temperature sufficiently far above $10^6$ K so that a new plasma regime, yielding longer confinement times might, perhaps, be established (see paper #27). Another result of importance to the fusion program was the high effectiveness of the Model C divertor, which reduced the level of oxygen, the dominant impurity in the heated plasma, by about two orders of magnitude.

Late in 1969, during Mel Gottlieb's directorship at PPPL, experiments in the Soviet Union with toroidal plasma devices confirmed the temperatures of order $10^7$ K suggested by earlier measures. In these axially symmetric devices, named Tokamaks by the Soviet scientists, the toroidal field, produced by coils threading the toroid, gave no rotational transform; a large plasma current around the toroid produced the poloidal field necessary for confinement and equilibrium. Since the electron temperature in Model C rarely exceeded $10^6$ K, it was decided to convert the device to a Tokamak. For this purpose the helical winding was removed, and a bigger vacuum tube was inserted in the large space thus available, permitting a substantial increase in $a_p$, the minor plasma radius. Subsequent experiments showed[3] that for $a_p \approx 5$ cm, the original Model C value, the new Tokamak gave plasmas with about the same temperature as found in Model C. As $a_p$ was increased to 13 cm, the temperature rose to about $10^7$ K. Apparently a chief advantage to a Tokamak, as compared to a stellarator of about the same overall size (and cost), is that it permits a larger $a_p$, yielding longer confinement times and higher T. With neutral beam injection added in due course, the temperature obtained in larger Princeton Tokamaks ($a_p$ up to 80 cm) rose to about $10^8$ K for times exceeding 0.1 second, as has been achieved with similar large toroidal devices elsewhere. Energetic research over the years has produced plasma conditions approaching those required for a power-producing fusion reactor.

The next step, to confine and heat a plasma under the same conditions necessary in an economic power-producing reactor, requires a device considerably larger than any now existing. The cost of such a "test reactor"—with no provision for recovering any fusion power released—would strain the research budget of a single nation. Hence, an International Thermonuclear Experimental Reactor (ITER)[4] is now being designed with joint sponsorship by the European Community, Japan, Russia, and the United States. In this device, the plasma toroid will have a cross-section measuring roughly 5 meters across horizontally and 10 meters vertically, with a circumference around the toroid center line of about 50 meters (major plasma radius, $R_p$, of 8 meters). One of the most difficult problems facing this program will be the political question of where ITER should be located. The design work is being divided up at a variety of sites in the different countries, an option hardly available for ITER itself.

ITER is planned as a Tokamak. As experimental devices, Tokamaks have the great advantage over stellarators that they tend to be more compact. Thus, as we saw at Princeton, for a fixed cost a Tokamak can have a substantially larger minor plasma radius, and a correspondingly longer confinement time. For power production, as compared with research, however, a stellarator has the significant advantage that it is well adapted to steady-state operation, and its greater length for a given $a_p$ may be less important; a greater cost for a larger power producing reactor may be acceptable if the operating cost per kilowatt generated is appreciably reduced. Experiments with stellarator-type devices in Germany and Japan show plasma confinement generally the same as in a Tokamak of the same minor radius, with sometimes much better plasma behavior. For an economic reactor these advantages of a stellarator may be compelling, despite the much greater experience that scientists and engineers have had with Tokamaks.

For this reason, both in Germany and in Japan plans are underway to construct stellarator-type systems of rather large dimensions—smaller than ITER but large enough to provide a more nearly definitive comparison with Tokamaks. The vacuum-tube minor radii will be about 50 cm in each of these two devices. In both, the external coils for the main confining field will be modified so that they produce also the desired helical field and the associated rotational transform. The Germans have adopted[5] the interesting approach of optimizing many features of their device to give the most effective performance according to the best present theories, even though not all these theories are, as yet, thoroughly established. The Japanese Large Helical Device[6] is already under construction. Results from these two experimental systems will be awaited with interest.

The road to useful power from fusion may be a long one, but the commanding importance of the goal continues to arouse strong commitments.

### REFERENCES

1. Bromberg, J. L.—Fusion: Science, Politics and the Invention of a New Energy Source, Cambridge, MA: MIT Press (1982).
2. Herman, R.—Fusion, the Search for Endless Energy, Cambridge: Univ. Press (1990).

3. Stodiek, W.—*Nuclear Fusion* **25**, 1161 (1985).
4. Rebut, P.-H. et al.—In: Proceedings, Fifteenth International Conference on Plasma Physics and Controlled Nuclear Fusion Research, Vienna: International Atomic Energy Agency (1994), paper CN-60/E-1-I-1.
5. Grieger, G. et al.—*Fusion Technology* **21**, 1767 (1992).
6. Iiyoshi, A. et al.—*Fusion Technology* **17**, 169 (1990).

## *Paper*

I should point out that my role on this panel is very similar to that of Rip Van Winkle. I haven't been asleep for twenty years, but from the standpoint of my knowledge of plasma physics I might as well have been. I've been off on other things, although I'm still deeply interested in plasma physics, both as a scientific discipline and as a source of great promise for future energy supplies. But I must confess I am not well informed as to what's been going on since I left the Plasma Physics Laboratory in about 1966.

It may be of some interest here to review the very earliest days of the fusion program in this country. We know, of course, that ignorance is bliss; in the first two years, when we didn't really know very much, we had optimistic views as to how a plasma could be contained and used for controlled release of fusion power. The theories of that time were based on the assumption that the plasma would be quiescent. A quiescent plasma is one which shows no fluctuations with passing time, which sits there, beautifully confined, and follows the dictates laid down by the theorist. As to physical plasmas, at least deep inside the sun where we cannot see them, they are thought to conform to this model.

In any case, the theory of a quiescent plasma is of interest partly because it can provide reasonably exact results. One knows where one is and, furthermore, such a theory sets idealized values of various quantities, providing a goal for our research efforts. In any case, for the first few years of the fusion program, the assumption of quiescent plasma was the basis of the theoretical work. I shall review here what results were achieved during those early years, since in a way they created the underpinning for much of what has been done, at least in theoretical research, since that time.

In the field of confinement, the analysis of toroidal systems is based on the rotational transform. Martin Kruskal has shown that if magnetic lines rotate as they go around the plasma, then to a very high approximation, these lines of force are essentially closed and generate magnetic surfaces. A similar conclusion follows for particle orbits. This result gave us a wonderful feeling in the early days that we could rely on the confinement of at least single particles in a magnetic field.

There were a number of problems which even then we were beginning to realize. Kruskal's result was only approximate; if you looked at real fields, they didn't always conform exactly to the ideal. And if you were not particularly careful in designing the magnetic field and constructing the necessary coils, some lines of force could enter the vacuum vessel at one point and finally escape from another. Furthermore, some particles did not go entirely around a toroidal device, but were reflected in certain areas. Such particles could escape more rapidly than predicted by the classical theory. We did not explore these topics in those early years as thoroughly as they were explored later, but I think we went far enough to indicate that, in all probability, these were not very serious practical problems for the confinement of the thermonuclear plasma, provided, of course, that the plasma remained quiescent.

We were certainly interested in the subject of heating in those days, because obviously you have to heat the plasma. Some of our practical friends would say, "Gosh, how do you ever heat the gas to a hundred million degrees, that sounds terribly difficult." But if you really believe that plasma confinement is good, heating is not that much of a problem. You can't quite just light a match under it and have the confined plasma get hotter and hotter and hotter, but there are many ways that you can pour energy into plasma. And if these heating methods do not impair the confinement, almost any one of them might be successful in heating a quiescent plasma up to thermonuclear temperatures.

Based on the assumption of quiescent plasmas, in about the third year of the program here at Princeton a number of us got together and worked out some of the engineering aspects of a full-scale fusion reactor. Obviously, this was a very preliminary investigation, but we looked at such problems as: the heat transfer into the divertor and the resultant heating of the divertor end plates; what nuclear reactions would be produced by neutrons generated in D–T reactions; how one could cool a fusion reactor with lithium or some compound flowing through pipes; how the tritium produced in the molten lithium could almost all be recovered; how one would inject droplets or jets of hydrogen back into the confinement vessel. We touched on these problems rather than solving them, but we went far enough in the analysis to reach the general conclusion that none of these problems seemed insuperable, and the cost estimates that were made were at least encouraging to us.

So at that time, while we were not quite foolish enough to put this in writing, we did have informal optimistic estimates that if the plasma remained quiescent, one might have some prototype operating in perhaps ten or twenty years. Since then, the time until a successful prototype reactor can be achieved has remained roughly constant. This estimate of the time required is perhaps a little less now in view of the exciting results that have been obtained recently.

I might mention that it soon became apparent in the experimental programs throughout the world that the real world of plasma physics was one of cooperative phenomena. The plasmas were not quiescent.

They developed the "shimmies." They were subject to all sorts of time-dependent phenomena, which has as their consequence that plasma was lost from the confinement vessel and struck the walls. There was some reason to believe from theory, and also from observation, that confinement became better at higher temperature, and so the stage was set for quite a few years of plasma research, characterized by a race between the rate of heating and the rate of plasma loss; the effort is to reach that plateau which the theorists have held out to us of high confinement under high temperature and relatively undisturbed conditions. To reach that promised land of good confinement, one has to heat the plasma through a region in which the losses are perhaps much greater. I have the impression that the promised land is virtually within sight. And on that basis one can be intelligently optimistic about the future, as I think many of us are.

# PART FIVE

# Miscellaneous

# A CONTRIBUTION TO THE MATHEMATICAL THEORY OF BIG GAME HUNTING

(WITH J. TUKEY ET AL.)*

(AMER. MATH. MONTHLY, 45, 446, 1938)

## *Commentary*

IN 1935–36, when I was a first-year graduate student at St. John's College, Cambridge University, the light banter among us included how to catch a lion in the desert; we delighted in devising ingenious methods, preferably based on the more recondite scientific laws or theorems. As I remember, some of the mathematics and physics faculty joined this challenging sport, using the most up-to-date principles.

During my two graduate-student years at Princeton, 1936–38, this question of how we might capture our lion continued to intrigue us. Among those who joined the fun were mathematicians Ralph Boas and Frank Smithies, both in mathematics, and John Tukey in statistics. If I recall correctly, it was John, with active expertise in many of the physical sciences, who pushed for a publishable paper, presenting a few of the more interesting techniques that we had encountered or concocted for this problem. There was some discussion among us—who should be listed as authors of this spoof, and which of us would actually submit the paper for publication?

In the end, the article was sent off to the American Mathematical Monthly, together with a letter signed by one E. S. Pondicherry (of the Royal Institute of Poldavia), the name proposed by John as a cover for our group. This letter explained the author's desire to publish his pioneering but possibly controversial work under a pen name, and proposed H. Pétard as a suitable pseudonym. I suppose it is seldom that a nonexistent individual has published in a serious professional journal under a pen name, especially one so appropriate for the subject.

It is a great pleasure for me to see this path-breaking paper reprinted here. As to recent progress in this field, I have no information, though some advances have surely been made over the years!

---

* In published version, author given is H. Pétard, Princeton, New Jersey.

# Paper

This little-known mathematical discipline has not, of recent years, received in the literature the attention which, in our opinion, it deserves. In the present paper we present some algorithms which, it is hoped, may be of interest to other workers in the field. Neglecting the more obviously trivial methods, we shall confine our attention to those which involve significant applications of ideas familiar to mathematicians and physicists.

The present time is particularly fitting for the preparation of an account of the subject, since recent advances both in pure mathematics and in theoretical physics have made available powerful tools whose very existence was unsuspected by earlier investigators. At the same time, some of the more elegant classical methods acquire new significance in the light of modern discoveries. Like many other branches of knowledge to which mathematical techniques have been applied in recent years, the Mathematical Theory of Big Game Hunting has a singularly happy unifying effect on the most diverse branches of the exact sciences.

For the sake of simplicity of statement, we shall confine our attention to lions (*Felis leo*) whose habitat is the Sahara Desert. The methods which we shall enumerate will easily be seen to be applicable, with obvious formal modifications, to other carnivores and to other portions of the globe. The paper is divided into three parts, which draw their material respectively from mathematics, theoretical physics, and experimental physics.

The author desires to acknowledge his indebtedness to the Trivial Club of St. John's College, Cambridge, England; to the M.I.T. chapter of the Society for Useless Research; to the F.o.P., of Princeton University; and to numerous individual contributors, known and unknown, conscious and unconscious.

## 1. Mathematical Methods

1. The Hilbert, or Axiomatic, Method. We place a locked cage at a given point of the desert. We then introduce the following logical system.

Axiom I. *The class of lions in the Sahara Desert is non-void.*

Axiom II. *If there is a lion in the Sahara Desert, there is a lion in the cage.*

Rule of Procedure. *If p is a theorem, and "p implies q" is a theorem, then q is a theorem.*

Theorem I. *There is a lion in the cage.*

2. The Method of Inversive Geometry. We place a *spherical* cage in the desert, enter it, and lock it. We perform an inversion with respect to the cage. The lion is then in the interior of the cage, and we are outside.

3. The Method of Projective Geometry. Without loss of generality, we may regard the Sahara Desert as a plane. Project the plane into a line, and then project the line into an interior point of the cage. The lion is projected into the same point.

4. The Bolzano-Weierstrass Method. Bisect the desert by a line running N–S. The lion is either in the E portion or in the W portion; let us suppose him to be in the W portion. Bisect this portion by a line running E–W. The lion is either in the N portion or in the S portion; let us suppose him to be in the N portion. We continue this process indefinitely, constructing a sufficiently strong fence about the chosen portion at each step. The diameter of the chosen portions approaches zero, as that the lion is ultimately surrounded by a fence of arbitrarily small perimeter.

5. The "Mengentheoretisch" Method. We observe that the desert is a separable space. It therefore contains an enumerable dense set of points, from which can be extracted a sequence having the lion as limit. We then approach the lion stealthily along this sequence, bearing with us suitable equipment.

6. The Peano Method. Construct, by standard methods, a continuous curve passing through every point of the desert. It has been remarked* that it is possible to transverse such a curve in an arbitrarily short time. Armed with a spear, we traverse the curve in a time shorter than that in which a lion can move his own length.

7. A Topological Method. We observe that a lion has at least the connectivity of the torus. We transport the desert into four-space. It is then possible‡ to carry out such a deformation that the lion can be returned to three-space in a knotted condition. He is then helpless.

8. The Cauchy, or Function Theoretical, Method. We consider an analytic lion-valued function $f(z)$. Let $\zeta$ be the cage. Consider the integral

$$\frac{1}{2\pi i} \int_C \frac{f(z)}{z - \zeta}\, dz,$$

where $C$ is the boundary of the desert; its value is $f(\zeta)$, i.e., a lion in the cage.[†]

---

* By Hilbert. See E. W. Hobson, The Theory of Functions of a Real Variable and the Theory of Fourier's Series, 1927, vol. 1, pp 456–57.

‡ H. Seifert and W. Threlfall, Lehrbuch der Topologie, 1934, pp. 2–3.

[†] *N.B.* By Picard's Theorem (W. F. Osgood, Lehrbuch der Funktionentheorie, vol. 1, 1928, p. 748), we can catch every lion with at most one exception.

9. THE WIENER TAUBERIAN METHOD. We procure a tame lion, $L_0$, of class $L(-\infty, \infty)$, whose Fourier transform nowhere vanishes, and release it in the desert. $L_0$ then converges to our cage. By Wiener's General Tauberian Theorem,[‡] any other lion, $L$ (say), will then converge to the same cage. Alternatively, we can approximate arbitrarily closely to $L$ by translating $L_0$ about the desert.[§]

## 2. METHODS FROM THEORETICAL PHYSICS

10. THE DIRAC METHOD. We observe that wild lions are, *ipso facto*, not observable in the Sahara Desert. Consequently, if there are any lions in the Sahara, they are tame. The capture of a tame lion may be left as an exercise for the reader.

11. THE SCHRÖDINGER METHOD. At any given moment there is a positive probability that there is a lion in the cage. Sit down and wait.

12. THE METHOD OF NUCLEAR PHYSICS. Place a tame lion in the cage, and apply a Majorana exchange operator[‖] between it and a wild lion.

    As a variant, let us suppose, to fix ideas, that we require a male lion. We place a tame lioness in the cage, and apply a Heisenberg exchange operator[¶] which exchanges the spins.

13. A RELATIVISTIC METHOD. We distribute about the desert lion bait containing large portions of the Companion of Sirius. When enough bait has been taken, we project a beam of light across the desert. This will bend right around the lion, who will then become so dizzy that he can be approached with impunity.

## 3. METHODS FROM EXPERIMENTAL PHYSICS

14. THE THERMODYNAMICAL METHOD. We construct a semi-permeable membrane, permeable to everything except lions, and sweep it across the desert.

15. THE ATOM-SPLITTING METHOD. We irradiate the desert with slow neutrons. The lion becomes radioactive, and a process of disintegration sets in. When the decay has proceeded sufficiently far, he will become incapable of showing fight.

---

[‡] N. Weiner, The Fourier Integral and Certain of its Applications, 1933, pp. 73–74.

[§] N. Weiner, *l. c.*, p. 89.

[‖] See, for example, H. A. Berthe and R. F. Bacher, Reviews of Modern Physics, vol. 8, 1936, pp. 82–229; especially pp. 106–7.

[¶] Ibid.

16. The Magneto-Optical Method. We plant a large lenticular bed of catnip (*Nepeta cataria*), whose axis lies along the direction of the horizontal component of the earth's magnetic field, and place a cage at one of its foci. We distribute over the desert large quantities of magnetized spinach (*Spinacia oleracea*), which, as is well known, has a high ferric content. The spinach is eaten by the herbivorous denizens of the desert, which are in turn eaten by lions. The lions are then oriented parallel to the earth's magnetic field, and the resulting beam of lions is focussed by the catnip upon the cage.

# PHYSICS OF SOUND IN THE SEA. I. TRANSMISSION, SUMMARY[†]

(WITH P. G. BERGMANN)

(IN: PHYSICS OF SOUND IN THE SEA, PART I TRANSMISSION, EDS. P. G. BERGMANN AND A. YASPAN, NEW YORK: GORDON & BREACH, CHAPTER 10, P. 236, 1968)

## *Commentary*

PAPER #29 describes some of the results obtained on the behavior of sound in the sea, a subject of considerable research during World War II. Scientists not previously acquainted with problems of underwater sound may be interested to read here how sound transmission in the deep ocean is so markedly affected by shadow zones, sound channels, and interference between direct and surface-reflected sound. For future research in this field, the discussion in the final section 10.5 may be of particular importance. In the long run, a better knowledge of specific mechanisms and processes involved in undersea sound transmission should be helpful in the most effective use of sonar for many purposes.

This commentary relates some of my personal wartime experiences in this scientific field. Early in 1942, shortly after the Japanese attack on Pearl Harbor, I accepted a staff position in New York City with Division 6 of the National Defense Research Committee (NDRC) to participate in their broad program of research concerned with subsurface warfare. During the first month, as I commuted daily from New Haven to Division 6 Headquarters in downtown New York, I remember thinking that I was fortunate indeed that this work, which seemed of some importance for the war effort, promised to be interesting scientifically as well.

The NDRC, organized under the aegis of the National Academy of Sciences, was carrying out, under civilian control, a general program of war research. The top management were mostly academic scientists; Division 6 was headed by Professor John Tate, a physicist from the University of Minnesota. Each division, after consultation with the

[†]Printed in 1946 for the U.S. Office of Scientific Research and Technology and distributed by the Office of Naval Research; reissued in 1968 by Gordon and Breach.

appropriate military officers, set up specific research programs, which were then carried out by private organizations (companies, universities, etc.) under contract with the NDRC. The staff at Division 6 headquarters provided liaison between top management at headquarters and the various research programs.

After several months my particular assignment was to provide liaison with the research being done on the behavior of underwater sound. The Woods Hole Oceanographic Institution at Woods Hole, Massachusetts, and the University of California Underwater Sound Laboratory at San Diego were the two principal organizations carrying out this important work. Largely because of the efforts by these two groups, the information on underwater sound transmission and scattering was much less rudimentary at the end of the war than at the beginning. As a staff adviser to NDRC Division 6, I had no direct authority over this research. My function was to suggest, not to direct. My quarterly visits to the Woods Hole and San Diego laboratories consumed much precious time of the scientists in these two groups, but may well have had a positive influence on the scientific programs. I certainly found these contacts engrossing and educational.

Particularly important during these years were my relationships with the Navy. My opposite number in the Bureau of Ships was Roger Revelle, an oceanographer from the Scripps Institution. A brilliant man, he was one of those who first warned, shortly after the war, that consumption of fossil fuels could produce global warming. As a Lieutenant Commander during the war, he was responsible for looking after the Navy's interest in basic underwater sound research. Hence, it was he who provided the necessary liaison with Division 6, NDRC. He would grill me on just what research was underway and why, and I would discuss with him how the Navy could best make use of the basic information which Division 6 of the NDRC was obtaining. Roger's clear, deliberate explanations were vital for presenting concepts and detailed suggestions to his superior officers.

Roger was thorough, unhurried, critical, and refused to accept anything he did not entirely understand and agree with. His comments frequently required many changes in the draft letters and reports I would bring down to him. As a result, delays were frequent. After a while I developed a technique for expediting action when official correspondence was required between NDRC and the Navy's Bureau of Ships. I would draft a letter to the Chief of the Bureau for Dr. Tate's signature, and go to Washington a few days later with a suggested reply. At the Bureau of Ships I would paw through Roger's pile of unanswered (largely routine) correspondence and pull out Dr. Tate's letter, which had been routed down to him. I would show this to Roger together with my draft of a suitable reply from the Navy. Roger would study the material carefully, pointing out several considerations which I had ignored. But a few hours later he had usually worked out a reply agreeable to us both, which he would then route through channels back up to the Chief of the Bureau for possible approval and signature. I would then return to New York to await the reply. Between the two of us, Roger and I were an effective pair, and I like to think we had some influence in this branch of naval activity.

I must confess that I had no yearning for combat experience. The closest I came to military activity was the day an Army two-engine plane accidentally smashed its way into the Empire State Building in New York. During the war, the Division 6 NDRC office had expanded and moved uptown, occupying all the 64th floor of this skyscraper. The cloud level that day was just about at our floor. Apparently, the pilot came down just below the clouds to see where he was, found the building straight ahead, pulled his stick up sharply to climb back up, didn't make it and crashed into the building some ten floors above us. The aviation gasoline set fire to two floors, which were soon completely gutted. The two motors plummeted to the ground, one down an elevator shaft, the other on to a building across the street. A couple of office workers on our floor were injured when they dove beneath their desks, thinking the impact was a bomb attack. We could not use elevators or stairs. We looked at the flames engulfing the two floors higher up and wondered.... Fortunately, the firemen, whose trucks jammed the street below, escorted us down the stairs a few hours later.

Paper #29 is a chapter from a four-part monograph on the physics of underwater sound, written at the war's end as part of the Summary Technical Report of Division 6, NDRC. I had a general responsibility for this monograph; most of the writing was done by the Sonar Analysis Group, which I had assembled to aid in our liaison work. This paper is the final summarizing Chapter 10 for the research described in Part I Transmission. Parts II, III, and IV deal with Reverberation, Reflections from Submarines and Surface Vessels, and Acoustic Properties of Wakes. The monograph as a whole was initially produced in 1946 by the Columbia University Press for the NDRC and its parent organization, the Office of Scientific Research and Technology; some two decades later it was reissued in three volumes by Gordon & Breach as part of their "Documents on Modern Physics" series.

# *Paper*[†]

Research on sound transmission during World War II was concerned almost exclusively with the investigation of sound fields which were operationally important. More than half of the experimental work was devoted to the sound field of standard echo-ranging transducers operating at frequencies around 24 kc. The purpose of the work was primarily to provide information which could be used to increase the effectiveness of Navy gear already in use on submarines and antisubmarine vessels. The instrumentation used for research differed as little as possible from standard operational gear; what modifications were made usually represented the minimum necessary for quantitative evaluation of the data obtained. Questions which did not seem important operationally, such as the physical cause of the observed attenuation of supersonic sound in the sea, received scant attention in these studies.

In the sections which follow, the essential results of these experiments on underwater sound transmission are summarized. Section 10.1 lists the definitions of the most important quantities used in describing underwater sound fields. Sections 10.2 and 10.3 summarize what is known concerning the average transmission of sound in the sea at various frequencies. In Section 10.4, data on the fluctuation and variation of seaborne sound are summarized. Finally, Section 10.5 provides a brief discussion of probable trends in future research on sound transmission.

## 10.1. Basic Definitions

### *Sound Pressure and Sound Field Intensity*

A sound wave in a fluid can be described conveniently in terms of the pressure disturbance which arises in the vicinity of a sound source, travels through the fluid, and is finally received by a hydrophone. The *instantaneous sound pressure* is the difference between the instantaneous value of the pressure at a chosen location and the mean or equilibrium pressure at the same point. The rms value of the instantaneous sound pressure is usually called the *rms sound pressure*. Usually, the average is carried out over a time interval which is long compared with the periods of the principal frequencies making up the sound signal. In the case of single-frequency sound, the average is extended over one period (or an integral number of periods). Unless specified otherwise, "sound pressure" as used in the technical literature is short for rms sound pressure. Except in the case of standing waves, the rms sound pressure is an excellent measure of the energy carried by the sound wave. At the present time, sound pressure values are uniformly reported in units of dynes per square centimeter.

The *sound field intensity* is defined as the averaged power carried by a sound wave per unit cross section of a wave front. The units in present use are watts per square centimeter. If the radii of curvature of the wave fronts are large compared with the wavelength, then the rms sound pressure and the sound field intensity are connected in excellent approximation by the formula

$$I = 10^{-7} \frac{p^2}{\rho c},\tag{1}$$

in which $p$ is the rms sound pressure, $\rho$ is the density of the fluid in grams per cubic centimeter, $c$ is the sound velocity in the fluid in centimeters per second, and $I$ is the sound field intensity.

### Sound Level

The sound field intensity is usually reported on a logarithmic scale. The most common scale for this purpose is the decibel scale. The quantity $L$,

$$L = 20 \log p \tag{2}$$

in which the rms sound pressure $p$ is expressed in units of dynes per square centimeter, is called the *sound pressure level* or simply the *sound level*. As defined by equation (2), $L$ is the sound level in decibels above a standard which corresponds to a sound pressure of 1 dyne per sq cm. In the past, sound levels were frequently reported in decibels above 0.0002 dyne per sq cm.

### Source Level

The source level is a measure of the power output of a sound source on the decibel scale. Briefly, it is the sound level due to a point source at a distance of 1 yd, in decibels above 1 dyne per sq cm. If a point source is located in a homogeneous, nondissipative medium which is infinitely extended in all directions, the intensity of the sound field is inversely proportional to the square of the distance from the source,

$$1 = \frac{F}{r^2}.\tag{3}$$

This law is called the inverse square law. In terms of the sound level, equation (3) becomes

$$L = S - 20 \log r. \qquad (4)$$

In these equations, $F$ and $S$ are constants which depend on the power output of the source, and $r$ denotes the distance (slant range) from the source. That $S$ is the source level as defined above can be verified by setting $r$ equal to 1 in equation (4).

For real sound sources in real media, equations (3) and (4) are not everywhere valid. Because of the finite extension of an actual sound source, the inverse square law fails at ranges of the order of the dimensions of the source. Because of absorption of the sound in the medium and because of scattering and reflection from bounding surfaces, it fails at very long ranges. However, there is frequently an intermediate range interval for which equation (4) holds. If there is such an interval, then the constant $S$ is considered the source level, even though $S$ may not be the actual sound level at a distance of 1 yd.

For a highly directional sound source, such as a standard echo-ranging transducer, the definition of the source level is further specified by the condition that the sound measurements are to be carried out on the axis, that is, the radial line of greatest sound field intensity.

### Transmission Loss and Transmission Anomaly

The *transmission loss H* at the range $r$ is defined by the formula

$$H(r) = S - L(r), \qquad (5)$$

where $S$ is the source level, and $L$ is the sound level defined by equation (2). The transmission loss defined in this way measures the drop of the sound level with increasing distance from the source and has the virtue of being independent of the particular power output of the source. Other parameters of the source, such as operating frequency and directivity pattern, are known to affect the value of the function $H(r)$. The units of $H$ are decibels.

The *transmission anomaly A* is the deviation of the transmission loss from that functional behavior demanded by the inverse square law of spreading. The defining equation for $A(r)$ is

$$A(r) = H(r) - 20 \log r = S - L(r) - 20 \log r. \qquad (6)$$

The transmission anomaly vanishes if the inverse square law of spreading is satisfied, and it is positive if the sound level drops off more rapidly than $20 \log r$. Large positive transmission anomalies, therefore, correspond to poor sound conditions.

In sound transmission work, it has been customary to train the projector in a horizontal plane on the receiving hydrophone, but not to tilt the acoustic axis away from the horizontal. Hence, measured transmission anomalies will be large for a close deep hydrophone beneath the sound beam.

In supersonic transmission work, it has been found that when successive signals are transmitted a few seconds apart over the same transmission path, the received sound intensity is subject to irregular fluctuations. Reported transmission anomalies always represent values which have been obtained by averaging over a number of signals received during a brief period so that much of this fluctuation is smoothed out.

### Variance of Amplitudes

The standard deviation of the individual pressure amplitudes in a sample of signals, divided by the average pressure amplitude for the sample, is called the *variance of amplitudes* for the sample. This variance is used as a measure of the fluctuation of received sound intensity. Observed values of the variance are summarized in Section 10.4 below.

### Deep and Shallow Water

Water is effectively deep when bottom-reflected sound is much weaker than the direct sound; otherwise, the water is effectively shallow. Over the continental shelf (depth less than 100 fathoms) the water is effectively shallow for most situations. Away from the continental shelf, the ocean is always deep when sharply directional sound is used (as in echo-ranging at supersonic frequencies), but may be shallow when listening at audible frequencies to a target at long range.

### 10.2. Deep-Water Transmission

The transmission loss in the open ocean depends on the way the velocity of sound changes with position in the sea, since velocity gradients distort the sound beam. These velocity gradients change with time and location, but in any localized region at any given time depend primarily on depth and relatively little on horizontal position within that region. Changes in sound velocity in deep water closely follow changes in water temperature; the effect of pressure changes is relatively slight and usually need not be considered except for transmission to great depths.

The following subsections tell of the transmission anomalies expected for various common temperature-depth distributions in the ocean.

### Isothermal Water

When the top 50 ft of the ocean are isothermal, transmission anomalies are determined by two major effects, absorption and surface reflection.

#### SONIC FREQUENCIES

At low sonic frequencies, sound is reflected from the sea surface in somewhat the same way as from a flat, perfectly reflecting mirror. The partial cancellation of direct and surface-reflected sound reduces the sound intensity at long range near the surface. The transmission anomaly at any range may be computed from the equation

$$A = -10 \log\left(1 - 2\gamma_a \cos \frac{4rh_1 h_2}{R\lambda} + \gamma_a^2\right), \qquad (7)$$

where $h_1$ is the depth of the sound source, $h_2$ is the depth of the receiving hydrophone, $R$ is the range from source to hydrophone, and $\lambda$ the wavelength. The quantity $\gamma_a$, called the *effective reflection coefficient* of the surface, is a semi-empirical parameter; its average value for different frequencies is given in Table 1.

Absorption has little effect on sound transmission at frequencies below 2,000 c.

#### HIGH SONIC AND SUPERSONIC FREQUENCIES

At frequencies above 2,000 c, the value of $\gamma_a$ to be used in equation (7) is seldom greater than 0.5 in the open sea and is frequently so small that image interference can scarcely be said to exist. Absorption plays

TABLE 1
Effective Reflection Coefficient of the Surface

| Frequency in cycles | 200 | 600 | 1,800 |
|---|---|---|---|
| $\gamma_\epsilon$ | 0.8 | 0.7 | 0.5 |

TABLE 2
Attenuation Coefficient in the Sea

| Frequency in kc | 20 | 24 | 30 | 40 | 50 | 60 | 80 | 100 | 500 | 1,000 |
|---|---|---|---|---|---|---|---|---|---|---|
| $a$ in db per kiloyard | 3 | 4 | 6 | 10 | 13 | 18 | 26 | 35 | 150 | 300 |

an increasingly important role as the frequency increases. The transmission anomaly $A$ may be computed from the relation

$$A = \frac{ar}{1,000},$$ (8)

where $r$ is the range in yards and where $a$ is the attenuation coefficient in decibels per kiloyard. Average values of $a$ at a number of frequencies are given in Table 2. At frequencies above 1,000 kc, the attenuation coefficient is about three times the value predicted from the viscosity of the water. At frequencies of 24 kc and below, $a$ is more nearly 100 times this theoretical value.

### Thermocline below Isothermal Layer

When sound from an isothermal layer passes at grazing angle into a thermocline or temperature layer, where the temperature decreases sharply with increasing depth, the sound rays are bent downward and become more spread out. The increased distance between sound rays in and below the thermocline reduces the sound intensity; this phenomenon is known as *layer effect*. The transmission anomaly below the thermocline, at ranges out to 4,000 yd, may be computed from the equation

$$A = 5 \log\left(1 + \frac{2\Delta c - r^2}{c_0 h_1^2}\right) + \frac{ar}{1,000},$$ (9)

where $\Delta c$ is the change in sound velocity in the top 30 ft of the thermocline. (If several thermoclines lie above the hydrophone or if the gradient in the thermocline increases with depth, $\Delta c$ is the velocity change in the 30-ft interval giving the maximum value of $\Delta c / h_1^2$.) $c_0$ is the sound velocity in the surface layer; $h_1$ is the height of the sound projector above the top of the thermocline, that is, above the top of the 30-ft interval in which $\Delta c$ is measured; $r$ is the range from projector to hydrophone. The last term on the right is taken over from equation (8); the values of the attenuation coefficient used are given in Table 2.

Equation (9) has been checked in detail at 24 kc only, but presumably gives an approximate indication of the anomalies expected below the thermocline at all frequencies above a few hundred cycles. At increasing depths below the thermocline, the anomaly decreases somewhat, the decrease being most marked for the shallower thermoclines.

### TEMPERATURE GRADIENTS NEAR SURFACE

When temperature gradients are present in the top 50 ft of the ocean, the transmission loss from a projector at 16 ft to a distant hydrophone is correlated with the following variables: the sharpness and depth of the gradients (for practical purposes, the decrease of temperature from the surface down to 30 ft); and $D_2$, the depth at which the temperature is 0.3 F less than the surface temperature. For a deep hydrophone, the temperature gradients at intermediate depths are also of importance.

### SHARP SURFACE GRADIENTS

When the temperature change in the top 30 ft is more than $1/100$ times the surface temperature, the sound beam is bent downward by the decrease of sound velocity with increasing depth. The plot of transmission anomaly against range usually shows three different regions as follows:

1. *The direct sound field* from the projector out to the shadow boundary. The anomaly within the direct sound field is primarily the result of absorption, and equation (8) is applicable.

2. *The near shadow zone.* Beyond the shadow boundary, the sound intensity decreases very rapidly for some distance. Representative values for this decrease are 50 db per kyd at 25 kc and about one-third this at 5 kc. These coefficients of attenuation in the shadow zone are apparently about half the values estimated from the theory of diffraction by a smooth velocity gradient. The range to the shadow boundary increases with depth in accordance with ray theory, but seems to be systematically somewhat less than predicted.

3. *The far shadow zone.* With standard echo-ranging gear and pulses 100 msec long, the transmission anomaly of scattered sound at ranges of several thousand yards is about 50 db. Thus when the transmission anomaly of the direct or diffracted sound in the shadow zone exceeds about 50 db, the observed sound is scattered sound, with an anomaly which does not depend strongly on further increases in range. This scattered sound is incoherent. For short pulses the intensity of this scattered sound is proportional to the pulse length; it becomes negligibly small for explosive sound.

To predict the anomalies expected under given temperature conditions, it is simplest to use curves of average anomalies for such conditions. Since unexplained deviations are frequently found between individual anomalies and the predictions of ray theory, use of average curves gives results about as accurate as the more elaborate methods. An example of this approach is figure 40 of Chapter 5, where average curves are given for different values of $D_2$, the depth at which the temperature is 0.3 F less than the surface temperature.

WEAK SURFACE GRADIENTS

When the temperature change in the top 30 ft is less than 1/100 of the surface temperature, but gradients are present in the top 50 ft., the division of the sound field into the three regions described previously is usually not observed. Since a small change in such temperature conditions may lead to a large change of transmission anomaly, the observed anomalies are highly variable and can neither be compared with theory nor predicted practically with much accuracy. Average anomalies for different values of $D_2$ are given in Figure 49 in Chapter 5 for a shallow hydrophone. For a deep hydrophone, below the thermocline, equation (9) may be used for approximate results.

*Sound Channels*

When the velocity of sound above and below the sound source is appreciably greater than the velocity at the source, the sound rays which leave the source with small inclinations will propagate out indefinitely without surface or bottom reflections, bending back and forth but always remaining within some fixed interval of depths.

SURFACE SOUND CHANNELS

When the sound projector lies below a sharp negative gradient and above a sharp positive gradient, sound channel effects should be marked, with regions of alternately high and low anomaly found out to considerable ranges. When a sharp gradient lies just above the projector and a layer of nearly isothermal water 100 ft or more in thickness lies below, ray theory predicts that the sound bent back up by the positive velocity gradient in the isothermal layer should be focused at shallow depths and long ranges, thus giving anomalously high intensities. Observations made under these conditions show that the transmission anomaly on a shallow hydrophone is sometimes as much as 40 db less than normal over a narrow range interval several thousand yards away. The details of these observed effects are not in good agreement, however, with the exact predictions of ray theory.

At a depth of several thousand feet there is usually a deep sound channel. The effect of pressure on sound velocity increases the velocity at greater depths, and a thermocline usually present closer to the surface increases the velocity at shallower depths. Sound of frequency less than 200 cycles, for which the absorption is very low, has been observed to propagate out for several thousand miles in such a deep channel. With small explosive charges, the arrivals of the different pulses agree with the different rays predicted theoretically. The largest number of arrivals, with the highest observed intensity, occur just before the observed sound stops entirely; these last arrivals are the rays coming almost straight along the axis of the channel.

## 10.3. SHALLOW-WATER TRANSMISSION

In shallow water, the transmission of underwater sound is determined primarily by the character of the bottom, and by the frequency of the transmitted sound. The state of the sea is a much more important factor than in deep water. Temperature gradients are of secondary importance. There are two situations in which sound conditions do not differ appreciably from those found in deep water: (1) soft MUD bottom; (2) strong positive velocity gradients (PETER pattern) below a directional sound source. In both these cases, transmission is very nearly the same as in deep water with the same thermal conditions.

### Sonic Frequencies

Most of the information on the transmission of sonic sound in shallow water was obtained in harbor surveys. The data obtained may be summarized as follows.

No systematic difference was found between different types of bottoms, with the exception of soft MUD, which turned out to be a poor reflector. All other bottoms apparently reflect equally well.

Transmission over sloping bottoms in the presence of downward refraction tends to be poor, in agreement with theoretical predictions.

Over flat bottoms, at ranges greater than the water depth and out to several thousand yards, average sound transmission can be best represented by an inverse 1.5th power law of spreading plus an attenuation which appears to increase roughly linearly with the frequency up to about 20 kc. The transmission loss is, thus, given roughly by the formula

$$H = 15 \log r + \frac{1}{4,000}(f - 2)r + C, \tag{10}$$

where $r$ is the range in yards, $f$ is the frequency in kilocycles, and $C$ is a constant independent of the range $r$.

### Twenty-four Kilocycles

In moderately shallow water, and in the presence of any bottom but MUD and soft SAND-AND-MUD, the transmission anomaly can usually be represented in fairly good approximation by a straight line. For wind forces 0 to 2, transmission anomalies increase with the range at a rate of 5 db per thousand yards over STONY and SAND bottoms, and at a rate of 6 db per thousand yards over ROCK bottoms. About half of all the runs carried out yield values which differ from these average values by no more than 2 db per kyd.

The following special results are also worth noting. (1) For heavy seas, transmission is somewhat worse than for light seas. For wind force 3, about 1 db per kyd should be added to the attenuation coefficients given above. (2) Over sloping bottoms and in the presence of negative gradients, transmission is poor. The transmission anomaly may increase with the range at a rate exceeding 10 db per thousand yards. (3) In shallow isothermal water, transmission is at least as good as in deep isothermal water. (4) In very shallow water (5 fms deep), a series of experiments carried out over SAND gave very poor transmission; the anomaly increased at the rate of about 16 db per kyd.

### High Supersonic Frequencies

As far as is known, transmission in shallow water at high supersonic frequencies is similar to that at 24 kc, except for greatly increased absorption losses; transmission anomalies in the presence of negative gradients are linear and their slopes are somewhat higher than those in deep isothermal water.

### 10.4. FLUCTUATION AND VARIATION

The transmission loss measured at any instant in the ocean will usually differ from the value found several seconds earlier. This rapid change of sound level is called fluctuation. Measured transmission losses and transmission anomalies are averaged to smooth out this fluctuation.

Fluctuation is invariably observed in the transmission of single-frequency supersonic signals transmitted over a path at least 100 yd long. Fluctuation is negligible over transmission paths of the order of 5 yd. Little is known concerning fluctuation over intermediate path lengths. For frequencies of 5 kc and less, fluctuation appears to be less pronounced than at frequencies of 10 kc and higher. The summary which follows is concerned only with the fluctuation of supersonic signals transmitted over paths at least 100 yd in length.

## Variance with Shallow Projector

For a projector at a depth of 16 ft, the direct sound from an echo-ranging projector cannot be distinguished from the surface-reflected sound. The fluctuation is large and inexplicably variable. Observed values of variance average 40 percent with an inter-quartile spread of about 20 percent. The variance at 24 kc is significantly correlated with the variance at 16 kc or 60 kc, the coefficient of correlation being about 0.7.

## Variance with Deep Projector

For a deep projector and a deep hydrophone, the direct signal can be resolved from the surface-reflected signal. The observed fluctuation of the direct signal is small; observed values of the variance at 24 kc lie between 5 and 10 percent and may result from the variability of the measuring equipment. The surface-reflected pulse is highly variable with a variance between 50 and 70 percent.

With explosive pulses, the direct sound can be resolved from the surface-reflected pulse even at shallow depths. The observed variance for the direct pulse is about 1 or 2 percent if the transmission path lies wholly in an isothermal layer, but up to 20 percent if part of the transmission path lies in the thermocline.

## Rapidity of Fluctuation

The time during which the sound level is not likely to change appreciably is also variable, but seems to increase with increasing range. At a fixed range of less than a few hundred yards, the transmission loss for a shallow sound projector changes by about 20 percent on the average during 0.5 sec. At a fixed range of several thousand yards in the direct sound field, the average time for a 20 percent change might be 2 sec; while in the shadow zone, this average time is likely to be nearer 0.02 sec.

## Variation

Slow changes in the (averaged) transmission of sound in the sea, which take place in several minutes and which cannot be explained in terms of observable changes in the vertical temperature pattern, are called variations. It has been found that at 24 kc the variation between two transmission runs about 20 minutes apart has an average value of about 4 db if only pairs of transmission runs are considered in which the bathythermograph pattern is significantly the same. This average value for the variation does not appear to depend significantly on range.

## 10.5. FUTURE RESEARCH

During World War II research on the transmission of underwater sound has been largely devoted to the empirical investigation of certain practical problems. A wealth of detailed information has been accumulated on the transmission loss of sound from a standard echo-ranging projector under conditions likely to be observed in practice. Although this information has been useful in subsurface warfare, it has not led to any complete understanding of the physical processes involved in underwater sound transmission. For example, the average attenuation in deep isothermal water near San Diego has been extensively measured, but the causes of this attenuation are completely unknown.

In the years to come, research in this field will problably change its character. The quest for empirical data on some particular situation has been carried about as far as usefulness requires, and future studies will most profitably be directed to a more fundamental investigation of the basic factors underlying the observed data of underwater sound transmission. Without such a reorientation of the basic research program, it will be impossible to predict the behavior of underwater sound under new and unexplored conditions. Suppose, for instance, that sound gear using a nondirectional supersonic projector were to be proposed. The transmission loss for the sound from such a system could not be predicted definitely from present data, which are all obtained with directional supersonic sources. To make such predictions would require some knowledge of the importance of the scattering of sound through small angles. Similarly, the attenuation of sound transmitted from a deep projector to a deep hydrophone cannot be predicted from the present empirical data taken with shallow projectors, but might be estimated if the basic causes of attenuation were known.

In principle, the answer to any practical question about underwater sound transmission could be obtained by a program of measurements planned wholly for the purpose of answering that question. When haste is required, this is frequently the quicker method. When time is available, however, such answers can most efficiently be provided by a broad program designed to yield a physical understanding of what is happening. Such a program makes it ultimately possible to answer not one but a large number of practical questions. Thus, in the long run, improved technology can best be based on a foundation of long-term fundamental research.

This final section gives a brief discussion of some of the basic physical factors that may be expected to be important in underwater sound transmission and also treats the type of observations that might be expected to give meaningful information on these different factors.

## Basic Factors

The wave equation, equation (27) of Chapter 2, presumably governs in good approximation the propagation of sound waves in the interior of the ocean. It appears reasonable at first to investigate solutions of the simple wave equation, taking account of the presence of velocity gradients in the sea and of the reflections from sea surface and sea bottom. If the results are in flagrant disagreement with observations, then the effects of the approximations entering into the derivation of the wave equation must be investigated in detail. Apart from the validity of the wave equation as such, it is known that the body of the ocean contains scatterers (their nature uncertain) which deflect a fraction of the sound energy from its original direction of propagation. Furthermore, the observed absorption at supersonic frequencies far exceeds the value predicted on the basis of viscosity alone, necessitating the assumption of additional dissipative processes.

The most important problems of underwater sound transmission may thus be summarized under the following four headings.

1. The effects of velocity gradients in the sea.
2. Absorption and scattering in the volume of the sea.
3. Surface reflection.
4. Bottom reflection.

Each of these topics is discussed in the following subsections.

### SOUND VELOCITY

The velocity of sound is known as a function of temperature, pressure, and salinity and thus can be calculated at any point in the ocean where these physical quantities are known. The refraction effects produced by smooth vertical changes of temperature have been extensively investigated theoretically, and the results are in general qualitative agreement with the observations. Since the agreement is not complete, however, other effects must also play an important part. While the pressure is known as a function of depth, changes in temperature and salinity over distances of a few feet have not been extensively measured, and the acoustic effects to be expected from such changes have not been thoroughly explored. Microstructure of temperature and perhaps also of salinity may have an important effect on sound transmission, especially when the smoothed vertical gradient of sound velocity is small. Also, microstructure probably accounts for some part at least of the observed fluctuation of transmitted sound.

### ABSORPTION AND SCATTERING

The attenuation observed in deep isothermal water is presumably the result of absorption, that is, some dissipative process which converts sound energy into heat. Since the attenuation observed at 24 kc exceeds by a factor of about 100 the value predicted on the basis of shear viscosity alone, the principal cause of the observed attenuation must be some other mechanism. Among the dissipative mechanisms considered are compression viscosity (which, however, certainly is not the principal factor at 24 kc), gas bubbles present in the water, fish bladders, plankton, and thermodynamically irreversible chemical reactions, such as the hydrolysis of dissolved salts.

Gas bubbles and other inhomogeneities would not only absorb but also scatter sound. That scatterers are present in the sea is known. Scattering may account for part of the attenuation of highly collimated beams and also is probably responsible for most of the sound observed in predicted shadow zones in the presence of negative velocity gradients.

All hypotheses concerned with the cause of the absorption of sound as well as with the role of volume scattering on sound transmission are at present largely speculative. Until further experimental and theoretical work has provided a scientific understanding of the mechanisms involved, it will not be possible to predict with confidence the attenuation under many different conditions.

### SURFACE REFLECTION

The change in density at the sea surface is known and is so large that for most practical purposes the density of air may be set equal to zero; that is, the surface is almost a perfect reflector of sound. The complexity of surface-reflected sound arises from the complicated form of the ocean surface. In principle, it is simply a mathematical problem to compute the sound reflected from any surface of known properties. In practice, observations are unquestionably required. A thorough understanding of this topic would be important in studies both of fluctuation and of the average transmission anomaly in the surface layer.

### BOTTOM REFLECTION

The ocean bottom may have a topography equally as complicated as the ocean surface. In addition, the relative change in the elastic parameters and in density across the interface is much less extreme than across the ocean surface, and the detailed values of these changes must be considered. Since the physical properties of the bottom may vary with position, both vertically and horizontally, the problem of bottom-reflected sound can be very complicated physically as well as mathematically. In certain regions, where the bottom is flat, and of uniform composition, the acoustic phenomena are perhaps capable of being understood. Bottom-reflected sound is obviously important in many situations, especially when the direct sound is weakened by temperature gradients.

## Methods

To understand the physics of underwater sound transmission, each problem must be given separate consideration. The following methods may be applicable, however, to the investigation of a considerable variety of problems.

### OCEANOGRAPHIC MEASUREMENTS

An important part of any basic research on sound in the sea must be the investigation of the physical properties of the medium in which the sound is transmitted. It is in terms of these properties that the acoustic data are presumably to be interpreted.

In the first place, detailed measurements of the factors influencing sound velocity seem desirable, especially temperature measurements showing the full detail actually present in the sea. In the second place, detailed measurements of the shape of the ocean surface are required before any attempt can be made to explain surface-reflected sound; in particular, statistical information on the spectrum of the surface water waves present during any interval seems desirable. In the third place, complete physical data on the ocean bottom (on topography, composition, porosity and compactness, etc.) are required to interpret physically the data on bottom-reflected sound. Finally, it may be necessary to make a variety of physical measurements on ocean water as part of the attempt to identify the cause of absorption.

### CONTROLLED ACOUSTIC MEASUREMENTS

The experimental techniques of underwater acoustics research will probably be developed in a number of directions. Greater emphasis may be expected on detailed accuracy of the acoustic data; probable errors of several decibels for a transmission anomaly can presumably be considerably reduced. Measurements involving smaller samples of the ocean may perhaps be anticipated with relatively complete oceanographic data obtained for the small samples investigated. Some such experiment might be devised for measuring the sound absorption in a relatively small volume. Another possible development is along the lines of multiple measurement, in which many different items are measured almost simultaneously. For example, the inclination of the wave front might be measured at the same time as its intensity with simultaneous recordings at a number of different frequencies. Increasing complexity of the necessary equipment may probably be anticipated.

It is possible that explosive sound may be useful as a research tool. Short explosive pulses provide resolution of the direct and reflected pulses even at nearly grazing angles and also reveal clearly any multiple ray paths that may be present. By means of Fourier analysis, it is possible also to obtain with explosive sound many of the results which could be obtained by simultaneous transmission of many single frequencies over the entire spectrum. Finally, the high sound intensities possible with explosive pulses can provide data at longer ranges than are possible with standard sound projectors. Thus, explosive sound would appear to be a valuable tool of underwater sound research, deserving wider application than it has had in the past.

Regardless of what specific technique is used, the primary requirement for any basic experiment is that it be devised to give answers to certain physical questions rather than to operational problems. To satisfy this requirement, the theory underlying each experiment must be studied in detail before each experiment is actually performed to make sure that the results obtained will be significant. Considerable ingenuity may be required to find means for isolating the effects of the different factors involved in order to investigate them separately. It is only by such carefully designed experiments that our general understanding of sound in the sea can be continually increased.

# H. N. RUSSELL, ASTRONOMER

(*Science*, **125**, 1133, 1957)

## *Commentary*

THIS obituary for Henry Norris Russell (1877–1957) discussed his creative scientific work. The scope and wide influence of his research have been described by his colleagues, students and others.[1] His personal qualities have also been discussed by many who knew him. This commentary treats my own contacts with Russell during nearly three decades.

In 1930, when I was fifteen and a student at Phillips Academy, Andover, I became interested in astronomy through reading popular books by Eddington and Jeans. Frederick M. Boyce, my physics and astronomy teacher, who launched me on my scientific career, offered to take me into Boston to hear Professor Russell give a popular talk on stellar evolution. I accepted with pleasureable anticipation. Russell gave a broad picture of his topic. His enthusiasm and his sweeping theories I found tremendously exciting. According to modern knowledge, his point of view, with stars moving down the main sequence, is incorrect. But in any case what appealed to me in his work was the bold generality of his theories and the pattern of his research; i.e., his primary emphasis on the physical interrelations—what is happening and how—explained in the simplest terms. He used mathematics only as needed to verify any deductions and to yield numerical results.

It was six years later that I saw Russell again, this time when I was a graduate student entering Princeton in 1936. For two years I listened to him with rapt attention in course lectures and in occasional small discussion groups. I made no attempt to conceal how much I enjoyed hearing him talk, and as a result he usually directed his lectures toward me. His encyclopaedic knowledge of observations and theories together with his contagious enthusiasm made him an ideal mentor for me, much more effective in this regard than Eddington had been in 1935–36. One colorful idiosyncrasy, which others have commented on, was his tendency to fall asleep during colloquia[2] and, very occasionally, even during his own talks.[3]

Russell usually spent a few months each year at the Mt. Wilson Observatory, where his theoretical interests had a substantial and important influence on various observational programs. Toward the end of my first year, he returned from one of these visits, and reported excitedly that Walter Adams had agreed that I could use for my Ph.D. thesis his spectrograms of several red supergiant stars. Since this material did not seem to offer much opportunity for testing broad theories, it did not thrill me particularly. However, I gathered from other graduate students that Russell did not often consult with his students on assignment of thesis topics.[3] So I spent many hours analyzing Adams's stellar spectra. As it turned out, my work on these stars led me to some interesting fountain models for their atmospheres (soon disproved by later data). More important, it introduced me to the Mt. Wilson Observatory, where in later years I was to spend many months as a visiting observer.

It was after 1947, when I had returned to Princeton as a colleague, that my wife and I came to know the Russells well. We adopted them into our own family as Uncle Henry and Aunt May. Together with our children we enjoyed various activities in their company. Visiting them in their old-fashioned, high-ceilinged house, which Uncle Henry occupied for most of his life, was always a memorable experience. We recall vividly the pre-Christmas celebrations in their formal parlor, with lighted candles burning on their 14-foot evergreen tree (and a bucket of water close by). Sometimes at the Christmas season they joined us to sing carols together; Uncle Henry knew by heart all the verses in all the carols. In the spring we would go for a picnic together by a country stream to see the mertensia in bud.

I had some concern that Russell might interfere with my administration of the Department, since he had a reputation for riding rather roughshod over his junior colleagues. This concern turned out to be completely groundless; if he made suggestions to me at various times, they must have been minor indeed, since I do not recall them. He was at all times friendly and encouraging to me; as a scientist and a person he was an inspiration to us all.

## REFERENCES

1. In Memory of Henry Norris Russell, Dudley Observatory Report No. 13, eds. A.G.D. Philip and D. H. DeVorkin, Albany, NY (1977).
2. Greenstein, J. L.—ibid., p. 87; Spitzer, L.—ibid., p. 4.
3. Green, L. C.—ibid., p. 79.

# 30

## *Paper*

The death of Henry Norris Russell on 18 February marks the passing of one of the most brilliant minds that has flourished in the modern scientific world. A scientist of truly remarkable breadth, he was for many years the leading theoretical astronomer in this country and a pioneer in the use of atomic physics for the analysis of stars.

Russell's astronomical career began, as it ended, at Princeton University, where he spent 62 of his 79 years. He entered Princeton with the class of 1897 and was graduated with the highest scholastic record of his generation. During the quarter of a century when Latin honors were awarded to high-standing seniors, he was the only Princeton undergraduate whose record was so outstanding that he was graduated *insigni cum laude*. His Ph.D. degree was obtained at Princeton three years later, under C. A. ("Twinkle") Young, his famous predecessor at the Princeton Observatory. After two years in Cambridge, England, he returned to Princeton University, where he rose rapidly up the academic ladder and became professor of astronomy in 1911, at the age of 33. The following year, he became director of the Princeton University Observatory, succeeding Young.

During his early years as an astronomer, Russell engaged in a number of programs, largely in such classical topics as stellar parallaxes, photographic positions of the moon, and celestial mechanics. In 1912 he turned to the analysis of stellar spectra, a subject in which he remained active throughout his career. The famous diagram of stellar luminosity against surface temperature, originated by Hertzsprung, was used so extensively by Russell in his physical analyses of stars and in his theories of stellar evolution that this basic plot is generally referred to as the "Hertzsprung-Russell diagram." In the same year he published his analysis of the light curves of eclipsing variable stars. These double stars, whose light varies periodically because one star eclipses the other, had been known for some time, but no simple method existed for interpreting the observed light curves quantitatively. With a characteristic genius for handling complicated data, Russell devised a convenient and useful method for extracting full information from such observations. The method did not long remain untried. "For then," as Russell later used to remark, "the Lord sent me Harlow Shapley." This young Ph.D. candidate at Princeton, who was destined to become Russell's best-known student, applied Russell's techniques to 87 double stars as his doctoral thesis. With the advent of photoelectric techniques, the observations are now much more accurate then they were 45 years ago, and more refined methods have been developed for certain types of binaries. However, Russell's method remains the standard one for a first interpretation of any eclipsing-variable light curve.

During World War I, Russell, like so many other scientists, became involved in military research. During the decade following the war, much of his time was spent in preparing a textbook on astronomy, in collaboration with his colleagues, R. S. Dugan and J. Q. Stewart. The first volume of this work was a revision of Young's *Manual of Astronomy*, but the second volume, dealing with modern astrophysics and stellar astronomy, was, for the most part, new. A great deal of effort went into the preparation of these two volumes. Virtually nothing was taken on faith, and much original research was carried on to provide material for the book. An indication of the thorough care that went into this textbook for undergraduates is the fact that, for almost three decades, "Russell, Dugan, and Stewart" remained a standard reference work, despite enormous advances in most astronomical fields.

It was, perhaps, the preparation of this general textbook that gave Russell the extraordinary knowledge of astronomy that was so characteristic of his mature years. An astronomer in any field could count on Russell's intelligent interest and helpful comments.

In 1924 he embarked on a research problem that was destined to take much of his time during the rest of his active life. In that year he published his first paper on the analysis of atomic spectra, a subject in which he became interested because of the pivotal importance of spectroscopy in astrophysics. During the next ten years he published twenty-five papers in this new and exciting field of basic physics. A pioneer in this field, he originated, with Saunders, a theory for the coupling between spin and orbital momentum in atomic electrons. He took great delight in unravelling complicated spectra and in determining the atomic energy levels involved.

The years 1928-29 may be regarded as the apex of Russell's professional career. He was then just over fifty years of age and at the peak of his powers. Although immersed in the task of spectroscopic research, he carried on several other important analyses at about this time. His analysis, with Adams, of stellar spectra, appeared in 1928, and his great work, *On the Composition of the Sun's Atmosphere*, appeared in 1929. In these papers he used the Saha ionization equation to determine the pressure and chemical composition of stellar and solar atmospheres. He concluded, in variance with accepted beliefs, that hydrogen was overwhelmingly the most abundant element in the solar atmosphere. For many elements his values of relative abundances are still the best available, and his assertion of the great abundance of hydrogen has now, after a prolonged controversy, been accepted as one of the most basic facts of cosmology. During this same period he also published an analysis of the rotation of the line of apsides in eclipsing variables, in which he pointed out a very important method—in fact, the only observational method—for obtaining the density distribution in stars.

During his later years, Russell continued active research in all these fields. In addition, he maintained his lively interest in stellar evolution and planetary origin. His little monograph, *The Solar System and Its Origin*, did not present any new and sweeping theory but was of great help in clarifying the field and stimulated several of his students to productive work in this area.

On so outstanding a scientist, honors were showered from all sides. Medals were awarded him at various times by the Royal Astronomical Society (England), by the French Academy, by the National Academy of Sciences, by the American Academy of Arts and Sciences, and by the Astronomical Society of the Pacific. Honorary degrees were awarded him by Yale, Harvard, and Princeton universities, among others. He was a member of many learned societies and past president of several. In 1946, Mexico presented him with its highest award for foreigners, the Order of the Aztec Eagle.

There are few men now living who remember Russell as a young man. Those who knew him in his later years remember him for his unbounded energy and his enthusiasm for ideas. It is characteristic of the man that he would frequently be so carried away in his graduate lectures that he would talk enthusiastically for an additional hour or two, carrying his fascinated audience into exciting new realms of research. He brought this same keenness and enthusiasm to all the many experiences to his full and active life—to his extensive travels, his wide reading of both prose and poetry, and his happy hours with his grandchildren. He would keep small children engrossed for hours with the paper boats, balls, birds, and animals that he constructed with facility, his long, dexterous fingers folding and creasing the paper with unerring speed. His knowledge was encyclopedic; it included facts and theories not only in all branches of science but also in such varied subjects as the Bible and the wild flowers of New Jersey.

A deeply religious man, he was convinced that there was no basic conflict between science and religion. For several years he organized among the graduate students regular discussions of the interrelations between science and religion and published a little book on the subject, *Fate and Freedom*. Through his vast and cogent writings, through his many and vital discussions with colleagues and students, and through the sheer force of this personality and giant intellect and their impact on all who knew him, the influence of Henry Norris Russell will continue through many generations to come.

# LETTER FROM L. SPITZER TO H. SHAPLEY, WITH ATTACHED LONG-RANGE PLAN

(UNPUBLISHED, 1946)

## *Commentary*

THE correspondence below represents a major milestone for me, since it led directly to my position at Princeton as successor to Henry Norris Russell. This Commentary discusses the background underlying my letter and explains how it happened that Professor Shapley, Director of the Harvard College Observatory, served as an intermediary between Princeton University and me.

I had known Harlow Shapley for almost a decade. In autumn 1937, he interviewed me in connection with my application for a National Research Council Post-Doctoral Fellowship; I had indicated a preference for a year of research at Harvard under Shapley's guidance. We spent an hour together at the Harvard Club in New York. I was greatly impressed by his friendly personality, his incisive mind, and his penetrating understanding of the many topics we discussed. During my stimulating and profitable year at Harvard, 1938–39, I much admired the effective way he ran the famous "Hollow-Square" discussions of contemporary astronomical problems. When at the end of that year Yale offered me an Instructorship in Physics, half-time, Shapley advised me to accept this position, with its tenure possibilities, rather than to hold out for a possible appointment in Harvard's prestigious Society of Fellows. "Tenure-track positions at first-rate institutions are not often available!" he said. I followed his advice, which in retrospect I believe was thoroughly sound.

In November 1945, Professor Russell invited me to visit Princeton to discuss with him and Hugh S. Taylor, Dean of the Graduate School, "the future of the Observatory." Ever since my days as a graduate student, the top position at Princeton—Director of the Observatory and successor to Russell as Research Professor—had a powerful appeal for me. So of course I expressed interest and went to Princeton for interviews. I was surprised that my discussions with Dean Taylor were mostly concerned with literature, history, culture, and the arts, but pleased that about a month later I received a definite offer of an Associate Professorship. One important development was that a few years earlier Professor Rosseland had come to Princeton from Norway (by way of the USSR and the Pacific, since Norway had been occupied by the Germans) to take the Endowed Research chair on Russell's retirement. Hence the position to which I aspired would not be available. However, in a friendly talk, Rosseland told me that his wife's poor health might soon lead him to resign from Princeton and return to Norway. In that case, there would be two astronomy positions available at a senior level (Russell's and the late Professor Raymond S. Dugan's). After further discussion with Rosseland, it appeared that my chances of being offered Russell's position, if it became available, might well be greater if I were back at Yale than if I were already at Princeton.

In the meantime, Yale, aware of my interest in building up a program of research and graduate education in astrophysics, was proposing to set up an Astrophysics Research Unit, sponsored jointly by the departments of physics and astronomy, and under my direction. While this rather complex arrangement was less desirable than the top astronomy position at Princeton, it seemed preferable to a secondary position there. So I turned down the Princeton offer and accepted that from Yale.

Practically my first action back at Yale was to urge Martin Schwarzschild to join me as a professor in the new Astrophysics Unit. As pointed out in the Commentary for Paper #1, the discussions which the two of us had enjoyed together at Harvard had given me an enormous respect for him. He turned down the Yale offer, largely because (for good reason) he had doubts in general as to the long-term stability of such unusual administrative structures. He left me with the impression that he might well accept a similar offer from the Princeton Astronomy Department if I were the director there. Naturally this episode strengthened my hope that Princeton might yet invite me to return as Director and Research Professor.

My wife and I were back in New Haven in January 1946, in time for the start of the second academic term. We had purchased a house, but refrained from completing some needed redecoration until we might know what the chances were at Princeton. At the AAS meeting in Madison, Wisconsin, that summer, I approached Shapley and told him frankly how I felt about Princeton. Then I asked him if he knew of any developments there. He informed me that Rosseland had indeed resigned, and that the Research Professorship had been offered to Chandrasekhar, who had accepted it. I remarked that we could now proceed to put new wallpaper in our dining room.

A few months later, in October 1946, Shapley phoned me concerning subsequent developments at Princeton. Chandrasekhar had been offered such a unique and distinguished position at Chicago that he had asked to be released from his commitment to Princeton. So again there were two senior positions in astronomy open. Shapley reported that Princeton had been glad to hear from him that I was still much interested in returning to Princeton. However, they did not wish to negotiate with me directly on this possibility, for fear that in the end I would simply improve, once again, my position at Yale. I understood why they might feel that way, in view of the promotions I had received at Yale following two earlier offers (Pittsburgh in 1944, Princeton in 1946). Hence I was glad to comply with Shapley's request that I indicate precisely the conditions under which I would feel committed to accept an offer from Princeton. My letter (Paper #31), together with the attached Long-Range Plan, gives my response. Part II of the Plan, giving some detailed cost estimates, has been omitted here.

So I was back again at Princeton in January 1947, for more discussions. My wife and I stayed with the Russells, who were very supportive. Uncle Henry had been informed of the various negotiations, though I believe he played no great part in them. All seemed to go well, and shortly after this visit, my Long-Range Plan was approved with no major changes. I was able to convince Yale that no counter-offer could keep me in New Haven. Martin and I each accepted formal offers from Princeton. By the summer of 1947 Spitzers and Schwarzschilds were officially attached to the Princeton University Observatory, well launched on a program of research and graduate education, supported during more than thirty years by our happy and successful collaboration.

*Paper*

New Haven, Conn.
November 2, 1946

Dr. Harlow Shapley
Harvard College Observatory
Cambridge 38, Massachusetts

Dear Dr. Shapley:

Several days ago you asked me to let you know the conditions under which I would be interested in accepting a position at Princeton. For many reasons, I believe that the chairmanship at Princeton offers very great opportunities of the sort which interest me, and I would definitely accept an offer from Princeton University if it were along the lines which I visualize, and which I describe below.

The most important aspect of the Princeton opening, from my point of view, is the general policy of the University administration toward the Astronomy Department. I have outlined a long-range plan for the Princeton Astronomy Department, and am enclosing a copy with this letter. While I would not anticipate much alteration in the major features of this plan, the specific details are of course quite tentative, and subject to considerable revision. You will, I hope, understand my unwillingness to accept the responsibility for the Princeton Department unless the administration were to approve the general outlines of this plan, and were willing to give it enthusiastic support. In particular, the suggested annual budget of $40,000, exclusive of heating, maintenance of the grounds and property, etc., is about the minimum required to put into effect the plans I should like to see materialize at Princeton. The approximate nature of this figure results from the fact that I do not know Princeton's precise salary scale. You will note that eventual promotions and salary increases for the relatively young staff envisaged would increase the funds required, and therefore within 10 to 15 years, about $50,000 would be needed.

In addition, there are two rather minor, but nevertheless important, ways in which this program could be supported. First, I would assume that as in the past the unspent funds of the Astronomy Department would be allowed to accumulate as a reserve, and that the funds thus accumulated under the previous Chairman would be available for the support of future astronomical research. In the second place, I would assume that the house and apartments which form an integral part of the Department offices on Prospect Street would be at the disposal of the Department. These would be particularly important during the next few years, when the housing shortage would otherwise make it difficult to persuade new men to come to Princeton. In later years, these quarters might be useful in providing room for possible expansion of office space, should this become necessary.

I assume that Princeton will have no objection, and will, in fact, encourage any support of astronomical research by outside funds. As noted in the accompanying plan, it is possible that funds from a governmental Science Foundation might be used to build up a somewhat larger theoretical group than would otherwise be possible. Incidentally I am serving as a consultant to the Douglas Aircraft Co. on astronomical and interplanetary problems, and hope that this will lead to a close connection on my part with the ultimate establishment of an extraterrestrial satellite observatory. I trust that Princeton would be sympathetic with this work.

I am assuming that my own appointment would carry the title, Chairman of the Department of Astronomy and Director of the Observatory, with the rank of full professor. For the initial appointment, the minimum salary of a full professor would be wholly adequate. A strong factor in the appeal of the position at Princeton would be, of course, the honour of holding the professorship which Professor Russell has made so distinguished; my willingness to come to Princeton is based on the assumption that the administration considers my research sufficiently promising to warrant an informal understanding that I would be awarded this professorship a certain interval after my arrival in Princeton as head of the Department—say, in seven years.

My own respect for astronomy at Princeton in general and for Professor Russell in particular is so profound that it would be a great personal pleasure for me to come to Princeton under almost any conditions. The very strong support which astrophysics enjoys at Yale, however, would make it very difficult for me to leave New Haven, with its opportunities for effective research and growth, unless the corresponding opportunities at Princeton are at least as great.

If the authorities at Princeton would like to discuss these proposals with me, I shall be very glad to visit Princeton in the near future. Naturally I should appreciate receiving your reaction and that of the Princeton administration to these ideas.

With very best wishes,

Sincerely yours,
*Lyman Spitzer, Jr.*

# PROPOSED LONG-RANGE PLAN FOR
# THE PRINCETON UNIVERSITY OBSERVATORY

BY LYMAN SPITZER, JR.

Princeton University is justly known as one of the world's leading centers of theoretical astronomy. This reputation has been built up over a considerable period of years, and should be preserved. The plan presented here is devised to continue this historical tradition in the field of theoretical astrophysics, and at the same time to preserve a balanced department by maintaining research in an observational field that is an integral part of the Princeton tradition—precise photometry of variable stars. In the first section below the scientific aspects of the plan are discussed, while the cost estimates are presented in the second section. It should be emphasized that in detail this plan is to be regarded as somewhat flexible, since its execution would naturally depend on the availability of qualified personnel as well as on the facilities at Princeton.

## 1. SCIENTIFIC PROGRAM

It is proposed that the primary effort in astronomy at Princeton continue in the field of theoretical astrophysics, with three men of professorial rank in this field—the Director, Professor Stewart, and an additional man. Dr. Martin Schwarzschild would be an excellent choice for this third position, and there is reason to believe he might accept an offer of this type. If he were not available, and if no one of similar calibre could be found, a temporary Visiting Professor could be brought in, possibly a new man every year. On Professor Stewart's retirement, in some 15 years, it is assumed that his place would be taken by another theoretical astrophysicist with wide abilities and broad training.

To keep theory in touch with current observational problems, it is planned that in the near future the two new members of the permanent staff would each spend one academic term out of every four in a major observational center such as the Mount Wilson Observatory. It is understood that staff members would continue to receive their usual stipend from Princeton while carrying out research at other observatories in this manner. Such an arrangement would provide, at very moderate expense to Princeton, the observational facilities afforded by the world's largest telescopes. It is believed that the material obtained in these trips could also be used by Princeton graduate students, in keeping with the Princeton tradition.

In addition, it is projected that each year one of the world's leading theoretical astrophysicists would be brought to Princeton for half an academic year. Such visits would also be in the best Princeton tradition, and would strengthen Princeton's reputation as a center of theoretical astrophysics.

To preserve a balanced department, giving a broad training in astronomy, and to make effective use of the Princeton observational equipment, a competent observational astronomer should be included in the staff. The traditional program of photometric measurements on variable stars, inaugurated by Professor Dugan, should be continued if possible, but with the more accurate photoelectric techniques replacing the older visual methods. An assistant professor is contemplated for this work in the immediate future. This is not a very long-term solution, however, and ultimately a permanent appointment for the right man will be required. This appointment might be that of an Associate Professor, or possibly that of a Research Associate with permanent tenure. If a young graduate, trained in photoelectric methods, is brought to Princeton, and is advanced gradually, about eight years could elapse before a permanent appointment in this field would be required.

In addition to the professorial staff, two instructors would be envisaged. One would normally be in the field of theoretical astrophysics; this would be a strictly temporary appointment, and after three years at most would not be renewed. This post could probably not be filled at the present time. In a few years, however, a considerable number of students should be emerging from graduate schools, and the most outstanding young theorists should profit from a few years of research, with some teaching experience, in an active astrophysical center such as Princeton. The other instructorship would sometimes be filled by a theoretical astrophysicist, sometimes by a worker in a more observational field. As already indicated, in the near future such a man might be a variable star observer, trained in photoelectric techniques, under consideration for an ultimate permanent appointment at Princeton. In the more distant future this position might be filled in some years by an outstanding young observational man, serving as assistant to the observing astronomer of professorial rank. More often, however, this post would be given to another young theorist.

Such a staff as outlined would serve as a center or focus of an active research group. A number of graduate students should be attracted each years by such a stimulating department. If a governmental Science Foundation is set up, and if such a Foundation decides to support theoretical astrophysics on a substantial scale, the Astronomy Department at Princeton would make an ideal focus for such support. To cross-fertilize the different fields of astronomy, and to keep theorists and observationalists in touch with each other, it would be desirable to bring scientists from other institutions to Princeton from time to time for joint consideration of the major problems under study by astrophysicists. Thus the establishment of a considerable number of Visiting Professorships, financed by some governmental research unit, seems a definite possibility. Such a development would again be in harmony with the historic Princeton tradition.

# DREAMS, STARS, AND ELECTRONS

## (ANNU. REV. ASTRON. ASTROPH. 27, 1, 1989)

*Commentary*

PORTRAIT OF MY FATHER

Root. Branch.
Wind. Sun.
Snowdrift. Seed-drift.

Tent peaks. Mountains.
Trail. Slope. Heavy pack.
Show me the dance.

Morning. Cold.
Blow on bud of flame.
Cookfire.

Twilight. Race dark.
Down. Through trees.
Grip my hand.

Ask crucial questions.
Straight to the bone.

Bring Saturday's opera
out under trees.
I learn music.
What you carry inside.

Pure Science.
Poetry by heart.
Legacy of melancholy.
Delight in sweets.

Love of fountains.
Dream of Versailles.
Order my first Vouvray.
What do seeds know of wind
or what pushes them?
Look to the sky.
See beyond it.

Courtesy,
a kind of cloak.
Imp in eclipse.

Always, eyebrows
drooping like elm branches.

LYDIA SPITZER

## *Paper*[†]

"What were you doing today, Daddy?" my children often asked me. I would reply, "I was thinking what I would do if I were an electron." Much of my professional work has been based on an ability to visualize a physical system and how it operates and thus to predict its behavior without benefit of mathematics—though mathematics provides a much needed check. This particular facility has led me to work on a variety of physical processes important in astrophysics.

Another characteristic that has greatly affected my research is an attraction to spectacular and difficult projects that strike me as important. Three major such projects, which provided long-term goals for my career, concerned the theory of star formation, a large general-purpose space telescope, and generation of useful power from fusion energy. The objective of the first was to understand the sequence of complex processes occurring; those of the last two involved engineering as well as scientific problems.

This paper deals with my various research activities, especially those that now seem relevant to further work in astronomy. The personal characteristics described above are mentioned because they provide the unifying strands in what might otherwise seem a scattered, somewhat miscellaneous professional career. Some personal items relevant to my scientific life are included also in the subsequent text. In the words of Pooh-Bah, perhaps these will give verisimilitude to an otherwise bald and unconvincing narrative.

### EARLY YEARS

As a boy I enjoyed much of my school work, but a serious interest in science developed only after entering Phillips Academy, Andover, in 1929, at age fifteen, for two precollege years. There I was introduced to physics by Frederick Boyce, who had a unique capability for making this topic both clear and exciting. It was a wonderful experience to find that by understanding a few simple principles one could master an entire area of physics. In response to a request from several of us, Mr. Boyce gave an astronomy course, supplementing the books by Jeans and Eddington that I had been reading and leaving me permanently fascinated with astrophysics.

---

[†] The latter portion of this article, describing my research activities in a number of specific fields, has been omitted here.

At the end of my stay at Andover I became involved in my mind with the first of the big projects that have appealed to me. This concerned the development of a global transportation system, based on the electromagnetic propulsion of levitated cars in evacuated tubes. Others have suggested such ideas. My concept was to integrate intercity travel with local transportation, so that one could get into a car on a high floor of an office building, dial the local or long distance number desired, and get out some minutes later at the desired floor in one's own city or in some remote center. I spent so much time considering various details of such a system that my parents, seriously concerned for my sanity, extracted a promise from me to stop thinking about this enterprise!

From Andover I went on to Yale, where astronomy (as well as the global transportation system) was put aside while majoring in physics. Under Leigh Page I took a general survey course in theoretical physics and found a keen aesthetic pleasure from his elegantly organized mathematical presentations. A limitation of my own in such areas became painfully evident during my final year at Yale, when I took a final oral examination on a physics thesis I had written for some independent work. Said one of my examiners, "Your second equation, Mr Spitzer, states that the power $P$ equals $ir^2$. Could you derive this for us, please?" Swiftly going through the elementary steps, I suddenly realized that I had inadvertently written $ir^2$ instead of the familiar $i^2r$, and that all the subsequent equations in my thesis were incorrect! My mind tends to make inversions of this sort, and while I have learned to ferret out most of them, they still dog my work from time to time.

In graduate school, first at Cambridge, England, (as a Henry Fellow) in 1935–36 and then at Princeton, I returned to astrophysics, partly because I was so enchanted by the beautiful theories that S. Chandrasekhar had presented at a series of informal evening seminars in his Trinity College rooms, and partly because I was so impressed by the physical lucidity of several contemporary papers by Bengt Strömgren. Studying under Henry Norris Russell was also a unique experience; his enormous knowledge, physical insight, and unfailing enthusiasm made him a very stimulating mentor. The lectures by Edward Condon were particularly helpful in extending my own insight into various aspects of the quantum theory.

My thesis topic, suggested by Russell, was an analysis of high-dispersion spectra that Walter Adams had obtained for three cool supergiant stars. I proposed a fountain model to explain the observed dependence of radial velocity on excitation potential. While later work by others disproved this model, the research was educational for me.

During my postdoctoral year at Harvard (as a National Research Council Fellow) in 1938–39, I found tremendously stimulating the scientific discussions among the active scientists whom Harlow Shapley. Donald Menzel, and Bart Bok had assembled at the Observatory. In particular, the free-ranging exchange of views at Shapley's "hollow square" discussion groups and at the evening seminars organized by Bok at his home gave all of us many exciting, important problems to work on. As pointed out below, my subsequent research in stellar dynamics and in star formation and interstellar matter grew out of these discussions.

In this active ferment of astronomical ideas Martin Schwarzschild was an imposing figure. He greatly impressed me by his insight, originality, and forcefulness. I felt strongly that he would be an ideal colleague if events should make it possible for us to be in the same institution.

## YALE AND PRINCETON

A brief discussion of my years at these two institutions will provide a backdrop for a discussion of my research work there. At Yale, where I spent two and a half years before shifting to war research in New York and Washington, and another year and a half afterward, I was alone much of the time as far as astrophysics was concerned. This did not worry me, since the impetus I had received at Princeton and Harvard would suffice for some years, and informal get-togethers of astronomers in the region from Washington to Cambridge were frequent.

My initial appointment in 1939 was as an instructor in the physics department, which gave me a welcome chance to broaden my background in this basic field, as well as to continue some research I had begun at Harvard. Under the influence of Henry Margenau I published a few papers on the collisional broadening of spectral lines, giving relatively exact solutions in certain idealized cases. When I returned to Yale after four years of underwater sound research, I had a joint appointment in physics and astronomy and concentrated my research on problems related to interstellar matter.

In 1947 I was offered a professorship at Princeton and also the directorship of the Observatory, succeeding Russell. I soon accepted; ever since my graduate student days I had felt this would be an ideal position. For this offer I am greatly indebted to three of this country's greatest astronomers—to Russell for his support, to Chandrasekhar for not accepting the earlier offer to him of Russell's professorship, and to Shapley for informing Princeton of my long-standing interest in this particular post and then serving as a helpful intermediary between me and the Princeton administration. In addition to his widespread other activities, Shapley ran sort of an informal employment bureau for young astronomers. Following the tradition established by Russell, I held the directorship (together with the departmental chairmanship) for some three decades; my brilliant colleague Jerry Ostriker succeeded me in 1979, three years before my retirement from the faculty (compulsory then at age 68).

One of my most important achievements at Princeton was to persuade Martin Schwarzschild to join me there. We worked closely and effectively together, his wisdom in practical affairs supplementing my willingness to get absorbed occasionally with administrative details. Our joint objective, when we both moved to Princeton in 1947, was to build a significant graduate program in theoretical astrophysics, with some emphasis also on observations. In support of this latter purpose the Princeton administration agreed to send Schwarzschild and me to the western US (in practice, usually the Mt. Wilson Observatory) for observing, each of us in alternate years for one semester. This arrangement certainly encouraged observational work by the two of us and our students. After fifteen years, the shift of our observational research plans to space telescopes forced us to drop, with great reluctance, these regular observing trips to the West Coast.

# List of published papers
## by Lyman Spitzer, Jr.

### BOOKS

*Physics of Sound in the Sea* (Columbia Univ. Press) 1946; reprinted 1968 (Gordon & Breach, New York) (Editor).

*Physics of Fully Ionized Gases* (Interscience Publishers, New York), 105 pages (1956).

*Physics of Fully Ionized Gases*, 2d ed. (Interscience Publishers, New York), 170 pages (1962).

*Diffuse Matter in Space* (Interscience Publishers, New York), 262 pages (1968).

*Physical Processes in the Interstellar Medium* (J. Wiley, New York), 318 pages (1978).

*Searching Between the Stars* (Yale Univ. Press), 179 pages (1982) (based on Silliman Lectures, 1978).

*Dynamical Evolution of Globular Clusters* (Princeton Univ. Press), 180 pages (1987).

### TECHNICAL ARTICLES

Noncoherent Dispersion and the Formation of Fraunhofer Lines, Monthly Notices of the Royal Astronomical Society, vol. 96, pp. 794–807 (1936).

New Solutions of the Equation of Radiative Transfer, Astrophysical Journal, vol. 87, pp. 1–8 (1938).

A Contribution to the Mathematical Theory of Big Game Hunting (with J. Tukey et al. under pen name of H. Pétard), American Mathematical Monthly, vol. 45, pp. 446–47 (1938).

The Hubble-Tolman Analysis of Dispersion of Nebular Magnitudes, Observatory, vol. 61, pp. 104–6 (1938).

Spectra of M Supergiant Stars, Astrophysical Journal, vol. 90, pp. 494–540 (1939), Contribution from the Mt. Wilson Observatory, No. 619.

A General Solution for the Equations of Natural Line Breadth, Physical Review, vol. 55, pp. 361–64 (1939).

Stark-Effect Broadening of Hydrogen Lines, I. Single Encounters, Physical Review, vol. 55, pp. 699–708 (1939).

Stark-Effect Broadening of Hydrogen Lines, II. Observable Profiles, Physical Review, vol. 56, pp. 39–47 (1939).

The Dissipation of Planetary Filaments, Astrophysical Journal, vol. 90, pp. 675–88 (1939).

The Stability of Isolated Clusters, Monthly Notices Royal Astronomical Society, vol. 100, pp. 396–413 (1940).

Impact Broadening of Spectral Lines, Physical Review, vol. 58, pp. 348–57 (1940).

The Dynamics of the Interstellar Medium. I. Local Equilibrium, Astrophysical Journal, vol. 93, pp. 369–79 (1941).

The Dynamics of the Interstellar Medium. II. Radiation Pressure, Astrophysical Journal, vol. 94, pp. 232–44 (1941).

The Dynamics of the Interstellar Medium. III. Galactic Distribution, Astrophysical Journal, vol. 95, pp. 329–44 (1942).

Notes on the Theory of Noncoherent Scattering, Astrophysical Journal, vol. 99, pp. 1–7 (1944).

Acoustic Properties of Gas Bubbles in a Liquid, Office of Scientific Research and Development Report No. 1705 (Columbia Univ. Div. of War Research: New York, NY), 62 pages (1943).

Recent Progress in Astrophysics, Astrophysical Journal, vol. 99, pp. 107–15 (1944).

The Magnetic Field of the Galaxy, Physical Review, vol. 70, pp. 777–78 (1946).

Astronomical Advantages of an Extra-Terrestrial Observatory, PROJECT RAND, pp. 71–75 (1946); reprinted in Astronomy Quarterly, vol. 7, pp. 131–42 (1990).

The Temperature of Interstellar Matter, I, Astrophysical Journal, vol. 107, pp. 6–33 (1948).

The Formation of Cosmic Clouds, Harvard Observatory Monograph No. 7, pp. 87–108 (1948).

The Distribution of Interstellar Sodium, Astrophysical Journal, vol. 108, pp. 276–309 (1948).

The Terrestrial Atmosphere above 300 km, *The Atmospheres of the Earth and Planets*, ed. G. P. Kuiper (Univ. of Chicago Press: Chicago, IL), pp. 213–49 (1949).

Interstellar Polarization, Galactic Magnetic Fields and Ferromagnetism (with J. W. Tukey), Science, vol. 109, pp. 461–62 (1949).

The Temperature of Interstellar Matter. II, Astrophysical Journal, vol. 109, pp. 337–53 (1949).

Upper Limits on the Abundances of Interstellar Li and Be, Astrophysical Journal, vol. 109, pp. 548–50 (1949).

On the Origin of Heavy Cosmic-Ray Particles, Physical Review, vol. 76, p. 583 (1949).

Equivalent Widths of Interstellar Calcium Lines (with I. Epstein and Li Hen), Annales d'Astrophysique, vol. 13, 147–63 (1950).

The Temperature of Interstellar Matter. III (with M. P. Savedoff), Astrophysical Journal, vol. 111, pp. 593–608 (1950).

Perturbations of a Satellite Orbit, Journal of the British Interplanetary Society, vol. 9, pp. 131–36 (1950).

The Acceleration of Dust Grains by Supernovae (with B. Wolfe, P. McR. Routly, and A. S. Wightman), Physical Review, vol. 79, pp. 1020–21 (1950).

The Electrical Conductivity of an Ionized Gas (with R. S. Cohen and P. McR. Routly), Physical Review, vol. 80, pp. 230–38 (1950).

On the Interpretation of Measured Solar Wavelengths, Monthly Notices of the Royal Astronomical Society, vol. 110, pp. 216–19 (1950).

A Proposed Stellarator, Project Matterhorn Report PM-S-1, Princeton University, NY0-993 (1951).

Physical Properties of the Interstellar Gas, in *Problems of Cosmical Aerodynamics*, pp. 31–40 (1951).

Stellar Populations and Collisions of Galaxies (with W. Baade), Astrophysical Journal, vol. 113, pp. 413–18 (1951).

The Density of Molecules in Interstellar Space (with D. R. Bates), Astrophysical Journal, vol. 113, pp. 441–63 (1951).

A Theory of Interstellar Polarization (with J. W. Tukey), Astrophysical Journal, vol. 114, pp. 187–205 (1951).

Report of the Standing Committee on Problems of the Upper Atmosphere, Transactions American Geophysical Union, vol. 32, p. 757 (1951).

The Possible Influence of Interstellar Clouds on Stellar Velocities (with M. Schwarzschild), Astrophysical Journal, vol. 114, pp. 385–97 (1951).

On the Difference in Chemical Composition between High- and Low-Velocity Stars (with M. Schwarzschild and R. Wildt), Astrophysical Journal, vol. 114, pp. 398–406 (1951).

Continuous Emission from Planetary Nebulae (with J. L. Greenstein), Astrophysical Journal, vol. 114, pp. 407–20 (1951).

Interplanetary Travel between Satellite Orbits, Journal of the American Rocket Society, vol. 22, pp. 92–96 (1952).

A Comparison of the Components in Interstellar Sodium and Calcium (with P. McR. Routly), Astrophysical Journal, vol. 115, pp. 227–43 (1952).

The Distribution of Interstellar Sodium. II (with J. B. Oke), Astrophysical Journal, vol. 115, pp. 222–26 (1952).

Equations of Motion for an Ideal Plasma, Astrophysical Journal, vol. 116, pp. 299–316 (1952).

Internal Motions within Interstellar Clouds (with A. Skumanich), Astrophysical Journal, vol. 116, pp. 452–54 (1952).

Transport Phenomena in a Completely Ionized Gas (with R. Härm), Physical Review, vol. 89, pp. 977–81 (1953).

The Possible Influence of Interstellar Clouds on Stellar Velocities. II (with M. Schwarzschild), Astrophysical Journal, vol. 118, pp. 106–12 (1953).

On the Evolution of Stars and Chemical Elements in the Early Phases of a Galaxy (with M. Schwarzschild), Observatory, vol. 73, pp. 77–79 (1953).

Behavior of Matter in Space (Henry Norris Russell Lecture, Dec. 27, 1953), Astrophysical Journal, vol. 120, pp. 1–17 (1954).

Problems of the Stellarator as a Useful Power Source (with D. J. Grove, W. E. Johnson, L. Tonks, and W. F. Westendorp), Project Matterhorn Report PM-5-14, Princeton University, NYO-6047 (1954).

Acceleration of Interstellar Clouds by O-Type Stars (with J. H. Oort), Astrophysical Journal, vol. 121, pp. 6–23 (1955).

Abundance of Interstellar Beryllium (with G. B. Field), Astrophysical Journal, vol. 121, pp. 300–305 (1955).

The Distribution of Interstellar Sodium and Calcium (with D. A. Lautman), Astrophysical Journal, vol. 123, pp. 363–66 (1956).

On a Possible Interstellar Galactic Corona, Astrophysical Journal, vol. 124, pp. 20–34 (1956).

Astrophysical Research with an Artificial Satellite, Earth Satellites as Research Vehicles, Franklin Institute of Pennsylvania Monograph No. 2, pp. 69–78 (1956).

On the Determination of Air Density from a Satellite, in Scientific Uses of Earth Satellites, ed. J. A. van Allen (Univ. of Michigan Press: Ann Arbor, MI), pp. 99–108 (1956).

Star Formation in Magnetic Dust Clouds (with L. Mestel), Monthly Notices of the Royal Astronomical Society, vol. 116, pp. 503–14 (1956).

Influence of Fluid Motions on the Decay of an External Magnetic Field, Astrophysical Journal, vol. 125, pp. 525–34 (1957).

Cooperative Phenomena in Hot Plasmas, Nature, vol. 181, pp. 221–22 (1958).

Disruption of Galactic Clusters, Astrophysical Journal, vol. 127, pp. 17–27 (1958).

Evaporation of Stars from Isolated Clusters (with R. Härm), Astrophysical Journal, vol. 127, pp. 544–50 (1958).

The Stellarator Concept, Physics of Fluids, vol. 1, pp. 253–64 (1958).

Mass Exchange with the Interstellar Medium and the Formation of Type I Stars, Semaine d'Etude sur le Probleme des Populations Stellaires, Académie Pontificale des Sciences, Scripta Varia, No. 16, pp. 445–58 (1958).

Theoretical Problems of Stellar Magnetism, in Electromagnetic Phenomena in Cosmical Physics, IAU Symposium No. 6, ed. B. Lehnert (Cambridge, University Press), pp. 169–81 (1958).

General Summary, Third Symposium on Cosmical Gas Dynamics, IAU Symposium No. 8, Reviews of Modern Physics, vol. 30, pp. 1102–3 (1958).

The Stellarator Program, Progress in Nuclear Energy Series, XI, Vol. 1, Plasma Physics and Thermonuclear Research (Pergamon Press: New York) pp. 107–24 (1959).

A High Dispersion Photoelectric Spectrophotometer (with J. B. Rogerson and J. D. Bahng), Astrophysical Journal, vol. 130, pp. 991–1002 (1959).

Interstellar Research with a Spectroscopic Satellite (with F. R. Zabriskie), Publications of the Astronomical Society of the Pacific, vol. 71, pp. 412–20 (1959).

Space Telescopes and Components, Astronomical Journal, vol. 65, pp. 242–63 (1960).

Problems in Magneto-Fluid Dynamics, Reviews of Modern Physics, vol. 32, pp. 696–700 (1960).

Particle Diffusion across a Magnetic Field, Physics of Fluids, vol. 3, pp. 659–61 (1960).

The Far Ultraviolet Line Spectrum of a B2 Star (with J. E. Gaustad), Astrophysical Journal, vol. 134, pp. 771–76 (1961).

Princeton Observatory Work and Plans in Connection with Satellite Spectroscopy (with J. B. Rogerson), Mémoires Société Royale Sciences, Liège, Tome IV, pp. 86–88 (1961).

The Beginnings and Future of Space Astronomy, American Scientist, vol. 50, 473–84 (1962).

Star Formation and Magnetic Fields, in *The Distribution and Motion of Interstellar Matter in Galaxies*, ed. L. Woltjer (W. A. Benjamin, Inc.: New York), pp. 98–107 (1962).

Star Formation, in *Origin of the Solar System*, ed. R. Jastrow and A.G.W. Cameron (Academic Press: New York), pp. 39–53 (1963).

Theoretical Equivalent Widths of the Interstellar $H_2$ Lines (with K. Dressler and W. L. Upson, II), Proceedings of the Astronomical Society of the Pacific, vol. 76, pp. 387–98 (1964).

On the Evolution of Galactic Nuclei (with W. C. Saslaw), Astrophysical Journal, vol. 143, pp. 400–419 (1966).

Line Spectra of Delta and Pi Scorpii in the Far Ultraviolet (with D. C. Morton), Astrophysical Journal, vol. 144, pp. 1–12 (1966).

Controlled Nuclear Fusion Research, Review of Experimental Results, Plasma Physics and Controlled Nuclear Fusion Research, International Atomic Energy Agency, vol. 1, pp. 3–11 (1966).

Magnetic Alignment of Interstellar Grains (with R. V. Jones), Astrophysical Journal, vol. 147, pp. 943–64 (1967).

Thermal Deformations in a Satellite Telescope Mirror (with B. Boley), Journal of the Optical Society of America, vol. 57, pp. 901–13 (1967).

On the Evolution of Galactic Nuclei. II (with M. E. Stone), Astrophysical Journal, vol. 147, pp. 519–28 (1967).

Photomultiplier Tube Pulses Induced by Gamma Rays (with K. Dressler), Review of Scientific Instruments, vol. 38, pp. 436–38 (1967).

Heating of H I Regions by Energetic Particles (with M. G. Tomasko), Astrophysical Journal, vol. 152, pp. 971–86 (1968).

Dynamics of Interstellar Matter and the Formation of Stars, in *Stars and Stellar Systems*, vol. VII, ed. B. M. Middlehurst and L. H. Aller (Univ. of Chicago Press: Chicago), pp. 1–63 (1968).

Astronomical Research with the Large Space Telescope, Science, vol. 161, pp. 225–29 (1968); see also *Manned Laboratories in Space*, ed. S. F. Singer (Dordrecht: Reidel), pp. 88–98 (1969).

Absorption Lines Produced by Galactic Halos (with J. N. Bahcall), Astrophysical Journal Letters, vol. 156, pp. L63–L65 (1969).

Equipartition and the Formation of Compact Nuclei in Spherical Stellar Systems, Astrophysical Journal Letters, vol. 158, pp. L139–L143 (1969).

Heating of H I Regions by Energetic Particles. II. Interaction between Secondaries and Thermal Electrons (with E. H. Scott), Astrophysical Journal, vol. 158, pp. 161–71 (1969).

Optical Space Astronomy, Position Paper of the Astronomy Missions Board, NASA, ed. R. O. Doyle (NASA SP-213), pp. 46–63 (1969).

Scientific Uses of the Large Space Telescope, National Academy of Sciences *Ad Hoc* Committee on the Large Space Telescope, 47 pages (1969).

Random Gravitational Encounters and the Evolution of Spherical Systems. I. Method (with M. H. Hart), Astrophysical Journal, vol. 164, pp. 399–409 (1971).

Random Gravitational Encounters and the Evolution of Spherical Systems. II. Models (with M. H. Hart), Astrophysical Journal, vol. 166, pp. 483–511 (1971).

Random Gravitational Encounters and the Evolution of Spherical Systems. III. Halo (with S. L. Shapiro), Astrophysical Journal, vol. 173, pp. 529–47 (1972).

Orientation of Rotating Grains (with E. M. Purcell), Astrophysical Journal, vol. 167, pp. 31–62 (1971).

Dynamical Evolution of Dense Spherical Star Systems, in: Pontificiae Academiae Scientiarum Scripta Varia, no. 35, pp. 443–75 (1971).

Random Gravitational Encounters and the Evolution of Spherical Systems. IV. Isolated Systems of Identical Stars (with T. X. Thuan), Astrophysical Journal, vol. 175, pp. 31–61 (1972).

On the Evolution of Globular Clusters (with R. A. Chevalier and J. P. Ostriker), Astrophysical Journal Letters, vol. 176, pp. L51–L56 (1972).

Random Gravitational Encounters and the Evolution of Spherical Systems. V. Gravitational Shocks (with R. A. Chevalier), Astrophysical Journal, vol. 183, pp. 565–81 (1973).

Spectrophotometric Results from the *Copernicus* Satellite. I. Instrumentation and Performance (with J. B. Rogerson, J. F. Drake, K. Dressler, E. B. Jenkins, D. C. Morton, and D. G. York), Astrophysical Journal Letters, vol. 181, pp. L97–L102 (1973).

Spectrophotometric Results from the *Copernicus* Satellite, II. Composition of Interstellar Clouds (with D. C. Morton, J. F. Drake, E. B. Jenkins, J. B. Rogerson, and D. G. York), Astrophysical Journal Letters, vol. 181, pp. L103–L109 (1973).

Spectrophotometric Results from the *Copernicus* Satellite. III. Ionization and Composition of the Intercloud Medium (with J. B. Rogerson, D. G. York, J. F. Drake, E. B. Jenkins, and D. C. Morton), Astrophysical Journal Letters, vol. 181, pp. L110–L115 (1973).

Spectrophotometric Results from the *Copernicus* Satellite. IV. Molecular Hydrogen in Interstellar Space (with J. F. Drake, E. B. Jenkins, D. C. Morton, J. B., Rogerson and D. G. York), Astrophysical Journal Letters, vol. 181, pp. L116–L121 (1973).

Spectrophotometric Results from the *Copernicus* Satellite. V. Abundances of Molecules in Interstellar Clouds (with E. B. Jenkins, J. F. Drake, D. C. Morton, J. B. Rogerson, and D. G. York), Astrophysical Journal Letters, vol. 181, pp. L122–L127 (1973).

Spectrophotometric Results from the *Copernicus* Satellite. VI. Extinction by Grains at Wavelengths between 1200 and 1000 Å (with D. G. York, J. F. Drake, E. B. Jenkins, D. C. Morton, and J. B. Rogerson), Astrophysical Journal Letters, vol. 182, pp. L1–L6 (1973).

Rotational Excitation of Interstellar $H_2$ (with W. D. Cochran), Astrophysical Journal Letters, vol. 186, pp. L23–L28 (1973).

On the Theory of $H_2$ Rotational Excitation (with E. G. Zweibel), Astrophysical Journal Letters, vol. 191, pp. L127–L130 (1974).

History of the Large Space Telescope (AIAA 12th Aerospace Sciences Meeting, Washington, DC. Jan. 30–Feb. 1, 1974), pp. 3–6 (1974).

Column Densities of Interstellar Molecular Hydrogen (with W. D. Cochran and A. Hirshfeld), Astrophysical Journal Supplement No. 266, vol. 28, pp. 373–89 (1974).

Random Gravitational Encounters and the Evolution of Spherical Systems. VI. Plummer's Model (with J. M. Shull), Astrophysical Journal, vol. 200, pp. 339–42 (1975).

Random Gravitational Encounters and the Evolution of Spherical Systems. VII. Systems with Several Mass Groups (with J. M. Shull), Astrophysical Journal, vol. 201, pp. 773–82 (1975).

Dynamical Theory of Spherical Stellar Systems with Large N, in *Dynamics of Stellar Systems*, IAU Symposium No. 69, ed. A. Hayli (Reidel: Dordrecht), pp. 2–26 (1975).

Ultraviolet Studies of the Interstellar Gas (with E. B. Jenkins), Annual Review of Astronomy and Astrophysics, vol. 13, pp. 133–64 (1975).

The Formation of the Nuclei of Galaxies. I. M31 (with S. D. Tremaine and J. P. Ostriker), Astrophysical Journal, vol. 196, pp. 407–11 (1975).

Components in Interstellar Molecular Hydrogen (with W. A. Morton), Astrophysical Journal, vol. 204, pp. 731–49 (1976).

Hydrogen Molecules in Interstellar Space (George Darwin Lecture, Dec. 12, 1975), Quarterly Journal Royal Astronomical Society, vol. 17, pp. 97–120 (1976).

Note on the Collapse of Magnetic Interstellar Clouds (with T. Ch. Mouschovias), Astrophysical Journal, vol. 210, pp. 326–27 (1976).

High-Velocity Interstellar Clouds, Comments on Astrophysics, vol. 6, pp. 177–87 (1976).

Interstellar Matter Research with the *Copernicus* Satellite (Karl Schwarzschild Lecture, Sept. 16, 1975), Mitteilungen der Astronomischen Gesellschaft, no. 38, pp. 27–39 (1976).

Russell and Theoretical Astrophysics, *In Memory of Henry Norris Russell*, IAU Symposium No. 80, Dudley Observatory Report No. 13, pp. 3–8 (1977).

Disorientation of Interstellar Grains in Suprathermal Rotation (with T. A. McGlynn), Astrophysical Journal, vol. 231, pp. 417–24 (1979).

History of the Space Telescope, Quarterly Journal Royal Astronomical Society, vol. 20, pp. 29–36 (1979).

Random Gravitational Encounters and the Evolution of Spherical Systems. VIII. Clusters with an Initial Distribution of Binaries (with R. D. Mathieu), Astrophysical Journal, vol. 241, pp. 618–36 (1980).

Ultraviolet Spectra of Stars, in *Space Science Comes of Age*, ed. P. A. Hanle and Von Del Chamberlain (Smithsonian Institution Press: Washington, DC), pp. 2–13 (1981).

Panel on Dynamic vs. Static Models of the Interstellar Medium—Introductory Remarks, NRAO Workshop on Phases of ISM, pp. 1–5, May 11, 1981.

Acoustic Waves in Supernova Remnants, Astrophysical Journal, vol. 262, pp. 315–21 (1982).

Introduction to the Orion Symposium, Annals of the New York Academy of Sciences, vol. 395, pp. 1–7 (1982).

Interstellar Abundances of Oxygen and Nitrogen (with D. G. York, R. C. Bohlin, J. Hill, E. B. Jenkins, B. D. Savage, and T. P. Snow), Astrophysical Journal Letters, vol. 266, L55–L59 (1983).

A Survey of Ultraviolet Interstellar Absorption Lines (with R. C. Bohlin, J. K. Hill, E. B. Jenkins, B. D. Savage, T. P. Snow, and D. G. York), Astrophysical Journal Supplement, vol. 51, 277–308 (1983).

The Interstellar Gas as Viewed by Copernicus, in *Topics in Plasma-, Astro- and Space Physics*, ed. G. Haerendel and B. Battrick, pp. 57–66 (1983).

Dynamics of Globular Clusters, Science, vol. 225, pp. 465–72 (1984).

Scattering of Shock Waves by a Spherical Cloud (with S. Ikeuchi), Astrophysical Journal, vol. 283, pp. 825–32 (1984).

Average Density along Interstellar Lines of Sight, Astrophysical Journal Letters, vol. 290, pp. L21–L24 (1985).

Precollapse Evolution of Globular Clusters, in *Dynamics of Star Clusters*, I.A.U. Symposium No. 113, ed. J. Goodman and P. Hut (Reidel: Dordrecht), pp. 109–37 (1985).

What Next? Priorities in Theorem and Observations, in *Dynamics of Star Clusters*, I.A.U. Symposium no. 113, ed. J. Goodman and P. Hut (Reidel: Dordrecht), p. 499 (1985).

Abundances of Interstellar Atoms from Ultraviolet Absorption Lines (with B. D. Savage and E. B. Jenkins), Astrophysical Journal, vol. 301, pp. 355–79 (1986).

Clouds between the Stars, Crafoord Lecture, submitted to Royal Swedish Academy of Sciences, October 1985; published in *Physica Scripta*, T11, pp. 5–13 (1985).

Spectroscopy with *Copernicus* and the Edwin P. Hubble Observatory, in *Vistas in Astronomy*, vol. 29, pp. 143–50 (1986).

Ultraviolet Absorption Studies of the Interstellar Gas, Publications of the Astronomical Society of the Pacific, vol. 100, pp. 518–23 (1988).

Dreams, Stars and Electrons, Annual Review of Astronomy and Astrophysics, vol. 27, pp. 1–17 (1989).

Hot Gas in Interstellar Space, Matematisk-fysiske Meddelelser, 42:4, Royal Danish Academy of Sciences and Letters, pp. 157–77 (1990).

Theories of the Hot Interstellar Gas, Annual Review of Astronomy and Astrophysics, vol. 28, pp. 71–101 (1990).

Highly Ionized Atoms Toward HD93521 (with E. L. Fitzpatrick), Astrophysical Journal Letters, vol. 391, pp. L41–L44 (1992).

The Interstellar Abundances of Tin and Four Other Heavy Elements (with L. M. Hobbs, D. E. Welty, D. C. Morton, and D. G. York), Astrophysical Journal, vol. 411, pp. 750–55 (1993).

Composition of Interstellar Clouds in the Disk and Halo I. HD93521 (with E. L. Fitzpatrick), Astrophysical Journal, vol. 409, pp. 299–318 (1993).

Composition of Interstellar Clouds in the Disk and Halo II. $\gamma^2$ Velorum (with E. L. Fitzpatrick), Astrophysical Journal, vol. 427, pp. 232–58 (1994).

Composition of Interstellar Clouds in the Disk and Halo III. HD149881 (with E. L. Fitzpatrick), Astrophysical Journal, vol. 445, pp. 196–210 (1995).
Highly Ionized Interstellar Atoms—Heated, Cooled or Mixed? Astrophysical Journal Letters, vol. 458, pp. L29–L32 (1996).

## POPULAR ARTICLES

The Encounter Theory Falls, The Sky, vol. 5, pp. 6–7 (1941).
Astronomy at Princeton, Princeton Alumni Weekly, vol. 48, pp. 5–7 (March 19, 1948).
The Formation of Stars, Physics Today, vol. 1, pp. 6–11 (1948); reprinted in Smithsonian Institution Report for 1949, pp. 153–60.
The Formation of Stars, Astronomical Society of the Pacific Leaflet No. 241, 8 pages (April 1949).
Newton Lacy Pierce (obituary), Popular Astronomy, vol. 58, pp. 425–27 (1950).
The Birthplace of Stars, Research Review, Office of Naval Research (1951).
The Birth of Stars from Interstellar Clouds, Journal of the Washington Academy of Sciences, vol. 41, pp. 309–18 (1951); see also Scientia, vol. 46, pp. 1–6 and 42–49 (1952).
Horizons in Astronomy, American Scientist, vol. 43, pp. 323–30 (1955).
Voyages Interplanetaires, Extrait du Bulletin "Ciel et Terre" de la Société Belge d'Astronomie, de Météorologie et de Physique du Globe, Bruxelles LXXI Année, pp. 57–64 (1955).
Travel Through Space, Princeton Alumni Weekly, vol. 55, pp. 12–15 (March 18, 1955).
H. N. Russell, Astronomer (obituary), Science, vol. 125, pp. 1133–34 (1957).
The Stellarator, Scientific American, vol. 199, pp. 28–35 (October 1958).
Flying Telescopes, Bulletin of the Atomic Scientists, vol. XVII, pp. 191–94 (1961).
Experimental Plasmas, Physics Today, pp. 33–39 (Dec. 1965).
Robert E. Danielson (obituary) (with M. Schwarzschild), Icarus, vol. 30, pp. 601–3 (1977).
The Space Telescope, American Scientist, vol. 66, pp. 426–31 (1978).
Le Milieu Interstellaire: un Maillon Important de l'Evolution de l'Univers (with J. Audouze and A. Vidal-Madjar), Le Monde, Paris, pp. 19–20 (June 11, 1980).
The Space Telescope (with J. N. Bahcall), Scientific American, vol. 247, no. 1, pp. 40–51 (July 1982).
Interstellar Matter and the Birth and Death of Stars, Mercury, vol. 12, pp. 142–43, 146–49 (1983).
Perspectives on the Past, Present and Future of Fusion Research, Journal of Fusion Energy, vol. 7, pp. 221–23 (1988).
Bengt Strömgren (obituary), Physics Today, vol. 41, pp. 112, 114 (March 1988).
Harnessing the Sun, New York Times, OP-ED letter, p. 23 (December 11, 1993).

# INDEX

This list contains chiefly references to substantial discussions of those topics most likely to interest readers. Items mentioned in the text with no elaboration and with no relation to these particular topics are, for the most part, not included here.

LYMAN SPITZER, JR. is Charles A. Young Professor of Astronomy Emeritus and Senior Research Astronomer at Princeton University.

JEREMIAH P. OSTRIKER is Charles A. Young Professor of Astronomy and Provost at Princeton University.